# LIGNOCELLULOSE BIOTECHNOLOGY

## Future Prospects

# LIGNOCELLULOSE BIOTECHNOLOGY

## Future Prospects

*Editors*

**Ramesh Chander Kuhad**

Department of Microbiology
University of Delhi
South Campus, New Delhi, India

**Ajay Singh**

Department of Biology
University of Waterloo
Waterloo, Ontario, Canada

**I.K. International Publishing House Pvt. Ltd.**

NEW DELHI • MUMBAI • BANGALORE

*Published by*

I.K. International Publishing House Pvt. Ltd.
S-25, Green Park Extension
Uphaar Cinema Market
New Delhi - 110 016 (India)
E-mail : ik_in@vsnl.net

ISBN 81-88237-58-2



5 4 3 2



Published by Krishan Makhijani for I.K. International Publishing House Pvt. Ltd., S-25, Green Park Extension, Uphaar Cinema Market, New Delhi - 110 016 and Printed by Rekha Printers Pvt. Ltd., Okhla Industrial Area, Phase II, New Delhi - 110 020.

# Foreword

It gives me immense pleasure to write the foreword for *Lignocellulose Biotechnology: Future Prospects* edited by Ramesh Chander Kuhad and Ajay Singh. My interest in lignocellulose biotechnology started in early seventies, when it was realized that if Indian livestock have to produce to this optimal level the only source of carbohydrates was from plant cell wall through the crop residues. They have to become a staple diet of all farm livestock as they are the only source of carbohydrates available to them. The crop residues can only become available source of energy if microbes can be used to release carbohydrates after breakdown of lignin ring in the plant cell wall.

This initial interest in the area of renewable energy resources and agro-industrial waste utilization became intense as new tools of biotechnology became available. Energy is the most fundamental requirement in our life. Due to increasing use of the conventional energy sources such as fossil fuels, hydro and nuclear energy, the supply of such resources is likely to be limited in future. The potential to use lignocellulosic biomass for energy production derives from its position as the most abundant and renewable organic material in the biosphere, accounting for approximately 50% of world biomass. Potential feedstock may include food crops, crop residues and woody biomass. The low cost and chemical composition of crop residues make them attractive as feedstock for production of energy, food, feed and industrial chemicals.

It is difficult to accurately estimate the quantities of agricultural waste and crop residues produced worldwide. According to an estimate of Food and Agriculture Organization, about three billion tons of cereal straws are produced annually in the world. Annual volumes of wastes from pulse crops, oilseeds and plantation crops amount to an almost a billion tons per annum. Sugarcane bagasse can also be a desirable feedstock for the production of industrial chemicals and bioethanol. Brazil and India are the world's two largest sugarcane growers with production of 300 and 285 million tons per year, respectively. Estimates of total residues produced in the US are 375 million dry tons per year. This estimate only includes corn, small grains, sorghum, rice and sugarcane crop residues and does not include soybean and cotton field residues which are difficult to collect.

However, there are numerous challenges in dealing with any waste processing technology. Since type and availability of agricultural and forestry wastes vary with geographic region, climate, environmental conditions, agricultural and processing practices, generic processes for bioconversion of lignocellulosic wastes to industrial products must be versatile and robust as well as cost effective. The main factor that affects the biotechnology of commodity products is the high cost of current processing technology rather than the cost of raw materials. Challenges to development of a cost competitive process may be grouped in terms of converting lignocellulose into reactive intermediates by overcoming the recalcitrance and converting reactive intermediates into useful products, using all the sugars and byproduct value recovery of non-sugars.

*Lignocellulose Biotechnology: Future Prospects*, discusses a wide range of topics related to the fundamental and applied aspects of lignocellulose utilization, processing and biotechnology. The book contains a range of topics including biodiversity of lignocellulolytic microorganisms and their enzymes, molecular biology relevant to biodegradation, characterization of lignocellulolytic enzymes, bioconversion of cellulosic material to produce enzymes, animal feed, bioethanol and xylitol and industrial applications of cellulases and xylanases. Chapters dealing with industrial applications address current biotechnological approaches in lignocellulose bioconversion to value added products.

The book has contributions from the scientists from different disciplines including microbiology, biochemistry, molecular biology, genetics and biochemical engineering with diverse backgrounds; universities, government laboratories and industry. The editors Prof. Kuhad and Singh have done a commendable job in bringing together a range of excellent papers from these experts in the field of lignocellulose biotechnology and shaping them to provide a unique piece of work in the form of this book. I am confident that the book should prove to be very useful to the students, teachers, scientists and engineers in the disciplines of biotechnology, life sciences, microbiology, biochemistry, and environmental sciences and engineering.

**Prof. P.N. Bhat**
Former Deputy Director General
(Animal Science), ICAR, New Delhi, India

# Preface

The residues from agricultural crops and agro- and forest-based industrial processes are abundant and along with grasses and weeds constitute the major part of the biomass available on earth representing an important energy and material resource. Life on earth depends on photosynthesis, resulting in production of plant biomass having cellulose as the major component. Lignocellulose, the major component of biomass, makes up about half of the matter produced by photosynthesis. The carbon cycle is closed primarily as a result of the action of cellulose-utilizing microorganisms present in soil and the guts of animals. The plant biomass consists of three types of polymer—cellulose, hemicellulose, and lignin—that are strongly bonded by non-covalent forces and by covalent cross-linkages. In nature, cellulose, lignocellulose and lignin are major sources of plant biomass, and therefore their recycling is indispensable for the carbon cycle. Each polymer is degraded by a variety of microorganisms through production of hydrolytic and oxidative enzymes that work synergistically.

Several biological methods for utilizing or recycling of lignocellulosic biomass have been suggested, including composting and their use as raw material for the production of ethanol as an alternative combustible seem to be the most economically feasible. The general use of alternative, environment-friendly technologies that introduce lignocellulose-degrading enzymes at different stages of pulp and paper manufacture as a pretreatment to pulping (biopulping), bleaching (biobleaching), or wastewater treatment has allowed considerable electrical power savings and a reduction of pollutants in the waste water from these industries. In addition, pretreatment of agricultural wastes with ligninolytic fungi enables their use as raw material for paper manufacturing. The use of microorganisms or their enzymes to enhance the de-inking of recycled fibres and the release of toners from office wastes is another promising field that is under research.

Many research groups worldwide are involved in studying different aspects of natural biodegradation of these compounds. Our current knowledge is quite advanced regarding enzymology, physiology, biochemistry, and molecular biology of lignocellulolytic microbes. Bioprocesses utilizing enzymes and microorganisms are being explored for their biotechnological applications. Microbiological conversion of lignocellulosic residues to various industrially useful products like hydrolytic and oxidative enzymes, organic acids, ethanol, fine chemicals, microbial lipids, protein and animal feed can be achieved through the efficient utilization of all the three fractions of biomass—cellulose, hemicellulose and lignin. Production of ethanol and other alternative fuels from lignocellulosic biomass can reduce urban air pollution, decrease the release of carbon dioxide in the atmosphere, and provide new options for agricultural wastes utilization and recycling. Biopulping and biobleaching are leading to cleaner and more efficient pulp and paper manufacture.

Despite the progress achieved, more effort is needed for lignocellulosic enzymes and/or microorganisms to have significant industrial impact.

*Lignocellulose Biotechnology : Future Prospects* addresses a wide range of topics related to the basic, applied and biotechnological aspects of lignocellulose bioconversion. The book is divided into four major sections: (1) Overview of Lignocellulose Biotechnology; (2) Lignocellulolytic Microorganisms and Enzymes; (3) Lignocellulose Bioconversion; and (4) Biotechnological Applications. Topics include lignocellulose resource management, diversity of lignocellulolytic microorganisms, fundamental and applied aspects of lignocellulolytic enzymes and bioconversion of lignocellulosic residues for production of enzymes, feed and chemicals. Several chapters cover applications of lignocellulolytic microbes and enzymes in food, pulp and paper chemical and textile industries.

The book contains contributions from leading scientists with diverse backgrounds—universities, institutions and industry who are involved in the microbiological, molecular, biochemical and biotechnological aspects of lignocellulose biodegradation and utilization research. This book should be of immense use to the students of biotechnology, microbiology, biochemistry, environment sciences, food sciences and animal feed sciences. This book will certainly be of equal interest to the teachers, scientists and engineers, whether in academia, industry or government directly involved in the research or want to learn about biomass utilization and lignocellulosic biotechnology.

We are grateful to all the authors for their excellent contributions.

**Ramesh Chander Kuhad**
**Ajay Singh**

# Contents

# PART I

---

## OVERVIEW OF LIGNOCELLULOLYTIC BIOTECHNOLOGY

# 1

# Lignocellulolytic Microorganisms, their Enzymes and Possible Biotechnologies based on Lignocellulolytic Microorganisms and their Enzymes

**R.C. Kuhad[1], S. Kuhar[1], M. Kapoor[1], K.K. Sharma[1] and A. Singh[2]**

[1]*Department of Microbiology, University of Delhi, South Campus, New Delhi 110021, India*

[2]*Department of Biology, University of Waterloo, Waterloo, Ontario, Canada*

Lignocellulose, the major component of biomass, makes up about half of the matter produced by photosynthesis. The plant biomass consists of cellulose, hemicellulose and lignin, which are strongly bonded by non-covalent cross linkages. Each of the lignocellulosic component is degraded by a variety of microorganisms through production of various hydrolytic and oxidative enzymes. The research advancement about lignocellulolytic microorganisms and their enzymes have proved their potential in animal feed, biotransformation, bioremediation, production of chemicals, stabilization of food beverages and paper manufacture etc. This chapter deals with diversity of lignocellulolytic microorganisms and their enzymes. The possible biotechnologies based on these organisms and enzymes are also discussed.

## INTRODUCTION

Lignocellulosics in the form of agricultural and forestry residues are the most abundant and inexhaustible or renewable natural resource. There are great possibilities for employing biotechnology if the lignocellulosics is taken as the raw material. Lignocellulose is the major structural component of woody plants and non-woody plants such as grass. It consists of lignin, hemicellulose and cellulose. The chemical properties of the components of lignocellulosics make them a substrate of enormous biotechnological value. Improvement in many processes related to lignocellulose biotechnology has appeared in the past few years.

The huge amounts of residual plant biomass considered "waste", which is generally burnt, can potentially be converted into different value added products including biofuels, chemicals, and cheap carbon sources for fermentation, improved animal feeds and human nutrients by cellulose-degrading microorganisms. Enzymes produced by these microbes also have significant applications in various industries including chemicals, fuel, food, brewery and wine, animal feed, textile and laundry, pulp and paper and agriculture. Microbial degradation of lignocellulosic residues and potential biotechnological applications are discussed in this chapter.

## Composition of Lignocellulosic Biomass

The composition of lignocellulosic biomass varies; the major components being cellulose (35-50%), followed by hemicellulose (20-35%) and lignin (10-25%) (Sun and Cheng, 2002). Proteins, essential oils and ash make up the remaining fraction. The structure of these materials is very complex and native biomass is generally resistant to enzymatic hydrolysis. Cellulose fibres are embedded in a lignin polysaccharide matrix. Xylan may play a significant role in the structural integrity of cell walls by both covalent and non-covalent associations (Saha, 2003). The composition of hardwoods and softwoods are significantly different. The lignin content of softwoods is generally higher than that of hardwoods, whereas hemicellulose content of hardwoods is higher than the softwoods.

The cellulose molecule is a homopolymer of glucose units linked with β-1,4-glucosidic units, although the true repeating stereochemical unit of cellulose is cellobiose (β-1,4-D-glucosyl-D-glucose). Glucose and cellodextrins are the products when cellulose is hydrolyzed (Gilbert *et al.,* 1983). The cellulose molecule is a polymer with a degree of polymerization of up to about 15,000. The degree of crystallinity of cellulose may vary from 0% (amorphous) to approximately 100% (Beguin *et al.,* 1994). Crystalline cellulose is highly resistant to microbial attack and enzymatic hydrolysis, whereas amorphous cellulose is degraded at a much faster rate (Eriksson *et al.,* 1990).

Hemicelluloses are easily hydrolyzed short chains of branched heteropolysaccharides composed of both hexoses and pentoses (Whistler *et al.,* 1970). D-Xylose and L-arabinose are the major constituents of the pentosans (xylans), while D-glucose, D-galactose and D-mannose are the constituents of the hexosans (mannans). The major hemicellulose components in softwood are mannan-based, and those in hardwood are xylan-based. The principal sugar components of these hemicellulose heteropolysaccharides are: D-xylose, D-mannose, D-glucose, D-galactose, L-arabinose, D-glucuronic acid, 4-O-methyl-D-glucuronic acid, D-galactouronic acid, and to a lesser extent, L-rhamnose, L-fucose and various O-methylated sugars. They usually have degree of polymerization of 100 to 200 (Kuhad *et al.,* 1997a).

The mannan hemicelluloses, galactoglucomannans and glucomannans, in softwoods and hardwoods, are both branched heteropolysaccharides. Their backbones are built up of 1, 4-linked β-D-glucopyranose and β-D-mannopyranose units. The mannose and glucose units in the backbone are partially substituted in C-2 and C-3 positions by acetyl groups (Eriksson *et al.,* 1990).

Xylan is the most abundant hemicellulose and it comprises 15-30% of annual plants, 20-25% of hardwoods, and 7-12% of softwoods. Xylans have homopolymeric backbone chains of 1, 4-linked β-D-xylopyranose units. Besides xylose, xylans may contain arabinose, glucuronic acid or its 4-O-methyl ether, and acetic, ferulic, and *p*-coumaric acids. The frequency and composition of the branches are dependent on the source of xylan. The backbone consists of *O*-acetyl, α-L-arabinofuranosyl, α-1, 2-linked glucuronic or 4-O-methylglucuronic acid substituents (Kuhad *et al.,* 1997a).

Lignins are highly branched polymeric molecules consisting of phenyl-propane-based monomeric units linked together by different types of bonds, including alkyl-aryl, alkyl-alkyl, and aryl-aryl ether bonds. The molecular weight of lignins may be 100 kDa or more. The relative proportions of three cinamyl alcohol precursors incorporated into lignin, i.e., *p*-coumaryl alcohol, coniferyl alcohol and sinapyl alcohol, vary not only with the plant species but also with the plant tissues and location of the lignins within the plant cell wall. Lignin is found in the highest concentration in the middle lamella, but is most abundant in the secondary walls of the vascular plants. The hydrolysable linkages in lignins are suggested to be of two types: β-aryl ether and α-aryl ether (Adler, 1977). The predominant β-aryl ether type bond is more resistant to cleavage. Under mild hydrolytic conditions, the cleavage of the ether bond is exclusively restricted to the α-aryl ether type (Kirk *et al.,* 1987).

Plant cell walls contain three kinds of structural proteins, which differ in amino acid composition are hydroxyproline-rich glycoproteins, glycine-rich proteins and proline-rich proteins, have been found in plants (Cassab and Varner, 1988). All are considered to be important structural components of plant cell walls.

Plant cell walls also contain extraneous materials, including extractives and non-extractives (Ladisch *et al.,* 1983). The extractives can be broadly divided into three groups, i.e., terpenes (isoprene polymers), resins (fatty acids, alcohol, resin acids) and phenols (tannins). In addition, low molecular weight carbohydrates, alkaloids, gums, and various other cytoplasmic constituents are present (Akin *et al.,* 1995). The non-extractives make up 0.2-0.8% of the dry weight and include inorganic components such as silica, carbonates, oxalates and non-cell wall substances such as starch, pectin and protein (McDonald, 1969).

## Microorganisms and their Lignocellulolytic Enzymes

A diverse spectrum of lignocellulolytic microorganisms, mainly fungi and bacteria have been isolated and identified over the years and this list still continues to grow rapidly. Despite the impressive collection of lignocellulolytic microorganisms only a few have been studied extensively. The exact mechanism by which lignocellulose is degraded enzymatically is still not fully understood but significant advances have been made to gain insight into the lignocellulolytic microorganisms, their genes and various enzymes involved in the process.

## Microorganisms Producing Cellulose-degrading Enzymes

A large number of microorganisms including fungi, bacteria and actinomycetes have the ability to produce cellulose-degrading enzymes. But few of them produce all the necessary enzymes for degradation of crystalline cellulose. Fungi are the most studied organisms with respect to degradation of cellulose and production of cellulolytic enzymes. Different categories of microorganisms with lignocellulolytic activity are briefly described in this section. Classification of cellulases capable of hydrolyzing 1, 4-β-glucosidic bonds is provided in Table 1.

**Table 1. Classification of cellulases as enzymes capable of hydrolyzing 1,4-b-glucosidic bonds**

| *Type and origin of the enzyme* | *EC number* | *Family* |
|---|---|---|
| Fungal endo-1,4-β-mannanases, and aerobic and anaerobic bacteria endo-glucanases | 3.2.1.78<br>3.2.1.4 | 5/A1 |
| Endoglucanases of actinomycetes, and aerobic and anaerobic bacteria, as well as nematodes | 3.2.1.4 | 5/A2 |
| Exo-1,3-β-glucanases;<br>Endoglucanases/1,3-1,4-β-glucanases and<br>1,3-β-glucanases (yeast, *Clostridium*) | 3.2.1.58<br>3.2.1.4/73<br>3.2.1.39 | 5/A3 |
| Endoglucanases and mannanases of actinomycetes, aerobic and anaerobic bacteria, and anaerobic fungi | 3.2.1.4<br>3.2.1.78 | 5/A4 |
| Endoglucanases of filamentous fungi and aerobic bacteria | 3.2.1.4 | 5/A5 |
| Endo-1,6-β-glucanases | 3.2.1.75<br>3.2.1.123 | 5 |

Soft-rot fungi mainly degrade the polysaccharides. The best known of these producing a complete set of cellulases is *Trichoderma viride* (Bisaria *et al.,* 1989). Other well-known fungi causing soft-rots are *Aspergillus niger*, *Chaetomium cellullolyticum*, *Fusarium oxysporium*, *Neurospora crassa*, *Penicillium pinophilum* (Singh and Hayashi, 1995).

Brown-rot fungi causing brown rot in plants degrade cellulose rapidly but the enzyme system seems to operate differently from those of soft-rot fungi and white-rot fungi. *Poria placenta*, *Lanzites trabeum*, *Tyromyces palustris* and *Coniophora puteana* are brown-rot fungi among the most studied for their cellulolytic activities (Eriksson *et al.,* 1990).

White-rot group of fungi cause white-rot in plants is rather heterogeneous, but these fungi have, in common, the ability to degrade lignin as well as other lignocellulosic components (Eriksson, 1981). The most studied white-rot fungi were *Phanerochaete chrysosporium* that was first isolated from wood chip piles. Other important white-rot fungi are *Sporotrichum thermophile* and *Coriolus versicolor.*

Rumen fungi secrete cellulases, often described as a complex of enzymes that, by acting together, solubilize cellulose efficiently (Wood, 1991). Some fungal species of relevance are *Sphaeromonas communis*, *Neocallimastrix frontalis*, *N. patriciarum* and *Piromyces communis.*

Cellulolytic enzyme systems of bacteria are not directly comparable to those of fungi. Bacteria often produce cellulases in small amounts and degradation of cellulose seems to take place by a cluster of multi-enzyme complexes, which are difficult to disrupt without loss of total activity as well as of the individual components (Lamed *et al.,* 1993; Rabinovich, 2002a). The most studied bacteria with respect to their cellulase systems are species of *Clostridium*, *Cellulomonas*, *Bacillus* and *Pseudomonas.* Other bacteria producing cellulosome like entities include *Acetovibrio cellulolyticus*, *Bacterioides cellulosolvens*, *Butyrivibrio fibrisolvens*, *Fibriobacter succinogenes*, *Ruminococcus albus*, *R. flavefaciens* and *Thermomonospora curvata* (Lamed *et al.,* 1993).

Extracellular cellulases of actinomycetes degrade cellulose by mechanisms similar to that of fungi, i.e., non-associated enzymes excreted into the culture medium. Some of important members of actinomycetes responsible for the degradation of cellulose are mesophilic species of *Streptomyces* and thermophilic species of *Thermomonospora* and *Thermoactinomyces* (Stutzenberger *et al.,* 1986; Calza *et al.,* 1985).

## Microorganisms Producing Hemicellulose-degrading Enzymes

Hemicellulose-degrading system is widespread as they are produced by fungi, bacteria from terrestrial and marine environments, rumen microorganisms, yeasts and marine algae. Fungi, yeasts and some bacteria secrete hemicellulolytic enzymes extracellularly, although cell wall bound and intracellular hemicellulolytic enzymes have also been reported (Dekker, 1985; Kuhad *et al.,* 1997a). Table 2 lists some important microorganisms producing hemicellulose-degrading enzymes. Only a few yeast species such as some species of *Aureobasidium*, *Cryptococcus* and *Trichosporon* are known to produce xylan-degrading enzymes and are therefore identified as xylanase producers (Eriksson *et al.,* 1990).

**Table 2. Important microorganisms producing hemicellulose-degrading enzymes**

| *Enzyme* | *Bacteria* | *Fungi* |
|---|---|---|
| β-1,4-Xylanases | *Bacillus pumilus* | *Aspergillus niger* |
| | *Bacillus subtilis* | *Fusarium oxysporum* |
| | *Clostridium acetobutylicum* | *Aspergillus wentii* |
| | *Cellulomonas uda* | *Trichoderma koningii* |
| | *Streptomyces xylophagus* | *Neurospora crassa* |
| β-1,4-Xylosides | *Clostridium thermocellum* | *Aspergillus niger* |
| | *Bacillus pumilus* | *Corticium rolfsii* |
| | *Acetovibrio cellulolyticus* | *Penicillium wortmanni* |
| | | *Trichoderma reesei* |
| α-Arabinosidase | *Streptomyces pupurascens* | *Aspergillus niger* |
| | *Bacillus subtilis* | *Corticium rolfsii* |
| | *Ruminococcus albus* | *Trichoderma reesei* |
| α-Glucuronidase | | *Trichoderma reesei* |
| | | *Agaricus bisporus* |
| | | *Pleurotus ostreatus* |
| Esterases | *Fibrobacter succinogenes* | *Aspergillus niger* |
| | *Bacteroides cellulosolvens* | *Aspergillus phoenicus* |
| | *Costridium thermocellum* | *Trichoderma reesei* |
| β-1,4-Mannanases | *Aerobacter aerogenes* | *Aspergillus niger* |
| | *Bacillus subtilis* | *Thielavia terrestris* |
| | *Caldocellum saccharolyticum* | *Trichoderma reesei* |
| | *Streptomyces lividans* | *Paecilomyces variotii* |
| β-1,4-Mannosidase | *Bacillus subtilis* | *Aspergillus niger* |
| | | *Aspergillus awamori* |
| | | *Thielavia terrestris* |
| | | *Polyporus sulphureus* |
| α-Galactosidase | *Bacillus subtilis* | *Aspergillus niger* |
| | *Cellulomonas sp.* | *Sclerotium rolfsii* |
| | | *Aspergillus tamarii* |
| | | *Mortierella vinacea* |

## Microorganisms Producing Lignin-degrading Enzymes

Lignin is degraded to different extents by different microorganisms, of which wood-rotting fungi are the most effective, white rot fungi in particular (Eriksson *et al.,* 1990).

Soft-rot fungi efficiently degrade wood polysaccharides but degrade lignin slowly and incompletely (Janshekar *et al.,* 1983). They have been observed more commonly on hardwoods than on softwoods.

Some soft-rot fungi with lignin degrading ability are *Chaetomium globosum*, *Daldinia concentrica*, *Lecythophora hoffmannii*, *Petrillidium boydii* and *Pialophora mutabilis*.

Brown-rot fungi prefer coniferous wood. These fungi cause extensive degradation of the polysaccharides and only a limited degradation of lignin. These are some brown-rot fungi, which have the capability of degradation of lignin *Fomitopsis pinicola*, *Gleophyllum trabeum*, *Poria placenta*, *Lentinus lepideus*, *Pholiota adipose* and *Tyromyces palustris* (Kuhad *et al.,* 1997a).

White-rot fungi are the most effective wood-rotting fungi. These predominanly degrade wood from deciduous trees but coniferous wood trees are also degraded (Eriksson *et al.,* 1990). Most white-rot fungi degrade wood by a simultaneous attack on the lignin, cellulose and hemicelluloses, but a few are rather specific lignin degraders (Blanchette *et al.,* 1992). There are a number of white-rot fungi studied for the ligninolytic activity, some of which are *Coriolus versicolor*, *Cyathus bulleri*, *Phlebia radiata*, *Pycnoporus sanguineus*, *Pleurotus ostreatus*, *Phanerochaete chrysosporium*, *Cereporiopsis subvermispora* and *Pleurotus eryngii*.

Bacteria generally cause a low percentage of mineralization of lignin in lignocellulosic materials. A number of gram-negative bacterial species belonging to the genera *Pseudomonas*, *Xanthomonas*, *Acinetobacter*, *Achromobacter*, *Agrobacterium*, *Aerobacter* and *Erwinia* (Zimmerman, 1990).

Actinomycetes have been reported to mineralize lignin successfully, although not as fast or comprehensively as white-rot fungi. The degradation of lignin results in the release of lignin-rich, water-soluble fragments called acid precipitable polymeric lignin (APPL) (Eriksson *et al.,* 1990). Some of the most active lignin degraders among actinomycetes are *Streptomycetes badius*, *S. cyaneus* and *Thermomonospora mesophila* (Kuhad *et al.,* 1997a).

## Biodegradation of Lignocellulosic Biomass

Of the three components, lignin is the most recalcitrant to degradation whereas cellulose is more resistant to hydrolysis as compared to hemicellulose. Alkaline and acid hydrolysis methods have been used to degrade lignocellulose. Weak acids tend to remove lignin but result in poor hydrolysis of cellulose whereas strong acid treatment occurs under relatively extreme corrosive conditions of high temperature and pH, which necessitate the use of expensive equipment.

Microorganisms such as fungi and bacteria both secrete cellulases, endoglucanases and exoglucanases, which act synergistically to degrade cellulose. Endoglucanases hydrolyze β-1,4-glucosidic bonds in a random fashion over a cellulose chain. As a result, there is a rapid decrease in chain length and a slow increase in reducing end groups. Cellobiohydrolases degrade cellulose by splitting off cellobiose from the non-reducing end of the chain (Rabinowich *et al.,* 2002b). β-Glucosidases, secreted by cellulolytic organisms, catalyze the hydrolysis of water-soluble cellodextrins as well as alkyl- and aryl-β-D-glucosides.

Endoxylanase, the most widely studied and best-characterized xylanolytic enzyme, attacks the xylan backbone to produce both substituted and non-substituted shorter oligomers, xylobiose and xylose. β-xylosidase is employed to convert oligomers to xylose and acts in concert with endoxylanases, α-glucuronidase, α-arabinosidase and acetyl-xylan esterase to achieve total hydrolysis of xylans to monosaccharides (Biely, 1985). Endomannases hydrolyze β-D-mannose and a series of mannose oligosaccharides of DP 2-6 (Ishihara, 1980). Table 3 lists important hemicellulose degrading enzymes.

**Table 3. The major hemicellulases and their classification**

| *Enzymes* | *Substrates* | *EC number* | *Family* |
|---|---|---|---|
| Exo-β-1,4-xylosidase | β-1,4-Xylooligomers xylobiose | 3.2.1.37 | GH 3<br>39<br>43<br>52<br>54 |
| Endo-b-1,4xylanase | β-1,4-Xylan | 3.2.1.8 | GH 5<br>8<br>10<br>11<br>43 |
| Exo-β-1,4-mannosidase | β-1,4-Mannooligomers mannobiose | 3.2.1.25 | GH 1<br>2<br>5 |
| Endo-β-1,4-mannanase | β-1,4-Mannan | 3.2.1.78 | GH5<br>26 |
| Endo-α-1,5-arabinanase | α-1,5-Arabinan | 3.2.1.99 | GH43 |
| α-L-arabinofuranosidase | β-Arabinofuranosyl (1→ 2) or (1→ 3) xylooligomers α-1,5-arabinan | 3.2.1.55 | GH3<br>43<br>51<br>54<br>62 |
| α-glucuronidase | 4-O-Methyl-α-glucuronic acid (1→ 2) xylooligomers | 3.2.1.139 | GH 67 |
| α-galactosidase | α-Galactopyranose (1→ 6) mannooligomers | 3.2.1.22 | GH 4<br>27<br>36<br>57 |
| Endo-galactanase | β-1,4-Galactan | 3.2.1.89 | GH 53 |
| β-Glucosidase | β-Glucopyranose (1→ 4) mannopyranose | 3.2.1.21 | GH 1<br>3 |
| Acetyl xylan esterases | 2-or 3-o-Acetyl xylan | 3.1.1.72 | CE 1<br>2<br>3<br>4<br>5<br>6 |
| Acetyl mannan esterase | 2-or 3-O-Acetyl mannan | 3.1.1.6 | CE 1 |
| Ferulic and p-coumaric acid esterases | 2-or 3-O-Acetyl mannan | 3.2.1.73 | CE 1 |

Lignin peroxidase (ligninase) catalyzes a large variety of reactions, e.g. cleavage of β-O-4 ether bonds and $C_{\alpha}-C_{\beta}$ bonds in dimeric lignin model compounds, the basis for the depolymerization reactions catalyzed by LiP. The enzyme also catalyzes decarboxylation of phenylacetic acids, oxidation of aromatic $C_{\alpha}$-alcohols to $C_{\alpha}$-oxo compounds, hydroxylation, quinone formation, and aromatic ring opening (Haemmerli *et al.,* 1987). Manganese peroxidase is involved in the oxidation of phenols and phenolic ring structures (Wariishi *et al.,* 1992). Laccase catalyzes demethoxylation reactions of terminal phenolic units. It can also degrade β-dimers and β-O-4 dimers via α-oxidation of alkyl-aryl cleavage (Youn *et al.,* 1995), and $C_{\alpha}-C_{\beta}$ cleavage. Laccase has also been shown to catalyze the cleavage of aromatic rings in a similar way to LiP (Kawai *et al.,* 1998). Other enzymes of importance in lignin degradation are

$H_2O_2$-producing enzymes, NAD(P)H:quinone oxidoreductase, aryl alcohol dehydrogenase and probably also, CDH(CBQ). Lignin-degrading enzymes are shown in Table 4.

**Table 4. Ligninolytic enzymes**

| *Enzyme* | *Substrate* | *EC number* |
|---|---|---|
| Lignin peroxidase | Veratryl alcohol<br>$H_2O_2$ | 1.11.1.14 |
| Manganese peroxidase | Vanillyl acetone<br>Phenol red<br>3-methyl-2-benzothiazoline (MBTH) and<br>3-dimethylaminobenzoic acid (DMAB)<br>$H_2O_2$ | |
| Laccase | Syringaldehyde<br>2,2'-azino-bis-(3-ethylbenzothiazoline)-6-sulphonate (ABTS)<br>MBTH and DMAB | 1.10.3.2 |
| Intracellular and extracellular $H_2O_2$-producing enzymes | | |
| Intracellular enzymes | | |
| Glucose-1-oxidase, glucose-2-oxidase, methanol oxidase and fattyacyl-CoA oxidase | Glucose<br>Methanol | |
| Extracellular enzymes | | |
| Glyoxal oxidase | | |
| Aryl-alcohol oxidase | Aromatic alcohols | |
| Oxidoreductases | | |
| NAD(P)H:quindone oxidoreductase | 2,6-dichlorophenol-indophenol (DCPIP)<br>3-methyl-5-t-butyl-benzoquinone | |
| Aryl alcohol dehydrogenase | (MTBBQ) Cellobiose | |
| CDH(CBQ) | Cytochrome c'Cellobiose | |

## Biotechnological Applications

In general the bioprocessing technology of raw materials or their constituents into bioproducts entails three steps: process design, system optimization and model development. Processing involves the use of biocatalysts, whole microorganisms or their enzymes from other organisms to synthesize or transform raw materials into new products, recover/purify such bioproducts and subsequently any needed downstream modifications.

Bioconversions of lignocellulosic materials into useful, higher value products normally require multi-step processes which include:

(i) Pretreatment (mechanical, chemical or biological) (Grethlein, 1984; Grethlein and Converse, 1991).

(ii) Hydrolysis of the polymers to produce readily metabolizable molecules (e.g. hexose or pentose sugars).

(iii) Bio-utilization of these molecules to support microbial growth or to produce chemical products.

(iv) The separation and purification.

Lignocellulose degradation is of great importance for biotechnological conversion of lignocellulosic materials into value added products. There is some excellent and comprehensive literature (Wong and Saddler, 1992a,b; Kuhad *et al.,* 1993, 1997ab; Bhat, 2000; Beg *et al.,* 2001; Sun and Cheng, 2002; Beauchemin *et al.,* 2001, 2003; Subramaniyan and Prema, 2002) available on the various bioproducts from lignocellulose-degrading microbes and their applications. Lignocellulolytic enzymes are used in various industries such as pulp and paper, textile, detergents, fodder, bioconversion, food, environment, chemical and pharmaceutical.

Technologies are currently available for all steps in the bioconversion of lignocelluloses to ethanol and other chemical products. However, these technologies must be improved and new technologies developed to produce renewable biofuel and other bioproducts at prices, which can compete with current production costs.

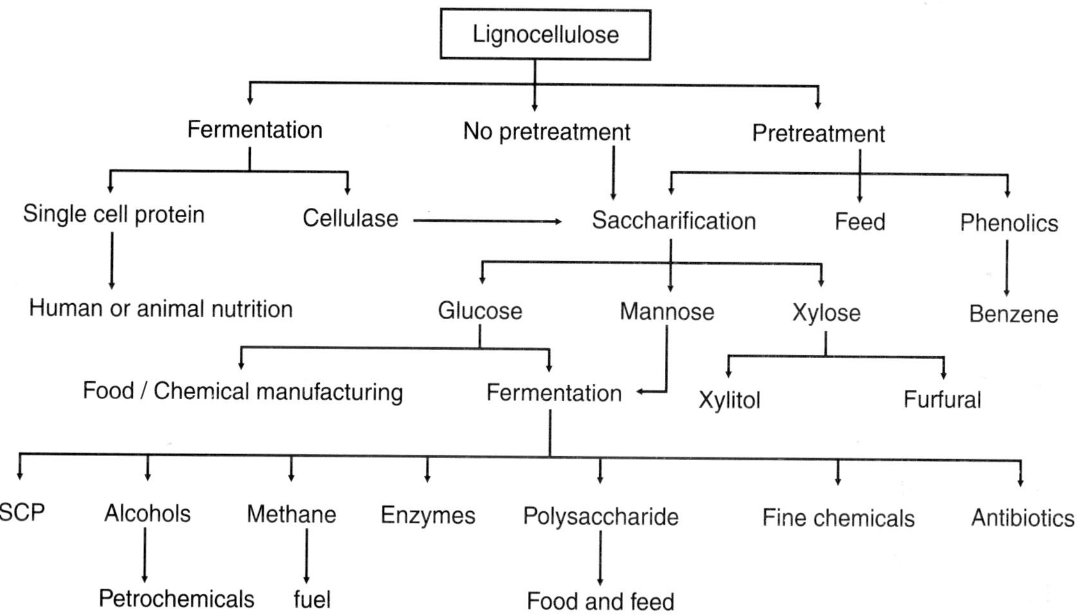


**Fig. 1.** Generalized process stages in lignocellulose bioconversion into value-added bioproducts

## Pulp and Paper Industry

Cellulases alone, or in combination with a few other enzymes such as xylanases, have been found useful for deinking different types of paper waste. Conventional deinking uses surfactants to float the toner followed by high temperatures to aggregate it, and then vigorous dispersion for size reduction. Those processes are expensive and energy-consuming. Cellulases and xylanases release toner particles facilitating flotation and subsequent steps. Endoglucanases and endoxylanases have been demonstrated to effectively deink selected old newsprint waste and to improve optical and strength properties of paper from enzymatically deinked pulp. The enzymatic deinking process also improves freeness (water drainage of the recycled fibres) compared to chemically deinked paper. Enzymatic deinking represents one of the few promising biotechnological applications of cellulases in the pulp and paper industry (Kim *et al.,* 1991). This enzymatic method has been scaled up and has proven to be economically feasible (Heise *et al.* 1996).

One of the approaches currently being developed for the application of biotechnology to the pulp and paper industry is the use of hemicellulose-hydrolysing enzymes for biobleaching processes. These enzymes are being used in the treatment of cellulosic pulps to remove residual xylans in the production of dissolving pulps. Xylanases, mannases and their accessory enzymes are able to degrade the hemicellulose portion in pulps selectively without affecting the cellulose.

Hemicellulose-degrading enzymes have also been studied for enzymatic debarking and pulp refining to reduce energy demands in mechanical pulping processes. In an enzymatic beating process, the enzymes are added to bleach pulp fibres to increase external fibrillation and thus improve papermaking properties. Mill trials of this concept have been carried out successfully using a commercial *T. reesi* enzyme called Liftase. Xylanases probably enhance the process of fibrillation as well as facilitate enzymatic deinking of recycled fibres in combination with cellulases. Dissolving pulps are used to produce cellulose derivatives such as acetates, cellophanes, and rayons. Hemicellulose contaminants lead to colour and haze in the product and to poor cellulose derivatization.

The organic chlorine byproducts formed during pre-bleaching are toxic and persistent. The incorporation of xylanase pre-bleaching also permits the use of lower chlorine charges and lower charges of unselective bleaching chemicals such as peroxide and ozone. Xylanase treatment may reduce the chemical loading required in caustic extraction of xylan from kraft pulps. Among the new bleaching technologies, the use of enzymes to specifically modify the pulp by environmentally safe processes was first proposed by Viikari *et al.* (1994). A patented Lignozyme process is effective in delignifying wood in a pilot pulp and paper process (Call and Mucke, 1997)

Lignin peroxidase (LiP), Manganese peroxidase (MnP) and laccase, produced by white rot fungi, have all been studied for use in pulp bleaching. Different approaches such as the use of suitable redox mediators for enhanced laccase delignification and brightening of kraft pulp have been worked out (Arias *et al.,* 2003). Laccase requires a higher amount of $H_2O_2$ and a low molecular mass electron carrier, while MnPs require a low and controlled supply of $H_2O_2$ along with Mn(II) ions and a suitable chelator such as oxalate or malonate (Onysko, 1993).

## Textile Industry

Cellulases have been successfully used for the biostoning of jeans and biopolishing of cotton and other cellulosic fabrics. Enzymatic stonewashing allows up to 50% higher jean load and provides the desired soft finish (Lange, 1993). Enzymatic biopolishing is a novel biological finishing process for textiles made from cellulosic fibres such as cotton, rayon etc. The controlled treatment with acidic cellulases improves softness and water absorbency of the fibres, strongly reduces the tendency for pill formation and provides a clearer surface structure with less fuzz (Sato *et al.,* 1990).

There are some very useful applications of xylanases in pretreatment of low-quality jute fibre, which amounts to about one-third of total world jute production. The low-quality jute fibre is harder due to its higher content of hemicellulose, posing difficulties in proper spinning in the mill for effective utilization. The spinning process therefore needs pretreatment (or softening), which allows the selective removal of xylan without affecting the fibre-strength (Hoq *et al.,* 1992) for this process, pretreatment with pure and thermostable xylanase is attractive.

The cleaning power of cellulase-containing detergents shows remarkable improvement (Hoshino, 1991). Alkaline cellulase from *Bacillus* sp. can interact selectively with cellulose in the interior of fibres and remove soil in the interfibre spaces in the presence of usual detergent ingredients.

## Bioconversion

Enzymatic saccharification of lignocellulytic materials produces monosaccharides, which subsequently can be fermented to a variety of products such as ethanol, other alcohols, organic acids, single cell protein, and lipids (Singh and Hayashi 1995; Kuhad *et al.* 1997a). Bioconversion of glucose is a well-documented technology (Vallander *et al.,* 1990). Bioethanol can be produced by using agricultural residues such as cereal straws and corn stalks. It is an advanced new transport fuel with a unique combination of attributes including low life-cycle greenhouse gas emissions, a high level of sustainability, seamless integration into the existing transport system and has the potential to have a large-scale impact (Ward and Singh, 2002). Bioethanol is also cost-effective way to reduce greenhouse gas emissions and gasoline use in transport, especially when compared to vehicle solutions. Its manufacturing process does not consume fossil fuels, but rather uses plant byproducts to create the energy to run the process. Recently, a patented process EcoEthanolä to make cellulose ethanol uses an enzyme hydrolysis to convert the biomass into sugars. These sugars are fermented and distilled into ethanol fuel using conventional ethanol distillation technology.

The key challenge in lignocellulose biotechnology is to develop a robust enzyme production system that can produce adequate amounts of highly efficient cellulase to make biomass conversion more rapid and less expensive. Cellulase yield and productivity above 150 FPU/g cellulose and 55 FPU/L/h, respectively, have been achieved using Solka floc as substrate (Himmel *et al.,* 1999). It has been estimated that a bioethanol process would require about 11 million FPU (19 kg, 42 lbs) of cellulase to process 1 ton of biomass (1000 lbs cellulose) to 84 gallons of ethanol at an enzyme loading of 25 FPU/g cellulose (Himmel *et al.,* 1997). It costs US$ 3/gallon of ethanol ($0.8/L) for commercial enzyme formulations and $0.5/gallon ($0.13/L) for less processed formulations (Lynd *et al.* 1999). To achieve total cost competitiveness, enzyme costs must be reduced to less than $0.07/gallon of ethanol, requiring a 10-fold increase in specific activity or production efficiency (Himmel *et al.,* 1999).

Commercial bioethanol production facilities could be located in feedstock producing areas and provide a solution for the surplus residue, while at the same time, create substantial economic opportunities in these areas. The U.S. Department of energy, and particularly, the National Renewable Energy Laboratory (NREL), runs a major bioethanol programme. Detailed analyses of bioprocesses for conversion of biomass to ethanol indicate a trend toward lower costs with greater consolidation of the hydrolysis and fermentation steps, namely, in cost terms: consolidated bioprocessing < simultaneous saccharification and fermentation < separate hydrolysis and fermentation (Lynd *et al.,* 1999).

Complete cellulolytic and hemicellulytic enzyme systems are required to achieve maximum hydrolysis of complex substrates to yield monomeric sugars. Sugars obtained in saccharification processes can be further utilized for the production of protein-rich biomass (single cell protein), and various processes have been developed for this purpose. Dilute acid pretreatment at a relatively low temperature to minimize the formation of inhibitory compounds, followed by enzymatic saccharification of the cellulosic portion, is an excellent workable process for generating fermentable sugars from lignocellulosic biomass. It is essential that a xylose tolerant β-xylosidase, along with other xylan-degrading enzymes, is developed in order to make enzymatic saccharification of any xylan substrate a commercial success (Saha, 2003).

The utilization of hemicellulose sugars is essential for efficient and cost-effective conversion of lignocellulosic material to fuel ethanol. Various waste and underutilized lignocellulosic agricultural residues can serve as low-cost feedstocks for production of fuel ethanol. In 2002, over 2 billion gallons of ethanol was produced, mainly by fermenting corn starch. Any hemicellulose containing lignocellulose generates a mixture of sugars upon pretreatment alone or in combination with enzymatic hydrolysis. The yeasts *Pachysolen tannophilus, Pichia stipitis*, and *Candida shehate* have the capability to ferment xylose to ethanol. Some bacteria, such as *E. coli, Klebsiella, Erwinia, Lactobacillus, Bacillus,* and *Clostridia*, can utilize mixed sugars but produce no, or only a limited quantity of, ethanol. Several microorganisms have been genetically engineered to overproduce ethanol from mixed sugar substrates. Various recombinant strains, e.g. *E. coli* K011, *E. coli* SL40, *E. coli* FBR3, *Zymomonas* CP4 (Pzb4) and *Sacchaaromyces* 1400 (Plnh32) have been evaluated for the fermentation of mixed sugar substrates (Saha, 2003).

Xylitol, a five-carbon sugar alcohol, used as a natural food sweetner (Parajo *et al.,* 1998; Roberto *et al.,* 2003; Saha, 2003), is currently produced by chemical reduction in alkaline conditions of the xylose derived mainly from wood hydrolyzate. Xylitol is used as a natural food sweetener. The demand for xylitol in the food and pharmaceutical industries has created a strong market for the development of low cost xylitol production process. Various xylose-rich hemicellulosic materials can serve as abundant and cheap feedstocks for production of xylitol by fermentation. There is a strong interest in creating recombinant yeast that can produce xylitol from glucose.

A product of hemicellulosic hydrolyzate, 2,3-Butanediol is a valuable chemical feedstock because of its application as a solvent, liquid fuel, and as a precursor of many synthetic polymers and resins. Dehydration of 2,3-butanediol yields the industrial solvent methyl ethyl ketone, which is much more suited as a fuel because of its much lower boiling point. Further dehydration yields 1,3-butanediene, which is the starting material for synthetic rubber and is also an important monomer in the polymer industry (Maddox, 1996). Other value added products from hemicellulose hydrolyzates include lactic acid, which is used in the food, pharmaceutical and cosmetic industries. It is a component of biodegradable plastic polylactate, the market for which is expected to grow significantly (Saha, 2003). Furfural is used in the manufacture of furfural-phenol plastics, varnishes and pesticides (Montane *et al.,* 2002).

Hemicellulases can be applied for the enzymatic synthesis of oligosaccharides (Palcic, 1999). Specifically, high yields of different oligosaccharides can be obtained using engineered retaining glycoside hydrolases, coined glycosynthases, in which the nucleophilic residue has been replaced. When using activated glycosyl donors of the opposite anomeric configuration together with suitable acceptors, these mutant enzymes can efficiently synthesize oligosaccharides, without hydrolyzing the resulting products. Since the first demonstration by Withers and co-workers on β-glucosidase, this technique has been applied successfully with many other glycoside hydrolases (Fairweather *et al.,* 2002).

Controlled degradation of lignin particularly by white-rot fungi or actinomycetes could give valuable aromatic products. Some laccases from *Coriolus hirsutus* and *C. zonatus* have been characterized for the synthesis of conducting polyaniline. Polyaniline is one of the most important conducting polymers which may be used as active component of organic light-weight batteries, microelectronics, optical display, bioanalyses etc. (Karamyshev *et al.,* 2003). The recombinant laccases from *Myceliophthora thermophila, Coprinus cinereus, Rhizoctonia solani* were used for the study of naphthol oxidation. Naphthol is widely used for dyes, medicinal organics and synthetic perfumes production. The compounds produced during oxidative coupling of naphthol may be useful as new polymeric materials (Kulys *et al.,* 2003).

Currently a number of organic acids, amino acids, vitamins and a number of bacterial and fungal polysaccharides such as xanthan gum are produced by fermentation using glucose as the fermentation substrate with the help of hemicellulose-degrading enzymes.

## Feed

Feed enzymes for ruminants have received considerable attention recently, because of their potential to improve animal performance. Cellulases are also used to improve silage of cattle feed, i.e. to produce high quality silage from grasses containing only small amounts of water-soluble carbohydrates. Cellulases, used as feed additives alone or with proteases have been shown to significantly improve body weight gain in piglets and to improve the quality of pork meat (Oshida *et al.,* 1992).

Pretreatment of agricultural silage and grain feed by xylanases has been reported to improve its nutritional value (Beg *et al.,*2001). This is the rapidly emerging thrust area for hemicellulases as animal feed additive for monogastrics like poultry and pigs which unlike the ruminants do not have xylanolytic microflora in their digestive tracts. Xylanases reduce viscosity and increase absorption by breaking down the non-starch polysaccharides (NSPs) in high fibre rye- and barley- based feeds. Most commercial xylanases are produced by *Trichoderma, Bacillus, Aspergillus, Penicillium, Aureobasidium* and *Talaromyces* spp. (Godfrey *et al.,* 1996).

Possibility of converting lignocellulosic substrate to more digestible and nutritious feedstuff for ruminants has concentrated research into upgrading animal feeds by solid-state fermentation processes. A large number of white-rot fungi have been studied for solid-state bioconversion of lignocellulosic materials because of their ability to degrade lignin (Akin *et al.,* 1993). *P. chrysosporium, Cyathus* sp., *Coriolus hirsutus, Dichomitus squalens* and *Pleurotus* spp. are the extensively studied organisms. The mode of attack was mineralization of lignin leaving carbohydrates *in situ.* In a study using *P. sajor-caju* for fermentation of wheat straw, 20% increase in *in vitro* digestibility was achieved using sterilized straw. Cinnamic acids are involved in lignin-polysaccharide linkages and their degradation could increase lignin extractability. Breakdown of β-*O*-4 and β-1 dimers has been investigated extensively and simultaneous degradation of condensed and uncondensed lignin and biphenyl structures has been demonstrated.

Most of the low-quality feedstuff contains high concentrations of cellulose, small amounts of protein and fat, and relatively high ash contents when compared with high-quality feedstuffs. However, they contain large amounts of total digestible nutrients and energy values. The nutritional value of these low-quality feedstuffs can be increased significantly if they are effectively processed before feeding. Considerable potential exists in upgrading fodder using solid-state fermentation employing ligninolytic fungi. Apart from improving digestibility and feed value of straw, the mycostraw obtained after solid-state fermentation has potential for biogas production. After microbial removal of lignin, the straw components are more accessible for anaerobic digestion. Mycostraw may constitute a technical substrate for many biotechnological processes (Kuhad *et al.,* 1993).

Lignocellulosic waste from pulp and paper industries as well as wastes from dairy and agricultural industries is the potential substrate for use in the production of SCP for feed and food purposes. The two possibilities of converting cellulose to SCP are:

1. Growing the microbe directly on cellulosic substrate for microbial protein production.
2. Production of extracellular cellulase enzyme that can be used for saccharification of cellulosic substrates.

Spent sulfite liquor (available as a byproduct of paper pulping) is widely used as a carbohydrate source and contains about 3% fermentable sugars. The organisms used for SCP include bacteria *Cellulomonas* and *Lactobacillus*; molds *Aspergillus, Penicillium, Chaetomium* and *Polyporus*; the yeasts *Saccharomyces, Candida* and *Rhodotorula*, and the algae *Spirulina* and *Chlorella* (Kuhad *et al.,* 1993; 1997b).

## Food

Fungal cellulases have been shown to improve the yield of sugars from unmodified malts in the brewing process. In production of alcohol from cassava, addition of *Trichoderma* cellulases increases ethanol yields by conversion of cellulose to glucose. Cellulases used to improve and speed up colour extraction by pectinases from the skin of fruits also have an important application as a part of an enzyme complex in the maceration of fruits and vegetables.

Currently, the most commercially viable process for bioconversion of agricultural lignocellulosic waste is composting to provide substrate for growth of edible fungal fruiting bodies. Fungal fruiting bodies are rich in protein. *Agaricus, Lentinus, Volvariella and Pleurotus* are the dominant taxa cultured for human consumption. In Europe, the organism of interest is *Agaricus bisporus.* In the Far East, a wide variety of mushrooms, especially the oyster mushroom, *Pleurotus ostreatus*, and shitake are cultivated.

Xylanases are particularly useful for applications in food processing (Wong and Saddler, 1992). Enzymatic pretreatment of cereal and millet flours using xylanase is an attractive development as it can reduce subsequent processing times in fermentation of traditional foods. Xylanase have also been used for clarifying juices and wines, for extracting coffee and plant oils, for improving nutritional properties of silage, for macerating plant cell walls, for producing food thickeners and for providing different textures to bakery products (Subramaniyam and Prema, 2002). They are used in the baking industry for improving desirable texture, loaf volume and shelf-life of bread. A xylanase, Novozyme 867, has shown excellent performance in the wheat separation process (Christopherson *et al.,* 1997).

Xylanases have also been proposed for use in the processing of cereals. Addition of xylanase causes loss in wheat flour dough water absorption and consistency, as well as changes in loaf texture (McCleary, 1986). Xylanases have also been used to enhance the recovery of starch from wheat flours. The major industrial use of α-galactosidases is in the hydrolysis of raffinose in the sugarbeet industry

(Linden, 1982). Limited endo-hydrolysis or removal of side groups may be used for the modification of polymers, e.g., galactoglucomannans used as thickeners in the food industry.

White-rot fungi are cultivated on lignin and cellulose containing substrates such as cotton stalks, wheat/rice straw etc. As the white rot fungi can decompose lignocellulose efficiently without chemical or biological pretreatment, a large variety of lignocellulosic wastes can be utilized and recycled for the production of edible mushrooms. It offers well-proven, low-cost profitable technology of a good quality human food (Cohen *et al.,* 2002). Wine and beer stabilization is one of the main applications of laccase in the food industry (Minussi *et al.,* 1998; Minussi *et al.* 2002). These are also involved in the fruit juice processing, i.e. in the extraction and stabilization of juices (Piacquadio *et al.,* 1997).

## Environment

Environmental pollutants, such as phenols used in detergents, dyes and agricultural chemicals, and aromatic hydrocarbons found in gasoline, paints, polyurethanes and waterproof leisure wear, are known to damage the environment and affect all the living organisms. New approaches have been developed to combat environmental contaminants. This includes establishing ways to degrade these compounds using bacteria and fungi as well as genetically modifying such microorganisms.

The demand for fuel ethanol is expected to rise as a safer alternative to methyl tertiary butyl ether (MTBE), the most common additive of gasoline used to provide cleaner combustion. MTBE is a common groundwater contaminant. Lignocellulosic materials are cost effective substrates for bioethanol production as discussed above in the bioconversion section.

Ligninolytic cutures of *Phanerochaete chrysosporium* and *Trametes versicolor* have been reported to degrade PCB, anthracene, fluorine, phenanthrene, benzo[a]pyrene and dichloroaniline (Morgan *et al.,* 1991; Field *et al.,* 1992; Collins *et al.,* 1996; Baldrian *et al.,* 2000). White rot fungi are thus gaining interest because of their capability to degrade a wide variety of natural and synthetic materials and environmentally persistent organopollutants (Pointing 2001). Fungi are capable of degrading large spectrum of azo, anthracene and triphenylmethane dyes (McMullan *et al.,* 2001, Tekere *et al.,* 2001). White-rot fungi are the most widely studied microorganisms for dye decolorization/degradation (Reddy 1995). The ability of these fungi to degrade such a range of organic compounds results from the relatively non-specific nature of their ligninolytic enzymes, such as lignin peroxidase (LiP), manganese peroxidase (MnP) and laccase (Edens *et al.,* 1999). Laccases and peroxidases can convert a range of azo dyes including the ones carrying a phenolic substituent in para-position to the azo bond and additional methyl- or methoxy-substituents in 2- or 2,6-position in relation to the hydroxy group and certain complex azo dyes (Chivukula and Renganathan 1995; Schliephake *et al.,* 2000) and a wide variety of synthetic dyes making them suitable for the treatment of wastewater from the textile industry (Murugesan *et al.,* 2003; Zille *et al.,* 2003).

The dehalogenation ability of fungal laccase was demonstrated by the formation of dechlorinated C-C linked dimers during the oligomerization of 2-hydroxy-5-chlorobiphenyl and 3-chloro-4-hydroxybiphenyl. Including the nonhalogenated dimers, both monohalogenated and dihalogenated dimers were present (Schultz *et al.,* 2001). Laccase from *Chaetomium thermophilum* has been suggested to be involved in the humification process during composting as this enzyme can remain active for a long period of time at high temperatures and alkaline pH values (Chefetz *et al.,* 1998).

## Chemical Industry

Many substances like bisphenol A (BPA), nonylphenols (NP), phthalic esters (PAEs) have potential to disrupt the hormonal system and are called endocrine disrupters (EDs). As the most of natural hormones and artificial EDs are phenolic compounds and they are therefore good substrates for the ligninolytic enzymes. Recently, lignin-degrading enzymes have been applied to the detoxification of endocrine-disrupting chemicals (Tanaka *et al.,* 2003). Alkyl phenols, bisphenols, and natural and synthetic estrogens adsorbed on soil can be treated with laccases from different white-rot basidiomycetes (Tanaka *et al.,* 2001). Vanillin and gallic acid, the two frequently used chemicals in the pharmaceutical and chemical industries (Howard *et al.,* 2003), are known to be synthesized by lignocellulolytic fungus *Phanerochaete chrysosporium* (Ribbons, 1987).

## Biosensors

A biosensor is a device that detects, transmits and records information regarding a physiological or biochemical change. A number of biosensors containing laccase have been developed for immunoassays, glucose determination, aromatic amines and phenolic compound determinations (Minussi *et al.,* 2002). A carbon paste biosensor modified with a crude enzymatic extract of the *Pleurotus ostreatus* fungi as a laccase source is proposed for catecholamine determination in pharmaceutical formulations. This enzyme catalyzes the oxidation of adrenaline or dopamine in the corresponding quinones and the current obtained in the electrochemical reduction of each of the products is related to the concentration of these catecholamines in the sample solution. Catecholamines play an important role in the central nervous system as neurotransmitters. They also affect the regulation of blood pressure and metabolic processes. They possess important intrinsic pharmacological properties and are used for the correction of hemodynamic disorders associated with shock episodes (Leite, 2003).

## CONCLUSION

The energy and environmental crises which the world is experiencing is forcing us, among other things, to re-evaluate the efficient utilization or finding alternative uses for natural, renewable resources, especially organic "waste", using clean technologies. Developing countries are still grappling with socio-economic issues including meeting the massive energy-shortage demands, food security and developing technological solutions in the agriculture, agro-processing and other related manufacturing sectors. Lignocellulose biotechnology offers significant opportunities to developing countries for addressing some of the issues highlighted since most of the technology is based on the utilization of the readily available residual plant biomass considered "waste" or "residue" to produce numerous value-added products. Enzyme-produced sugar-syrups, xylitol, and glucose and fructose sweeteners can replace conventional sugar.

The high cost of cellulase is a major barrier to bioethanol industry from biomass mediated by enzymatic hydrolysis. Although, commercially available cellulase preparations are currently being employed in non-biomass conversion applications such as food processing, detergents and textiles, currently these markets present a much higher price for cellulase than can be afforded by any potential bioethanol process.

The technology of lignocellulose bioconversion, from a capital costs investment perspective, is an attractive technology since its biodegradation could follow solid-state fermentation comparable to silage or mushroom production, thus making such technology suitable for farms and small industrial plants without the need for large engineering infrastructure. It is also important to emphasize that in order for lignocellulose biotechnology to make meaningful impact on developing countries; suitable bioconversion processes need to be developed on a much wider scale and these countries should begin to pull their meager resources and biological science expertise in a cooperative and integrated manner towards modern, advance genomics and proteomics technologies for identifying novel lignocellulolytic enzymes and engineering enzymes with improved activities suitable for industrial-scale application.

## REFERENCES

Adler, E. (1977). Lignin chemistry: past, present and future. *Wood Sci Technol* 11, 69.

Akin, D.E., Rigsby, L.L., Sethuraman, A., Morrisson, W.H., Gamble, G.R., Eriksson, K.-E.L. (1995). Alterations in structure, chemistry, and biodegradability of grass lignocellulose treated with the white rot fungi *Ceriporiopsis subvermispora* and *Cyathus stercoreus. Appl Environ Microbiol* 61,159.

Akin D.E., Sethuraman A., Morrison W.H., Martin S.A. and Eriksson K.E. (1993). Microbial delignification with white rot fungi improves forage digestibility. *Appl Environ Microbiol* 59, 4272-4282.

Arias M.E., Arenas M., Rodriguez J., Soliveri J., Ball A.S. and Hernandez M. (2003). Kraft pulp biobleaching and mediated oxidation of a nonphenolic substrate by laccase from *Streptomyces cyaneus* CECT 3335. *Appl Environ Microbiol* 69, 1953-1958.

Baldrian P., Wiesche C., Gabriel J., Nerud F. and Zadrazil F. (2000). Influence of cadmium and mercury on activities of ligninolytic enzymes and degradation of polycyclic aromatic hydrocarbons by *Pleurotus ostreatus* in soil. *Appl Environ Microbiol* 66, 2471-2478.

Beauchemin, K.A., Colombatto, D., Morgavi, D.P. and Yang, W.Z. (2003). Use of Exogenous Fibrolytic Enzymes to Improve Feed Utilization by Ruminants. *J. Anim. Sci.* 81, E37-E47.

Beauchemin, K.A., Morgavi, D.P., Mcallister, T.A., Yang, W.Z. and Rode, L.M. (2001). Recent Advances in Animal Nutrition, Nottingham University Press, p. 296-322.

Beg,Q.K., Kapoor, M., Mahajan, L. and Hoondal, G.S. (2001). Microbial xylanases and their industrial applications: a review. *Appl Microbiol Biotechnol* 56, 326-338.

Beguin P. and Aubert J.P. (1994). The biological degradation of cellulose. *FEMS Microbiol Rev* 13, 25.

Bhat, M.K. (2000). Cellulases and related enzymes in biotechnology. *Biotechnol Adv* 18, 355-383.

Biely P. (1985). Microbial xylanolytic systems. *Trends Biotechnol* 3, 286.

Bisaria, V.S. and Mishra, S. (1989). Regulatory aspects of cellulase biosynthesis and secretion. *Crit Rev Biotechnol* 9, 61.

Blanchette, R.A. Burnes, T.A., Eerdmans, M.M. and Akhtar, M. (1992). Evaluating isolates of *Phanerochaete chrysosporium and Ceriporiopsis subvermispora* for use in biological pulping processes. *Holzforschung* 46,109.

Call, H.P. and Mucke, I. (1997). History, overview and applications of mediated lignolytic systems, especially laccase-mediator-systems (Lignozymeâ -process). *J Biotechnol* 53,163-202.

Calza, R.E., Irwin, D.C. and Wilson, D.B. (1985). Purification and characterization of two b-1-4-endoglucanases from *Thermomonospora fusca. Biochemistry.* 24, 7797.

Cassab, G.I. and Varner, J.E. 1988. Cell wall proteins. *Annu Rev Plant Physiol Plant Mol Biol* 39, 321.

Chefetz B., Chen Y. and Hadar Y. 1998. Purification and characterization of laccase from *Chaetomium thermophilum* and its role in humification. *Appl Environ Microbiol* 64, 3175-3179.

Chivukula, M. and Renganathan V. (1995). Phenolic azo dye oxidation by laccase from P*yricularia oryzae. Appl Environ Microbiol* 60, 4374-4377.

Christopherson, C., Anderson, E., Jokobsen, T.S., Wagner, P. (1997). Xylanases in wheat separation. *Starch* 49, 5-12.

Cohen, R., Persky, L., Hadar, Y. (2002). Biotechnological applications and potential of wood degrading mushrooms of the genus *Pleurotus*, *Appl Microbiol Biotechnol* 58, 582-594.

Collins, P.J., Kotterman, M.J.J., Field, J.A. and Dobson, A.D.W. (1996). Oxidation of anthracene and benzo(a)pyrene by laccase from *Trametes versicolor. Appl Environ Microbiol* 62, 4563.

Dekker, F.H.R. (1985). In: Biosynthesis and Biodegradation of wood components, (ed.), T. Higuchi Academic Press, Pronto 521-526.

Edens, W.A., Goins, Y.Q., Dooley, D. and Henosn, J.M. (1999). Purification and chraracterization of secreted laccase of *Gaeumannomyces graminis* var. *tritici. Appl Environ Microbiol* 45, 1741-1747.

Eriksson, K-E.L., Blanchette, R.A., Ander, P. (1990). Microbial and enzymatic degradation of wood and wood components, Springer, Berlin Heidelberg NewYork.

Eriksson, K-E.L. (1981). Fungal degradation of wood components. *Pure Appl Chem* 53, 33.

Fairweather, J.K., Faijes, M., Diguez, H. and Planas, A. (2002). Specifity studies of bacillus 1,3-1,4-beta-glucanases and application to glycosynthase-catalyzed transglycosylation. *Chembiochem* 3, 866-873.

Field, J.A., deJong, E., Feijoo-Costa, G. and deBont J.A.M. (1992). Biodegradation of polycyclic aromatic hydrocarbons by new isolates of white rot fungi. *Appl Environ Microbiol* 58, 2219.

Gilbert, I.G. and Psao, G.T. (1983) Annu Rep Ferm Proc 6, 323.

Godfrey, T., West, S., editors. (1996). Industrial enzymology: the application of enzymes in industry. New York: Macmillan.

Grethlein, H.E., Converse, A.O. (1991). Characterization of fungal cellulases. *Biores Technol* 36, 37-50.

Grethlein, H.E. (1984). Pretreatment for enhanced hydrolysis of cellulosic biomass. *Biotech Adv* 2, 43-62.

Haemmerli, S.D., Schoemaker, H.E., Schmidt, H.W.H., Leisola, M.S.A. (1987). Oxidation of veratryl alcohol by the lignin peroxidase of *Phanerochaete chrysosporium.* FEBS Left. 220, 149-154.

Heise, O.U., Unwin, J.P., Klungness, J.H., Fineran, W.G., Sykes, M., Abubakr, S. (1996) Industrial scale up of enzyme-enhanced deinking of nonimpact printed toners. *Tappi* J 79, 207–212.

Himmel, M.E., Ruth, M.F. and Wyman C.E. (1999). Cellulase for commodity product from cellulosic biomass. *Curr Opinion Biotechnol* 10, 358.

Himmel, M.E., Adney, W.S., Baker, J.O., Elander, R., McMillan, J.D., Nieves, R.A., Sheehan, J., Thomas, S.R., Vinzant, T.B. and Zhang, M. (1997) In: Woodward, J. and Saha, B. (eds.) Fuels and Chemicals from Biomass, ACS Series 666, Washington, DC: American Chemical Society, p 2-45.

Hoq, M.M., Alam, M., Mohiuddin,G. and Gomes, I. (1992). Enzyme biotechnology in developing countries: production and application of thermostable xylanase for jute fibre upgradation. In: Michael, R.L., and Bose, A. (eds.) Harnessing biotechnology for the 21st century, American Chemical Society, Washington, pp 568-573.

Hoshino, E. (1991). Hyomen. 29, 709.

Howard, R.L., Abotsi, E., Jansen van Rensburg, E.L., and Howard, S. (2003). Lignocellulose biotechnology: issues of bioconversion and enzyme production. *Afr J Biotechnol* 2, 602-619.

Ishihara, M. and Shimizu, K. (1980). *Mok Gakk* 26, 811.

Janshekar, H. and Fiechter, A. (1983). *Adv Biochem Eng / Biotechnol*27, 119.

Karamyshev A.V., Shleev S.V., Koroleva O.V., Yaropolov A.I. and Sakharov I.Y. 2003. Laccase-catalyzed synthesis of conducting polyaniline. *Enz MicrobTechnol* 33, 556-564.

Kim, T.J., Ow, S. and Eom, T.J. (1991). Enzymatic deinking of wastepaper. *Tappi Pulping Conference*, p 1023.

Kirk, T.K. 1987. Biochemistry of lignin degradation by *P. chrysosporium*, Proc of the Fed of the Microbiol Soc Symp, p 43.

Kuhad, R.C. and Singh, A. (1993). Lignocellulose biotechnology: Current and future prospects. *Crit Rev Biotechnol* 13, 151-172.

Kuhad, R.C., Singh, A. and Eriksson, K-E.L. (1997a). Microorganisms and enzymes involved in the degradation of plant fibre cell walls. *Adv Biochem Eng Biotechnol* 57, 45-125.

Kuhad, R.C., Singh, A., Tripathi, K. K., Saxena, R.K., and Eriksson, K.-E.L. (1997b) Microorganisms as an alternative source of protein. *Nutr Rev* 55, 65-75.

Kulys, J., Vidziunaite, R. and Schneider, P. (2003). Laccase-catalyzed oxidation of naphthol in the presence of soluble polymers. Enz Microb Technol 32, 455-463.

Ladish, M.R., Lin, K.W., Valoch, M. and Tsao, G.T. (1983). Enz Microb Technol. 35:156.

Lamed, R., Bayer, E. A. (1993). In: Genetics, biochemistry, and ecology of lignocellulose degradation. (eds) Shimada, K., Ohmiya, K., Kobayashi,Y., Hoshinos, S., Sakka, K., Karita, S. UNI, Tokyo, p.1.

Lange, N.K. (1993). In: Suominen, P., Reinikainen T(eds) *Trichoderma reesei* cellulases and other hydrolases. Foundation for Biotechnical and Industrial Fermentation Research, Espoo, Finland, p 263.

Leite, (2003). Determination of catecholamines in pharmaceuticals formulations using a biosensor modified with a crude extract of fungi laccase. *J Braz Chem Soc* 14, 297-303.

Linden, J.C. (1982). *Enz Microb Technol* 4, 130.

Lynd, L.R., Wyman, C.E., and Gerngross, T.U. (1999). Biocommodity engineering. *Biotechnol Prog* 15, 777-793.

Maddox, I.S. (1996). Microbial production of 2, 3-butanediol. In: Roehr, M. (ed) Biotechnology. VCH, Weinheim, 6, 269-291.

McCleary, B.V., Gibson, T.S., Allen, H., Gomes, T.C. (1986). *Starch* 38: 433.

McDonald, R. (1969). The pulping of wood, McGraw Hill, New York. p 34.

McMullan, G., Meehan, C., Conneely, A., Kirby, N., Robinson, T., Nigam, P., Banat, I. M., Marchant, R. and Smyth, W.F. (2001). Mini-review: microbial decolorisation and degradation of textile dyes. *Appl Microbiol Biotechnol* 56, 81-87.

Minussi, R.C., Pastore, G.M. and Duran, N. (1998). Phenol removal and stabilization of red wine in the presence of laccase from *Trametes versicolor*. In Brazilian Meeting on the Chemistry of food and beverages. Sao Carlos-Brazil. Abstr. 10.

Minussi, R.C., Pastore, G.M. and Duran, N. (2002). Potential applications of laccase in the food industry. *Trends Food Sci Technol* 13, 205-216.

Montane, D., Salvado, J., Torras, C. and Farriol X. (2002). High-temperature dilute-acid hydrolysis of olive stones for furfural production. *Biom Bioene* 22, 295-304.

Morgan, P.S., Lewis T. and Watkinson R.J. (1991). Comparison of abilities of white rot fungi to mineralize selected xenobiotic compounds. *Appl Microbiol Biotechnology* 34, 693.

Murugesan, K. and Kalaichelvan, P.T. (2003). Synthetic dye decolourization by white rot fungi. *Ind J Exp Biol* 41, 1076-1087.

Oshida, T., Sakata, R., Inomata, T., Fukuyasu, T., Kohzakil, K., Kuwano, K. (1992). Effect of enzyme additives on pig production for environment protection in livestock industry. $38^{th}$ ICoMST, Clermont-Ferrand, France, p 113.

Onysko, K.A. (1993). Biological bleaching of chemical pulps: a review. *Biotechnol Adv* 11, 179-198.

Palcic, M.M. (1999). Biocatalytic synthesis of oligosaccharides. *Curr Opin Biotechnol* 10, 616-624.

Parajo, J.C., Dominquez, H.D. and Dominquez, J.M. (1998). Biotechnological production of xylitol. Part 1: Interest of xylitol and fundamentals of its biosynthesis. Biores. Technol. 65, 191-201.

Piacquadio, P., De Stefano, G., Sammartino, M. and Sciancalepore, V. (1997). Phenols removal from apple juice by laccase immobilized on $Cu^{2+}$ -chelate regenerable carrier. *Biotechnol Technol* 11: 515-517.

Pointing, S. B. (2001). Feasibility of bioremediation by white rot fungi. *Appl Microbiol Biotechnol* 57, 20-33.

Rabinovich,M. L., Melnik, M. S., Bolobova, A. V. (2002a). Microbial cellulases: A review. *Appl Biochem Microbiol* 38, 305-321.

Rabinovich, M.L., Melnik, M.S., Bolobova, A.V. (2002b). The structure and mechanism of action of cellulolytic enzymes. *Biochemistry* (Moscow) 67, 850-871.

Reddy, C.A. (1995). The potential of white rot fungi in the treatment of pollutants. *Curr Opin Biotechnol* 6, 320-328.

Ribbons, R.W. (1987). Chemicals from lignin. *Phil Trans R Soc Lond Ser A* 321, 485-494.

Roberto, I.C., Mussatto, S.I. and Rodrigues, R.C.L.B. (2003). Dilute-acid hydrolysis for optimization of xylose recovery from rice straw in a semi-pilot reactor. *Indust Crops* Prod. 17, 171-176.

Saha, B.C. (2003). Hemicellulose bioconversion. *J Ind Microbiol Biotechnol* 30, 279-291.

Sato, T. and Daimon, K. (1990). *Kakogijyutu.* 26, 183.

Schliephake, K., Maninwaring, D.E., Lonegran, G.T., Jones, I.K. and Baker, W.L. (2000) Transformation and degradation of the disazo dye Chicago Sky Blue by a purified laccase from *Pycnoporus cinnbarinus. Enz Microb Technol* 27, 100-107.

Schultz, A., Jonas, U., Hammer, E. and Schauer, F. (2001). Dehalogenation of chlorinated hydroxybiphenyls by fungal laccase. *Appl Environ Microbiol* 67, 4377-4381.

Singh, A. and Hayashi, K. (1995). Microbial cellulases: protein architecture, molecular properties and biosynthesis. *Adv Appl Microbiol* 40, 1-44.

Sun, Y. and Cheng, J. (2002). Hydrolysis of lignocellulosic material from ethanol production: a review. Biores Technol. 83, 1-11.

Stutzenberger, F.J. and Kehler, R. (1986). Cellulase biosysnthesis by *Thermonospora curvata J Appl Bacteriol* 61, 225.

Subramaniyam, S. and Prema, P. (2002). Biotechnology of microbial xylanases: Enzymology, Molecular Biology and Application. *Crit Rev Biotechnol* 22, 33-64.

Tanaka T., Nose M., Endo A., Fujii T. and Taniguchi M. (2003). Treatment of nonylphenol with laccase in a rotating reactor. *J Biosci Bioeng* 96, 541-546.

Tanaka T., Tonosaki T., Nose M., Tomidokoro N., Kadomura N., Fujii T. and Taniguchii M. (2001). Treatment of model soils contaminated with phenolic endocrine-disrupting chemicals with laccase from *Trametes* sp. in a rotating reactor. *J Biosci Bioeng* 92, 312-316.

Tekere, M., Mswaka, A.Y., Zvauya, R. and Read, J.S. (2001). Growth, dye degradation and ligninolytic activity studies on Zimbabwean white rot fungi. *Enzyme Microb Technol* 28, 420-426.

Vallander, L., Eriksson, K-E. L. (1990). *Adv Biochem Eng/Biotechnol* 42, 63.

Viikari, L., Kantelinen, A., Sundquist, J. and Linko, M. (1994). Xylanase in bleaching: from an idea to the industry. *FEMS Microbial Rev* 13, 335-350.

Ward, O.P., Singh, A. (2002) Bioethanol technology: developments and perspectives. *Adv Appl Microbiol* 51, 53-80.

Wariishi, H., Valli, K. and Gold, M. H. (1992). *J Biol Chem* 267, 23688.

Wong, K.K.Y. and Saddler, J.N. (1992a). Applications of hemicellulases in the food, feed, and pulp and paper industries. In Coughlan P.P., Hazlewood G.P. (eds) Hemicellulose and Hemicellulases. Portland Press, London, pp. 127-143.

Wong, K.K.Y. and Saddler, J.N. (1992b). *Trichoderma* xylanases: their properties and applications. In: Xylans and their xylanases. Visser *et al.* (eds) Elsevier, Amsterdam, pp 171-186.

Wood, T.M. (1991). Fungal cellulases. In: Biosynthesis and biodegradation of cellulose. Haigler *et al.* (eds) Macel Dekker Inc., New York, pp 491-534.

Youn, H-D., Hah, Y.C. and Kang, S-O. (1995). Role of laccase in lignin degradation by white-rot fungi. *FEMS Microbiol Lett* 132, 183.

Zille A., Tzanov T., Gubitz G. M. and Cavaco-Paulo A. (2003). Immobilized laccase for decolorization of reactive black 5 dyeing effluent. 25, 1473-1477.

Zimmermann, W. (1990). Degradation of lignin by bacteria. *J Biotechnol* 13, 119-130.

# 2

# Integrated Biosystems for Lignocellulosic Waste Management

**Murarilal and BMJ Pereira**

*Department of Biotechnology, Indian Institute of Technology Roorkee, Roorkee – 247 667 Uttaranchal, India.*

Scientists engaged in lignocellulosic research have developed biotechnologies for generating single products like sugars, ethanol, biofuels, high value biochemicals, etc., either by employing microbes or enzymes in bioconversion reactions. Considerable effort has also been put into making the process economically viable by optimizing the process conditions. However, for total mineralisation of the complex biomass with zero waste and zero emissions, several biotechnologies are to be combined in a specific sequence so as to evolve an integrated biosystem. Lignocellulose is one part of the complex biomass that requires to be released before being employed in a bioprocess. Indiscriminate and incomplete utilization of lignocellulosic material from total biomass will amount to tremendous wastes in resources and energy. The trend now is to integrate individual bioprocesses and come out with biotechnologies that are multi-product adhering to the zero waste/ zero emission concepts. Such an integrated approach would not only be ecofriendly and yield substantial profits but also ensure total recycling of biomass, complete mineralisation and minimum pollution. In this chapter we suggest how to blend biotechnologies to be in tune with nature's way of balancing carbon, nitrogen and sulfur cycles that can sustain plant, animal and human life.

## INTRODUCTION

Large amounts of lignocellulosic wastes pose an environmental pollution problem and unfortunately both developed and developing countries use the easiest option to get rid of this bulk by burning the biomass. The available technology suggests that biomass containing lignocellulosic materials represents an abundant and inexpensive source of sugars that can be microbiologically or enzymatically converted into industrial products (Howard *et al.,* 2003). Effective bioconversion requires an intimate understanding of the composition of the starting material that varies with the source from which it is derived. Lignocellulose is a complex of polysaccharide micro fibers most often formed by cellulose and hemicellulose filaments and covered by lignin layers. The lignin stabilizes the complex structure by protecting the polysaccharides against attack by hydrolytic enzymes and other external factors (Leonowicz *et al.,* 1999). The key technologies that are required for successful bioconversion of lignocellulosic biomass are therefore: 1. *Delignification* to liberate cellulose and hemicellulose from their complex with lignin, 2. *Depolymerization* of the carbohydrate polymers to produce free sugars, 3.*Bioutilization* of mixed sugars (glucose, xylose, arabinose, mannose and galactose) by fermentation, 4. *Downstream processing* for separation and purification of products/ byproducts and 5. *Recycling* of the byproducts/ wastes generated.

## PROBLEMS WITH LIGNOCELLULOSIC BIOMASS UTILIZATION

Lignin degradation is an oxidative process and needs the presence of oxygen at a partial pressure equal to that in the natural atmosphere (Reid and Seifert, 1980). In order to promote the delignification of lignocellulosic substrates it is essential to maximize the rate and specificity of lignin molecule degradation while avoiding polysaccharide consumption (Kerem and Hadar, 1995). Manganese promotes selective removal of lignin by white-rot fungi by increasing the activity of lignolytic enzymes (Perez and Jeffries, 1992). Perhaps, the Mn (III) ions oxidized by manganese peroxidases produce free radicals that preferentially degrade the aromatic structures present in the lignocellulose. Flavonoids are also reported to improve lignin degradation through an increased production of laccase in *Pleurotus ostreatus* (Ardon *et al.,* 1998). In short, effective biological delignification is possible if there is coordination between the lignin-degrading microorganisms, their ecophysiological requirements and optimal bioreactor/ bioprocess design.

The hydrolysis of lignocellulosic waste by chemical or enzymatic treatments or a combination of both, releases several simple sugars that could form the starting material for bioprocesses. Unfortunately, several other degradation products also turn out during the reaction that prove harmful to microbial fermentation. For example, acid hydrolysis of lignocellulosic materials produces several inhibitory compounds, such as sugar and lignin degradation products, compounds derived from the lignocellulosic structure and heavy metal ions. Their toxicity is a major factor limiting bioconversion processes that utilize hydrolysates. Pentose sugars degrade to furfurol and hexoses form hydroxymethyl furfural, both toxic compounds that inhibit cells, affect specific growth rates and thus considerably decrease the yield of the bioprocess. Similarly, a large variety of compounds (aromatic, polyaromatic, phenolic and aldehydic) are released from lignin.

Several other inhibitory products such as raw material extractives (acidic resins, taninic and terpenic acids, acetic acid etc.) are also discharged into hydrolysates. Sometimes, the products cause the internal pH to drop thereby modulating cell activity and even causing death. In brief, the fermentation of non-detoxified lignocellulosic hydrolysates shows slow kinetics with limited yield and productivity compared with fermentation of commercial sugars / pure substrates. Thus, to decide the kind of microorganisms to be employed and the kind of products that could be produced, it is essential to monitor the composition of hydrolysates. The identification of these compounds together with the degree of inhibition caused to the microorganisms would help to decide about the detoxification method so as to improve the overall efficiency of the fermentative processes.

## DETOXIFICATION METHODS

To start with, it would be ideal if the formation of inhibitors during hydrolysis were avoided. A combination of mechanical and enzymatic treatments for the biodegradation of lignocellulosic material could achieve this, primarily due to the fact that enzymes are substrate-specific. But when chemical hydrolysis is resorted to, a number of byproducts are formed and detoxification by pretreatment and neutralization of pH are essential before lignocellulosic hydrolysates are used for microbial fermentation. A variety of biological, physical and chemical techniques have been proposed to reduce the concentration of the inhibitors in lignocellulose hydrolysates (Mussatto and Roberto, 2004). However, it must be emphasized that there is no universal method for detoxification of lignocellulosic hydrolysates since the efficiency of any detoxification method depends both on the composition of the raw material used and the hydrolysis conditions employed. For example, vacuum evaporation is a method that could reduce the content of volatile compounds such as acetic acid, furfural and vanillin (Rodrigues *et al.,* 2001). But this procedure increases the concentration of non-volatile toxic compounds and consequently the degree of fermentation inhibition. Precipitation of toxic compounds, ionization of inhibitors by changing pH (Martinez *et al.,* 2000, 2001) and physical adsorption on substances like activated charcoal (Mussatto and Roberto, 2001) could be low-cost, yet effective, additional routes of detoxification. An alternative tactic would be to develop species of microorganisms able to resist inhibitors so that they could be employed in bioconversion processes. Besides naturally occurring microorganisms, some thermophilic anaerobes and recombinant bacteria have been tried for direct microbial conversions. The latest trend is the use of new metabolically engineered microbial species that can better tolerate inhibitors and decrease the need to detoxify hydrolysates (Taherzadeh *et al.,* 2000). But it may still become necessary to convert toxic compounds into products that do not interfere with metabolism.

## DIVERSITY OF BIOPRODUCTS

Several technologies are currently available for bioconversion of lignocellulose to a whole range of chemicals and/or value added products. These technologies continue to receive the attention of scientists and with the passage of time, it is expected that newer and improved quality products would be delivered at incredibly low prices. To make the product even more competitive, the industrial units are concentrating on agricultural and other cheap sources of raw material. Simultaneously, improvements in technologies are being attempted at all the levels of production, namely, process design, system optimization and model development. The industrial applications of some of the products generated through bioconversion of lignocellulosic material are consolidated in Table 1. The ever-increasing demand for bioproducts leads to over utilization of biomass that ultimately culminates in the accumulation of wastes. We need to address this issue using clean and ecofriendly technologies and one way to achieve this would be through integrated biosystems.

**Table 1. Some industrially important products derived from lignocellulosic material**

| *Starting material* | *Products* | *Application* | *Reference* |
|---|---|---|---|
| Biomass | Lignocellulose | Pulp & Paper | Breen and Singleton, 1999. |
| Lignocellulose | Lignin | Process fuel | Crawford *et al.,* 1984; Kuhad and Singh, 1993. |
| Lignin | Aminolignin | Sorption active material for dyes, heavy metals, phenol, enterosorbtion | Verkhanovsky *et al.,* 1982a; 1982b; Dizhbite *et al.,* 1999. |
| Lignin | Vanillin<br>Gallic acid | Herbicide, anti-foaming agents, anti-microbial agent, drugs | Ribbons, 1987.<br>Delgenes *et al.,* 1998; |
| Hemicellulose (xylose) | Xylitol | Sweetener, Teeth hardening Anti-microbial agent | Parajo *et al.,* 1998. |
| Hemicellulose (arabinose, xylose, etc.) | Furfurol | Furfural-phenol plastics, varnishes, pesticides | Gravitis *et al.,* 2004. |
| Cellulose (glucose) | Ethanol | Chemical feed stock, octane enhancer, petrol additive | Kuhad and Singh, 1993; Ingram and Doran, 1995; Lee, 1997; Kuhad *et al.,* 1997; Chandrakant and Bisaria, 1998; Millati *et al.,* 2005 |
| Lignocellulose | Residual lignified product from ethanol production | Co-product for industrial feedstock, nutritional antioxidants, ultraviolet absorbers, resins. | Anderson *et al.,* 2005 |

## COMPOSTING OF LIGNOCELLULOSIC WASTES

The microbiological process of composting lignocellulosic materials arises due to the combined activities of a succession of a mixed microbial population. During this microbial succession, each type of microbe is suited to one specific condition of short duration and is only active in decomposing one type or group of organic matter present (Wong *et al.,* 1990). Ultimately, organic material is converted into carbon dioxide, humus and heat by compost microorganisms. Although it would appear that lignin could be degraded or extensively transformed during composting, the lignin loss in compost environment is hardly due to extensive mineralisation. Lignin degradation in composts is regulated by temperature, the original lignin content and the thickness of material. The organisms most efficient at mineralising lignin do not survive the thermophilic phase of composting and thus cannot play any significant role in mineralisation (Tuomela *et al.,* 2000). However, from literature it is evident that lignin could be degraded by thermophilic microorganisms. The idea is not merely to degrade lignin but to make sure that the process yields another product of commercial value. We suggest that the compost be utilized first for mushroom cultivation so that the spent compost that retains high lignin content can still be available for generating biofuels.

## UTILIZATION OF LIGNOCELLULOSIC COMPOST FOR MUSHROOM CULTIVATION

Agro-wastes have been used since ages for cultivation of mushrooms on a commercial scale with varying degrees of success. In one study, the mycelium extension rates were evaluated on seven different substrates: wheat straw, cotton, gin-trash, peanut shells, poplar sawdust, oak sawdust, corncobs and olive press-cake (Zervakis *et al.,* 2001). The mycelium extension rates (linear growth and colonization rates) were found to be the highest on cotton gin-trash, peanut shells and poplar sawdust for *Pleurotus spp.* and *A. aegerita.* Wheat straw, peanut shells and particularly cotton gin-trash supported fast growth of *V. volvacea,* whereas wheat straw was the most suitable substrate for *L. edodes* and *A. auricula-judae.* Supplemented oak sawdust and olive press-cake were poor substrates for most species examined, while almost all strains performed adequately on corncobs.

The utilization of compost for growing and harvesting mushrooms increases their nutritional value and is considered a means of value addition (Fermor, 1993; Basak *et al.,* 1996). Today, mushroom cultivation is seen as a direct utilization of their ecological role in the bioconversion of solid wastes generated from industry and agriculture into edible biomass, which could also be regarded as a functional food or as a source of drugs and pharmaceuticals (Chiu *et al.,* 2000) The possibility of utilizing industrial, urban, and other wastes for the growth of a product which is directly edible by humans is fascinating. However, it should be borne in mind that many wastes containing toxic substances, like, heavy metals and agrochemicals could reach the food chain and produce adverse effects on human health (Bressa *et al.,* 1988).

## RECOVERY OF LIGNOCELLULOSE-DEGRADING ENZYMES FROM SPENT MUSHROOM COMPOST

Spent mushroom compost, a bulky solid waste generated from the mushroom industry, can no doubt be exploited as a soil fertilizer and as a prospective bioremediating agent (Chiu *et al.,* 2000). But before doing so, we may add another bioprocess that can assist in greater utilization of lignocellulosic wastes. Since the compost is first produced by the activity of a consortium of microorganisms following which the mushrooms are grown, it is obvious that the spent compost contains several ectoenzymes of both bacterial and fungal origin (Alam *et al.,* 1994). Analysis of extracts from spent compost reveals a range of lignocellulose degrading activities like peroxidases, the xylan-debranching enzymes (acetylesterase, arabinofuranosidase) and the cellulose degrading activities (endoglucanase, cellobiohydrase and β-glucosidase). The composition, recovery, activity and stability of the enzyme cocktails from compost extracts suggest that they could be exploited for preparing high-grade animal fodder (Ball and Jackson, 1995). Thus, even after growing mushroom, there is still scope to use the compost to extract enzymes which is yet another product of commercial value.

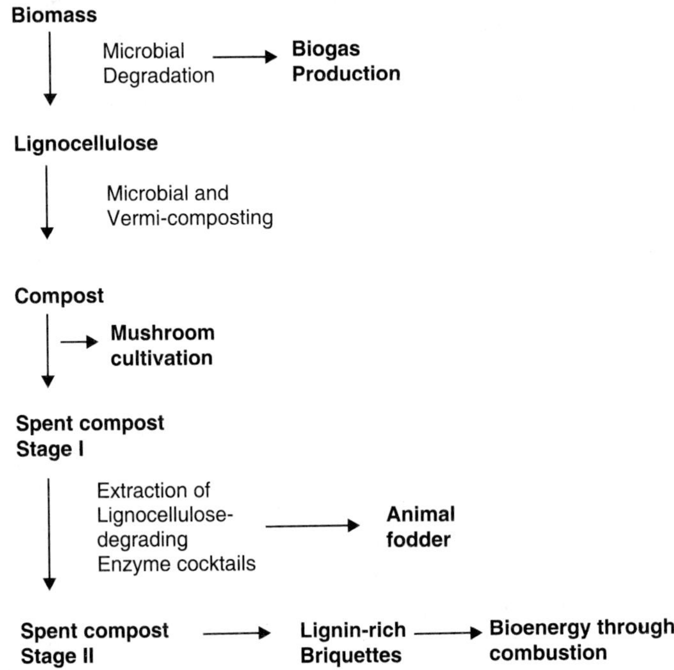


**Fig. 1.** A commercially viable pathway for integrating biosystems towards attaining zero waste.

## ENERGY PRODUCTION FROM LIGNOCELLULOSIC WASTES

Biomass combustion is the oldest and most popular method for energy production. Besides combustion, other technologies of the thermal conversion of biomass for energy production include gasification, liquefaction and pyrolysis. The advantages of the combustion method for waste treatment are an approximately 10-fold decrease in waste volume, the reduction of the risk for soil and water pollution and the recovery of the heat formed. Although the direct combustion of biomass for energy production is undoubtedly, rather attractive, a considerably higher economic effect can be reached by a multistage biomass conversion. A considerable share of bioenergy is now substantially contributing to the total energy balance of many countries owing to the high energy yielding combustion of lignin-containing wastes. After calculating the economics including labour and water, the feasibility of large-scale lignocellulose-based bioenergy production in a renewable intensive global energy scenario (RIGES) has been aggressively advocated (Berndes *et al.,* 2001). It has been estimated that the energy of 1 kg of lignin is roughly equivalent to that of 0.6 kg of oil. At present special attention is paid to the production of fuel briquettes and granules with and without binders. Briquettes and granules are also produced using energy additives (coal siftings, mazut, organic materials etc). When we talk about integrated biosystems for energy production while preparing these briquettes and granules, we need to think in terms of mixing lignocellulosic wastes with activated sludge of water treatment plants. These technologies have a significant advantage since the products obtained may be used as the energetical or technological fuel while in the combustion method only the energetical use of heat from the raw material is possible. The analysis of energy consumption patterns reveals that the world is heading towards utilization of bioenergy. There is already a move to mix ethanol with petrol in different proportions to sustain and meet our ever-increasing energy demands. With the price of crude oil touching US $ 61.50 a barrel there is no longer a need to increase the competitiveness of biofuel through tax subsidies.

## PRODUCTION OF ANIMAL FEEDS

Due to the phenomenal expansion in the agro-industrial activity in recent times we are now faced with a grave situation of accumulating lignocellulosic residues. This has resulted in the development of several processes to not only find ways to recycle such material but also convert them into products of economic value. Most of the agro-industrial byproducts are poor in nutrients such as proteins and vitamins but rich in fibres with low digestibility and therefore cannot be directly consumed. Microbial bioconversion of these wastes is seen as a practical and promising avenue to transform them into palatable animal feed with a simultaneous increase in their nutritional value. Microorganisms grow fast and produce protein in large amounts. Further, the growth of microbes on lignocellulosic wastes is able to furnish all the hydrolytic enzymes needed for preparation of feeds. However, it may be cautioned that some microbial products can also contain toxic metabolites harmful to animal and human health and therefore the risk should be properly evaluated. Nevertheless, the diversity of the material and the microorganisms employed permit a range of products to be produced suitable for different categories of animals (Villas-Boas *et al.,* 2002).

Genetic manipulation has no doubt resulted in higher levels of productivity in domestic and farm animals. In order to match their growth, production and reproduction levels there is now a demand for feeds of not only higher nutrients but also balanced compositions. The pressure to minimize cost and improve quality has forced the feed industry, to use substrates that were not used in the feed production in the past. They are constantly, evaluating, processing and using new ingredients through advanced technologies. The time has now come to define the type of bioconversion process since the procedures for obtaining high quality feed for improved and non-improved animals are not the same. Byproducts that were thus far being wasted are now incorporated into diets with astonishing results. The additives improve the level of essential aminoacids, vitamins and minerals. Some prove to be metabolic modifiers, anti-microbial agents, probiotics etc., enhancing growth and preventing/protecting against diseases (Wenk, 2000). Thus, animal feed production is yet another sector where lignocellulosic wastes can generate revenue without polluting the environment.

### Bio-refinery

This is a recent concept that separates plant biomass (lignocellulosic materials) into its building blocks (phenols and sugars) and is a methodological equivalent to oil refinery (Gravitis and Della

Senta, 2001). Bio-refinery technologies could generate value-added products that might range from basic food ingredients to complex pharmaceuticals and from simple building materials to complex industrial composites. The principal advantage of bio-refinery in comparison with oil refinery and other fossil fuel consumption systems is that biomass is renewable and a sustainable resource. To provide a complete wood and non-wood plant biomass refinery into value added chemicals, materials and biofuels, an integrated processing scheme has been recently proposed (Gravitis *et al.,* 2004). It allows producing a variety of products such as furfural, acetic acid, glucose, bioethanol, levoglucosan, microcrystalline cellulose, fibre materials, plastics, carbon materials, charcoal activated carbon and others (Tengerdy and Szakacs, 2003). A part of the lignocellulose bulk produces gas (Ekwenchi *et al.,* 1990) and the unused part of lignin are burnt in a boiler house to allow biomass processing with energy. Biogas has a slightly higher calorific value than natural gas and can be used to run generators for electricity production as well as cooking in households. It is advisable to resort to biogas production through anaerobic digestion to remove the ever-increasing infectious disease outbreaks especially in tropic developing countries. The strict anaerobic conditions required for a successful methane production kills most pathogens responsible for infectious diseases. This can not only remove the health hazard but also produce energy and promote aquaponics. The solids can be used as fertilizers directly or through composting which will ultimately improve and regenerate the microbial population in the soil, crucial for natural cycles of matter.

Structuring diverse industries into integrated clusters would essentially mean that the waste of one industry becomes the raw material for the next so that we are left with zero emissions (no waste in the air, water and soil). In addition, what is treated as waste in one chain of production is converted into value-added inputs in another. Products or part of them are also used as intermediary products for further processing into products of higher added value. The net result being the manufacturing line becomes a series of production cycles and recycling systems. Such an integration of various production systems facilitates increasing the diversity of value added products, competitiveness and solves environmental problems (Chang, 1987).

## From Concept to Implementation

Several projects have been undertaken to fully exploit renewable resources and replace monoculture/ mono-product farming with a multi-product system all over the globe. Each of these is tailored to the demands of the society in which it operates and differs from country to country. The Montfort Boy's Town school in Fiji has set up an excellent fully operational integrated biosystem for us to observe and emulate. The established fish farming industry and its main economic resource, sugar were the base upon which the integration was designed and built. The system has now evolved to comply with a near zero waste/zero emission level.

Tropical countries have a great opportunity to use their abundant lignocellulosic wastes to their advantage using the biotechnologies enumerated above. Several strategies that are in use for converting these wastes into useful products and byproducts have been reviewed and discussed. No doubt, ongoing research is going to ensure that the processes become more and more profitable. But it would be ideal if a scheme for clustering of processes is developed wherein multiple products are generated from a single raw material. We propose here an integrated model system (Figure 2) for the sugarcane-growing belts of India, which is self-explanatory. In this scheme we have shown how the biomass and the lignocellulosic wastes could be effectively recycled with minimum energy losses and pollution. The importance is the integration of production and management of waste that is simple, yet, flexible according to the needs, specific situations and constraints. Such an integrated approach would not only lead to proper and total utilization of lignocellulosic wastes but also promote the evolution of green and cleaner technologies with zero emission.

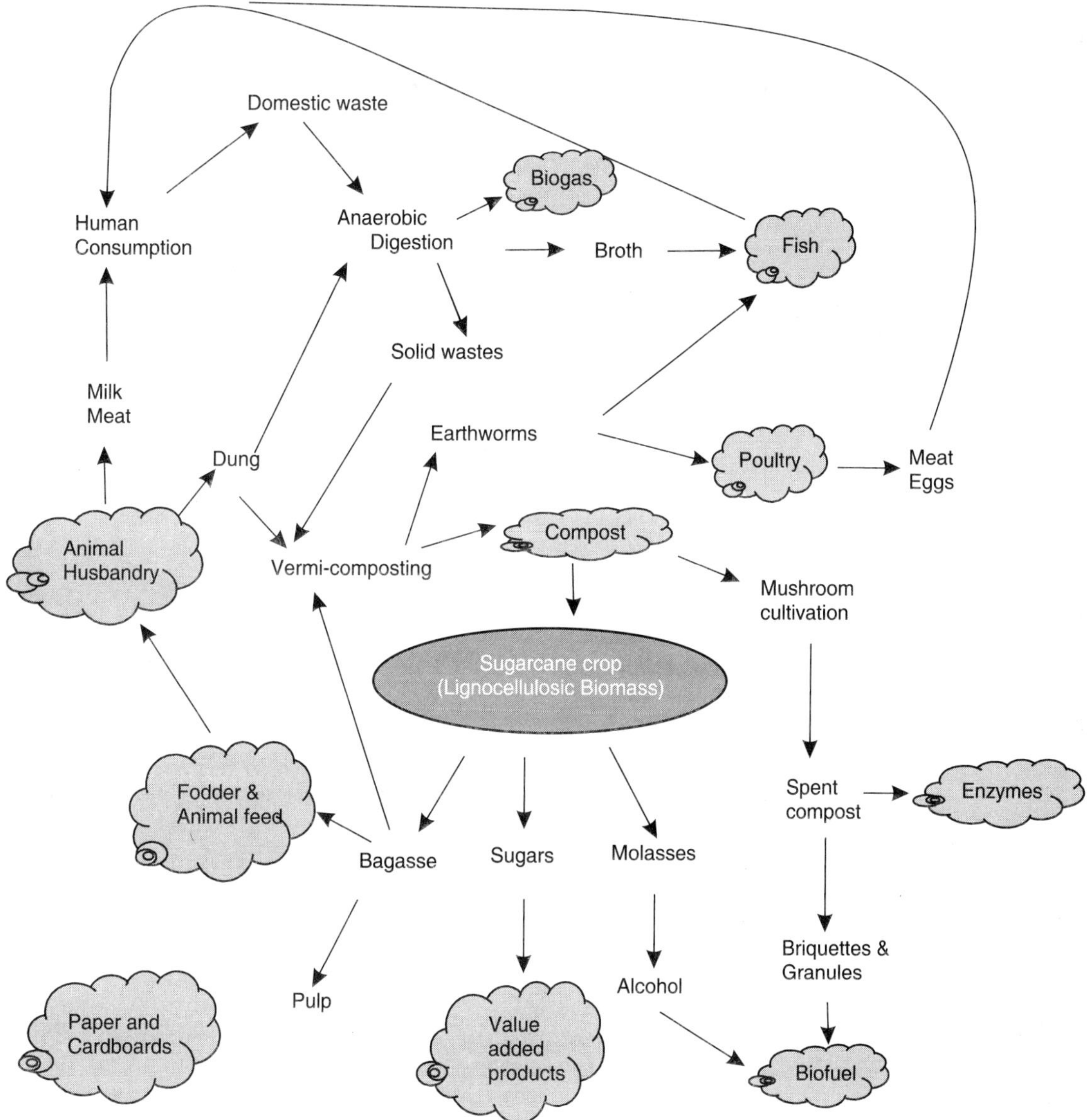


**Fig. 2.** A proposal of a multi-product integrated biosystem that could be adopted for sugarcane growing belts of the Indian subcontinent.

## REFERENCES


Alam, M., Gomes, I., Mohiuddin, G. and Hoq, M.M. (1994). Production and characterization of thermostable xylanases by *Thermomyces lanuginosus* and *Thermoascus aurantiacus* grown on lignocelluloses. *Enzyme and Microbial Technology,* 16, 298-302.

Anderson, W.F., Peterson, J., Akin, D.E. and Morrison, W.H. (2005). Enzyme pretreatment of grass lignocellulose for potential high-value co-products and an improved fermentable substrate. *Applied Biochemistry and Biotechnology,* 124, 303-310.

Ardon, O., Kerem, Z. and Hadar, Y. (1998). Enhancement of lignin degradation and laccase activity in *Pleurotus ostreatus* by cotton stalk extract. *Canadian Journal of Microbiology,* 44, 676-680.

Ball, A.S. and Jackson, A.M. (1995). The recovery of lignocellulose-degrading enzymes from spent mushroom compost. *Bioresource Technology,* 54, 311-314.

Basak, M.K., Chanda, S., Bhaduri, S.K., Mondal, S.B. and Nandi, R. (1996). Recycling of jute waste for edible mushroom production. *Industrial Crops and Products,* 5, 173-176.

Berndes, G., Azar, C., Kaberger, T. and Abrahmson, D. (2001). The feasibility of large-scale lignocellulose-based bioenergy production. *Biomass and Bioenergy,* 20, 371-383.

Breen, A. and Singleton, F.L. (1999). Fungi in lignocellulose breakdown and biopulping. *Current Opinion in Biotechnology,* 10, 252-258.

Bressa, G., Cima, L. and Costa, P. (1988). Bioaccumulation of Hg in the mushroom *Pleurotus ostreatus. Ecotoxicology and Environmental Safety,* 16, 85-89.

Chandrakant, P. and Bisaria, V.S. (1998). Simultaneous Bioconversion of Cellulose and Hemicellulose to Ethanol. *Critical Reviews in Biotechnology,* 18, 295-331.

Chang, S.T. (1987). Microbial biotechnology — Integrated studies on utilization of solid organic wastes. *Resources and Conservation,* 13, 75-82.

Chiu, S.W., Law, S.C., Ching, M.L., Cheung, K.W. and Chen, M.J. (2000). Themes for mushroom exploitation in the 21st century: Sustainability, waste management, and conservation. *Journal of General and Applied Microbiology,* 46, 269-282.

Crawford, D.L., Pometto, A.L. and Crawford, R.L. (1984). Production of useful modified lignin polymers by bioconversion of lignocellulose with *Streptomyces. Biotechnological Advances,* 2, 217-232.

Delgenes, J.P., Escare, M.C., Laplace, J.M., Moletta, R. and Navarro, J.M. (1998). Biological production of industrial chemicals, i.e. xylitol and ethanol, from lignocelluloses by controlled mixed culture systems. *Industrial Crops and Products,* 7, 101-111.

Dizhbite, T., Zakis, G., Kizima, A., Lazareva, E., Rossinskaya, G., Jurkjane, V., Telysheva, G. and Viesturs, U. (1999). Lignin — a useful bioresource for the production of sorption-active materials. *Bioresource Technology,* 67, 221-228.

Ekwenchi, M.M., Akunwanne, B.U., Okeke, N.R. and Ekpenyong, K.I. (1990). Gaseous fuel production from fungal lignocellulose degradation. *Fuel,* 69, 1569-1572.

Fermor, T.R. (1993). Applied aspects of composting and bioconversion of lignocellulosic materials: An overview. *International Biodeterioration and Biodegradation,* 31, 87-106.

Gravitis, J. and Della Senta, T. (2001). Global prospects of substituting oil by biomass. In: *World Forests, Markets and Policies,* Eds Palo, M., Uusivuori, J. and Mery, G., Kluwer Academic Publishers, Amsterdam Chapter 2. Pp 23-39.

Gravitis, J., Zandersons, J., Vedernikov, N., Kruma, I. and Ozols-Kalnins, V. (2004). Clustering of bio-products technologies for zero emissions and eco-efficiency. *Industrial Crops and Products* 20, 169-180.

Howard, R.L., Abots, E., Jansen van Rensburg, E.L. and Howard, S. (2003). Lignocellulose biotechnology: issues of bioconversion and enzyme production. *African Journal of Biotechnology,* 2, 602-619.

Ingram, L.O. and Doran, J.B. (1995). Conversion of cellulosic materials to ethanol. *FEMS Microbiology Reviews,* 16, 235-241.

Kerem, Z. and Hadar, Y. (1995). Effect of manganese on preferential degradation of lignin by *Pleurotus ostreatus* during solid-state fermentation. *Applied and Environmental Microbiology,* 61, 3057-3062.

Kuhad, R.C. and Singh, A. (1993). Enhanced production of cellulases by *Penicillium citrinum* in solid state fermentation of cellulosic residue. *World Journal of Microbiology and Biotechnology,* 9, 100-101.

Kuhad, R.C., Singh, A. and Eriksson, K.E. (1997). Microorganisms and enzymes involved in the degradation of plant fiber cell walls. In: *Advances in Biochemical Engineering Biotechnology,* Ed. Eriksson, K.E., Springer-Verlag, Germany. Pp 46-125.

Lee, J. (1997). Biological conversion of lignocellulosic biomass to ethanol. *Journal of Biotechnology,* 56, 1-24.

Leonowicz, A., Matuszewska, A., Luterek, J., Ziegenhagen, D., Wojtas-Wasilewska, M., Cho, N.S., Hofrichter, M. and Rogalski, J. (1999). Biodegradation of lignin by white rot fungi. *Functional and General Biology,* 27, 175-185.

Martinez, A., Rodriguez, M.E., York, S.W., Preston, J.F. and Ingram, L.O. (2000). Effect of $Ca(OH)_2$ treatments ('Overliming') on the composition and toxicity of bagasse hemicellulose hydrolysates. *Biotechnology and Bioengineering,* 69, 526-536.

Martinez, A., Rodriguez, M.E., Wells, M.L., York, S.W., Preston, J.F. and Ingram, L.O. (2001). Detoxification of dilute acid hydrolyzates of lignocellulose with lime. *Biotechnology Progress,* 17, 287-293.

Millati, R., Edebo, L. and Taherzadeh, M.J. (2005). Performance of *Rhizopus*, *Rhizomucor*, and *Mucor* in ethanol production from glucose, xylose, and wood hydrolyzates. *Enzyme and Microbial Technology,* 36, 294-300.

Mussatto, S.I. and Roberto, I.C. (2001). Hydrolyzate detoxification with activated charcoal for xylitol production by *Candida guilliermondii. Biotechnology Letters,* 23, 1681-1684.

Mussato, S.I. and Roberto, I.C. (2004). Alternatives for detoxification of diluted-acid lignocellulosic hydrolyzates for use in fermentative processes: a review. *Bioresource Technology,* 93, 1-10.

Parajo, J.C., Dominguez, H. and Dominguez, J.M. (1998). Biotechnological production of Xylitol. Part 3: Operation in culture media made from lignocellulosic hydrolysates. *Bioresource Technology,* 66, 25-40.

Perez, J. and Jeffries, T.W. (1992). Roles of manganese and organic acid chelators in regulating lignin degradation and biosynthesis of peroxidases by *Phanerochaete chrysosporium. Applied and Environmental Microbiology,* 58, 2402-2409.

Reid, I.D. and Seifert, K.A. (1980). Lignin degradation by *Phanerochaete chrysosporium* in hyperbaric oxygen. *Canadian Journal of Microbiology,* 26, 11658-11671.

Ribbons, R.W. (1987). Chemicals from lignin. *Phil Trans Royal Society of London Ser. A,* 321, 485-494.

Rodrigues, R.C.L.B., Felipe, M.G.A., Almeida e Silva, J.B., Vitolo, M. and Gomez, P.V. (2001). The influence of pH, temperature and hydrolysate concentration on the removal of volatile and non-volatile compounds from sugarcane bagasse hemicellulosic hydrolysate treated with activated charcoal before and after vacuum evaporation. *Brazilian Journal of Chemical Engineering,* 18, 299-311.

Taherzadeh, M.J., Niklasson, C. and Liden, G. (2000). On-line control of fed-batch fermentation of dilute-acid hydrolysates. *Biotechnology and Bioengineering,* 69, 330-338.

Tengerdy, R.P. and Szakacs, G. (2003). Bioconversion of lignocellulose in solid substrate fermentation. *Biochemical Engineering Journal,* 13, 169-179.

Tuomela, M., Vikman, M., Hatakka, A. and Itavaara, M. (2000). Biodegradation of lignin in a compost environment: A review. *Bioresource Technology,* 72, 169-183.

Verkhanocsky, V.G., Nikiforov, A.F. and Pushkarev, V.V. (1982a). Sorption of synthetic cation dyes by hydrolysis lignin. *Khimya Technologiya Vody,* 4, 29-31.

Verkhanocsky, V.G., Skorokhodov, V.I., Nikiforov, A.F. and Akhmanaeva, G.B. (1982b). Extraction of heavy metals from aqueous solutions by sorption by hydrolysis lignin. *Zhur Prikl Khim,* 56, 2606-2608.

Villas-Boas, S.G., Esposito, E. and Mitchell, D.A. (2002). Microbial conversion of lignocellulosic residues for production of animal feeds. *Animal feed Science and Technology,* 98, 1-12.

Wenk, C. (2000). Recent advances in animal feed additives such as metabolic modifiers, antimicrobial agents probiotics, enzymes and highly available minerals-review. *Asian Australian Journal of Animal Science,* 13, 86-95.

Wong, P., Au, F., Fong, W. and Chan, K. (1990). Bacterial flora isolated from spent mushroom composts with special reference to cellulolytic bacteria. *Microbios,* 61, 153-167.

Zervakis, G., Philippoussis, A., Ioannidou, S. and Diamantopoulou, P. (2001). Mycelium growth kinetics and optimal temperature conditions for the cultivation of edible mushroom species on lignocellulosic substrates. *Folia Microbiologica (Praha),* 46, 231-234.

# PART II

## LIGNOCELLULOLYTIC MICROORGANISMS AND ENZYMES

# 3

# Biodiversity of Ligninolytic Fungi

**Sarika Kuhar[1], Mukesh Kapoor[1], Rajeev Kapoor[1], Krishna Kant Sharma[1], Ajay Singh[2] and R.C. Kuhad[1]**

*[1]Department of Microbiology, University of Delhi, South Campus, New Delhi-110021, India*
*[2]Department of Biology, University of Waterloo, Waterloo, Ontario, Canada*

Besides three major classes of wood-rots (soft-rot, brown-rot and white-rot), their different enzymatic profile still makes the pattern of delignification complex and the biodiversity of fungi so wide. Only white-rot has the potential to completely degrade all three major components of wood, thus making them ecologically most important to study their diversity in detail. For germination and growth, they require several essential conditions, which are listed as source of infection, suitable substrate (food), moisture, oxygen and suitable temperature. Various ligninolytic enzymes produced by white-rot fungi work synergistically to decay the wood efficiently resulting in the establishment of composite system. White-rot fungi has been classified into six main groups depending on the production of different combinations of enzymes and into two main groups depending upon the degradation of three wood polymers cellulose, hemicellulose and lignin at different rates and extent. To study the wood decay in detail, scanning electron microscopy was used as powerful tool to directly observe and select microorganisms with a capacity to preferentially degrade lignin. Thus reviewing biodiversity of microorganism's especially white-rot fungi, organisms with required characteristics would be preferred for their exploitation in industrial biotechnology.

## INTRODUCTION

The nature has maintained the biodiversity of fungi so broad that different wood-decaying fungi are present with three major classes of wood-rots and then their different enzymatic profile, making the pattern of delignification complex. They require different conditions for their growth and to cause wood decay. Also, various ligninolytic enzymes work in synergism to decay the wood efficiently resulting in the establishment of composite system. With all the decades, these factors are conserved in the nature so that different fungi adapt in accordance to conditions existing in the ecosystem and complete their task of recycling the carbon of lignified tissues as white-rot fungi can only completely mineralize it (Kuhad *et al.,* 1997; Aro, 2005). These factors will be important to understand the diversity of ligninolytic fungi and their application in industry. In this chapter, biodiversity of ligninolytic fungi is reviewed including different classes of wood-rot, conditions favourable for growth of fungi, enzymology of ligninolytic system, classification of white-rot fungi based on enzyme profile, synergistic action of ligninolytic enzymes for wood decay, different patterns of wood decay and microscopic features of wood decay.

## WOOD-ROT AND ITS TYPES

By colonizing dead or dying tree trunks and stumps, the fungus utilizes one or the other cell wall constituents preferentially, resulting in the wood decay known as wood-rot. These include three types of wood-rot i.e., soft-rot, brown-rot and white-rot which are based on the component utilized and the characteristics of the decayed wood (Table 1). Of these only white-rot has the potential to completely degrade all three major components of wood, thus making them ecologically most important to study their diversity in detail. These fungi mainly belong to Ascomycetes, Deuteromycetes or Basidiomycetes groups (Kuhad *et al.,* 1997).

**Table 1. Types of Wood rots**

| *Type* | *Family to which Wood rot belongs* | *Texture of Wood after attack by Wood rot* | *Components of Wood Degraded* |
|---|---|---|---|
| Soft-rot | Ascomycetes and Deuteromycetes | Wood remains wet, spongy, pitted and brownish. Usually grows on surface of wood and are least specialized among wood rots | Prefer carbohydrates and modify lignins to a limited extent |
| Brown-rot | Basidiomycetes | Wood remains fibrous and brownish | Primarily carbohydrates degraded, leaving behind brownish modified lignin but doesn't degrade it |
| White-rot | Basidiomycetes and Ascomycetes | Wood remains fibrous and whitish as cellulose mainly is left behind | All components of wood degraded |

**Soft-rot** causes degradation of cellulose and hemicellulose but only partially digests lignin. They are particularly prevalent at the early stages of wood decay and in conditions of high moisture and increased nitrogen content (Blanchette, 1995). Wood affected by soft rot may appear wet, spongy, or pitted. There are over 300 species of known soft-rots, which are either the members of Ascomycetes or Deuteromycetes (Kuhad *et al.,* 1997) with *Chaetomium*, *Paecilomyces* and *Fusarium* (Rayner and Boddy, 1988; Eriksson *et al.,* 1990; Blanchette, 1995) as typical examples.

The fungi capable of degrading cellulose and hemicellulose but are unable to digest the lignin component of wood belong to **Brown-rot** category. In this case, the lignin remains more or less intact and becomes modified with brown and crumbly matrix appearance (Eriksson *et al.,* 1990; Blanchette, 1995; Highley and Dashek, 1998). Unlike the soft-rots, the brown-rot fungi are relatively few in number, comprising only 6% of all wood-rotting Basidiomycetes (Gilbertson 1980; Rayner and Boddy, 1988; Kuhad *et al.,* 1997). *Poria*, *Polyporus*, *Coprinus* are some examples of brown-rot fungi (Buswell and Odier, 1987; Rayner and Boddy, 1988; Eriksson *et al.,* 1990; Straatsma *et al.,* 1994; Blanchette, 1995; Dix and Webster, 1995).

**White-rot fungi** are capable of degrading all the major components of wood (cellulose, hemicellulose, and lignin). The fungi causing white-rots mostly belong to Basidiomycetes (Table 2), but some Ascomycetous fungi have been also identified to cause wood decay (Eriksson *et al.,* 1990) such as *Xylaria hypoxylon* and *X. polymorpha* commonly known as candle snuff fungus and dead man's finger, respectively (Deacon 1997; Liers *et al.,* 2006). *Ustulina vulgaris* (Kuhad *et al.,* 1997) have been reported to aid in lignin degradation. Other laccase producing ascomycetes are *Rhizoctonia solani* (Wahleithner *et al.,* 1996), *Aspergillus nidulans* (Law and Timberlake, 1980), *Podospora anserina, Neurospora crassa* (Tamura and Inoue 1989) and *Gaeumannomyces graminis* var. *tritici* (Edens *et al.,* 1999). The role of Ascomycetous fungi in wood degradation has yet to be explored in details. Basidiomycotina attack either hardwood or softwood, while Ascomycotina probably degrade only hardwood (Kirk and Farrell, 1987).

**Table 2. Different wood-decaying fungi with their family and ligninolytic enzyme profile**

| *S. No.* | *Wood Rot* | *Family* | *Enzymes Produced* | *Reference* |
|---|---|---|---|---|
| 1. | *Armillaria mellea* (Honey fungus, Bootlace fungus) | Marasmiaceae | MnP/Lac | Robene-Soustrade *et al.,* 1992 |
| 2. | *Armillaria ostoyae* | Marasmiaceae | MnP/Lac | Robene-Soustrade *et al.,* 1992 |
| 3. | *Auricularia sp.* (Wood ear, jew's ear) | Auriculariaceae | MnP/Lac | Hofrichter and Fritsche, 1997; Saparrat *et al.,* 2002 |
| 4. | *Bjerkandera adusta* | Hapalopilaceae | LiP/MnP | Kimura *et al.,* 1990; Kaal *et al.,* 1993; Pickard *et al.,* 1999; Moreira *et al.,* 2000a,b; Wang *et al.,* 2001 |
| 5. | *Ceriporiopsis subvermispora* | Hapalopilaceae | MnP/Lac | Ruttimann *et al.,* 1992a,b; Lobos *et al.,* 1994; |
| 6. | *Cerrena maxima* | Polyporaceae | Lac | Koroleva *et al.,* 2002a |
| 7. | *Chrysonilia sitophila* | - | LiP/MnP | Duran *et al.,* 1987a,b Ferrer *et al.,* 1992 |
| 8. | *Clitocybula dusenii* | Tricholomataceae | MnP/Lac | Ziegenhagen and Hofrichter, 1998; Scheel *et al.,* 2000; Nuske *et al.,* 2002 |
| 9. | *Coriolopsis gallica* | Polyporaceae | Lac | Pickard *et al.,* 1999b |
| 10. | *Coriolopsis occidentalis* | Polyporaceae | LiP | Nerud *et al.,* 1989 |
| 11. | *Coriolopsis polyzona* | Polyporaceae | MnP/Lac | Nerud *et al.,* 1991 |
| 12. | *Coriolopsis rigida* | Polyporaceae | MnP/Lac | Saparrat *et al.,* 2002 |
| 13. | *Cyathus bulleri* | Nidulariaceae | LiP/MnP/Lac | Dhawan and Kuhad 2002 |
| 14. | *Cyathus stercoreus* | Nidulariaceae | MnP/Lac | Orth *et al.,* 1993; Dhawan and Kuhad 2002 |
| 15. | *Daedalea flavida* | | Lac | Arora and Gill, 2001 |
| 16. | *Daedaleopsis confragosa* | Polyporaceae | LiP/MnP | De Jong *et al.,* 1992 |
| 17. | *Dichomiotus squalens* (Red rot) | Polyporaceae | MnP/Lac | Nerud *et al.,* 1991; Orth *et al.,* 1993 ; Périé and Gold 1999; |
| 18. | *Fomes fomentarius* | Polyporaceae | LiP | Varela *et al.,* 2000 |
| 19. | *Fomes lignosus* | Polyporaceae | LiP | Huoponen *et al.,* 1990; Waldner *et al.,* 1988 |
| 20. | *Fomes sclerodermeus* | Polyporaceae | MnP/Lac | Papinutti *et al.,* 2003 |
| 21. | *Ganoderma applanatum* | Ganodermataceae | LiP | Varela *et al.,* 2000 |
| 22. | *Ganoderma australe* | Ganodermataceae | LiP | Varela *et al.,* 2000 |
| 23. | *Ganoderma lucidum* | Ganodermataceae | LiP/MnP/Lac | Orth *et al.,* 1993; D'Souza *et al.,* 1999 |
| 24. | *Ganoderma valesiacum* | Ganodermataceae | MnP/Lac | Nerud *et al.,* 1991 |
| 25. | *Grifola frondosa* | Meripilaceae | MnP/Lac | Orth *et al.,* 1993 |
| 26. | *Irpex lacteus* | Steccherinaceae | LiP/MnP/Lac | Novotny *et al.,* 2000; Máximo and Costa-Ferreira, 2004 |
| 27. | *Junghuhnia separabilima* | - | LiP/Lac | Vares et al 1992 |
| 28. | *Lentinula edodes* (Shiitake) | Marasmiaceae | MnP/Lac | Daniel *et al.,* 1990; Forrester *et al.,* 1990; Orth *et al.,* 1993; Hatvani and Mecs 2001, 2002; Nagai *et al.,* 2003 |

| | | | | |
|---|---|---|---|---|
| 29. | *Lentinus degener* | Polyporaceae | LiP | Varela *et al.*, 2000 |
| 30. | *Lentinus tigrinus* | Polyporaceae | MnP/Lac | Moreira *et al.*, 2000 |
| 31. | *Marasmius quercophilus* | Tricholomataceae | MnP/Lac | Tagger *et al.*, 1998; Dedeyan *et al.*, 2000; Klonowska *et al.*, 2001 |
| 32. | *Nematoloma frowardii* | Strophariaceae | LiP/MnP/Lac | Hofrichter *et al.*, 1999; Nuske *et al.*, 2002 |
| 33. | *Oudemansiella radicata* | Marasmiaceae | LiP/Lac | Huttermann *et al.*, 1989 |
| 34. | *Panus tigrinus* | Tricholomataceae | MnP/Lac | Maltseva *et al.*, 1991; Golovleva *et al.*, 1993; Fenice *et al.*, 2003 |
| 35. | *Peniophora gigantea* | Peniophoraceae | LiP | Varela *et al.*, 2000 |
| 36. | *Perenniporia medulla-panis* | Polyporaceae | MnP | Orth *et al.*, 1993 |
| 37. | *Phanerochaete sordida* | Phanerochaetaceae | MnP | Moreira *et al.*, 2000 |
| 38. | *Phanerochaete chrysosporium* | Phanerochaetaceae | LiP/MnP | Glenn *et al.*, 1983; Tein and Kirk, 1983; Glenn and Gold, 1985; Schoemaker and Leisola, 1990; Kaal *et al.*, 1995; Gill and Arora 2003; Moldes *et al.*, 2003; Grgic and Perdih 2003 |
| 39. | *Phanerochaete flavido-alba* | Phanerochaetaceae | LiP/MnP | Hamman *et al.*, 1999; Varela *et al.*, 2000 |
| 40. | *Phellinus gilvus* | Hymenochaetaceae | MnP/Lac | Sapparat *et al.*, 2002 |
| 41. | *Phellinus igniarius* | Hymenochaetaceae | Lac | Szklarz *et al.*, 1989 |
| 42. | *Phellinus pini* | Hymenochaetaceae | LiP/MnP | Blanchette *et al.*, 1989; Bonnarme and Jeffries, 1990 |
| 43. | *Phlebia brevispora* | Meruliaceae | LiP/MnP/Lac | Perez and Jeffries, 1990; Rüttiman *et al.*, 1992b; Arora and Rampal, 2002 |
| 44. | *Phlebia fascicularia* | Meruliaceae | Lac | Arora and Gill, 2001; Arora and Rampal, 2002 |
| 45. | *Phlebia floridensis* | Meruliaceae | Lac | Arora and Gill, 2001; Arora and Rampal, 2002 |
| 46. | *Phlebia ochraceofulva* | Meruliaceae | LiP/Lac | Vares *et al.*, 1993 |
| 47. | *Phlebia radiata* | Meruliaceae | LiP/MnP/Lac | Hatakka *et al.*, 1987; Saloheimo *et al.*, 1989; Karhunen *et al.*, 1990; Niku-Paavola *et al.*, 1988, 1990; Moreira *et al.*, 2000; Arora and Gill 2001; Arora and Rampal 2002 |
| 48. | *Phlebia subserialis* | Meruliaceae | MnP | Bonnarme and Jeffries 1990 |
| 49. | *Phlebia tremellosa* | Meruliaceae | LiP/MnP/Lac | Ralph *et al.*, 1996; Vares *et al.*, 1994 Robinson *et al.*, 2001b |
| 50. | *Pleurotus eryngii* | Pleurotaceae | LiP/MnP | Martinez *et al.*, 1996; Munoz *et al.*, 1997a,b; Heinfling *et al.*, 1998a,b; Orth *et al.*, 1993 |
| 51. | *Pleurotus florida* | Pleurotaceae | LiP/Lac | Huttermann *et al.*, 1989; Das *et al.*, 1999 |
| 52. | *Pleurotus ostreatus* (Oyster mushroom) | Pleurotaceae | LiP/MnP/Lac | Waldner *et al.*, 1988; Sarkar *et al.*, 1997; Palmieri *et al.*, 1997; Ha *et al.*, 2001; Morais *et al.*, 2001b; Reddy *et al.*, 2003; |
| 53. | *Pleurotus pulmonarius* | Pleurotaceae | MnP/Lac | Masaphy and Levanon, 1992; Orth *et al.*, 1993; De Souza *et al.*, 2002 |
| 54. | *Pleurotus sajor-cajo* | Pleurotaceae | LiP/MnP/Lac | Boyle *et al.*, 1992; Hatakka, 1994; Tuor *et al.*, 1995; Chagas and Durrant, 2001; Reddy *et al.*, 2003 |
| 55. | *Pleurotus sapidus* | Pleurotaceae | MnP/Lac | Orth *et al.*, 1993 |
| 56. | *Polyporus brumalis* | Polyporaceae | LiP/Lac | Huttermann *et al.*, 1989 |

| | | | | |
|---|---|---|---|---|
| 57. | *Polyporus ciliatus* | Polyporaceae | MnP/Lac | Moneira *et al.*, 2000 |
| 58. | *Polyporus osteiformis* (Brown rot fungus) | Polyporaceae | LiP | Dey *et al.*, 1991 |
| 59. | *Polyporus pinsitus* | Polyporaceae | LiP/Lac | Huttermann *et al.*, 1989 |
| 60. | *Polyporus platensis* | Polyporaceae | LiP/Lac | Huttermann *et al.*, 1989 |
| 61. | *Polyporus varius* | Polyporaceae | LiP | Huttermann *et al.*, 1989 |
| 62. | *Polyporus versicolor* | Polyporaceae | MnP | Orth *et al.*, 1993 |
| 63. | *Pycnoporus cinnabarinus* | Polyporaceae | Lac | Lomascolo *et al.*, 2002 |
| 64. | *Pycnoporus sanguineus* | Polyporaceae | LiP/MnP/Lac | Pointing *et al.*, 2000; Lomascolo *et al.*, 2002 |
| 65. | *Rigidoporus lignosus* | Meripilaceae | MnP/Lac | Galliano *et al.*, 1991; Cambria *et al.*, 2000 |
| 66. | *Stereum hirsutum* | Stereaceae | MnP/Lac | Nerud *et al.*, 1991; Moneira *et al.*, 2000 |
| 67. | *Trametes cingulata* | Polyporaceae | MnP/Lac | Orth *et al.*, 1993; Tekere *et al.*, 2001 |
| 68. | *Trametes elegans* | Polyporaceae | MnP/Lac | Tekere *et al.*, 2001 |
| 69. | *Trametes gibbosa* | Polyporaceae | LiP/MnP/Lac | Nerud *et al.*, 1991 |
| 70. | *Trametes hirsuta* (Hairy stereum) | Polyporaceae | LiP/MnP/Lac | Nerud *et al.*, 1991; Koroleva *et al.*, 2002a,b; Rosales *et al.*, 2002 |
| 71. | *Trametes pocas* | Polyporaceae | MnP/Lac | Tekere *et al.*, 2001 |
| 72. | *Trametes trogii* | Polyporaceae | LiP/MnP/Lac | Levin and Forchiassin, 2001; Levin *et al.*, 2002 |
| 73. | *Trametes versicolor* (Turkey tail) | Polyporaceae | LiP/MnP/Lac | Blanchette *et al.*, 1989; Huoponen *et al.*, 1990; Black and Reddy, 1991; Jonsson *et al.*, 1992; Johansson *et al.*, 1993; Hatakka, 1994; Couto *et al.*, 2002d, 2003; Gill and Arora 2003; |
| 74. | *Trametes villosa* | Polyporaceae | MnP | De Jong *et al.*, 1992 |

Overlapping habits have been found in the three main eco-physiological groups of fungi (wood-decaying, mycorrhiza-forming and litter-decomposing fungi, LDF). Wood decay fungi such as *Hypholoma* spp. (*Nematoloma* spp.), *Pleurotus* spp., *Armillaria* spp. are also capable of colonizing soil in contact with wood debris (Dix and Webster 1995). Other litter decomposing fungi can also grow on straw (e.g. *Stropharia rugosoannulata*) that is favoured only by wood-decay fungi (Haselwandter *et al.,* 1990). *Volvariella volvacea*, a litter-decomposing fungi belonging to Pluteaceae family, secretes ligninolytic enzymes (laccase) (Chen *et al.,* 2004). *Agaricus bisporus* is another LDF which secretes laccases and manganese peroxidases (Bonnen *et al.,* 1994). The ability to break down lignin and cellulose enables some of the LDF to function as typical white-rot fungi in soil (Hofrichter, 2002). Diversity of all these groups of fungi is broad and hence they work together explaining their importance in the carbon cycle (Dix and Webster, 1995).

## CONDITIONS FAVOURABLE FOR FUNGI TO CAUSE WOOD DECAY

Decay of wood will depend on the conditions to which it is exposed and whether these conditions are favourable for fungal growth. There are several essential conditions for germination and growth such as source of infection, suitable substrate (food), moisture, oxygen and suitable temperature. Some other factors such as light, fungal interactions, pH, level of nitrogen etc. also play an important role in either inhibiting or increasing the growth of decay fungi.

## SOURCE OF INFECTION

If healthy wood is in contact with infected wood, then disease can easily spread to the sound wood by normal growth of the fungal hyphae from the decayed wood, even without the production of spores. Infection may also spread directly from soil to wood in contact with it, because most soils contain organic matter inhabiting microbes. Regardless of any contact between healthy wood and infected materials, the interspaces can be bridged by airborne spores. These airborne spores are produced during the fruiting phase, which under favourable conditions germinate on the wood surface and produce filaments called hyphae to invade the wood (Carll and Highley, 1999).

## SUBSTRATE

Only wood decay fungi (especially white-rot fungi) are able to utilize the complex organic substrates. Wood contains only small amounts of easily usable substrates such as simple sugars and starch. The principal components of wood are **cellulose** (40-50% dry weight, composed of long, straight chains of β-1,4-linked glucose), **hemicelluloses** (25-40%, consisting of mixed polymers of glucose, mannose, xylose, arabinose, etc.) and **lignin** (20-35%; a complex 3-dimensional polymer consisting of three types of phenyl-propane units). Out of which, lignin is highly resistant to microbial and enzymatic attack, and it encrusts the other wall materials or forms chemical complexes with them, preventing them from being degraded easily (Kuhad *et al.,* 1997). Lignin degradation by white-rot fungi is faster than that of any other organisms and they are responsible for most of the lignin decomposition in nature. However, the growth substrate is not only lignin, but also hemicelluloses and cellulose (Kirk and Farrell, 1987; Blanchette, 1995). It has been demonstrated that at least one of the polysaccharide in wood must be degraded simultaneously with lignin (Kuhad *et al.,* 1997). In cellulaseless mutants, hemicellulose was also removed when lignin was degraded (Gold and Alic, 1993). This preference may be due to the fact of close spatial relationship of lignin and hemicellulose within the cell wall (Kirk and Cullen, 1998). Also, host specificity of wood rotters has been found to influence decay pattern (Tuor *et al.,* 1995). With change of tree species, some white-rot fungi are even capable to show different patterns of decay and the examples are *Ganoderma applanatum* and *Ischnoderma resinosum.* Some fungi have a preference for certain cell types. Earlywood cells appeared to be delignified by *Dichomitus squalens* and *Inonotus dryophilus*, whereas a preference for latewood was depicted by *Phellinus pini* (Otjen and Blanchette, 1982; Blanchette, 1984).

## MOISTURE

The development of fungi on wood is largely controlled by the moisture content. Wood moisture content is defined as the weight of water in wood that is expressed as a fraction or more commonly as a percentage of the weight of oven dry wood. It is necessary for the substratum to be moist and the humidity of the surrounding atmosphere to be high for the germination of fungal spores. Moisture conditions conducive to decay are around 30% moisture content, which is above the fibre saturation point (refers to as the moisture content at which cell walls are completely saturated but there in no free water in the cell cavities according to Tiemann in 1906). This moisture content actually depends on the species of fungi that has attacked the wood and the kind of wood. Also it should be well below the waterlogged conditions. Moisture content when present below than 20% in the wood, the fungal development is completely inhibited (Panshin and De Zeeuw, 1964; Scheffer and Verrall, 1973; Zabel and Morrell, 1992).

## OXYGEN

All wood-rotting fungi require some air for growth, and many species die quickly if they are deprived of it. The air-moisture balance in the cells within the wood, therefore, is a most important factor controlling the susceptibility of wood to decay. Fungi need oxygen for the oxidation of sugars, which they use for growth and the supply of energy. When cell spaces are completely filled with water, as they are when wood is submerged in water, the air supply is cut off and growth is stopped.

## TEMPERATURE

Temperature is one of the most important factors affecting fungal growth (Tuomela *et al.,* 2000). The growth of wood-rotting fungi is affected by temperature in the way that most wood-inhabiting fungi would develop only between 15°C and 40°C (Humphrey and Siggers, 1933). There are variations in the response to temperature, and for each species there is an optimum temperature at which growth is most rapid. The optimum temperature generally lies between 25°C and 30°C (Dix and Webster, 1995). But some Basidiomycotina also grows well at elevated temperatures. *Phanerochaete chrysosporium* has an optimum temperature of 36-40°C (Mouchacca, 1997). *Ganoderma colossum* is another white-rot fungus, which is still capable of growing at 45°C and has an optimum temperature of 40°C (Adaskaveg *et al.,* 1990; 1995).

## OTHER FACTORS

In addition to the essential factors for the growth of wood-rotting fungi, there are other factors that may have an effect on growth. Light usually has a retarding influence, and exposure to intense light such as bright sunlight can kill the mycelium of some species, most species grow more vigorously in subdued light, but some do not grow normally in total darkness.

Interactions among the different fungi come into play when conditions are favourable for microbial invasion and hence different species of wood decay fungi may inhabit the same wood. Interactions may range from antagonism to co-existence and synergism among the different fungi (Zabel and Morrell, 1992).

The acidity and alkalinity of the wood can also have an effect. Most wood-rotting fungi flourish on materials that are slightly acidic, and very few can tolerate alkaline conditions (Dix and Webster, 1995).

Wood typically has a very low content of nitrogen (usually less than 0.1% by weight). This constraint has been overcome mainly by wood-rotting fungi. Indeed it has been found that low nutrient nitrogen level is often a prerequisite for lignin degradation (Eriksson *et al.,* 1990; Dix and Webster, 1995).

## ENZYMOLOGY OF LIGNINOLYTIC SYSTEM

Lignin is an integral cell wall constituent, which provides plant strength and resistance to microbial and enzymatic degradation (Argyropoulos and Menachem, 1997). This resistance has resulted in the development of lignin-degrading system, which faces major three challenges. Firstly, the lignin polymer is large, ligninolytic systems have to be extracellular. Secondly, the structure of lignin is composed of inter-unit carbon-carbon and ether bonds, hence oxidative degradative mechanism is required instead of hydrolytic. Thirdly lignin is stereo-irregular; therefore the system must be much less specific in order to degrade lignin. Only white-rot fungi have evolved to meet these challenges and act nonspecifically via production of extracellular oxidative enzymes (Kirk and Cullen, 1998). The three important extracellular oxidative ligninolytic enzymes (lignin peroxidase, manganese peroxidase and laccases) have been discussed here. Recently versatile peroxidase has also been reported.

## LIGNIN PEROXIDASE (LiP)

LiP (EC 1.11.1.14) has been isolated from several white-rot fungi as shown in Table 2. The extracellular N-glycosylated LiP with molecular masses between 38 and 47 kDa have acidic pIs and pH optima (Gold and Alic, 1993). Its active site contains heme moiety which is dependent on $H_2O_2$ and operates via a typical peroxidase catalytic mechanism. As compared to typical peroxidases, LiP are more powerful oxidants and consequently oxidize not only phenolic compounds, but also a variety of nonphenolic lignin structures and other aromatic ethers that resemble the basic structural unit of lignin (Kersten *et al.,* 1990; Barr and Aust, 1994a). It catalyzes a one-electron oxidation of nonphenolic aromatic compounds, forming the aryl cation radical (Schoemaker *et al.,* 1985), suggesting that oxidized intermediates of the enzyme possess a very high redox potential. The mechanism of LiP catalytic action on lignin is still uncertain, because it has not been clear whether LiP can oxidize lignin through a direct interaction or through radical mediation. The Veratryl (3,4-dimethoxybenzyl) alcohol (VA), cation radical has been proposed as a radical mediator to oxidize polymeric substrates with which LiP presumably cannot interact directly (Hammel and Moen, 1991; Schoemaker *et al.,* 1994; Barr and Aust, 1994b). However, it has been shown that ferrocytochrome *c* is oxidized by LiP/$H_2O_2$ in the absence of VA, suggesting that LiP is capable of oxidizing polymeric substrates at the protein surface by a long-range electron transfer mechanism (Schoemaker *et al.,* 1994; Wariishi *et al.,* 1994). LiP from *P. chrysosporium* has been isolated for study of binding properties against a synthetic lignin (dehydrogenated polymerizate, DHP). The kinetic and spectral studies strongly suggest that LiP is capable of oxidizing lignin directly at the protein surface by a long-range electron transfer process (Johjima *et al.,* 1999).

LiP may not be essential for the attack on lignin as several highly active WRF and litter-decaying fungi (e.g., *C. subvermispora, Dichomitus squalens, Panus tigrinus, Rigidosporus lignosus, Armillaria spp.*) do not excrete this enzyme (Galliano *et al.,* 1991, Hatakka, 1994, Maltseva *et al.,* 1991, Périé and Gold, 1991; Robene-Soustrade *et al.,* 1992; Ruttimann *et al.,* 1992a, b; Lobos *et al.,* 1994; Fenice *et al.,* 2003). Notwithstanding the problem of role of LiP in ligninolysis, LiP can efficiently result in Cα-Cβ cleavage and extracellular cleavage of aromatic rings that is characteristic of ligninolysis by white-rot fungus.

## MANGANESE PEROXIDASE (MnP)

MnP is one of the most common lignin degrading peroxidases produced by the majority of wood decaying fungi and by many litter-decomposing fungi (Hofrichter, 2002). Table 2 depicts the various white-rot fungi producing MnP (EC 1.11.1.13). It is an extracellular heme containing peroxidase with a requirement for $Mn^{2+}$ as its reducing substrate. These are glycosylated proteins with an iron protoporphyrin IX (heme) prosthetic group (Glenn and Gold, 1985; Nie *et al.,* 1999). Their molecular weights are slightly larger than LiPs and range between 32 and 62.5 kDa (Hofrichter, 2002) and these are secreted in multiple isoforms (Urzua *et al.,* 1995). Like LiPs, they too have a conventional peroxidase catalytic cycle, but with Mn (II) as the substrate (Gold *et al.,* 1989). The $Mn^{3+}$ formed is highly reactive and act as diffusible oxidants that can act at some distance from the enzyme. Mn(II) must be chelated by bidentate chelating organic acids such as oxalate, glycolate or malate (Cui and Dolphin, 1990; Kishi *et al.,* 1994), These chelators are also produced by the white-rot fungus (Galkin *et al.,* 1998; Mäkelä *et al.,* 2002; Ürek and Pazarlioglu, 2005). The redox potential of the MnP-Mn system is lower than that of LiP and preferably oxidizes phenolic substrates producing phenoxyl radicals (Vares, 1996). These radicals can further react with the eventual release of $CO_2$. The phenoxyl radicals formed subsequently cleave Cα-Cα or alkyl-phenyl bonds causing depolymerization to smaller intermediates including quinones and hydroxyl quinines (Kuhad *et al.,* 1997). Purified or crude MnP has been used in cell-free systems (*in vitro*) and shown to oxidize not only lignin (Hofrichter *et al.,* 2001) but also chlorolignins (Lackner *et al.,* 1991), and synthetic lignin compounds (Wariishi *et al.,* 1991).

## Laccases

Laccases (EC 1.10.3.2), an N-glycosylated extracellular blue copper oxidase, are widespread in nature (Mayer and Staples, 2002). They are produced by most of the white-rot fungi (Table 2). Some Ascomycetous fungi, as described previously, also produce laccases. The studied laccases from white-rot fungi are 40–90 kDa (Reinhammar, 1984; Call and Mûcke, 1997) and have acidic pIs and pH optima. It contains four copper atoms in the active site (as $Cu^{2+}$ in the resting enzyme) that are distributed among different binding sites, and are classified into three types with differential specific characteristic properties (Messerschmidt, 1997; McGuirl and Dooley, 1999). These catalyze four consecutive 1-electron oxidations, then transfer the electrons to molecular oxygen, reducing it to water, and returning the enzyme to its native state (Kuhad *et al.,* 1997). Nonphenolic lignin-related substrates can be also oxidized in the presence of certain auxiliary substrates: 3-hydroxyanthranilic acid (3-HAA) (Eggert *et al.,* 1996b), 2,2'-azinobis-(3-ethylbenzthiazoline-6-sulphonate) (ABTS) (Bourbonnais and Paice, 1990), 1-hydroxybenzotriazole (1-HBT) (Bourbonnais *et al.,* 1997), Polyoxometalates (Balakshin *et al.,* 2001) etc. Eggert and coworkers (1996a) showed that the white-rot fungus *Pycnoporus cinnabarinus*, which produces laccases but no detectable MnP or LiP, thus showing the importance of laccases in degradation of lignin. It also secretes a laccase substrate, 3-HAA, which apparently is a natural intermediary for oxidation of nonphenolic lignin substructures (Eggert *et al.,* 1996b).

## VERSATILE PEROXIDASE

Recently versatile peroxidases have been reported which have high affinity for $Mn^{2+}$, hydroquinones and dyes, and also oxidize veratryl alcohol, dimethoxybenzene and lignin dimers (Heinfling *et al.,* 1998c; Caramelo *et al.,* 1999). The deduced sequences seem to have higher identity with *Phanerochaete chrysosporium* lignin peroxidase than manganese peroxidase but molecular model revealed the presence of $Mn^{2+}$ binding site (Ruiz-Dueñas *et al.,* 2001). This ability of a peroxidase has been reported first in *Pleurotus eryngii* (Martinez *et al.,* 1996) and then in others also such as *P. pulmonarius* (Camarero *et al.,* 1996), *P. ostreatus* (Sarkar *et al.,* 1997), *Bjerkandera adusta* (Heinfling *et al.,* 1998), *Bjerkandera sp.* (Mester *et al.,* 1998) and *Geotrichum candidum* (Kim and Shoda, 1999).

## CLASSIFICATION OF WHITE-ROT FUNGI BASED ON ENZYME PROFILE

Overview of the white-rot fungi investigated and the oxidative enzyme activities as published in the literatures has been given in Table 2. Hattaka, 1994 divided the white-rot fungi into four groups depending on the production of different combinations of enzymes. Tuor and coworkers in 1995 classified the white-rot fungi in five main groups. But the report of versatile peroxidase has led to the division into six main groups, which are listed as follows.

First group includes white-rot fungi expressing LiP, MnP and laccase. It includes the best known fungi, namely, *Cyathus bulleri*, *Ganoderma lucidium*, *Phlebia brevispora*, *P. radiata*, *Pleurotus sajor-cajo*, *Trametes gibbosa*, *T. hirsuta* and *T. versicolor.* However, *Phanerochaete chrysosporium* is generally not included in this group as it is considered not to produce laccase. Yet laccase has been reported in *P. chrysosporium* under specific conditions by some workers (Srinivasan *et al.,* 1995).

Second group consists of fungi producing MnP and laccases but no detectable levels of LiP have been reported. These also include strong lignin degraders such as *Ceriporiopsis subvermispora*, *Dichomitus squalens*, *Lentinula edodes* and many others.

Third group also comprised effective lignin degraders, which produces LiP and MnP. *P. chrysosporium, P. flavido-alba, Phellinus pini* are some of the examples, which list in this group.

Fourth group (LiP-laccase group) consists of fungi with much lower lignin degradation capability (Hattaka, 1994). Some white-rot fungi that come in this group are *Oudemansiella radicata*, *Phlebia ochraceofulva*, *Pleurotus florida*, *Polyporus pinsitus* and others.

Fifth group comprised fungi producing any one out of the three major extracellular oxidative enzymes i.e. LiP, MnP and laccases. *Pycnoporus cinnabarinus* degrades lignin very efficiently with laccase as the only phenol oxidase. Other examples categorizing in this group are *Cerrena maxima* secreting laccases, *Fomes fomentarius* producing LiP and others.

Sixth group includes those fungi which have been reported recently to produce versatile peroxidases. This group includes *Pleurotus eryngii, P. pulmonarius, P. ostreatus, Bjerkandera adusta* and *Bjerkandera* sp.

The last group (seventh group) consists of white-rot degraders but neither of the oxidative enzymes has been reported. White-rot fungi coming in this category are *Ceriporia metamorphosa, Mycoacacia nothofai, Phellinus pseudopunctatus, Hypholoma capnoides* and the list goes on. Efforts are needed to explore and to study the diverse ligninolytic enzyme system of these white-rot fungi.

## SYNERGISTIC ACTION OF LIGNINOLYTIC ENZYMES FOR WOOD DECAY

Lignin degradation is a complex process and the enzymes have probably synergistic effects on each other (Tuomela *et al.,* 2000).

## MANGANESE PEROXIDASE (MnP) and LACCASES

Galliano *et al.,* (1991) demonstrated in concert action of two extracellular oxidative enzymes, namely, MnP and laccases which are secreted from *Rigidoporus lignosus* into the culture medium. It was found that *in vitro* neither laccase nor MnP alone was able to act efficiently upon radioactive Hevea lignins. However, solubilization was extensive when laccases and MnP were added together at the same time. Thus explaining the possibility of ligninolytic enzymes working together *in vivo* conditions too. Along with that, glucose oxidase was taken into consideration and when put into the reaction mixture, lignin solubilization was enhanced. This may be due to preventing of repolymerization of the radicals formed by these oxidative enzymes. The working of glucose oxidase with other ligninolytic enzymes has been explained previously. Nevertheless, with these experiments done *in vitro* gives the probablity of cooperation not only among the three main ligninolytic enzymes but also with other enzymes secreted by the fungus.

## GLUCOSE OXIDASE and LACCASES

In lignin-degrading cultures of the white-rot fungus *P. chrysosporium*, glucose oxidase has been shown to be predominant source of $H_2O_2$ (Eriksson *et al.,* 1986). Others white-rot fungus to form hydrogen peroxide are *G. applanatum*, *X. frustulatus*, *T. versicolor* and others (Highley, 1987). To check the production of hydrogen peroxide whether is linked or not, to ligninolytic activity of white-rot fungi, hydrogen peroxide negative and positive isolates of *P. chrysosporium* (ME-446 and ME-461, respectively) and *G. applanatum* (MAD-708 and 7823-S respectively) were compared. The hydrogen peroxide negative isolates were completely unable to decay pine and sweetgum while hydrogen peroxide positive isolates readily decayed wood. Hence, circumstantial evidence of hydrogen peroxide involvement in degradation of wood by white-rot fungi is provided by this observation. Although, additional degradative studies with hydrogen peroxide positive and negative isolates are required (Highley, 1987).

Quinonoid intermediates appeared *in situ* from chromogenic acid by *T. versicolor* laccase. Since glucose oxidase can reduce quinones to phenol, it was considered to affect quinonoid intermediates appeared as a result of laccase action on lignin (Green, 1977). Taking this into consideration, an experiment was conducted to find cooperation between fungal laccase and glucose oxidase by Szklarz and Leonowicz (1986). When glucose oxidase was added to the reaction mixture containing laccase and high molecular weight fraction, depolymerization was more extensive than in the experiment without glucose oxidase. In the case of low molecular weight fraction, the polymerization process was disturbed as the two enzymes operated in concert. In other words, depolymerization was greatly enhanced. Glucose oxidase preferentially takes up quinonoid oxygen (radicals) generated by laccase and their immediate reduction prevents their repolymerization as well as inhibition of laccase. This greatly enhances the efficiency of depolymerization. Hence, it was considered that the same mechanism might be operating in the nature where the various enzymes excreted by the fungus may operate synergistically to transform lignocellulose into easily utilized carbon sources.

## ARYL ALCOHOL OXIDASE (AAO) and LIGNINOLYTIC ENZYMES

Many white-rot fungi secrete aryl alcohol oxidase along with other major ligninolytic enzymes e.g. *T. versicolor* (Farmer *et al.,* 1960), *P. sajor-caju* (Bourbonnais and Paice, 1988), *P. ostreatus* (Sannia *et al.,* 1991), *P. eryngii* (Delgado *et al.,* 1992) and *B. adusta* (Muheim *et al.,* 1990). The combined action of AAO and intracellular dehydrogenases reduces aromatic aldehydes and acids which ultimately lead to the formation of redox cycle, providing the hydrogen peroxide needed for activities and the generation of hydroxyl ion (Guillén and Evans, 1994). In another experiment, conducted for the first time, the production of hydroxyl radical was found to be a synergistic action of fungal laccase and AAO (Guillén *et al.,* 2000). This radical is the strongest oxidant found in white-rot fungal cultures and is one of the low molecular weight agents, participating in the initial attack of lignocellulose when ligninolytic enzymes cannot penetrate plant cell wall (Backa *et al.,* 1993; Evans *et al.,* 1994). Fenton reaction (1) is considered to be involved in the production of hydroxyl ion by white-rot fungi (Barr *et al.,* 1992).

$$Fe^{2+} + H_2O_2 \longrightarrow Fe^{3+} + {}^{\cdot}OH + OH^- \quad \text{.........(1)}$$

Aromatic aldehyde redox cycling can provide the mechanism by which *P. eryngii* could generate the superoxide anion radical required for the extracellular $H_2O_2$ production via the superoxide driven Fenton reaction (2).

$$O_2^{\cdot -} + H_2O_2 \longrightarrow {}^{\cdot}OH + OH^- + O_2 \quad \text{........(2)}$$

A series of reactions, (reactions 1 to 7) as depicted in Figure 1, were established to explain the mechanism involved in the cycle (Guillen *et al.,* 2000). During the oxygen activation (reaction 1), methoxyhydroquinones was oxidized by laccases to semiquinones, which auto-oxidized $O_2^{\cdot -}$ (reaction 2). During this reaction of semiquinone auto-oxidation, hydrogen peroxide was even found to be produced in the absence of both ferric ion and AAO/anisyl alcohol system. Reaction 3 included dismutation of $O_2^{\cdot -}$. Hence in this condition, $H_2O_2$ levels produced was considered to be used to determine the extent of semiquinone auto-oxidation and the total amount of produced by the laccase. Ferrous ion production was caused by both $O_2^{\cdot -}$ and semiquinone radicals as depicted in reactions 4 and 5 respectively. As the extent of $O_2^{\cdot -}$ production to that of ferric ion reduction remained constant, therefore, ferric ion reduction was mainly due to semiquinone radicals. So, the formation of hydroxyl radical was because of semiquinone driven fenton reaction (reaction 5 and 6) as compared to $O_2^{\cdot -}$ driven one (reaction 4 and 6). From the comparison of ferrous ion and hydrogen peroxide levels, it was inferred that hydrogen peroxide was the limiting reagent for hydroxyl production. In reaction 7, hydrogen peroxide was produced by AAO taking anisyl alcohol as electron donor. The results obtained determined the cooperation of AAO with laccase to prevent repolymerization of lignin-derived radicals and to produce hydroxyl radical. Well, research is required to determine their occurrence in other white-rot fungi.

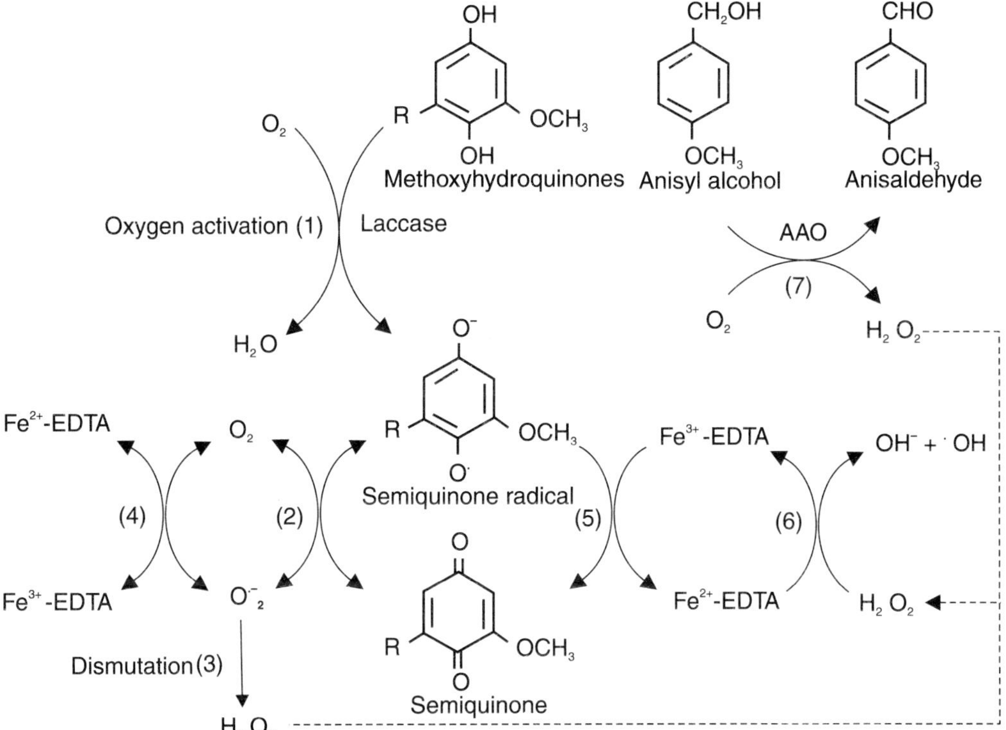


**Fig. 1:** Model to represent the synergistic action of laccase and AAO for the production of hydroxyl radical.

## CELLOBIOSE DEHYDROGENASES (CDH) AND LACCASES

CDH are a group of oxidative enzymes that are widespread in fungi, that can utilize cellulose (Eriksson *et al.,* 1993 ; Ander, 1994). CDH has been suggested to play an important role in lignocellulose degradation (Costa-Ferreira *et al.,* 1994 ; Henriksson *et al.,* 1995; Ander and Marzullo, 1996) but the exact function(s) is yet to be explored. *Phanerochaete chrysosporium* (Bao *et al.,* 1993),

*Trametes versicolor* (Roy *et al.,* 1996; Levin *et al.,* 2005), *Heterobasidion annosum* (Hüttermann and Noelle, 1982) and *Schizophyllum commune* (Fang *et al.,* 1998) are some examples of white-rot fungi from which CDH have been purified. It has been suggested that CDH may have function in both lignin degradation and cellulose degradation but the two processes are often studied as if they have taken place independently (Temp and Eggert, 1999). Therefore, synergistic effects between the two pathways are poorly understood. The generation of active hydroxyl radicals which initiate Fenton's reaction to degrade wood components (Kremer and Wood, 1992 ; Wood, 1994) and prevention of repolymerization of radicals generated during the oxidation of lignin by phenol oxidases are some of the functions considered to be performed by CDH (Eriksson *et al.,* 1993; Ander, 1994). Temp amd Eggert (1999) tried to find the interaction between laccase and CDH during pigment synthesis in the white-rot fungus *Pycnoporus cinnabarinus* (Figure 2). Here laccase mediates the formation of a characteristic orange-red pigment, cinnabarinic acid (CA) and serves as an antimicrobial agent (Eggert *et al.,* 1995 ; Eggert, 1997). The precursor of CA, 3-hydroxyanthranilic acid (3-HAA) aid the laccase during oxidation of lignin by acting as co-oxidant, also known as mediators (Eggert *et al.,* 1996b). Therefore in a natural system, recycling of co oxidants would be prerequisite. It was found that some laccase oxidation products of 3-HAA functions as *in vivo* electron acceptors for *P. cinnabarinus* CDH. Laccase oxidizes 3-HAA to CA in the absence of cellulose but in the presence of cellulose, CDH is produced, catalyzing the reduction of 3-HAA oxidation intermediates. If a suitable electron donor (preferably cellobiose) is available, the cycle is maintained indefinitely which prevents the accumulation of CA (Figure 2). CA do not form until fruiting bodies are produced. This novel interaction between the two enzymes has given the evidence which supports the hypothesis that

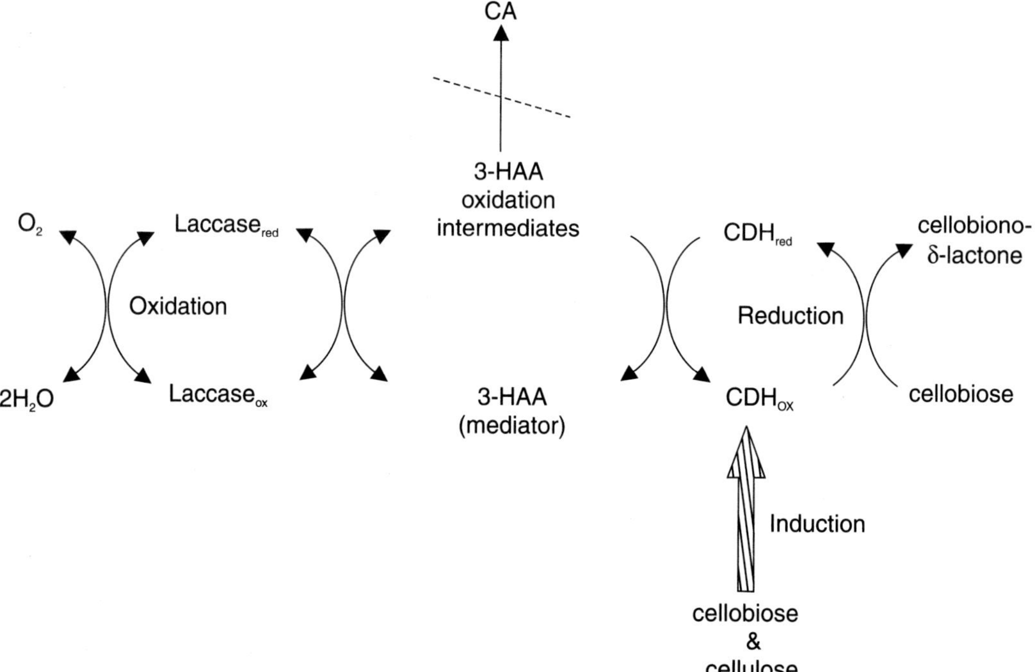


**Fig. 2 :** Model to show the interaction of laccases and CDH in production of pigment cinnabarinic acid

CDH are involved not only in cellulose degradation but also act in concert with other ligninolytic enzymes in lignin degradation.

## SIMULTANEOUS AND SELECTIVE MODE OF WOOD DECAY

On the basis of degradation of three wood polymers cellulose, hemicellulose and lignin at different rates and extent, the white-rot fungi have been divided into classes of simultaneous or nonselective degraders of lignin along with wood polysaccharides and/or selective or sequential lignin degraders (Table 3). In selective delignification, lignin and hemicelluloses are removed ahead of the cellulose, whereas simultaneous degraders remove hemicellulose and lignin more or less simultaneously (Eriksson *et al.,* 1990; Blanchette, 1995). Fungi vary greatly among species and among strains within a species, in causing the two types of decay.

**Table 3. Selective and Simultaneous Degraders of Wood**

| *S. No.* | *White-rot fungi* | *Reference* |
|---|---|---|
| | **Selective Degraders** | |
| 1. | *Pleurotus ostreatus* | Becker and Sinitsyn, 1993; Martinez *et al.,* 1994; Giardina *et al.,* 1996 |
| 2. | *Pleurotus eryngii* | Martínez *et al.,* 1996; Muñoz *et al.,* 1997 |
| 3. | *Dichomitus squalens* | Eriksson *et al.,* 1990; Périé *et al.,* 1996 |
| 4. | *Phlebia radiata* | Lundell and Hatakka, 1994 |
| 5. | *Pycnoporus cinnabarinus* | Eggert *et al.,* 1996b |
| 6. | *Bjerkandera adusta* | Blanchette, 1984 |
| 7. | *Cerrena unicolor* | Blanchette, 1984 |
| 8. | *Ganoderma lobatum, G. oregonense* and *G. tsugae* | Blanchette, 1984 |
| 9. | *Hapalophilus croceus* | Blanchette, 1984 |
| 10. | *Inonotus dryadeus, I. dryophilus, I. ludovicianus, I. rheades, I. texanus* and *I. tomentosus* | Blanchette, 1984 |
| 11. | *Laurilia sulcata* | Blanchette, 1984 |
| 12. | *Perenniporia medulla-panis* and *P. subacida* | Blanchette, 1984 |
| 13. | *Phellinus fastuosus, P. nigrolimitatus, P. pini* and *P. viticola* | Blanchette, 1984 |
| 14. | *Xylobolus frustulatus, X. subpileatus* | Blanchette, 1984 |
| | **Simultaneous Degraders** | |
| 1. | *Trametes versicolor* | Hatakka, 1994; Levin *et al.,* 2005 |
| 2. | *Fomes fomentarius* | Blanchette, 1984 |
| 3. | *Ganoderma curtissi* | Blanchette, 1984 |
| 4. | *Phellinus robustus* and *P. igniarius* | Blanchette, 1984 |
| | **Both Selective and Simultaneous Degraders** | |
| 1. | *Heterobasidion annosum* | Eriksson *et al.,* 1990; Blanchette, 1995 |
| 2. | *Ganoderma applanatum* | Blanchette, 1984; Dill and Kraepelin, 1988 |
| 3. | *Ischnoderma resinosum* | Blanchette, 1984 |

Classification based on these criteria has proved to be difficult since many white-rot fungi illustrate both types of decay while acting on the same substrate (Blanchette, 1984; Eriksson *et al.,* 1990). *In vivo* selectivity of lignin degradation may be governed by substrate characteristics of a tree species, environmental conditions (temperature, humidity, microclimate, nitrogen content of the substrate) and variation among isolates. These conditions have already been discussed previously. These factors may affect directly or indirectly the white-rot fungi to act on lignin simultaneously or selectively. But still the occurrence of different pattern of decay is as yet unclear.

## MICROSCOPIC FEATURES OF WOOD DECAY

It was over a century ago that Hartig in 1878 suggested by histological examination that some white-rot fungi might preferentially degrade lignin from wood. Afterwards, it was scanning electron microscopy (SEM) that became powerful tool to directly observe and select microorganism with a capacity to preferentially degrade lignin, thus giving the details of wood decay (Blanchette, 1984; Otjen and Blanchette, 1982; 1984). It easily creates demarcations between two modes of wood decay. In simultaneous decay, erosion troughs are observed beneath the hyphae and hence there is localized degradation pattern that was apparent immediately around the hyphae. The cell walls become gradually thinner, often in a non-uniform manner with the decay and holes start appearing between cells as decay advances. The extent of decay can vary from cell to cell. It happens that some heavily attacked cells could occur adjacent to relatively sound cells (cells without any damage by decay). On the other hand, in selective mode of decay, cell walls more or less retain their morphology, but gradually lignin content reduces and glucose levels increase as the protective lignin layer has been removed. Also, there is removal of extensive amounts of lignin from the middle lamella region, when fungal hyphae were located in cell lumens (Blanchette, 1991; 1995).

Blanchette (1984) analyzed 26 different white-rot fungi for selectively removal of lignin from various coniferous and hardwood tree species with the help of scanning electron microscopy. In the results, the gross patterns of delignification appeared similar among the various samples investigated but through the use of SEM, differences were evident in the fine detail of the cell walls. For example, *Ischnoderma resinosum* and *Phellinus viticola*, after the removal of lignin from wood, left the cells with a relatively smooth appearance and slight alteration of the secondary wall was observed (Kawase, 1962; Eriksson, 1978; Blanchette, 1984). On the other hand, *Perenniporia subacida* and *Phellinus* sp. resulted in more obvious degradation of the remaining cell wall layers. Apparently, a distinct fibrillar structure was evident giving the cells rough appearance. In *Ganoderma lobatum*, the mottled and pocket rots that had zones of delignified cells (selective pattern of decay) also had a simultaneous white-rot decay pattern in other areas of same wood. Within the decayed wood, both selective and simultaneous white-rotted cells were always found to be separated spatially (Blanchette, 1984). *Xylobolus frustulatus* subsequently also remove cellulose that remained in delignified pockets (Otjen and Blanchette, 1984).

It has been demonstrated by Blanchette (1984) that hemicellulose is also removed in addition to lignin. Hence, at least one of the polysaccharides in wood must be degraded simultaneously with lignin. The preference for hemicellulose is due to the close spatial relationship of lignin and hemicellulose, encrusting the cellulose microfibrils within the cell wall (Kerr and Goring, 1975;

Koshijima *et al.,* 1989; Jeffries, 1990). This selective removal of lignin-hemicellulose matrix starting at the lumen surfaces is considered to expose additional enzyme-accessible surfaces, thus allowing enzymes to act into the walls (Kirk and Cullen, 1998). After the cells of wood have been delignified, it has been demonstrated that ribbon-like structures of cellulose fibrils could be seen which bond in their radial planes to form lamella.

SEM has elucidated large number of fungi with capacity to selectively remove lignin from wood. However, further research is needed to understand the mechanisms involved in lignin degradation during simultaneous as well as selective pattern of wood decay.

## CONCLUSION

The work done in the field of wood decay till now is not sufficient to determine the entire mechanism of wood decay. Hence, more research is needed to understand how different enzymes interact with each other so that the ligninolytic system is sustained in the nature. Also, the list of white-rot fungi included in classification of fungi based on enzymes profile keeps changing as more and more work on biodiversity of white-rot fungi has led to the addition of new fungi producing one or more different ligninolytic enzymes. We should focus on the biodiversity of microorganisms, especially white-rot fungi so that organisms with novel characteristics can be exploited for better catalytic efficiency and sustainable utility in industrial biotechnology.

## REFERENCES

Adaskaveg, J.E., Gilertson, R.L. and Blanchette, R. A. (1990). Comparative studies of delignification caused by *Ganoderma* species. *Applied and Environmental Microbiology,* 56, 1932-1943.

Adaskaveg, J.E., Gilertson, R.L. and Dunlap, M.R. (1995). Effect of incubation time and temperature on in vitro selective delignification of silver leaf oak by *Ganoderma colossum. Applied and Environmental Microbiology,* 61, 138-144.

Ander, P. (1994). The cellobiose-oxidizing enzymes CBQ and CBO as related to lignin and cellulose degradation — a review. *Enzyme and Microbial Technology,* 15, 1002-1007.

Ander, P. and L. Marzullo. (1996). Sugar oxidoreductases and veratryl alcohol oxidase as related to lignin degradation. *Journal of Biotechnology,* 53, 115-131.

Argyropoulos, D.S. and Menachem, S.B. (1997). Lignin In: *Advances in Biochemical Engineering Biotechnology,* Ed Erriksson, K. –E. L. Springer, Germany. Pp 127-158.

Aro, N., Pakula, T., Penttila, M. (2005) Transcriptional regulation of plant cell wall degradation by filamentous fungi. *FEMS Microbiology Review,* 29, 719–739.

Arora, D.S. and Gill, P.K. (2001). Effects of various media and supplements on laccase production by some white rot fungi. *Bioresource Technology,* 77, 89-91.

Arora, D.S. and Rampal, P. (2002). Laccase production by some *Phlebia* species. *Journal of Basic Microbiology,* 42(5), 295-301.

Backa, S., Gierer, J., Reitberger, T. and Nilsson, T. (1991). On the participation of hydroxyl radicals in the early stage of wood degradation by rot fungi. *Proceedimgs 6th ISWPC,* Vol. 2, Pp 269-272.

Backa, S., Gierer, J., Reitberger, T. and Nilsson, T., (1993). Hydroxyl radical activity associated with the growth of white-rot fungi. *Holzforsching,* 47, 181-187.

Balakshin M.Y., Evtuguin, D.V., Pascoal Neto, C. and Cavaco-Paulo, A. (2001). Polyoxometalates as mediators in the laccase catalyzed delignification. *Journal of Molecular Catalysis B: Enzymatic,* 16, 131-140.

Bao, W., S.N. Usha, and V. Renganathan. (1993). Purification and characterization of cellobiose dehydrogenase, a novel hemoflavoenzyme from the white-rot fungus *Phanerochaete chrysosporium. Archives of Biochemistry and Biophysics,* 300, 705-713.

Barr, D.P. & Aust, S.D. (1994b). Pollutant degradation by white-rot fungi. *Review of Environmental Contamination and Toxicology,* 138, 49-72

Barr, D.P. and Aust, S.D. (1994a) Mechanisms white-rot fungi use to degrade pollutants. *Environmental Science Technology,* 28, 78-87.

Barr, D.P., Shah, M.M., Grover, T.A. and Aust, S.D. (1992). Production of hydroxyl radical by lignin peroxidase from *Phanerochaete chrysosporium. Archives of Biochemistry and Biophysics,* 298, 480-485.

Becker, H.G. and Sinitsyn, A.P. (1993). Manganese peroxidase from *Pleurotus ostreatus*, the action on the lignin. *Biotechnology Letters,* 15, 289-294.

Black, A.K. and Reddy, C.A. (1991). Cloning and characterization of a lignin peroxidase gene from the white-rot fungus *Trametes versicolor. Biochemical and Biophysical Research Communication,* 179(1), 428-435.

Blanchette, R.A. (1991). Delignification by wood-decay fungi. *Annual Review of Phytopathology,* 29, 381-398.

Blanchette, R.A. (1995). Degradation of lignocellulose complex in wood. *Canadian Journal of botany,* 73, s999-s1010.

Blanchette, R.A., Abad, A.R., Farrell, R.L. and Leathers, T.D. (1989). Detection of lignin peroxidase and xylanase by immunocytochemical labeling in wood decayed by asidiomycetes. *Applied and Enviromental Microbiology,* 55, 1457-1465.

Blanchette, R.A. (1984). Screening wood decayed by white rot fungi for preferential lignin degradation. Applied and Environmental Microbiology 48(3) 647-653.

Bonnarme, P., and Jeffries, T.W. (1990). Mn(II) regulation of lignin peroxidases and manganese-dependent peroxidases from lignin-degrading white rot fungi. *Applied and Enviromental Microbiology,* 56(1), 210-217.

Bonnen, A.M., Anton, L.H. and Orth, A.B. (1994). Lignin-degrading enzymes of the commercial button mushroom *Agaricus bisporus. Applied Environmental and Microbiology,* 60, 960-965.

Bourbonnais, R. and Paice, M.G. (1988). Veratryl alcohol oxidases from the lignin degrading basidiomycete *Pleurotus sajor-caju. Biochemical Journal,* 255, 445-450.

Bourbonnais, R. and Paice, M.G. (1990). Oxidation of non-phenolic substrates-an expanded role for laccase in lignin peroxidase. *FEBS Letters,* 267, 99-102.

Bourbonnais, R., and M.G. Paice. (1992). Demethylation and delignification of kraft pulp by *Trametes versicolor* laccase in the presence of 2,2-azinobis-(3-ethylbenzthiazoline-6-sulphonate). *Applied Microbiology and Biotechnology,* 36, 823–827.

Bourbonnais, R., Paice, M.G., Freiemuth, B., Bodie, E. and Bornemann, S. (1997). Reactivities of various mediators and laccases with kraft pulp and lignin model compounds. *Applied and Environmental Microbiology,* 63, 4627-4632.

Boyle, C.D., Kropp, B.R. and Reid I.D. (1992). Solubilization and mineralization of lignin by white rot fungi. *Applied and Enviromental Microbiology,* 58, 3217-3224.

Buswell, J.A. and Odier, E. (1987). Lignin biodegradation. *CRC Critical Review of Biotechnology,* 6, 1-60.

Call, H.P. and Mûcke, I. (1997). Minireview, history, overview and applications of mediated ligninolytic systems, especially laccase-mediator-systems (Lignozym-Process). *Journal of Biotechnology,* 53,163– 202.

Cambria, M.T., Cambria, A., Ragusa, S., and Rizzarelli, E. (2000). Production, purification, and properties of an extracellular laccase from *Rigidoporus lignosus. Protein Expression and Purification,* 18, 141-147.

Carll, C.G. and Highley, T.L. (1999). Decay of wood and wood based products above ground in buildings. *Journal of Testing and Evaluation,* JTEVA 27 (2), 150-158.

Chagas, E.P. and Durrant, L.R. (2001). Decolorization of azo dyes by *Phanerochaete chrysosporium* and *Pleurotus sajorcaju. Enzyme and Microbial Technology,* 29, 473-477.

Chen, S., Ge, W. and Buswell, J.A. (2004). Biochemical and molecular characterization of a laccase from the edible straw mushroom, *Volvariella volvacea. European Journal of Biochemistry,* 271, 318–328.

Cooke, R.C. and Rayner, A.D.M. (1984). *Ecology of Saprotrophic Fungi.* Longman, London.

Costa-Ferreira, M., P. Ander, and J. Duarte. (1994). On the relationship between cellobiose dehydrogenase and cellobiose, quinone oxidoreductase under conditions where [$^{14}$C]DHP is mineralized by whole cultures of *Phanerochaete chrysosporium. Enzyme and Microbial Technology,* 16, 771-776.

Couto, S.R., Gundin, M., Lorenzo, M., and Sanroman, M.N. (2002d). Screening of supports and inducers for laccase production by *Trametes versicolor* in semi-solid-state conditions. *Process Biochemistry,* 38, 249-255.

Couto, S.R., Moldes, D., Liebanas, A., and Sanroman, A. (2003). Investigation of several bioreactor configurations for laccase production by *Trametes versicolor* operating in solid-state conditions. *Journal of Biochemical Engineering,* 15, 21.

Cui, F., and Dolphin, D. 1990. The role of manganese in model systems related to lignin biodegradation. *Holzforschung,* 44, 279–283.

Daniel, G., Pettersson, B., Nilsson, T. and Volc, J. (1990). Use of immunogold chemistry to detect Mn(II)-dependent and lignin peroxidases in wood degraded by the white rot fungi *Phanerochaete chrysosponium* and *Lentinula edodes. Canadian Journal of Botany,* 68, 920-933.

Das, N., Chakraborty, T.K., and Mukherjee, M. (1999). Role of potato extract in extracellular laccase production of *Pleurotus florida. Journal of Basic Microbiology,* 39(5-6), 299-303.

De Jong, E., De Vries, F.P., Field, J.A., Van der Zwan, R.P. and De Bont, J.A.M. (1992). Isolation and screening of basidiomycetes with high peroxidative activity. *Mycology Research,* 96, 1098-1104.

De Souza, C. G.M., Zilly, A., and Peralta, R.M. (2002). Production of laccase as the sole phenoloxidase by a *Schizophyllum commune* dikaryon. *Journal of General Microbiology,* 132, 2817-2826.

Deacon, J.W. (1997). *Modern Mycology, 3rd Edition.* Blackwell Scientific, Oxford. UK.

Dedeyan, B., Klonowska, A., Tagger, S., Tron, T., Iacazio, G., Gil, G. and Le Petit, J. (2000). Biochemical and Molecular Characterization of a Laccase from *Marasmius quercophilus. Applied and Environmental Microbiology,* 66, 925-929.

Delgado, G., Guillén, F., Martínez, M.J., GonŸalez, A.E. and Martínez, A.T. (1992). Light stimulation of aryl-alcohol oxidase activity in *Pleurotus eryngii. Mycology Research,* 96, 984-986.

Dhawan, S., and Kuhad, R.C. (2002). Effect of amino acids and vitamins on laccase production by the bird's nest fungus *Cyathus bulleri. Bioresource Technology,* 84, 35-38.

Dhawan, S., Lal, R., and Kuhad, R.C. (2003). Ethidium bromide stimulated hyper laccase production from bird's nest fungus *Cyathus bulleri. Letters in Applied Microbiology,* 36, 64-67.

Dill, I. and Kraepelin, G. (1988). Der Abbau von Lignin/Cellulose durch Weissfäule-Pilze,Einfluss spezifischer ökologischer Faktoren. *Forum Mikrobiologie,* 11/88, 484-489.

Dix, N.J. and Webster, J. (1995). *Fungal ecology.* Chapman and Hall, London, U. K.

D'Souza, T.M., Merritt, C.S. and Reddy, C.A. (1999). Lignin-Modifying Enzymes of the White Rot Basidiomycete *Ganoderma lucidum. Applied Environmental and Microbiology,* 65, 5307-5313.

Duran, N., Ferrer, I., and Rodriguez, J. (1987a). Ligninases from *Chrysonilia sitophila* (TFB-27441 strain). *Applied Biochemistry and Biotechnology,* 16, 157-167.

Duran, N., Rodriguez, J.,Ferraz, A., and Campos, V. (1987b). *Chrysonilia sitophila* (TFB-27441), a hyperlignolytic strain. *Biotechnology Letters,* 9(5), 357-360.

Edens, W.A., Goins, T.Q., Dooley, D., and Henson, J.M. (1999). Purification and characterization of a secreted laccase of *Gaeumannomyces graminis* var. *tritici. Applied Environmental and Microbiology,* 65(7), 3071-3074.

Eggert, C. (1997). Laccase is responsible for antimicrobial activity of *Pycnoporus cinnabarinus. Microbiology Research,* 152, 315-318.

Eggert, C., Temp, U. and Eriksson, K.-E. (1996a). The ligninolytic system of the white rot fungus *Pycnoporus cinnabarinus,* purification and characterization of the laccase. *Applied Environmental and Microbiology,* 62, 1151–1158.

Eggert, C., Temp, U., Dean, J.F.D. and Eriksson, K.-E. L. (1995). Laccase-mediated formation of the phenoxazinone derivative, cinnabarinic acid. *FEBS Letters,* 376, 202-204.

Eggert, C., Temp, U., Dean, J.F.D. and Eriksson, K.-E. L. (1996b). A fungal metabolite mediates oxidation of non-phenolic lignin model compounds and synthetic lignin by laccase. *FEBS Letters,* 391, 144-148.

Eriksson, K.-E.L. (1978). Enzyme mechanisms involved in cellulose hydrolysis by the rot fungus *Sporotrichum pulverulentum. Biotechnology and Bioengineering,* 20, 317-332.

Eriksson, K.-E. L., Habu, N. and Samejima, M. (1993). Recent advances in fungal cellobiose oxidoreductases. *Enzyme and Microbial Technology,* 15, 1001-1008.

Eriksson, K.-E.L., Blanchette, R.A. and Ander, P. (1990). Microbial and enzymatic degradation of wood and wood components. *Springer-verlag. Berlin. New York,* 397.

Eriksson, K.-E. L., Pettersson, B., Volc, J. and Musilek, V. (1986). Formation and partial characterization of glucose-2-oxidase, a $H_2O_2$ producing enzyme in *Phanerochaete chrysosporium. Applied Microbiology and Biotechnology,* 23 (3-4), 257-262.

Evans, C.S., Dutton, M.V., Guillen, F. and Veness, R.G. (1994). Enzymes and small molecular mass agents involved with lignocellulose degradation. *FEMS Microbiology Review,* 13, 235-240.

Fang, J., Liu, W. and Gao, P.J. (1998). Cellobiose dehydrogenase from *Schizophyllum commune*, purification and study of some catalytic, inactivation and cellulose binding properties. *Archives of Biochemistry and Biophysics,* 353, 37-46.

Farmer, V.C., Henderson, M.E.K. and Russell, J.D. (1960). Aromatic-alcohol-oxidase activity in the growth medium of *Polystictus versicolor. Biochemical Journal,* 74, 257-262.

Fenice, M., Sermanni, G.G., Federici, F., and D' Annibale, A. (2003). Submerged and solid- state production of laccase and Mn-peroxidase by *Panus tigrinus* on olive mill wastewater-based media. *Journal of Biotechnology,* 100, 77-85.

Ferrer, I., Esposito, E. and Duran, N. (1992). Lignin peroxidase from *Chrysonilia sitophila,* heat denaturation kinetics and pH stability. *Enzyme and Microbial Technology,* 14, 402-406.

Forrester, I.T., Grabski, A.C., Mishra, C., Kelley, B.D., Strickland, W. N., Leatham, G.F. and Burgess, R.R. (1990). Characteristics and N-terminal amino acid sequence of a manganese peroxidase purified from *Lentinula edodes* cultures grown on a commercial wood substrate. *Applied Microbiology and Biotechnology,* 33, 359-365.

Galkin, S., Vares, T., Kalsi, M., and Hatakka, A. (1998). Production of organic acids by different white-rot fungi as detected using capillary zone electrophoresis. *Biotechnology Technology,* 12, 267–271.

Galliano, H., Gas, G., Seris, J.L. and Boudet, A.M. (1991). Lignin degradation by *Rigidoporus lignosus* involves synergistic action of two oxidizing enzymes, Mn peroxidase and laccase. *Enzyme and Microbiology Technology,* 13, 478-482.

Gilbertson, R.L. (1980). Wood rotting fungi of North America. *Mycologica,* 72 (1), 1-49.

Gill, P.K., and Arora, D.S. (2003). Effect of culture conditions on manganese peroxidase production and activity by some white rot fungi. *Journal of Indian Microbiology and Biotechnology,* 30, 28-33.

Glenn, J.K., Gold, M.H. (1985). Purification and characterization of an extracellular Mn(II)-dependent peroxidase from the lignin-degrading basidiomycete, *Phanerochaete chrysosporium. Archives of Biochemistry and Biophysics,* 242, 329-341.

Glenn, J. K., Morgan, M. A., Mayfield, M. B., Kuwahara, M. and Gold, M. H. (1983). An extracellular $H_2O_2$-requiring enzyme preparation involved in lignin biodegradation by the white rot basidiomycete *Phanerochaete chrysosporium. Biochemical and Biophysical Research Communication,* 114, 1077-1083.

Gold, M.H. and Alic, M. (1993). Molecular biology of the lignin-degrading basidiomycetes *Phanerochaete chrysosporium. Microbiology Review,* 57, 605-622.

Gold, M.H., Wariishi, H. and Valli, K. 1989. Extracellular peroxidases involved in lignin degradation by the white-rot basidiomycete *Phanerochaete chrysosporium.* In *Biocatalysis in Agricultural Biotechnology,* Eds Whitaker, J.R. and Sonnet, P.E., Vol. ACS Symp. Ser. No. 389, The American Chemical Society, Washington, DC. Pp 128-140.

Green, T.R. (1977). Significance of glucose oxidase in lignin degradation. *Nature* (London). 268, 78.

Grgic, I., and Perdih, A. (2003). Stimulation of ligninolytic enzyme production in *Phanerochaete chrysosporium* by polyoxyalkanes. *Journal of Applied Microbiology,* 94, 360-368.

Guillen, F. and Evans C.S. (1994). Anisaldehyde and veratraldehyde acting as redox cycling agents for H(inf2) O (inf2) production by *Pleurotus eryngii. Applied Environmental and Microbiology,* 60, 2811-2817.

Guillen, F., Munoz, C., Gomez-Toribio, V., Martinez, M.J. and Martinez, A.T. (2000) Production of hydroxyl radical by the synergistic action of fungal laccase and aryl alcohol oxidase. *Archives of Biochemistry and Biophysics,* 383, 142-147.

Ha, H-C., Honda, Y., Wantanabe, T., and Kuwahara, M. (2001). Production in manganese peroxidase by pellet culture of the lignin-degrading basidiomycete, *Pleurotus ostreatus. Applied Microbiology and Biotechnology,* 55, 704-711.

Hamman, O.B., de la Rubia, T. and Martinez, J. (1999). The effect of manganese on the production of *Phanerochaete flavido-alba* ligninolytic peroxidases in nitrogen limited cultures. *FEMS Microbiology Letters,* 177, 137-142.

Hammel, K.E. and Moen, M.A. (1991). Depolymerization of a synthetic lignin invitro by lignin peroxidase. *Enzyme and Microbial Technology,* 13, 15-18.

Hartig, R. (1878). *Die zersetzungserscheinungen des holzes der nadelholzebaume und der eiche in forstlicher botanischer und chemischer richtung.* Springer-Verlag. Berlin.

Haselwandter, K., Bobbleter, O., and Read, D.J. (1990). Degradation of $^{14}$C-labelled lignin and dehydropolymer of coniferyl alcohol by ericoid and ectomycorrhizal fungi. *Archives of Microbiology,* 153, 352-354.

Hatakka, A., Kantelinen, A., Tervilä-wilo, A. and Viikari, L. (1987). Production of ligninases by *Phlebia radiata* in agitated cultures. In: *Lignin Enzymatic and Microbial Degradation,* Ed Odier, E., Symp. Intern. INRA Paris. Pp. 185-189.

Hattaka, A. (1994). Lignin-modifying enzymes from selected white-rot fungi, production and role in lignin degradation. *FEMS Microbiology Review,* 13, 125-135.

Hatvani, N., and Mecs, I. (2001). Production of laccase and manganese peroxidase by *Lentinus edodes* on malt-containing by-product of the brewing process. *Process Biochemistry,* 37, 491-496.

Hatvani, N., and Mecs, I. (2002). Effect of the nutrient composition on dye decolorisation and extracellular enzyme production by *Lentinus edodes* on solid medium. *Enzyme and Microbial Technology,* 30, 381-386.

Heinfling, A., Martinez, M. J., Martinez, A.T., Bergbauer, M. and Szewzyk, U. (1998a). Transformation of industrial dyes by manganese peroxidases from *Bjerkandera adusta* and *Pleurotus eryngii* in a manganese-independent reaction. *Applied Environmental and Microbiology,* 64, 2788-2793.

Heinfling, A., Ruiz-Duenas, F.J., Martinez, M.J., Bergbauer, M., Szewzyk, U. and Martinez, A.T. (1998b). A study on reducing substrates of manganese oxidizing peroxidases from *Pleurotus eryngii* and *Bjerkandera adusta. FEBS Letters,* 428, 141-146.

Heinzkill, M., Bech, L., Halkier, T., Schneider, P. and Anke, T. (1998). Characterization of Laccases and Peroxidases from Wood-Rotting Fungi (Family Coprinaceae). *Applied Environmental and Microbiology,* 64, 1601-1606.

Henriksson, G., Ander, P., Pettersson, B. and Pettersson, G. (1995). Cellobiose dehydrogenase (cellobiose oxidase) from *Phanerochaete chrysosporium* as a wood-degrading enzyme. Studies on cellulose, xylan, and synthetic lignin. *Applied Microbiology and Biotechnology,* 42, 790-796.

Highley, T.L. (1987). Efeect of carbohydrate and nitrogen on hydrogen peroxide formation by wood decay fungi in solid medium. *FEMS Microbiology Letters,* 48, 373-377.

Highley, T.L. and Dashek, W.V. (1998). Biotechnology in the study of brown- and white-rot decay. In: *Forest products biotechnology,* Eds Bruce, A. and John, W. London, Great Britain: Taylor and Francis. Pp 15-36.

Hofrichter, M. (2002). Review, Lignin conversion by manganese peroxidase (MnP). *Enzyme and Microbial Technology* 30, 454-466.

Hofrichter, M. and Fakoussa, R. (2001) Microbial degradation and modification of coal. In: *Lignin, Humic Substances and Coal,* Eds Steinbüchel, A. and Hofrichter, M., Vol. 1, Wiley-VCH, Weinheim, Germany. 393-427.

Hofrichter, M., Vares, T., Kalsi, M., Galkin, S., Scheibner, K., Fritsche, W., and Hataka, A. (1999). Production of manganese peroxidase and organic acids and mineralisation of $^{14}C$-labelled lignin ($^{14}C$-DHP) during solid-state fermentation of wheat straw with the white rot fungus *Nematoloma frowardii. Applied Environmental and Microbiology,* 65(5), 1864-1870.

Humphrey, C. and Siggers, P. (1933). Temperature relations of wood-destroying fungi. *Journal of Agriculture Research,* 47, 997-1008.

Huoponen, K., Ollikka, P., Kalin, M., Walther, I., Mantsala, P. and Reiser, J. (1990). Characterization of lignin peroxidase-encoding gene from lignin-degrading basidiomycetes. *Gene,* 89, 145-150.

Hüttermann, A., and Noelle, A. (1982). Characterization and regulation of cellobiose dehydrogenase in *Fomes annosus. Holzforschung,* 36, 283-286.

Hüttermann, A., Milstein, O., Nicklas, B., Trojanowski, J., Haars, A. and Kharazipour, A. (1989). Enzymatic modification of lignin for technical use. In, *Lignin–Properties and Materials,* Eds Glasser, W. G., Sarkanen, S., ACS Symposium Series Vol. 397, 361-370.

Jeffries, T.W. (1990). Biodegradation of lignin-carbohydrate complexes. *Biodegradation,* 1, 163-176.

Johansson, T., Welinder, K.G. and Nyman, P.O. (1993). Isozymes of lignin peroxidase and manganese(II) peroxidase from the white-rot basidiomycete *Trametes versicolor. Archives of Biochemistry and Biophysics,* 300, 57-62.

Johjima, T., Itoh, N., Kabuto, M., Tokimura, F. , Nakagawa,T., Wariishi, H. and Tanaka, H. (1999). Direct interaction of lignin and lignin peroxidase from *Phanerochaete chrysosporium. Proceedings of National Academy of Science,* U S A. 96 (5), 1989–1994.

Jonsson, L. and Nyman, P.O. (1992). Characterization of a lignin peroxidase gene from the white rot fungus *Trametes versicolor. Biochimie,* 74, 177-182.

Kaal, E.E.J., Field, J.A. and Joyce, T.W., (1995). Increasing ligninolyic enzyme activites in several white-rot basidiomycetes by nitrogen-sufficient media. *Bioresource Technology,* 53, 133-139.

Kaal, E.E.J., De Jong, E., and Field, J.A. (1993). Stimulation of ligninnolytic peroxidase activity by nitrogen nutrients in the white-rot fungus *Bjerkandera* sp. strain BOS55. *Applied Environmental and Microbiology,* 59(12), 4031-4036.

Kawase, K. (1962). Chemical components of wood decayed under natural conditions and their properties. *Hokkaido Daigaku Sapporo J. Fac. Agric.* 52, 186-245.

Kerr, A.J. and D.A.I. Goring. (1975). The ultrastructural arrangement of the wood cell wall. *Cellulose Chemistry and Technology,* 9, 563-573.

Kersten, P.J., Kalyanaraman, B., Hammel, K.E., Reinhammer, B. and Kirk, T.K. (1990). Comparison of lignin peroxidase, horse peroxidase and laccase in the oxidation of methoxybenzenes. *The Biochemical Journal,* 268, 475–480.

Kim, S.J. and Shoda, M. (1999). Purification and characterization of a novel peroxidase from *Geotrichum candidum* dec 1 involved in decolorization of dyes. *Applied Environmental and Microbiology,* 65(3), 1029-1035.

Kimura, Y., Asada, Y., and Kuwahara, M. (1990). Screening of basidiomycetes for lignin peroxidase genes using a DNA probe. *Applied Microbiology and Biotechnology,* 32, 436-442.

Kirk, T.K. and Cullen, D. (1998). Enzymology and Molecular Genetics of Wood Degradation by White-Rot fungi. In: *Environmentally Friendly Technologies for the Pulp and Paper Industry,* Eds Young, R.A. and Akhtar, M., John Wiley and Sons, 280.

Kirk, T.K. and Farrel, R.L. (1987). Enzymatic combustion, the microbial degradation of lignin. *Annual Review of Microbiology,* 41, 465-505.

Kishi, K., Wariishi, H., Marquez, L., Dunford, H.B., and Gold, M.H. (1994). Mechanism of manganese peroxidase compound II reduction. Effect of organic acid chelators and pH. *Biochemistry,* 33, 8694-8701.

Klonowska, A., Le Petit, J., and Tron, T. (2001). Enhancement of minor laccase production in the basidiomycete *Marasmius quercophilus* C30. *FEMS Microbiology Letters,* 200, 25-30.

Koroleva, O.V., Garilova, V.P., Stepanova, E.V., Lebedeva, V.I., Sverdlova, N.I., Landesman, E.O., Yavmetdinov, I.S., and Yaropolov, A.I. (2002a). Production of lignin modifying enzymes by co-cultivated white-rot fungi *Cerrena maxima* and *Coriolus hirsutus* and characterization of laccase from *Cerrena maxima. Enzyme and Microbial Technology,* 30, 573-580.

Koshijima, T. Watanabe, T. and Yaku, F. (1989). Structure and properties of the lignin-carbohydrate polymer as an amphipathic substance. In, *Lignin-properties and Materials,* Eds Glasser, W. G. and Sarkanen, S., ACS Symposium Series Vol. 397, 11-28.

Kremer, S.M., and Wood, P.M. (1992). Production of Fenton's reagent by cellobiose oxidase from cellulolytic cultures of *Phanerochaete chrysosporium. European Journal of Biochemistry,* 208, 807-814.

Kuhad, R.C., Singh, A., and Eriksson, K. E. (1997). Microorganisms and enzymes involved in the degradation of plant fiber cell walls. In: *Advances in Biochemical Engineering Biotechnology,* Ed Eriksson, K.E., Springer-Verlag, Germany. 46-125.

Lackner, R., Srebotnik, E., and Messner, K. (1991). Oxidative degradation of high molecular weight chlorolignin by manganese peroxidase of *Phanerochaete chrysosporium. Biochemical and Biophysical Research Communication,* 178, 1092-1098.

Law, D.J. and Timberlake, W.E. (1980). Developmental regulation of laccase levels in *Aspergillus nidulans.* Journal of Bacteriology 144 (2), 509-517.

Levin, L., and Forchiassin, F., and Ramos, A.M. (2001). Ligninolytic enzymes of the white-rot basidiomycete *Trametes trogii. Acta Biotechnoligica,* 21(2), 179-186.

Levin, L., and Forchiassin, F., and Ramos, A.M. (2002). Copper induction of lignin-modifying enzymes in the white-rot fungus *Trametes trogii. Mycologia,* 94(3), 377-383.

Levin, L., Forchiassin, F. and Viale, A. (2005) Ligninolytic enzyme production and dye decolorization by *Trametes trogii,* application of the Plackett–Burman experimental design to evaluate nutritional requirements. *Proceedings of Biochemistry,* 40, 1381-1387.

Liers, C., Ullrich, R., Steffen, K.T., Hattaka, A. and Hofrichter, M. (2006). Mineralization of (14)C-labelled synthetic lignin and extracellular enzyme activities of the wood-colonizing ascomycetes *Xylaria hypoxylon* and *Xylaria polymorpha. Applied Microbiology and Biotechnology,* 69(5), 533-539.

Lobos, S., Larrain, J., Salas, L., Cullen, D. and Vicuna, R. (1994). Isozymes of manganes-dependent peroxidase and laccase produced by the lignin-degrading basidiomycete *Ceriporiopsis subvermispora. Journal of General Microbiology,* 140, 2691-2698.

Lomascolo, A., Cayol, J.L., Roche, M., Guo, L., Robert, J.L., Record, E., Lessage-Meessen, L., Olliver, B., Sigoillot, J.C., and Asther, M. (2002). Molecular clustering of *Pycnoporus* strains from various geographic origins and isolation of monokaryotic strains for laccase hyperproduction. *Mycology Research,* 106(10), 1193-1203.

Lundell,T. and Hattaka, A. (1994). Participation of Mn (II) in the catalysis of laccase, manganese peroxidase and lignin peroxidase from *Phlebia radiata. FEBS Letters,* 348, 291-296.

Mäkelä, M., Galkin, S., Hatakka, A., and Lundell, T. (2002). Production of organic acids and oxalate decarboxylase in lignin-degrading white rot fungi. *Enzyme and Microbial Technology,* 30, 542-549.

Maltseva, O.V., Niku-Paavola, M.-L., Leontievsky, A.A., Myasoedova, N.M. and Golovleva, L.A. (1991). Ligninolytic enzymes of the white rot fungus *Panus tigrinus. Biotechnology and Applied Biochemistry,* 13, 291-302.

Martinez, A.T., Camarero, S., Guillen, F., Gutierrez, A., Munoz, C., Varela, E., Martinez, M.J., Barrasa, J., Ruel, K. and Pelayo, J. (1994). Progress in biopulping of nonwoody materials, chemical, enzymatic and ultrastructural aspects of wheat straw delignification with ligninolytic fungi from the genus *Pleurotus. FEMS Microbiology Review,* 13, 265-273.

Martinez, D., Larrondo, L., Putnam, N., Sollewijn Gelpke, M., Huang, K., Chapman, J., Helfenbein, K., Ramaiya, P., Detter, J., Larimer, F., Coutinoho, P., Henrissat, B., Berka, R, Cullen, D. and Rokhsar, D., (2004). Genome sequence of the lignocellulose degrading fungus *Phanerochaete chrysosporium* strain RP78. Nature Biotechnology 22, 695-700.

Martinez, M.J., Ruiz-Duenas, F.J., Guillen, F. and Martinez, A.T., (1996). Purification and catalytic properties of two manganese peroxidase isoenzymes from *Pleurotus eryngii. European Journal of Biochemistry,* 237, 424-432.

Masaphy, S. and Levanon, D. (1992). The effect of lignocellulose on lignocellulolytic activity of *Pleurotus pulmonarius* in submerged culture. *Applied Microbiology and Biotechnology,* 36, 828-832.

Máximo, C. and Costa-Ferreira, M. (2004) Decolourisation of reactive textile dyes by *Irpex lacteus* and lignin modifying enzymes. *Proceedings of Biochemistry,* 39, 1475–1479.

Mayer, A.M., Staples, R.C. (2002) Laccase, new functions for an old enzyme. *Phytochemistry,* 60, 551–565.

McGuirl, M.A., Dooley, D.M. (1999). Copper-containing oxidases. *Current Opinion in Chemical Biology,* 3, 138-144.

Messerschmidt, A. (1997). Spatial structures of ascorbate oxidase, laccase and related proteins, implication for catalytic mechanism. In, *Multicopper oxidases,* Ed Messerschmidt, A., Singapore, World Scientific, Pp 23-80.

Mester, T. and Field, J.A. (1998). Characterization of a novel manganese peroxidase-lignin peroxidase hybrid isozyme produced by *Bjerkandera* species strain BOS55 in the absence of manganese. *Journal of Biology and Chemistry,* 273, 15412-15417.

Moldes, D., Couto, S.R., Camesslle, C., and Sanroman, M.A. (2003). Study of the degradation of dyes by Manganese peroxidase of *Phanerochaete chrysosporium* produced in a fixed-bed bioreactor. *Chemosphere,* 51, 295-303.

Morais, H., Ramos, C., Forgacs, E., Cserhati, T., and Oliviera, J. (2001b). Lignin-modifying enzymes of *Pleurotus ostreatus* grown on agro-residues. *Acta Aliment,* 30(4), 363-372.

Moreira, M.T., Feijoo, G., and Lema, J.M. (2000a). Manganese peroxidase production by *Bjerkandera* sp. BOS55.1. Regulation of enzymatic production. *Bioprocess Engineering,* 23, 657-661.

Moreira, M.T., Torrado, A., Feijoo, G., and Lema, J.M. (2000b). Manganese peroxidase production by *Bjerkandera* sp. BOS55.2. Operation in stirred tank reactors. *Bioprocess Engineering,* 23, 663-667.

Mouchacca, J. (1997). Thermophilic fungi: Biodiversity and taxonomic status. *Cryptogamie Mycology,* 18, 19-69.

Muheim, A., Waldner, R., Leisola, M.S.A. and Fiechter, A. (1990). An extracellular aryl-alcohol oxidase from the white-rot fungus *Bjerkandera adusta. Enzyme and Microbial Technology,* 12, 204-209.

Munoz, C., Guillen, F., Martinez, A.T. and Martinez, M.J. (1997a). Laccase isoenzymes of *Pleurotus eryngii,* characterization, catalytic properties and participation in activation of molecular oxygen and $Mn^{2+}$ oxidation. *Applied Environmental and Microbiology,* 63, 2166-2174.

Munoz, C., Guillen, F., Martinez, A.T. and Martinez, M.J., (1997b). Induction and characterization of laccase in the ligninolytic fungus *Pleurotus eryngii. Current Microbiology,* 34, 1-5.

Nagai, M., Kawata, M., Watanabe, H., Ogawa, M., Saito, K., Takesawa, T., Kanda, K. and Sato, T. (2003). Important role of fungal intracellular laccase for melanin synthesis, purification and characterization of an intracellular laccase from *Lentinula edodes* fruit bodies. *Microbiology,* 149, 2455-2462.

Nerud, F. and Misurcova, Z. (1989). Production of ligninolytic peroxidases by the white-rot fungus *Coriolopsis occidentalis. Biotechnology Letters,* 11, 427-432.

Nerud, F., Zouchova, Z. and Misurcova, Z. (1991). Ligninolytic properties of different white-rot fungi. *Biotechnology Letters,* 13, 657-660.

Nie, G., Reading, N.S., and Aust, S.D. (1999). Relative stability of recombinant versus native peroxidases from *Phanerochaete chrysosporium. Archives of Biochemistry and Biophysics,* 365, 328–334.

Niku-Paavola, M.L., Karhunen, E., Salora, P. and Raunio, V. (1988). Ligninolytic enzymes of the white-rot fungus *Phlebia radiata. Journal of Biochemistry,* 254, 877-884.

Niku-Paavola, M.-L., Karhunen, E., Kantelinen, A. , Viikari, L., Lundell, T. and Hatakka, A. (1990). The effect of culture conditions on the production of lignin modifying enzymes by the white-rot fungus *Phlebia radiata. Journal of Biotechnology,* 13, 211-221.

Novotny, C., Erbanova, P., Cathaml, T., Rothschild, N., Dosoretz, C. and Sasek, V. (2000) *Irpex lacteus*, a white rot fungus applicable to water and soil bioremediation. *Applied Microbiology and Biotechnology,* 54, 850-853.

Nuske, J., Scheibner, K., Dornberger, U., Ullrich, R. and Hofrichter, M. (2002). Large-scale production of manganese-peroxidase using agaric white-rot fungi. *Enzyme and Microbial Technology,* 30, 556-561.

Orth, A.B., Royse, D.J. and Tein, M. (1993). Ubiquity of lignin-degrading peroxidases among various wood-degrading fungi. *Applied Environmental and Microbiology,* 59, 4017-4023.

Otjen, J. and R.A. Blanchette. (1982). Pattern of decay caused by *Inonotus dryophilus* (Aphyllophorales, Hymenochaetaceae) a white-pocket rot of oaks. *Canadian Journal of Botany* 60, 2270-2279.

Otjen, L. and R.A. Blanchette. (1984). *Xylobolus frustulatus* decay of oak, patterns of selective delignification and subsequent cellulose removal. *Applied Environmental and Microbiology,* 47, 670-676.

Palmieri, G., Giardina, P., Bianco, C., Scaloni, A., Capasso, A. and Sannia, G. (1997). A Novel White Laccase from *Pleurotus ostreatus*. Journal of Biology and Chemistry 272, 31301-31307.

Panchin, A. and De Zeeuw, C. (1964). *Textbook of wood technology,* Vol. 1, McGraw-Hill Book Co., New York.

Papinutti, V. L., Diorio, L. A. and Forchiassin, F. (2003). Production of laccase and manganese peroxidase by *Fomes sclerodermeus* grown on wheat bran. *Journal of Indian Microbiology and Biotechnology,* 30, 157-160.

Perez, J. and Jeffries, T.W. (1990). Mineralization of $^{14}C$-ring-labeled synthetic lignin correlates with the production of lignin peroxidase, not of manganese peroxidase or laccase. *Applied Environmental and Microbiology,* 56, 1806-1812.

Perie, F.H. and Gold, M.H. (1991). Manganese regulation of manganese peroxidase expression and lignin degradation by the white rot fungus *Dichomitus squalens. Applied Environmental and Microbiology,* 57, 2240-2245.

Perie, F.H., Reddy, G.V., Blackburn, N.J. and Gold, M.H. (1998). Purification and characterization of laccases from the white-rot fungi basidiomycete *Dichomitus squalens. Biochimica Biophysics Acta,* 353, 349-355.

Perie, F.H., Sheng, D., and Gold, M.H. (1996). Purification and characterization of two manganese peroxidase isozymes from the white-rot basidiomycete *Dichomitus squalens. Biochimica Biophysics Acta,* 1297, 139-148.

Pickard, M.A., Vandertol, H., Roman, R. and Vazquez-Duhalt, R. (1999b). High production of ligninolytic enzymes from white-rot fungi in cereal bran liquid medium. *Canadian Journal of Microbiology,* 45, 627-631.

Rayner, A.D.M. and Boddy, L. (1988). *Fungal decomposition of wood.* Wiley, Great Britain.

Reddy, G.V., Babu, P.R., Komaraih, P., Roy, K.R.R.M. and Kothari, I.L. (2003). Utilization of banana waste for the production of ligninolytic and cellulolytic enzymes by solid substrate fermentation using two *Pleurotus* species (*P. ostreatus* and *P. sajor-caju*). *Proceedings of Biochemistry,* 38, 1457-1462.

Reinhammar, B. (1984). Laccase. In, *Copper proteins and copper enzymes,* Ed Lontie, R., Boca Raton (FL), CRC Press, Vol. 3, 1-36.

Robene-Soustrade, I., Lung-Escarmant, B., Bono, J.J. and Taris, B. (1992). Identification and partial characterization of an extracellular manganese-dependent peroxidase in *Armillaria ostoyae* and *Armillaria mellea. European Journal of Forest Pathology,* 22, 227-236.

Rosales, E., Couto, S.R. and Sanroman, A. (2002). New uses of food waste, application to laccase production by *Trametes hirsuta. Biotechnology Letters,* 24, 701-704.

Roy, B.P., Dumonceaux, T., Koukoulas, A.A. and Archibald, F.S. (1996). Purification and characterization of cellobiose dehydrogenase from the white rot fungus *Trametes versicolor. Applied Environmental and Microbiology,* 62, 4417-427.

Ruiz-Dueñas, F.J., Camarero, S., Perez-Boada, M., Martinez, M.J. and Martinez, A.T. (2001). A new versatile peroxidase from *Pleurotus. Biochemical Society Transactions,* 29(2), 116-122.

Ruiz-Dueñas, F.J., Martinez, M.J. and Martinez, A.T. (1999). Molecular characterization of a novel peroxidase isolated from the ligninolytic fungus *Pleurotus eryngii. Molecular Microbiology,* 31, 223-235.

Rüttimann, C., Salas, L. and Vicuna, R. (1992a). Studies on the ligninolytic system of the white rot fungus *Ceriporiopsis subvermispora.* In, *Biotechnology in the Pulp and Paper Industry,* Eds Kuwahara, M. and Shimada, M., Uni. Publishers Ltd., Tokio, Pp. 243-248.

Ruttimann, C., Schwember, E., Salas, L., Cullen, D. and Vicuna, R. (1992b). Ligninolytic enzymes of the white rot basidiomycetes *Phlebia brevispora* and *Ceriporiopsis subvermispora. Biotechnology and Applied Biochemistry,* 16, 64-76.

Saloheimo, M., Barajas, V., Niku-Paavola, M.-L. and Knowles , J.K.C. (1989). A lignin peroxidase-encoding cDNA from the white rot fungus *Phlebia radiata*, characterization and expression in *Trichoderma reesei.* **Gene** 85, 343-351.

Sannia, P., Limongi, P., Cocca, E., Buonocore, F., Nitti, G. and Giardiana, P. (1991). Purification and characterization of a veratryl alcohol oxidase enzyme from the lignin degrading basidiomycete *Pleurotus ostreatus. Biochimica Biophysics Acta,* 1073, 114-119.

Saparrat, M.C.N., Guillén, F., Arambarri, A.M., Martínez, A.T. and Martínez M.J. (2002). Induction, Isolation and Characterization of Two Laccases from the White Rot Basidiomycete *Coriolopsis rigida. Applied Environmental and Microbiology,* 68, 1534-1540.

Sarkar, S., Martinez, A.T. and Martinez, M.J. (1997). Biochemical and molecular characterization of a manganese peroxidase isoenzyme from *Pleurotus ostreatus. Biochimica et Biophysica Acta,* 1339, 23-30.

Scheel, T., Höfer, M., Ludwig, S. and Hölker, U. (2000). Differential expression of manganese peroxidase and laccase in white-rot fungi in the presence of manganese or aromatic compounds. *Applied Microbiology and Biotechnology,* 54, 686-691.

Scheffer, T. and Verrall, A. (1973). Principles for protecting wood buildings from decay. *USDA Forest Service Research Paper FPL-190,* U.S. Forest Products Laboratory, Madison, WI.

Schoemaker, H.E. and Leisola, M.S.A. (1990). Degradation of lignin by *Phanerochaete chrysosporium. Journal of Biotechnology,* 13, 101-109.

Schoemaker, H.E., Harvey, P.J., Bowen, R.M. and Palmer, J.M. (1985). On the mechanism of enzymatic lignin breakdown. *FEBS Letters,* 183, 7-12.

Schoemaker, H.E., Lundell, T.K., Hatakka, A.I. and Piontek, K. (1994). The oxidation of veratryl alcohol, dimeric lignin models and lignin by lignin peroxidase-the redox cycle revisited. *FEMS Microbiology Review,* 13, 321-332.

Srinivasan, C., D'Souza, T. M., Boominathan, K. and Reddy, C.A. (1995). Demonstration of laccase in the white rot basidiomycete *Phanerochaete chrysosporium* BKM-F1767. *Applied Environmental and Microbiology,* 61, 4274-4277.

Stewart, P. and Cullen, D. (1999). Organization and differential regulation of a cluster of lignin peroxidase genes of *Phanerochaete chrysosporium. Journal of Bacteriology,* 181, 3427-3432.

Straatsma, G., Samson, R.A., Olijnsma, T.W., Op den camp, H.j.M., Gerrits, J.P.G. and van Grioncvon, L.J.L.D. (1994) Ecology of thermophilic fungi in mushroom compost, with emphasis on *Scytalidium thermophilum* and growth stimulation of *Agaricus bisporus* mycelium. *Applied Environmental and Microbiology,* 60, 454-458.

Szklarz, G. and Leonowicz, A. (1986). Cooperation between fungal laccase and glucose oxidase in the degradation of lignin derivatives. *Phytochemistry,* 25(11), 2537-2539.

Szklarz, G.D., Antibus, R.K., Sinsabaugh, R.L. and Linkins, A.E. (1989). Production of Phenol oxidases and Peroxidases by wood-rotting fungi. *Mycologia,* 8, 234-240.

Tagger, S., Perissol, C., Gil, G., Vogt, G. and Le Petit, J. (1998). Phenoloxidases of the white-rot fungus *Marasmius quercophilus* isolated from an evergreen oak litter (*Quercusilex* L.). *Enzyme Microbial and Technology,* 23, 372-379.

Tamaru H. and Inoue H. (1989). Isolation and characterization of a laccase-depressed mutant of *Neurospora crassa. Journal of Bacteriology,* 171(11), 6288-6293.

Tekere, M., Zvvauya, R. and Read, J.S., (2001). Ligninolytic enzyme production in selected sub-tropical white rot fungi under different culture conditions. *Journal of Basic Microbiology,* 41, 115-129.

Temp, U. and Eggert, C. (1999). Novel interaction between laccase and cellobiose dehydrogenase during pigment synthesis in the white-rot fungi *Pycnoporus cinnabarinus Applied Environmental and Microbiology,* 65(2) 389-395.

Tien, M. and Kirk, T.K. (1983). Lignin-degrading enzyme from the hymenomycete *Phanerochaete chrysosporium* Burds. *Science,* 221, 661-663.

Tien, M. and Kirk, T.K. (1984). Lignin-degrading enzyme from *Phanerochaete chrysosporium*, purification, characterization and catalytic properties of a unique $H_2O_2$-requiring oxygenase. *Proceedings of National Academy of Science,* USA. 81, 2280-2284.

Tuomela, M., Vikman, M., Hattaka, A. and Itävaara, M. (2000). Biodegradation of lignin in a compost environment, a review. *Bioresource Technology,* 72, 169-183.

Tuor, U., Winterhalter, K. and Fiechter, A. (1995). Enzymes of white-rot fungi involved in lignin degradation and ecological determinants for wood decay. *Journal of Biotechnology,* 41, 1-17.

Ürek, R.O. and Pazarlioglu, N.K. (2005) Production and stimulation of manganese peroxidase by immobilized *Phanerochaete chrysosporium. Proceedings of Biochemistry,* 40, 83-87.

Urzúa, U., Larrondo, L.F., Lobos, S., Larraın, J. and Vicuña, R. (1995). Oxidation reactions catalyzed by manganese peroxidase isoenzymes from *Ceriporiopsis subvermispora. FEBS Letters,* 371, 132–136.

Varela, E., Martinez, A.T. and Martinez, M.J. (2000). Southern blot screening for lignin peroxidase and aryl-alcohol oxidase genes in 30 fungal species. *Journal of Biotechnology,* 83, 245-251.

Vares, T. (1996). Ligninolytic enzymes and lignin-degrading activity of taxonomically different white-rot fungi. *PhD Thesis,* Vol. 44, Department of Applied Chemistry and Microbiology, University of Helsinki, Finland, Pp 67.

Vares, T., Lundell, T.K. and Hattaka, A.I. (1992). Novel heme-containing enzyme possibly involved in lignin degradation by the white-rot fungus *Junghuhnia separabilima. FEMS Microbiology Letters,* 99, 53-58.

Vares, T., Lundell, T.K. and Hattaka, A.I. (1993). Production of multiple lignin peroxidases by the white-rot fungus *Phlebia ochraceofulva. Enzyme and Microbial Technology,* 15, 664-669.

Wahleithner J.A., Xu F., Brown K.M., Brown S.H., Golightly E.J., Halkier T., Kauppinen S., Pederson A. and Schneider P. (1996). The identification and characterization of four laccases from the plant pathogenic fungus *Rhizoctonia solani. Current Genetics,* 29(4), 395-403.

Waldner, R., Leisola, M.S.A. and Fiechter, A. (1988). Comparison of ligninolytic activities of selected white-rot fungi. *Applied Microbiology and Biotechnology,* 29, 400-407.

Wang, Y., Vazquez-Duhalt, R. and Pickard, M.A. (2002). Purification, characterization and chemical modification of manganese peroxidase from *Bjerkandera adusta* UAMH 8258. *Current Microbiology,* 45, 77-87.

Wariishi, H., Sheng, D. and Gold, M.H. (1994). Oxidation of ferrocytochrome c by lignin peroxidase. *Biochemistry,* 33, 5545-5552.

Wariishi, H., Valli, K., and Gold, M.H. (1991). In vitro depolymerization of lignin by manganese peroxidase of *Phanerochaete chrysosporium. Biochemical and Biophysical Research Communication,* 176, 269-275.

Wood, P. M. (1994). Pathways of production of Fenton's reagent by wood-rotting fungi. *FEMS Microbiology Review,* 13, 313-320.

Zabel, R. and Morrell, J. (1992). *Wood Microbiology,* Academic press, San Diego, CA.

Ziegenhagen, D. and Hofrichter, M. (1998). Degradation of humic acids by manganese peroxidase from the white-rot fungus *Clitocybula dusenii. Journal of Basic Microbiology,* 38,289-299.

# 4

# Structural and Functional Aspects of Lignolytic Enzymes

**Meera Yadav and K.D.S. Yadav**

*Department of Chemistry, D.D.U. Gorakhpur University, Gorakhpur 273009, India*

Structural and functional studies on lignolytic enzymes i.e. lignin peroxidase, Mn-peroxidase, versatile peroxidase and laccase have been reviewed. All lignolytic enzymes have broad substrate specificities, which have been discussed. High-resolution crystal structures of representative lignolytic enzymes except versatile peroxidase are available. Utilising the crystal structure models of LiP, MnP and other peroxidases, model for the structure of versatile peroxidase has been suggested. All the structural features of the lignolytic enzymes have been discussed in relation to the functional aspects of these enzymes.

## INTRODUCTION

Lignocellulose is the second most abundant material in the biosphere (Souza *et al.,* 1999). It is seen as a renewable alternative to the depleting oil reserves, which constitutes the backbone of our present chemical industries. Lignin is a natural polymer consisting of arylpropyl unit linked primarily with β-O-4 bond but also with various other C-C and C-O linkages as shown in Fig. 1.

Lignin when attached to cellulose makes it resistant to microbial attack and chemical conversion of lignocellulose materials to feed stock for chemical industries requires drastic reaction conditions, which are expensive in energy and pollute the environment.

There is a need to develop low energy and ecofriendly technology for the conversion of lignocellulose to materials, which could be used as, feed stock for chemical industries. If lignin is removed from the lignocellulose, it becomes amenable to further conversions to useful chemicals. Lignolytic microorganisms and lignolytic enzymes delignify lignocellulosic materials and therefore there is a possibility of developing a technology of delignification of lignocellulosic materials based on lignolytic microorganisms/enzymes. Key to the development of such a technology is the understanding of molecular mechanism of delignification of lignocellulosic materials by lignolytic microorganisms/enzymes. This has been an active area of global research during the last three decades and a vast literature (Martinez, 2002) is available on this topic.

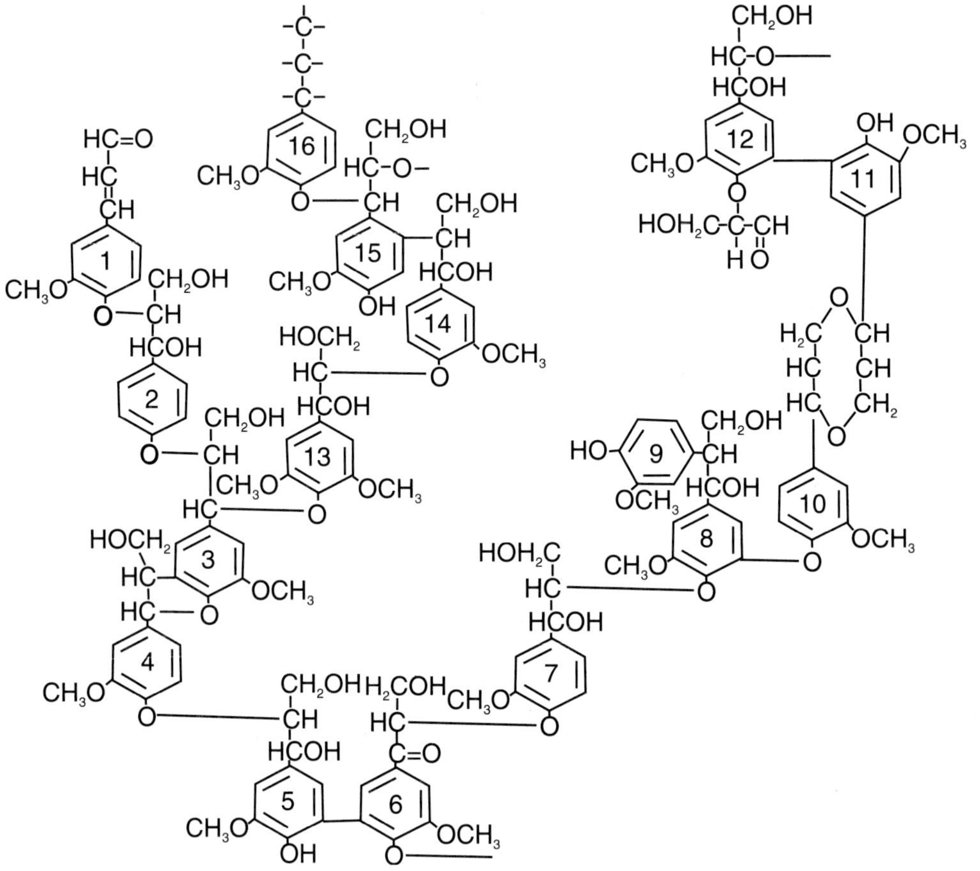

**Fig. 1:** Structure of lignin (Souza *et al.*, 1999)

There are clear evidences (Martinez, 2002) for the involvement of four enzymes lignin peroxidase [E.C.1.11.1.14], Mn-peroxidase [E.C.1.11.1.13], versatile peroxidase (VP) and laccase [E.C.1.10.3.2] – in lignin degradation. In addition, other supporting enzymes like $H_2O_2$ producing enzymes, glyoxal oxidase and aryl alcohol oxidase have also been implicated in lignin degradation (Martinez, 2002). In future, other enzymes (Su *et al.,* 2005; Mcpherson *et al.,* 2001) may add to the list of lignin biodegradation. However, the present article will cover only structural and functional aspects of lignin peroxidase, Mn-peroxidase, versatile peroxidase and laccase. The literature upto 2002 on lignin peroxidase, Mn-peroxidase and versatile peroxidase has been extensively reviewed by Martinez (Martinez, 2002), which covers structure, function as well as molecular biology of these enzymes. Authors are not in a position to add much to the information available in the review (Martinez, 2002) on lignin degrading haem peroxidases. Only recent findings on these enzymes will be included. However, some new structural information on laccases has been reported during recent years, which will be summarized in this article.

## LIGNIN PEROXIDASE

Ever since Tien and Kirk (1983) reported discovery of an extracellular enzyme from *Phanerochaete chrysosporium* which catalyses the cleavage of lignin model compounds (a) and (b) as shown in Fig. 2 and spruce and birch lignins in the presence of $H_2O_2$, the structural and functional studies on this enzyme have been an active area of research. The same group of workers (Tein and Kirk, 1984) purified and characterized the enzyme as a haem containing $H_2O_2$ requiring oxygenase of relative molecular mass 42KD which later has been proved to be a peroxidase (Kuila *et al.,* 1985). Kuwahara *et al.,* 1984 also independently reported the same enzyme from the culture filtrate of *Phanerochaete chrysosporium,* which was spectrally characterized by Anderson and coworkers (1985). The ligninase (as called initially) has spectral properties similar to those of hemoproteins (Martinez, 2002): catalase, horseradish peroxidase and c-type cytochromes and is a glycoprotein (Tein and Kirk, 1984).

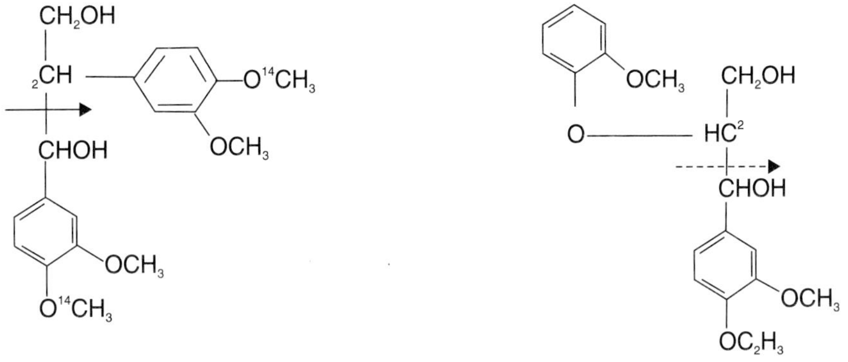

**Fig. 2:** Lignin model compounds (a) and (b) from Tein and Kirk (1983)

The enzyme catalyses a variety of oxidations (Tein and Kirk 1983):

(i) $C_\alpha$-$C_\beta$ cleavage of β-1 and β-O-4 lignin models as shown in Fig. 3 (ii) oxidation of benzyl alcohols; (iii) oxidation of phenols leading to radical coupling; (iv) hydroxylation of certain benzylic methelene groups; and (v) intradiol cleavage of phenyl glycol structures. All these reactions were discovered in studies with β-1 and β-O-4 lignin model compounds. The predominant reaction in oxidation of lignin models is the cleavage of the $C_\alpha$- $C_\beta$ bond (Hammel *et al.,* 1986).

Kersten and coworkers (1985) using EPR detected the formation of cation radicals in reaction mixtures containing methoxybenzenes, lignin peroxidase and $H_2O_2$ as shown in Fig 4.

This finding has formed basis for understanding the mechanisms of the various reactions catalysed by lignin peroxidase (Harvey *et al.,* 1986). Utilising these findings and a large number of experimental observations related to the function of lignin peroxidase, Harvey and coworkers (1986) have given an explanation, which covers most of the reactions catalysed by lignin peroxidase. The key to all the reactions is the formation of a radical cation intermediate, which can follow different paths depending on the nature of the substituent groups present on the radical cation intermediate and also on the prevailing reaction conditions. Initially, terms like oxidase (Tein and Kirk, 1983;

**Fig. 3.** $C_\alpha$ - $C_\beta$ cleavage of β-1 (A) and β-0-4 (B) lignin models compounds (Tein and Kirk, 1983).

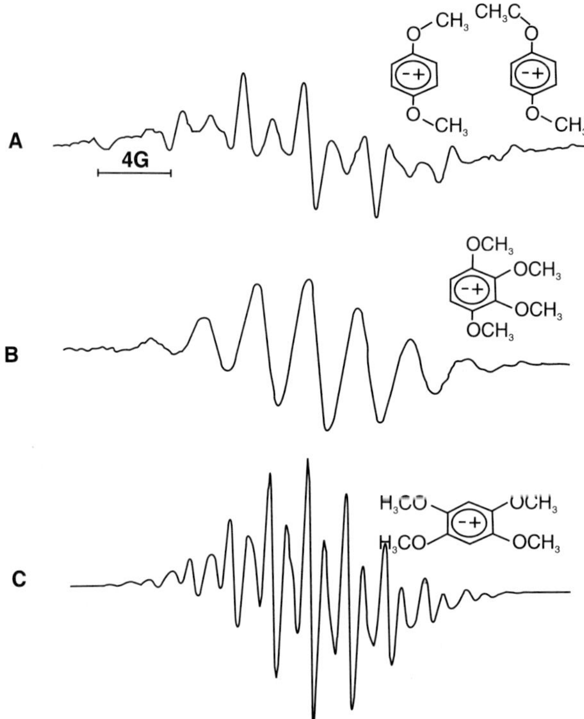


**Fig. 4.** ESR spectra of methoxybenzenes cation radicals (Kersten *et al.*, 1985)

Kuwahara *et al.,* 1984), Oxygenase (Tein and Kirk, 1984; Andersson *et al.,* 1985) and ligninase (Kuila *et al.,* 1985) were used for lignin peroxidase but Kuila and coworkers (1985) using Resonance Raman spectral studies showed that the active site of ligninase was similar to those of peroxidases. Once it was established that ligninases are peroxidases, the characterisation of enzyme intermediates compound I, II, III as observed in case of peroxidases were initiated.

Renganathan and Gold (1986) characterized the $H_2O_2$ oxidized forms of lignin peroxidase compound I, II and III using electronic absorption spectroscopy. The native enzyme contains high-spin ferric iron and has its Soret maximum at 407.6 nm and $\alpha$ and $\beta$ bands at 496 and 630 nm (Fig 5.)

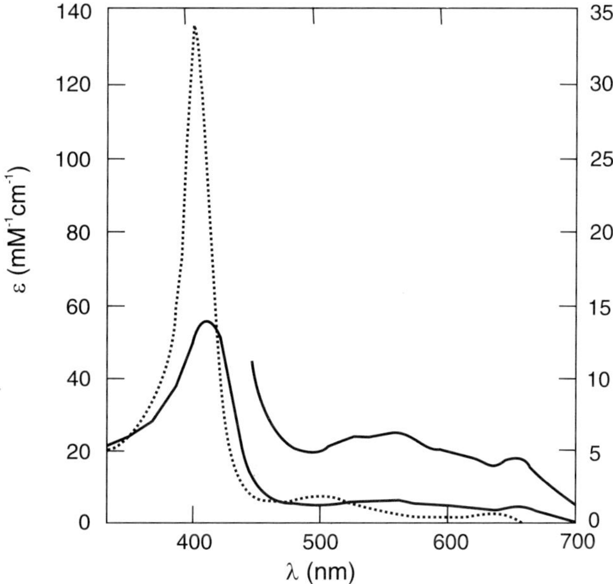


**Fig. 5.** Electronic absorption spectrum of the native enzyme (----) and of the compound I (—) from Renganathan and Gold., 1989.

Addition of 1 molar equivalent of $H_2O_2$ to the native enzyme in 20mM sodium phosphate, pH 6.0 yields compound I characterized by a Soret maximum at 408 nm with reduced intensity and by additional maxima at 550, 608 and 650 nm. Addition of 2 molar equivalent of $H_2O_2$ to the native enzyme yields compound II identified by absorption maxima at 420, 525 and 556nm (Fig 6). Addition of molar excess of $H_2O_2$ to the native enzyme yields compound III, which is characterized by absorption maxima at 419, 540 and 578nm. (Fig.7). These spectral characteristics are similar to those of horseradish peroxidase compound I, II and III. When 0.5 molar equivalent of veratryl alcohol, a two-electron substrate is added to lignin peroxidase compound I, it rapidly reduces to lignin peroxidase compound II. Further addition of 0.5 molar equivalent of veratryl alcohol to lignin peroxidase compound II, reduces it to the native ferric enzyme. These observations have indicated that like horseradish peroxidase compounds I and II, lignin peroxidase compounds

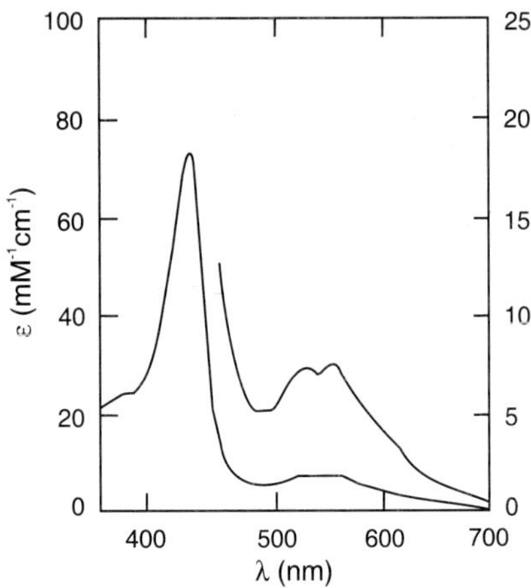


**Fig. 6.** Electronic absorption spectrum of LiP compound II (Renganathan and Gold, 1989)

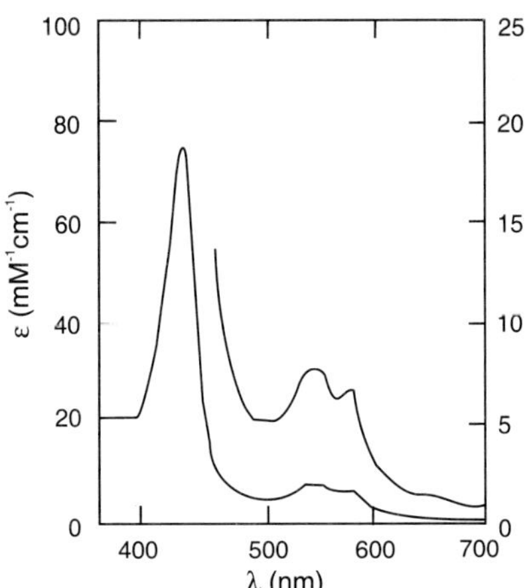


**Fig. 7.** Electronic absorption spectrum of LiP compound III (Renganathan and Gold, 1989)

I and II contain 2 and 1 oxidizing equivalent over the ferric resting state of the enzyme. It also has indicated that the dehydrogenation of veratryl alcohol by lignin peroxidase proceeds via two single-electron oxidation steps rather than through a single two-electron oxidation mechanism. The different enzyme intermediates could be depicted by Fig 8.

Lignin peroxidase compound III has been further characterized (Su *et al.,* 2005; Dunford and Stillman, 1976) and kinetics of its formation and decomposition have been studied (Renganathan and Gold, 1986; Wariishi and Gold, 1989).

Steady state kinetic studies on lignin peroxidase have shown Ping-Pong type mechanism as observed in the case of classical peroxidases (Mylranjan *et al.,* 1990). Transient state kinetic studies

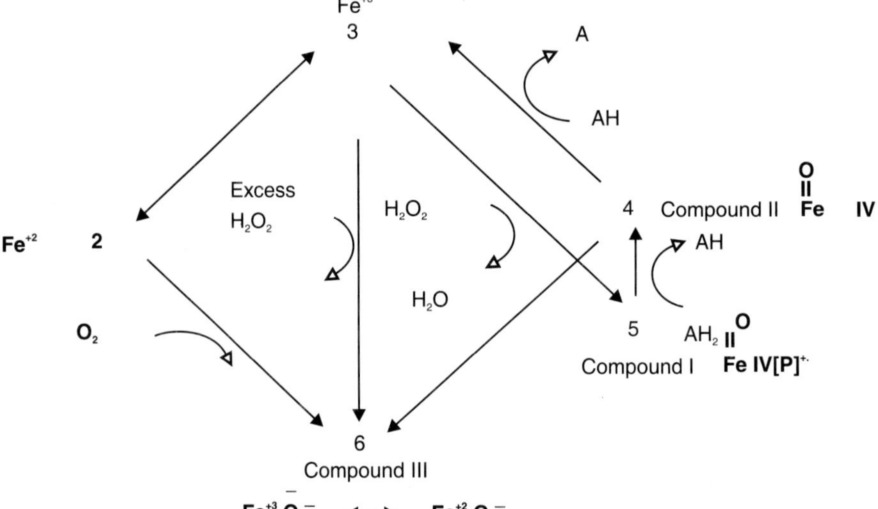


**Fig. 8.** Interrelationship between the five-redox states of lignin peroxidase (Renganathan and Gold, 1989). Reaction paths 3 → 5 → 4 indicate one catalytic cycle of the enzyme. $AH_2$ = substrate.

have shown the formation of compound I and compound II as observed in case of horseradish peroxidase (Mylranjan *et al.,* 1990; Wariishi and Gold, 1990).

The spectral characterisation of the enzyme intermediates of lignin peroxidase and steady state and transient state kinetic studies on lignin peroxidase and its enzyme intermediates have clearly established the similarity of lignin peroxidase to other peroxidases (Martinez, 2002). However, lignin peroxidase differs to other peroxidases in many respects particularly its compound I has higher oxidation potential than the oxidation potentials of compound I of other peroxidases. In order to understand the functional characteristics of lignin peroxidase, its three-dimensional structure was crucial. This crucial work has been done independently by two groups of research workers (Poulos *et al.,* 1993; Piotek *et al.,* 1993). When crystal structures of lignin peroxidase isoenzyme H8 (Poulos *et al.,* 1993) and H2 (Piotek *et al.,* 1993) were first published, the crystal structure of only cytochrome c peroxidase (Finzel *et al.,* 1984) in peroxidase family was available for comparison. The overall fold of lignin peroxidase as shown in Fig. 9 closely resembles that of cytochrome c peroxidase (Finzel *et al.,* 1984).

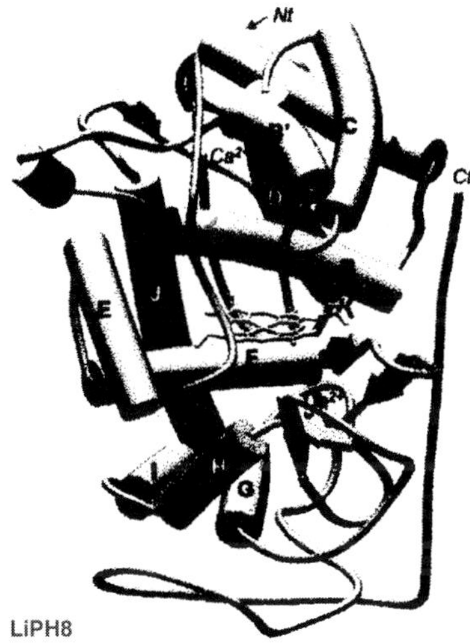


**Fig. 9.** Molecular structure of lignin peroxidase isozyme H8 (Poulos *et al.*, 1993)

The active sites as shown in Fig. 10 are also similar because both contain a proximal histidine haem ligand H176 hydrogen bonded to a buried aspartate residue D238 and both contain histidine H47 and arginine R43 residues in the distal peroxide binding pocket.

The most obvious difference in the active site is that whereas cytochrome c peroxidase has tryptophan residues located in the proximal and distal haem pockets, lignin peroxidase has phenylalanines. There are four other differences in the two structures. The haem pocket in cytochrome c peroxidase is open to solvent whereas in lignin peroxidase this opening is smaller. Lignin peroxidase has a carboxylate – carboxylate hydrogen bond important for haem binding that is not present in cytochrome c peroxidase. Lignin peroxidase contains 2 structural calcium ions as

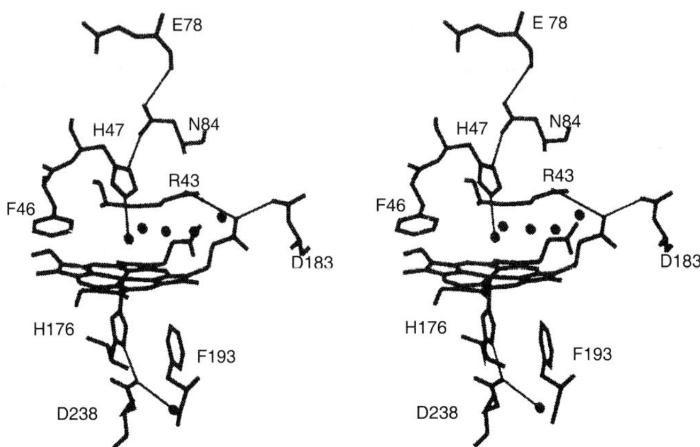


**Fig. 10.** Stereo view of the active site environment of LiP (Poulos *et al.*, 1993). The small dark spheres represent solvent molecules and the thin lines hydrogen bonds.

shown in Fig. 11 while cytochrome c peroxidase does not contain any calcium ion. The extra 49 residues in lignin peroxidase, which is not present in cytochrome c peroxidase, constitute the C terminal end of the molecule situated between the two-haem propionates.

Choinowski and co-workers (1999) have reported the crystal structure of lignin peroxidase isoenzymes H8 at 1.7 Å resolutions. They have observed that the surface residue tryptophan 171 in lignin peroxidase is stereospecifically hydroxylated at the C β-atom due to an autocatalytic process and have proposed the existence of a second substrate-binding site centered at Trp 171, which is

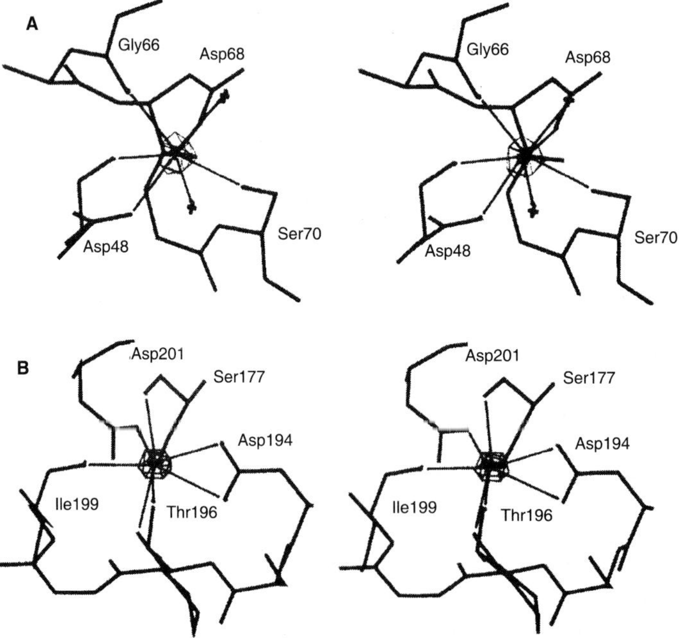


**Fig. 11.** Stereo view of the (a) "distal" and (b) "proximal" calcium site (Poulos *et al.*, 1993)

different from the classical haem edge site found in other peroxidases (Choinowski *et al.,* 1999). This point has been further confirmed by a number of experiments employing crystallography, site directed mutagenesis, protein chemistry, spin - trapping and spectroscopy (Blodig *et al.,* 2001; Pointek *et al.,* 2001). They have concluded that (i) Trp 171 is spectrospecifically hydroxylated at its C β – atom as the result of an autocatalytic process which occurs under turn over conditions in the presence of hydrogen peroxide, (ii) evidence for the formation of a Trp 171 radical intermediate has been obtained using spin trapping in combination with peptide mapping and protein crystallography, (iii) Trp 171 is very likely to be involved in electron transfer from natural substrates to the haem cofactor via long range electron transfer, (iv) mutagenetic substitution of Trp 171 in lignin peroxidase abolishes completely the oxidation activity of the enzyme for veratryl alcohol but not for artificial substrates, (v) structural changes due to mutation are marginal indicating that the lack of activity towards veratryl alcohol is due to the absence of the redox active indol side chain of Trp 171.

Mester and coworkers (2001) have studied the oxidation of a tetrameric non-phenolic lignin model compound by lignin peroxidase, which has shown that the substrate-binding site is well exposed. Mutation of the recombinant lignin peroxidase isoenzymes H8 in haem access channel, which was previously proposed to be the veratryl alcohol-binding site, had little effect on the oxidation of the tetramer. In contrast, mutation of a Trp residue (W 171 S) in the alternate proposed substrate-binding site completely inhibited the oxidation of the tetrameric model. These results are consistent with lignin peroxidase having an exposed active site capable of directly interacting with the lignin polymer without the involvement of low molecular weight mediators.

More information regarding the bonding of aromatic substrates in lignin peroxidases have come from the studies on the binding of aromatic substrates to versatile peroxidases (Martinez, 2002), which have been discussed in the section on versatile peroxidase.

## MANGANESE PEROXIDASE

Manganese peroxidase, MnP [E.C. 1.11.1.13] was first reported by Kuwahara and co-workers (1984) in the lignolytic culture of *P. chryosporium.* Since then, this enzyme has been reported in a number of fungal strains (Hatakka, 1994; Pelaez *et al.,* 1995) and it is characterised by oxidation of Mn(II) to Mn(III) (Glenn *et al.,* 1986). It has Mr of 46 KD. The haem Fe of the native enzyme is ferric, high spin and penta co-ordinate with a proximal histidine ligand and it oxidises a variety of organic compound but only in the presence of Mn (II). Kinetic and spectroscopic characterisation of oxidised intermediates, Mn-peroxidase compound (I) and (II) have indicated that the catalytic cycle of Mn-peroxidase is similar to that of lignin peroxidase. However, Mn-peroxidase is unique in its ability to oxidise Mn (II) to Mn (III) (Glenn *et al.,* 1986). The reactions involved in the Mn-peroxidase catalytic cycle are,

$$\text{MnP} + H_2O_2 \longrightarrow \text{MnP(I)} + H_2O$$

$$\text{MnP(I)} + \text{Mn}^{II} \longrightarrow \text{MnP(II)} + \text{Mn}^{III}$$

$$\text{MnP(II)} + \text{Mn}^{II} \longrightarrow \text{MnP} + \text{Mn}^{II} + H_2O$$

The enzyme generated Mn(III) is stabilized by chelators such as oxalate which are secreted by the fungi (Kuan and Tein, 1993; Hofrichter *et al.,* 1999). The Mn(III) chelator complex oxidises phenolic substrates such as lignin substructures (Wariishi *et al.,* 1989; Bao *et al.,* 1994; Tuor *et al.,* 1992; Lackner *et al.,* 1991; Hofrichter *et al.,* 2001; Wariishi *et al.,* 1991) and aromatic pollutants (Hofrichter *et al.,* 1998; Moreira *et al.,* 2001; Steffen *et al.,* 2003).

The crystal structure of Mn-peroxidase from *P. chryosporium* has been solved (Sundaramoorthy *et al.,* 1994). The overall structure as shown in Fig. 12 is similar to that of LiP from *P. chroysoporium.* It has same tertiary fold and share topology with other haem peroxidases (Martinez, 2002). However, MnP has a unique cation-binding site consisting of Glu-35, glu-39, Asp. 179, one of the haem propionates and two water molecules and this site has been proposed as Mn-binding site (Sollewijn Gelpke *et al.,* 1999) as shown in Fig.13.

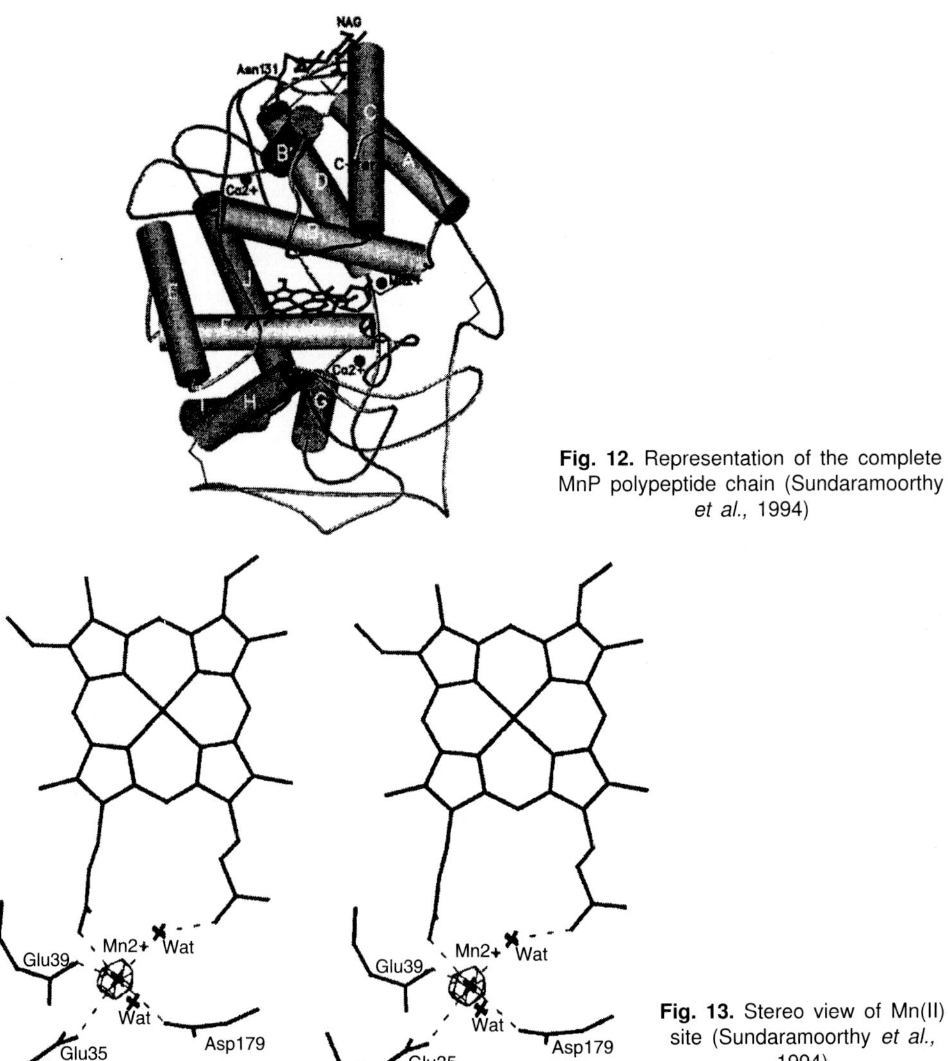


**Fig. 12.** Representation of the complete MnP polypeptide chain (Sundaramoorthy *et al.,* 1994)

**Fig. 13.** Stereo view of Mn(II) site (Sundaramoorthy *et al.,* 1994)

MnP also has two structural Ca ions as shown in Fig. 14. It also has two N-acetyl glucosamine residues, N-linked to Asn-131 that is readily visible in the electron density map. The active site as shown in Fig. 15 consisting of a proximal His-ligand H-bonded to an Asp-residue and a distal peroxide binding pocket consisting of a catalytic histidine and arginine is the same as in yeast cytochrome c-peroxidase. MnP differs in having five rather than four disulphide bonds.

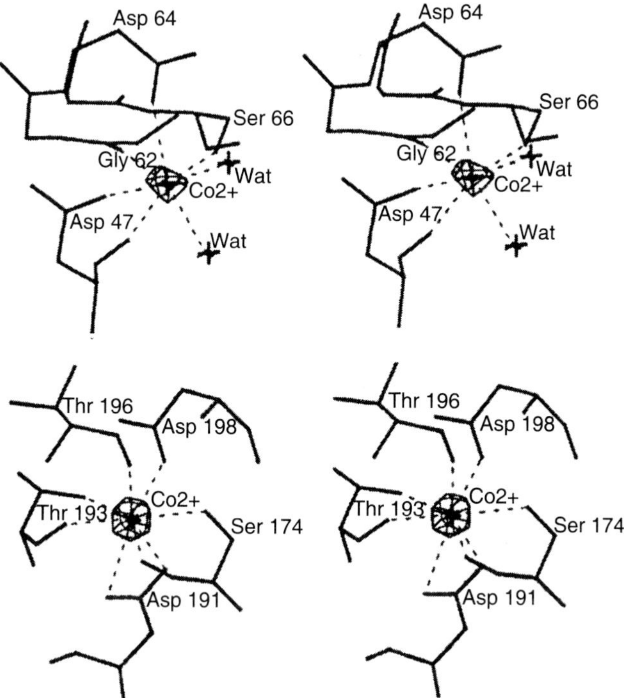


**Fig.14.** Stereo view of both $Ca^{2+}$ sites. Top, the distal site and bottom, the proximal site (Sundaramoorthy *et al.*, 1994)

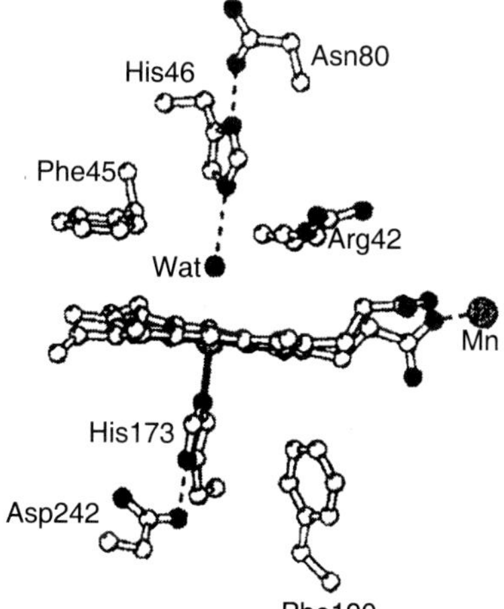


**Fig. 15.** Active site environment of MnP (Sundaramoorthy *et al.*, 1994)

The additional disulphide bond cys 341- cys 348 is located near the C-terminus of the polypeptide chain. The crystal structure also suggests that Arg-177 forms a salt bridge with Glu-35, thereby orienting this ligand for efficient Mn(II) binding (Sundaramoorthy *et al.,* 1994). Breaking this Arg-177-Glu-35-Glu-39 salt bridge by mutations such as R-177A and R-177K (Sollewijn Gelpke *et al.,* 1999) or E-35Q (Kishi *et al.,* 1996) affects the ability of the enzyme to adopt a close conformation upon Mn(II) binding resulting in an increase in the apparent Mn(II) binding constant. The disruption of the salt bridge between Arg-177 and Glu-35 by mutation of Arg-177 results in a decrease in Mn(II) affinity. This point has been further confirmed by using spectroscopic and kinetic analysis of four new Arg-177 mutants, namely, -R-177D, R-177E, R-177N and R-177Q. Unlike the mutants previously reported both the electron transfer rate and Mn(II) binding are affected in these mutants (Alic *et al.,* 1990).

The controversy regarding binding of Mn(II) to MnP in free hexa aqua form or oxalate complex form as raised by Warishii *et al* (1992) and Kuan *et al* (1993) has been finally settled by water proton relaxation studies (Baci *et al.,* 1998) which have indicated that oxalate did not bind to Mn(II) while it is co-ordinated to MnP but it binds to Mn(III) and removes it from the protein. Further information regarding these conclusions has come from the high-resolution crystal structure of MnP (Sundaramoorthy *et al.,* 2005). The new crystal structure of MnP complexed with Mn(II) provided a more accurate view of the Mn-binding site. The new features include possible partial protonation of Glu 39 in the Mn binding site and divalent Cd(II) binds Mn(II) binding site and exhibit octahedral hexacoordinate ligation geometry similar to Mn(II). Cd(II) also binds to a putative second weak metal binding site with tetrahedral geometry at the C terminus of the protein.

Unlike that of Mn(II) and Cd(II), co-ordination of trivariant Sm(III) at the Mn(II) binding site is octacoordinate. Sm(III) has been removed from MnP-Sm(III) crystal by soaking the crystal in oxalate and then has been reintroduced into the binding site. Thus direct comparison has been made between Sm(III) bound and metal free structures using the same crystal. Thus reversible binding of Sm(III) to the enzyme has provided a useful model for Mn(III) binding. Kinetic studies are required in this direction.

Several studies (Scott Reading and Steven, 2001) have been carried out to understand the structural stability of MnP. Glycoxylation of native peroxidases have been shown to play a role in stabilizing the protein against inactivation. Recombinant MnP has been found to inactivate at faster rate than native peroxidase. This trait was attributed to the lack of glycoxylation on the recombinant enzyme. Ca appears to have a critical role in stabilizing MnP against inactivation. Inactivation of native peroxidase has been correlated with the loss of atleast one Ca(II) from the protein. The Ca(II) has been proposed to be released from the distal Ca(II) binding site, resulting in the formation of a sixth ligand to the heme iron by a distal histidine leading to inactivation. Recent studies have provided evidence supporting the proposition that fungal peroxidases form a bis(histidile)complex when exposed to conditions that lead to inactivation. The addition of exogenous Ca(II) in each of the above studies has resulted in a restoration of enzyme activity. However, it has been demonstrated that the extent of recovery of activity was inversely dependent upon duration of exposure to extreme conditions suggesting that other conformational changes are occurring leading to irreversibly inactivated state. These observations have suggested that protein structure could be reinforced using disulphide bonds thereby preventing the protein from adopting an irreversible inactivated state after a prolonged exposure to extreme conditions. This has been accomplished successfully with the addition of disulphide bond near the distal Ca-binding site in Mn-peroxidase (Baci *et al.,* 1998).

Additional information regarding the binding and catalysis by MnP is available from studies on versatile peroxidase (Martinez, 2002), and have been discussed in the section on versatile peroxidase.

## VERSATILE PEROXIDASE

Versatile peroxidase (VP) combines the catalytic properties of the LiP and MnP and was reported first in *Pleurotus eryngii* (Martinez *et al.,* 1996) and then in other fungi (Comarero *et al.,* 1996; Fu *et al.,* 1997; Sarkar *et al.,* 1997; Heinfling *et al.,* 1998a; Giardiana *et al.,* 2000; Heinfling *et al.,* 1998b; Mester and Field, 1998). These peroxidases oxidise Mn(II) to Mn(III) (as does MnP), degrade the lignin model dimer veratryl glycerol-β-guaiacyl ether (used to demonstrate LiP activity) and oxidise veratryl alcohol and p-dimethoxy benzene to veratraldehyde and p-benzoquinone respectively (Ruiz-Duenas *et al.,* 2001). Moreover, they oxidise hydroquinones and substituted phenols that are substrates for plant peroxidases (Heinfling *et al.,* 1998a). Their pH optima for oxidation of Mn(II) (pH 5) and aromatic compounds and dyes (pH 3) differ being similar to those of optimal MnP and optimal LiP activities.VP have high affinity for Mn(II) ($K_m$ 19-20 μM), ferulic acid, α- naphthol and different hydroquinones ($K_m$ 6-22 μM), and dyes ($K_m$ 3-12) but affinity for veratryl alcohol ($K_m$ 3000 μM) and substituted phenols ($K_m$ 90-300 μM) are lower. VP has high affinity for $H_2O_2$ ($K_m$ 10 μM) also. VP has the highest activity on Mn(II) ($V_{max}$ 165-180 units/mg), moderate activities on different aromatic substrates including veratryl alcohol and 2,2′- azinobis-(3-ethyl benzothiazoline-6-sulphonate) (ABTS) ($V_{max}$ 12-40 units/mg), and lower activity on guaiacol and other dyes ($V_{max}$ 5-10 units/mg). The VP affinities for $H_2O_2$ and Mn(II) are higher than obtained for *Phanerochaete chrysosporium* peroxidases but veratryl alcohol affinity is lower than found for LiP. The maximal activities on Mn(II) and veratryl alcohol are also lower than obtained for MnP and LiP respectively (Ruiz-Duenas *et al.,* 2001).

A new isoenzyme of VP (termed as VP-PS1) has been reported (Caramelo *et al.,* 1999; Camarero *et al.,* 1999) from the lignocellulose culture of *Pleurotus eryngii* in addition to the VP isoenzyme (termed as VP PL) isolated from the liquid culture of the same fungal strain. The activities of VP-PS1 on Mn(II), 2,6-dimethoxy phenol and methoxy-p-hydroquinone are similar to those of VP – PL (Heinfling *et al.,* 1998a). However, it shows lower activities on veratryl alcohol, p-methoxy benzene and lignin model dimers suggesting a lower redox potential.

In VP-PS1 oxidation of methoxy-p-hydroquinone, a biphasic steady state kinetics with apparent saturation in both micromolar and millimolar ranges have been reported (Ruiz-Duenas *et al.,* 2001) as found for ABTS oxidation by VP-PL (Heinfling *et al.,* 1998a). Biphasic steady state kinetics has been reported also for LiP and S168W MnP variant oxidation of some substrates (Timofeevski *et al.,* 2000). The biphasic kinetics found for some substrates suggests the existence of more than one site for oxidation of the same molecule. The existence of different sites involved in oxidation of Mn(II) and other substrates by VP is supported also by non-competitive inhibition between oxidation of Mn(II) and high redox potential dyes in *Pleurotus eryngii* VP-PSI (Ruiz-Duenas *et al.,*

2001) (Fig. 16B) and VP-PLI (Heinfling *et al.,* 1998a). This suggests two sites with different affinities, which are not affected by the presence of the alternative substrate acting as inhibitor.

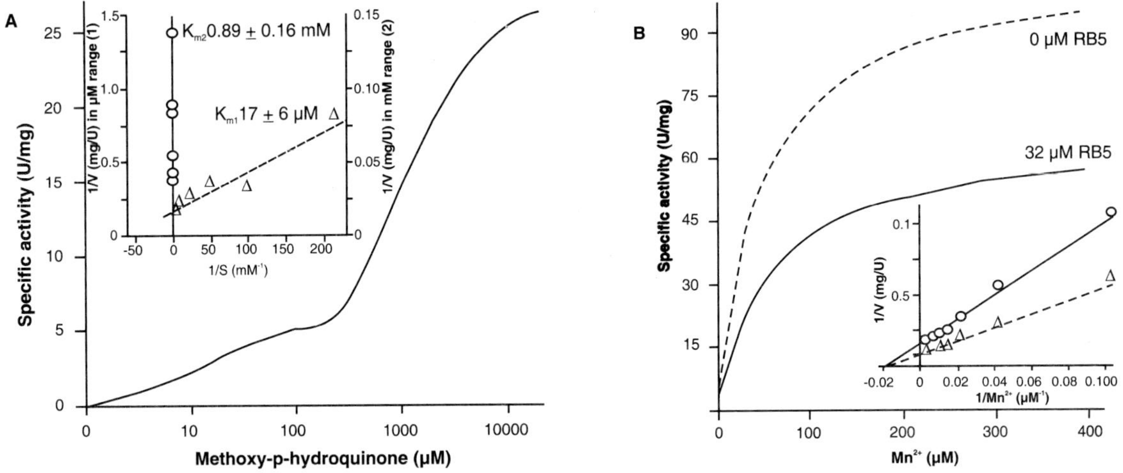


**Fig. 16.** Biphasic kinetics (Ruiz-Duenas *et al.,* 2001) for methoxy-p-hydroquinone oxidation by *Pleurotus eryngii* VP-PSI were obtained by Michaelis-Menten adjustment in the two steps (with substrate concentration on a logarithmic scale) resulting in two different apparent Km values from the Lineweaver-Burk plot (inset) (A): and non-competitive inhibition of $Mn^{2+}$ oxidation by the same peroxidase by 32 μM RBS, including Lineweaver-Burk plot (inset) (B).

At present no crystal structure is available for VP (Kamitsuji *et al.,* 2005). However, molecular models built using LiP and MnP and other peroxidases (Martinez, 2002) as templates show the presence of a manganese-binding site in the proximity of the haem ring on the side of the two-propionate groups in the manner similar to MnP (Ruiz-Duenas *et al.,* 2001). The multifunctional VP enzymes also possess an exposed tryptophan residue, which functions as a redox active site for veratryl alcohol oxidation (Caramelo *et al.,* 1999) as found in LiP. The current scientific information on VP is available in the reference (Kamitsuji *et al.,* 2005) and references cited therein. However, some important features are briefly described below.

## GENE CLONING AND PEROXIDASE SEQUENCE OF VP

The gene (pl), which encodes two peroxidases, isolated from liquid cultures of *Pleurotus eryngii* has been cloned and appeared as two alleles sharing 99% of the sequence. The VP allelic variants PL1 and PL2 differ only in three amino-acid residues (Ruiz-Duenas *et al.,* 1999c). The gene (PS1) encoding VP-PS1 produced in lignocellulose cultures of *Pleurotus eryngii* also has been cloned. Both *Pleurotus eryngii* VP genes (pl and ps1) contain 15 introns and encode proteins with 361 amino acids (including a 30 amino-acid signal peptide) in case of VP-PL, and 370 amino-acids (including a signal peptide of 31 amino acids) in case of VP-PS1, showing 74% sequence homology. It has been shown that *Pleurotus eryngii* VP is regulated at transcription level (Ruiz-Duenas *et al.,* 1999a; Irie *et al.,* 2001). VP PL2 has been expressed in *Aspergillus nidulans* with properties similar to wild VP (Ruiz-Duenas *et al.,* 1999b). It is interesting to note that the *Pleurotus eryngii* VP isoenzymes have higher sequence homology with *P. chrysosporium* LiP (upto 60% for VP-PL and 62% for VP-PS1) than MnP isoenzymes (upto 55% for VP-PL and 57% for VP-PS1) (Ruiz-Duenas *et al.,* 2001).

## VP MOLECULAR STRUCTURE

Utilising high sequence homology of *Pleurotus eryngii* isoenzymes with *P. chrysoporium* LiP and MnP and considering similar folding and helical topology of other peroxidases, molecular model for *Pleurotus eryngii* isoenzymes have been built by homology modeling (Peisch, 1996). Fig. 17 shows the predicted models (Ruiz-Duenas *et al.,* 1999c; Camarero *et al.,* 2000). Both include 12 predominantly α-helices; four disulphide bridges; two Ca(II) binding sites; several characteristic residues in haem pocket and putative substrate oxidising sites which are described below. The C-terminal tails of *Pleurotus eryngii* VP isoenzymes are a few residues shorter than *P. chrysosporium* LiP, where as *P. chrysosporium* MnP has a longer (18-26 residues) tail (Ruiz-Duenas *et al.,* 2001).

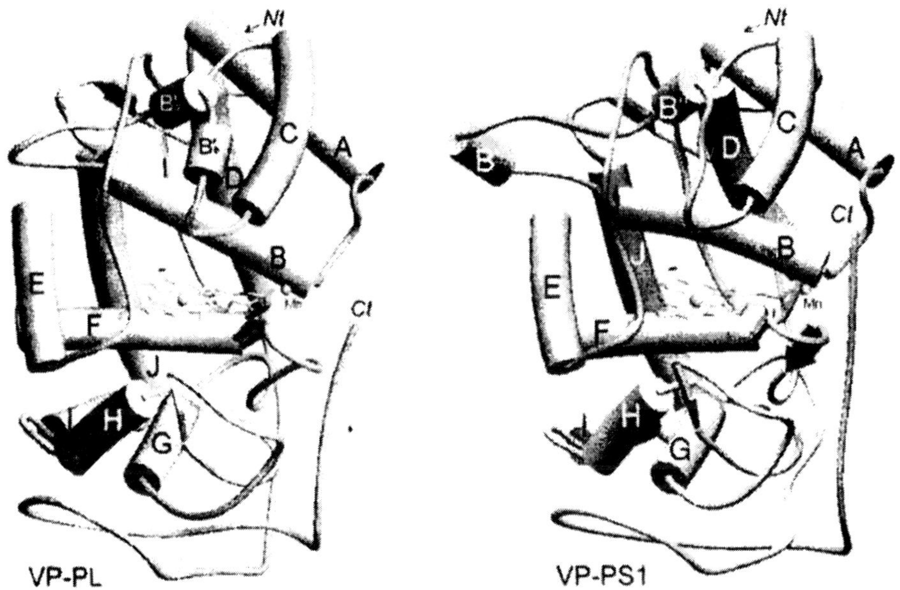


**Fig. 17.** Molecular models for *Pleurotus* VP (Ruiz-Duenas *et al.,* 2001)

## SUBSTRATE BINDING SITES IN VP

As shown in Fig. 18 a putative Mn-interaction similar to that of MnP (E35, E39 and D179) has been identified in *Pleurotus eryngii* VP-PL (E36, E40 and D175) and VP-PS1 (E36, E40 and D181) (Ruiz-Duenas *et al.,* 1999c; Camarero *et al.,* 2000). The three acidic residues delimit a small access channel in front of haem propionate. Mn enters this second haem access channel being coordinated by carboxylate oxygens of the internal propionate and the three acidic residues (its octahedral coordination being completed by two structural waters) and one electron is transferred to haem via

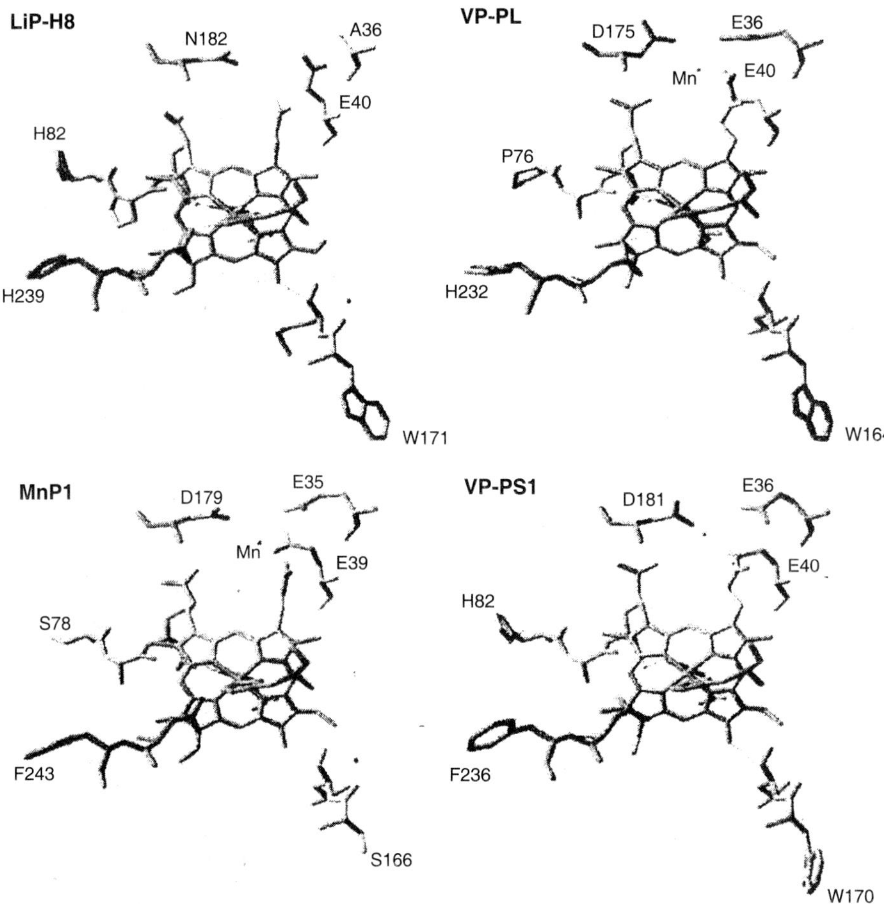


**Fig. 18.** Residues involved in electron transfer to haem in different ligninolytic peroxidases (Ruiz-Duenas *et al.*, 2001)

the propionate arm. No such site exists in LiP and therefore it cannot oxidise Mn(II). It is worth mentioning here that a Mn binding site has been created in *P. chrysosporium* LiPH8 isoenzyme by site directed mutagenesis (Mester and Tein, 2001) which oxidize both veratryl alcohol and Mn(II). Thus a versatile peroxidase has been prepared from lignin peroxidase. This can be considered a significant development in the area of versatile peroxidase research. The Mn(II) binding site in versatile peroxidase has also been investigated using NMR spectroscopy (Baci *et al.,* 2003) in the resting state and in the cyanide inhibited form of the enzyme which shed light on the factors affecting the function of this novel peroxidase. Interesting results regarding the binding site for $H_2O_2$ in case of versatile peroxidase from *Bjerkandera adusta* has been obtained using EPR Spectroscopy in which reaction of $H_2O_2$ with the enzyme in the absence of the reducing substrates has been studied (Pogni *et al.,* 2005). A tryptophan radical has been identified as a protein based radical at the moderate distance from the ferryl haem.

This finding has been further confirmed in another significant development in the versatile peroxidase research regarding the binding of aromatic substrates by Perez-Boada and co-workers (2005) where three long range electron transfer pathways have been investigated using site directed mutagenetic, spectroscopic and crystallographic studies. High-resolution crystal structures as well homology models for VP isoenzymes from the fungus *Pleurotus eryngii* have revealed three possibilities for long-range electron transfer for the oxidation of high redox potential aromatic compounds. The possible pathways have been thought to start either at Trp164 or His232 of isoenzyme VPL and at His82 or Trp170 of isoenzyme VPS1. These residues are exposed and less than 11Å apart from the haem. In order to test their functionalities, two single mutations (W164S and H232F) and one double mutation (W164/P76H) were introduced in VPL that (i) removed the two pathways in this isoenzyme and (ii) incorporated the absent putative pathway. Analysis of the variants has shown that Trp164 is required for oxidation of two high redox potential model substrates (veratryl alcohol and Reactive Blacks) whereas the two other pathways starting at His232 and His82 are not involved in long range electron transfer. None of the mutation affected Mn(II) oxidation, which would occur at the opposite side of the enzyme. Substitution of Trp164 by His also resulted in an inactive variant indicating that an indole side-chain is required for activity. It is proposed that substrate oxidation occurs via a protein-based radical. For the first time in a lignolytic peroxidase such an intermediate species has been detected using low temperature electron paramagnetic resonance of $H_2O_2$ activated VP and was found to exist at Trp164 as a neutral radical. The $H_2O_2$ activated VP was self reduced in the absence of reducing substrates. Trp164 is also involved in this reaction, which in W164S variant was blocked at the level of compound II. No hydroxylation of the Trp164 C-β atom has been observed in the analysis of VP crystal structures close to atomic resolution. This is in contrast to LiP Trp171 (Choinowski *et al.,* 1999). Analysis of the crystal structures of both peroxidases has shown difference in the environment of the protein radical forming residue that could affect its reactivity. These variations also explain the differences found for oxidation of some high redox potential aromatic substrates. Kamitsuji and coworkers (2005) have reported the involvement of Trp170 exposed on the enzyme surface of *Pleurotus ostreatus* VP which has 95.8% sequence homology with *Pleurotus eryngii* isoenzyme VP-PSI in the direct oxidation of polymeric substrates which supports the findings reported by Parez-Boada and co-workers (2005).

## LACCASE

Laccases, Polyphenol oxidases (E.C. 1.10.3.2), are multicopper enzymes belonging to the group of blue oxidases (Reinhammer and Malstrom, 1984; Solomon *et al.,* 1996; Messerschmidt *et al.,* 1997). They catalyse the oxidation of a variety of phenolic compounds as well as diamines and aromatic amines with concomitant reduction of molecular oxygen to water (Thurston, 1994). A general reaction scheme has been proposed as

$$4RH + O_2 \longrightarrow 4R + 2H_2O \qquad \text{.......(1)}$$

Laccases are widely distributed in higher plants (Huang *et al.,* 1991) and fungi (Thurston, 1994) and laccase-like activity has been observed in some bacteria (Martino *et al.,* 2002), insects (Thomas *et al.,* 1989) and wasp venom (Parkinson *et al.,* 2001).

Laccases have been implicated in lignin degradation (Have and Tennissen, 2001; Bourbonnais and Paice, 1990), in bleaching of pulp (Arias *et al.,* 2003; Geng and Xu, 2004), detoxification of textile dyes (Abadulla *et al.,* 2000; Baldrian, 2004) and in degradation of organic pollutants (Johannes and Majcherczyk, 2000; Schultz *et al.*, 2001) and in the roles played by lignolytic enzymes (Thurston, 1994). The biotechnological importance of laccases have increased after the discovery that the oxidizing reaction substrate range could be further extended in the presence of small readily oxidizable molecules called mediators (D' Acunzo and Galli, 2003). Laccases are considered the most promising enzymes for future industrial applications and a number of researchers have focussed their efforts on characterisation of laccases produced by lignin degrading fungi. The literature is vast and only the salient points are mentioned here.

Typical laccases contain four copper atoms per molecule of the enzyme (Huang *et al.,* 1991). Based on the redox, optical and magnetic properties, the coppers in laccase have been classified into three type I (T I) type II (T II) and type III (T III). The type I copper (blue copper) gives four charge transfer bands in the visible region, of which the band at 614nm originating from cys $\longrightarrow Cu^{2+}$ is the strongest ($\varepsilon$ =5800 $M^{-1}cm^{-1}$) and a very narrow hyperfine splitting (A11 = $4.3 \times 10^{-3}$ $cm^{-1}$) in the EPR spectrum. Type II copper (non-blue copper) is characterized by a weak absorption in visible region and a broad hyperfine splitting ($A_{11}$ = $20.6 \times 10^{-3}$ $cm^{-1}$) in the EPR spectrum. Type III copper has two copper centres is EPR non-detectable due to strong antiferromagnetic interaction through a hydroxide ion and exhibit a 330 nm shoulder band.

Spectroscopic studies on laccase combined with studies on other multicopper oxidases like ascorbate oxidase and ceruloplasmin have generated a description of active site concerned with the catalytic function of laccase (Huang *et al.,* 1991). It has been concluded that type I copper is four co-ordinated and has the tetrahedral distorted geometry. Type II copper is three co-ordinated by two imidazoles and water as ligands but shows EPR spectral features for a tetragonal or slightly tetrahedrally distorted geometry. Paired type III coppers with highly tetrahedrally hindered geometry are co-ordinated by three histidines and hydroxide or oxide ion as the bridging group. Type II and type III coppers are not bridged but located within 0.4 nm to form the trinuclear copper centre and thus bind and reduce dioxygen utilizing the electrons transferred from the type I copper site. A typical laccase catalysed reaction is as shown below in Fig. 19 (Ducros *et al.,* 1998).

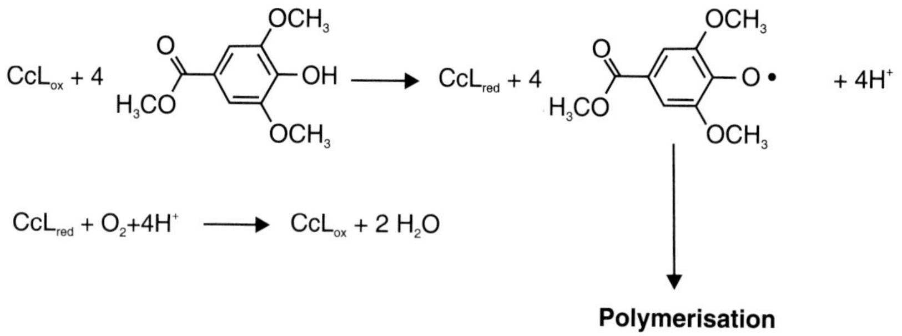

**Fig. 19:** A typical laccase catalyzed reaction, in this case the oxidation of methyl syringate (Ducros *et al.,* 1998). The substrate undergoes a single electron oxidation to form a free radical which subsequently undergoes polymerization. Molecular oxygen is used to regenerate the oxidized form of laccase. The stochiometry is four molecules of reducing substrate for every molecular oxygen, as shown in equation (1).

It has been proposed (Ducros *et al.,* 1998) that the mononuclear copper type I functions as the primary electron acceptor extracting electrons from the reducing substrate and delivering it to the trinuclear site where oxygen is reduced to water and the oxidized form of the enzyme is regenerated.

The oxygen centred radical bound to the trinuclear copper centre has been detected (Huang *et al.,* 1991) as intermediate during the reoxidation process of the reduced *Rhus vernicifera* laccase with dioxygen and has been characterized by using absorption, stopped flow and EPR spectroscopies and super conducting quantum interface devices measurement. On the basis of these studies following mechanism for dioxygen reduction to water by laccase has been proposed as shown in Fig. 20.

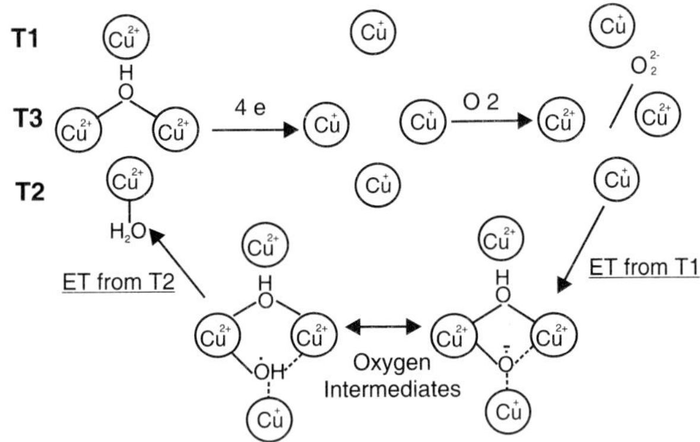


**Fig. 20.** The proposed reaction mechanism of dioxygen reduction to water by laccase (Huang *et al.,* 1991)

These conclusions are from the spectroscopic studies and need further confirmation by X-rays crystallographic and site directed mutagenetic studies.

Recently four fungal laccases-fungal laccase from *Coprinus cinereus* (Ducros *et al.,* 1998) [Protein Data Bank Codes 1A65(PDB) and 1HFU(PDB), fungal laccase from *Trametes versicolor* (Bertrand *et al.,* 2002; Piotek *et al.,* 2002); Protein Data Bank codes 1KYA(PDB) and 1CYC(PDB), fungal laccase from *Melanocarpus albomyces* (Hakulinen *et al.,* 2002); Protein Data Bank codes 1GEO(PDB)] and fungal laccase from *Rigidoporus lignosus* (Garavaglia *et al.,* 2004) have been structurally characterized in addition to one bacterial laccase cot A from *Bacillus subtilis* (Enguita *et al.,* 2003) [Protein Data Bank code 1GSK(PDB)].This has paved the way for structure function relationship studies in case of fungal laccases using molecular biology techniques. For detailed comparative structural information, interested readers are referred to the original papers (Ducros *et al.,* 1998; Bertrand *et al.,* 2002; Piotek *et al.,* 2002; Hakukinen *et al.,* 2002; Garavaglia *et al.,* 2004; Enguita *et al.,* 2003).

The ribbon 3D structure of *Trametes versicolor* laccase (Bertrand *et al.,* 2002) is shown in Fig. 21. It consists mainly of antiparallel β-barrets. There are three cupredoxin-like domains: the copper type I belongs to domain 3 and the trinuclear copper site (Cu2, Cu3a and Cu3b) is at the interface between the two other domains. This topology is essentially the same as that from the monomer of Zucchini ascordate oxidase (Reinhammer and Malstrom, 1984). The overall structure is very

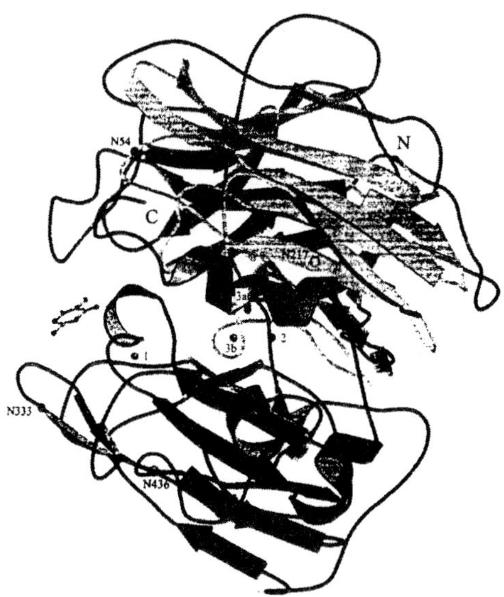


**Fig. 21.** The global fold of LacIIIb. Ribbon representation showing the secondary structures (Bertrand *et al.*, 2002)

close to that of the Cu2 depleted form of *Coprinus cinereus* laccase (Ducros *et al.*, 1998). The crystal structure (Bertrand *et al.*, 2002) of 2,5 - xylidine bound *Trametes versicolor* laccase has allowed unambiguous determination of both the mononuclear copper site of type I (T1) and the trinuclear copper site consisting of copper of type II (T2) and two coppers of type III (T3). Copper T1 is coordinated to histidine 395 and 458 and to Cys 453. The non-coordinated Phe463 residue is at a distance of 3.7 A from the copper. This phenylalanine is homologous to the axial methionine coordinated to the copper of the T1 site in ascorbate oxidase.

The trinuclear copper centre T2/T3 is coordinated to eight histidines (Figure 22 b) Histidines

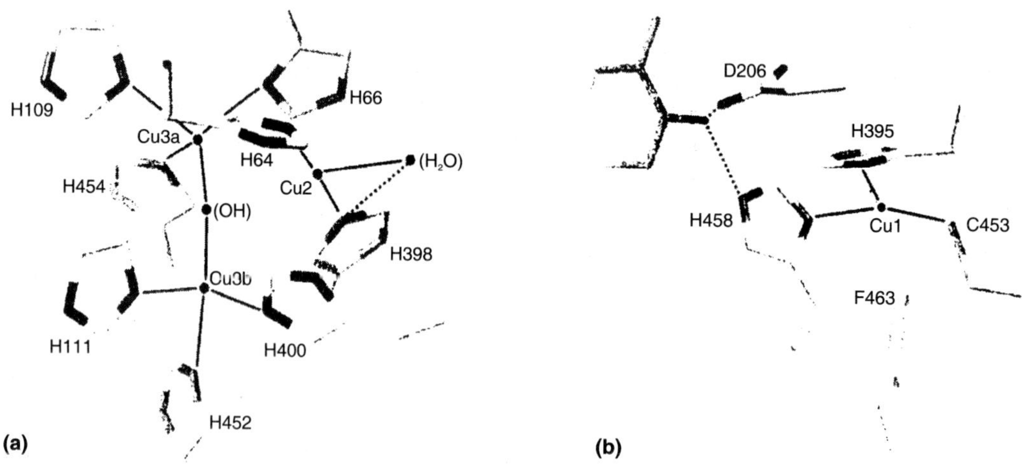


**Fig. 22.** The copper sites of LacIIIb and their environment (Bertrand *et al.*, 2002). (a) The trinuclear T2/T3 site. (b) The T1 site and the neighboring 2,5-xylidine ligand

66, 109 and 454 are coordinated to Cu 3a while histidines 111, 400 and 452 are coordinated to cu 3b. The bridging oxygen atom from the hydroxide known to bind Cu3a and Cu3b is equidistance from these two coppers. The oxygen-Cu3a distance is 1.8-1.9A, depending on the molecule of the asymmetric unit and the oxygen-Cu3b distance is 1.8-2.0A.This results in a Cu3a-Cu3b distance of 3.75A.Cu2 is coordinated to histidines 64 and 398 and to a water molecule from the solvent. The Cu2-water distance varies notably among the four different molecules of an asymmetric unit (2.3-2.9A). Recently a combination of spectroscopic and density functional (DFT) studies on a derivative of laccase where the T1 Cu has been replaced by a spectroscopically innocent $Hg^{2+}$ ion has led to a detailed geometric and electronic structure description of the resting trinuclear Cu cluster complementing crystallographic results (Quintanar *et al.,* 2005).

Garavaglia (Garavaglia *et al.,* 2004) has observed an unprecedented coordination arrangement for the copper type III pair in the structure of *Rigidoporus lignosus* laccase. Differently to that seen in all previous reported four copper laccases, the bridging oxygen ligand is not equidistant from the two type III coppers and therefore can no longer be considered bridging ligand. They have proposed that the Cu3b in their structure could represent the EPR active copper at the trinuclear site described for *Rigidoporus lignosus* laccase in solution (Bonomo *et al.,* 2004). These studies need further confirmation.

Though the mechanism of laccase catalysis is not fully understood, the current view is that the resting state of the enzyme has all the four copper ions in the oxidised form (Bonomo *et al.,* 2004). The catalysis begins with the oxidation of a single molecule of the reducing substrate by the copper of type I, with the subsequent transfer of the extracted single electron to the copper type II/type III cluster through a pathway consisting of a strictly conserved His-Cys-His motif. Successive one electron oxidation of the reducing substrate takes place at the copper type I site producing a fully reduced enzyme and the electrons are accumulated at the copper type II/type III site where the reduction of molecular oxygen takes place. This scheme is shown in Fig. 20. The catalysis at the copper type I site is reasonably understood, the catalytic steps taking place at the trinuclear copper site are still to be worked out in details.

Enguita *et al* (2004) have crystallized the cotA laccase from the endospore coat of *Bacillus subtilis* in presence of the non-catalytic co-oxidant 2, 2´-azinobis-(3-ethylbenzothia-zoline-6-sulfonate (ABTS) and the crystal structure has been solved. These studies indicated how ABTS in conjugation with laccases could act as an oxidative mediator towards non-phenolic moieties thus paving the way for further studies on the role of mediator molecules in the modification of lignins. Recently the screening of potential redox mediators for laccase of *Trametes hirsuta* has been performed using electrochemical and oxidation reactions of different mediators (Shumakovich *et al.,* 2005). It has been shown that N-hydroxy naphthalimide shows the same efficiency as ABTS as mediator. Further studies are needed in this direction.

It is obvious from the preceding discriptions that commendable advances have been made in understanding the enzymology and molecular biology of lignolytic enzymes, namely, lignin peroxidase, manganese peroxidase, versatile peroxidase and laccase during recent years. The development of an ecofriendly and energy inexpensive technology for delignification of lignocellulosic materials is yet to be developed. The requirement is that the developed technology should separate lignin and cellulose parts of lignocellulose without creating much destruction to them. The limitations of all natural lignolytic enzymes are that they are all non-specific. Thus there is a need for development of lignolytic enzymes specific for specific bonds. The other possibility is the development of synthetic catalysts utilising the structures of the active sites of the lignolytic enzymes. It is hoped that a suitable technology for delignification of lignocellulosic materials will become a reality in near future.

## ACKNOWLEDGEMENTS

Authors are thankful to Prof. R.C. Kuhad for the inspiration to write this article. Financial support of DST, Govt. of India, New Delhi, in the form of Women Scientist Fellowship SR/WOS-A/CS-27/2004 to Dr. Meera Yadav is thankfully acknowledged. The help rendered by Shri Jitendra Kumar Sharma in collecting scientific materials is also acknowledged.

## REFERENCES

Abadulla, E., Tzanov, T., Costa, S., Robra, Karl-Heinz, Cavaco-Pauls, A. and Giibitz, G.M.(2000). Decolorization and Detoxification of Textile Dyes with a Laccase from *Trametes hirsuta. Applied Environmental Microbiology,* 66, 3357-3362.

Albreeht Messerschmidt ed. (1997). Multicopper oxidases. *World Scientific Singapore.*

Alic M., Clark, E.K., Kornegay, J.R. and Gold, M.H. (1990). Transformation of *Phanerochaete chrysosporium* and *Neurospora crassa* with adenine biosynthetic genes from *Schizophyllum conmune. Current Genetic,* 17, 305-311.

Ambrosio, C., Scaloni, A. and Rizzi, M. (2004). The structure of *Rigidoporus lignosus* laccase containing a full complement of copper ions reveals an asymmetrical arrangement for the T3 copper pair. *Journal of Molecular Biology,* 342, 1519-1531.

Andersson, L.A, Rengnathan, V., Chiu, A.A., Loehr, T.M. and Gold, M.H. (1985). Spectral characterisation of diarylpropane oxygenase, a novel peroxide- dependent lignin-degrading heme enzyme. *Journal of Biological Chemistry,* 260, 6080-6087.

Arias, M.E., Arenas, M., Rodriquez, J., Soliveri, J., Ball, A.S. and Hernandez, M., (2003). Kraft pulp Biobleaching and mediated oxidation of a nonphenolic substrate by laccase from *Streptomyces cyaneus* CECT 3335 *Applied Environmental Microbiology,* 69, 1953-1958.

Baci, L., Bertini, I., Pozzo, L.D., Conte, R.D. and Tien, M.(1998). Monitoring the role of oxalate in manganese peroxidase. *Biochemistry,* 37, 9009-9015.

Baci, L., Comarero, S., Martinez, A.T., Martinez, M.J., Parez-Boada, M., Pierattelli, R. and Ruiz-Duenas, F.J.(2003). NMR study of manganese (II) binding by a new versatile peroxidase from the white-rot fungus *Pleurotus eryngii. Journal of Biological Inorganic Chemistry,* 8, 751-760.

Baldrian, P. (2004). Purification and characterization of laccase from the white rot fungus Daedalea quercina and decolorization of snthetic dyes by the enzyme. *Applied Environmental Microbiology,* 63, 560-563.

Bao, W., Eukushima, Y., Jensen, K.A., Moen, M.A. and Hammel, K. E. (1994). Oxidative degradation of nonphenolic lignin during lipid peroxidation by fungal manganese peroxidase. *FEBS Letters,* 354, 297-300.

Bertrand, T., Jolivalt, C., Briozzo, P., Caminade, E., Joly, N., Madzak, C. and Mougin, C. (2002). Crystal structure of a four copper Laccase complexed with an Aryl amine: Insight into substrate recognition and correlation with kinetics. *Biochemistry,* 41, 7325-7333.

Blodig, W., Smith, A.T., and Piontek, K. (2001). Crystal structures of pristine and oxidatively processed lignin peroxidase expressed in *Escherichia coli* and of the $W_{171F}$ variant that eliminates the redox active tryptophan 171. Implication for the reaction mechanism. *Journal of Molecular Biology,* 305, 851-861.

Bonomo, R.P., Castronovo, B.M.G. and Santoru, A.M.A (2004). A spectroscopic investigation of the interaction between nitrogen monooxide and copper sites of the fungal laccase from *Rigidoporus lignosus. Journal of Chemical Society Dolton Transaction,* 104-112.

Bourbonnais, R.E. and Parce, M.G., (1990). Oxidation of non-phenolic substrates- an expanded role for laccase in lignin biodegradation. *FEBS Letter,* 267, 99-102.

Cai, D. and Tien, M. (1990). Characterisation of the oxycomplex of lignin peroxidase from *Phanerochaete chrysosporium*: Equilibrium and Kinetic studies. *Biochemistry,* 29, 2085 – 2091.

Camarero, S., Ruiz-Duenas, F.J., Sarkar, S., Martinez, M.J. and Martinez, A.T. (2000). The doning of a new peroxidase found in lignocellulose cultures of *Pleurotus eryngii* and sequence comparision with other fungal peroxidases. *FEMS Microbiology Letters,* 191, 37-43.

Camarero, S., Sarkar, S., Ruiz-Duenas, F.J., Martinez, M.J. and Martinez, A.T. (1999). Description of a versatile peroxidase involved in the natural degradation of lignin that has both manganese peroxidase and lignin peroxidase substrate interaction sites. *Journal of Biological Chemistry,* 274, 10324-10330.

Caramelo, L., Martinez, M.J. and Martinez, A.T. (1999). A search for lignolytic peroxidases in the fungus *Pleurotus eryngii* involving α-keto-y-thiomethyl butyric acid and lignin model dimers. *Applied Environmental Microbiology,* 65, 916-922.

Choinowski, T., Blodig, W., Winterhalter, K.H., and Piontek, K. (1999). The crystal structure of lignin peroxidase at 1.7Å resolution reveals a hydroxy group on the C-β of tryptophan 171: A novel radical site formed during the redox cycle. *Journal of Molecular Biology,* 286, 809 – 827.

Comarero, S., Bockie, B., Martinez, M.J. and Martinez, A.T. (1996). Manganese-mediated lignin degradation by *Pleurotus pulmonarius. Applied Environmental Microbiology,* 62, 1070-1072.

D' Acunzo, F. and Galli, C. (2003). First evidence of catalytic mediation by phenolic compounds in the laccase- induced oxidation of lignin models. *Europian Journal of Biochemistry,* 270, 3634-3640.

Ducros, V., Brzowski, A.M., Wilson, K.S., Brown, S.H., Stergaard, P., Schneider, P., Yaver, D.S., Pedersen, A.H., and Davies, G.J. (1998). Crystal structure of the type- 2Cu depleted laccase from Coprinus cinereus at 2.2A resolution. *National Structural Biology,* 5, 310-316.

Dunford, H.B. and Stillman. (1976). *Coordination Chemistry Review,* 19, 187 – 251.

Enguita, F.J., Marcal, D., Martin, L.O., Grenha, R., Henriques, A.O., Lindley, P.F. and Carrondo, M.A. (2004). Substrate and dioxygen binding to the endospore coat laccase from *Bacillus subtilis. Journal of Biological Chemistry,* 279, 23472-23476.

Enguita, F.J., Martin, L.O., Henrique, A.O. and Carrondo, M.A. (2003). Crystal structure of a bacterial endospore coat component: A laccase with enhanced thermal stability properties. *Journal of Biological Chemistry,* 278, 19416-19425.

Finzel, B.C., Poulos, T.L., and Krant, T. (1984). Crystal structure of yeast Cytochrome c peroxidase refined at 1.7Å resolution. *Journal of Biological Chemistry,* 259, 13027 – 13036.

Fu, S.y., Yu, H.S. and Buswell, J.A.(1997). Effect of nutrient nitrogen and manganese on manganese peroxidase and laccase production by *Pleurotus sajor caju. FEMS Microbiology Letters,* 147. 133-137.

Garavaglia, S., Cambria, M. T., Miglio, M., Ragusa, S., Lacobazzi, V., Palmieri, F., D'

Geng, X., Li, K. and Xu, F. (2004). Investigation of hydroxamic acids as laccase mediators for bleaching. *Applied Microbiology Biotechnology,* 64, 493-496.

Giardina,P.,Palmieri, G., Fontanella, B., Rivieccio, V. and Sannia, G. (2000). Manganese peroxidase isozymes produced by *Pleurotus ostreatus* grown on wood sawdust. *Archives of Biochemistry and Biophysics,* 376, 171-179.

Glenn, J.K., Akileswaran, L. and.Gold, M.H. (1986). Mn(II) oxidation is the principal function of the extracellular Mn-Peroxidase from *Phanerochaete chrysosporium. Archives of Biochemistry and Biophysics,* 251, 688-696.

Hafrichter, M., Lundell, T. and Hatakka, A. (2001). Conversion of milled pine wood by manganese peroxidase from *Phlebia radiata. Applied and Environmental Microbiology,* 67, 4588-4593.

Hakulinen, N., Kiiskinen, Luara - Leena, Kruus, K., Saloheimo, M., Paananen, A., Koivula, A. and Rouvinen, J. (2002). Crystal structure of a laccase from *Melanocarpus albomyces* with an intact trinuclear copper site, *National structural Biology,* 9, 601-605.

Hammel, K.E., Tein, M., Kalyanaraman, B. and Kirk T.K. (1986). Mechanism of oxidative $C_\alpha$ - $C_\beta$ cleavage of lignin model dimmer by *Phanerochaete chrysosporium* ligninase: Stoichiometry and Involvement of free radicals. *Journal of Biological Chemistry,* 260, 8348 –8353.

Harvey, P.J., Schoemaker, H.E. and Palmer, J.M. (1986). Veratryl alcohol as a mediator and the role of radical cation in lignin biodegradation by *Phanerochaete chrysosporium FEBS Letters,* 195, 242 – 246.

Harvey, P.J., Schoemaker, H.E. and Palmer, J.M. (1987). Lignin degradation by white rot fungi *Plant Cell and Environment,* 10, 709 – 714.

Hatakka, A. (1994). Lignin-modifying enzymes from selected white-rot fungi-Production and role in lignin degradation. *FEMS Microbiology Review,* 13, 125-135.

Have, T.R. and Tennissen, P.J.M. (2001). Oxidative mechanism involved in lignin degradation by white-rot fungi. *Chemistry Review,* 101, 3397-3413.

Heinfling, A., Ruiz-Duenas, F.J., Martinez, M.J. Bergbauer, M., Szewzyk, U. and Martinez, A.T. (1998). A study on reducing substrates of manganese-oxidizing peroxidases from *Pleurotus eryngii* and *Bjerkandera adusta. FEBS Letters,* 428, 141-146.

Hemfling, A., Martinez, M.J., Martinez, A.T., Bergbuer, M. and Szewzyk, U. (1998). Purification and characterization of peroxidases from the dye-decolorization of peroxidases from the dye-decolorizing fungus *Bjerkander adusta.*. *FEMS Microbiology letter,* 165, 43-50.

Hofrichter, M., Scheibner, K., Schneegab, V. and Fritsche, W. (1998). Enzymatic combustion of Aromatic and Aliphatic compounds by MnP from *Nematoloma frowardii. Applied and Environmental Microbiology,* 64, 399-404.

Hofrichter, M., Vares, T., Kalsi, M., Galkin, S., Scheibner, K., Fritsche, W. and Hatakka, A. (1999). Production of MnP and organic acids and mineralization of $^{14}$C-labelled lignin ($^{14}$C-DHP) during solid-state fermentation of wheat-straw with the white rot fungus *Nematoloma frowardii. Applied and Environmental Microbiology,* 65, 1864-1870.

Huang, H.W., Zoppellaro, G., Sakurai, T. (1991). Spectroscopic and Kinetic Studies on oxygen centred radical formed during the four electron reduction process of dioxygen by *Rhus Vernicifera* laccase. *Journal of Biological Chemistry,* 274, 32718-32724.

Irie, T. Honda, Y., Watanabe, T. and Kuwahara, (2001). Homologous expression of recombinant manganese peroxidase genes in lignolytic fungus *Pleurotus ostreatus. Applied Micrbiology Biotechnology,* 55, 566-570.

Johannes, C., and Majcherczyk, A. (2000). Natural mediators in the oxidation of polycyclic Aromatic hydrocarbons by laccase mediator systems. *Applied Environmental Microbiology,* 66, 524 – 528.

Kamitsuji, H., Watanabe, T., Honda, Y. and Kuwahara, M. (2005). Direct oxidation of polymeric substrates by multifunctional manganese peroxidase isozymes from *Pleurotus ostreatus* without redox mediators. *Biochemical Journal,* 386, 387-393.

Kersten, P.J., Tien, M., Kalyanaraman, B. and Kirk, T.K. (1985). The ligninase of *Phanerochaete chrysosporium* generates cation radicals from methoxy benzenes *Journal of Biological Chemistry,* 260, 2609 – 2612.

Kishi, K., Kusters-van Someren, M., Mayfield, M.B., Sun, J., Loehr, T.M. and Gold, M.H. (1996). Characterization of manganese (II) binding site mutants of MnP. *Biochemistry,* 35, 8986-8994.

Kishi, K., Wariishi, H., Marquez, L., Dunford, H.B. and Gold, M.H. (1994). .Mechanism of manganese peroxidase compound II reduction: effect of organic acid chelators and pH. *Biochemistry,* 33, 8694-8701.

Kuan, I.C, Johnson, K.A. and Tien, M. (1993). Kinetic analysis of manganese peroxidase: The reaction with manganese complexes. *Journal of Biological Chemistry,* 268, 20064 – 20070.

Kuan, I.C. and Tien, M. (1993). Stimulation of Mn peroxidase activity: A possible role for oxalate in lignin biodegradation. *Biochemistry,* 90, 1242-1246.

Kuan, I.C., Johnson, K.A. and Tien, M. (1993). .Kinetic analysis of MnP: The reaction with manganese complexes. *Journal of Biological Chemistry,* 268, 20064-20070.

Kuila, D., Tien, M., Fee, J.A. and Ondrias, M.R. (1985). Resonance Raman spectra of extracellular ligninase: Evidence for a heme active site similar to those of peroxidases. *Biochemistry* 24, 3394 –3397.

Kuwahara, M., Gleen, J.K., Morgan M..A. and Gold, M.H. (1984). Separation and characterization of two extracellular $H_2O_2$ dependent oxidases from lignolytic cultures of *Phanerochaete chrysosporium. FEBS Letters,* 169, 247-250.

Lackner, R., Srebotnik, E. and Messner, K. (1991). Oxidative degradation of high molecular weight chlorolignin by MnP of *Phanerochaete chrysosporium. Biochemical and Biophysical Research Communication,* 178, 1092-1098.

Martinez, A.T. (2002). Molecular biology and structure-function of lignin-degrading heme-peroxidases. *Enzyme and Microbial Technology,* 30, 425-444.

Martinez, M.J., Ruiz-Duenas, F.J., Guillen, F. and Martinez, A.T. (1996). Purification and catalytic properties of two manganese peroxidase isoenzymes from *Pleurotus eryngii. European Journal of Biochemistry,* 237, 424-432.

Martino, L.O., Soares, C.M., Pereira, M.M., Teixeira, M., Costa, T., Jones, G.H. and Henriques, A.O. (2002). Molecular and Biochemical characterization of a highly stable bacterial laccase that occurs as a structural component of the *Bacillus subtiles* endospore coat. *Journal of Biological Chemistry,* 277, 18849-18859.

Mcpherson, M.J., Parsons, M.R., Spooner, R.K. and Wilmot, C.M. (2001). *Galactose oxidase in Handbook of Metalloproteins.* Eds Messerschmidt, A., Huber, R., Poulos, T. and Wieghardt, K, John Wiley and Sons Ltd, Chichester.

Mester, T. and Field, J.A. (1998). Characterization of a novel manganese peroxidase-lignin peroxidase Hybrid isozyme produced by *Bjerkandera* species strain BOS55. *Journal of Biological Chemistry* 273, 15412-15417.

Mester, T. and Tien, M. (2001).Engineering of a Mn-binding site in lignin peroxidase isozyme H8 from *Phanerochaete chrysosporium. Biochemistry Biophysics Research Communication,* 284, 723-728.

Mester, T., Balay, A.K., Baffoni, C.S., Banci, L., Jones, A.D., and Tien, M., (2001). Oxidation of a tetrameric nonphenolic lignin model compound by lignin peroxidase. *Journal of Biological Chemistry,* 276, 22985 – 22990.

Moreira, M.T., Palma, C., Mielgo, I., Feijoo, G.,.Lema, J.M. (2001). In vitro degradation of a polymeric dye (Poly R-478) by MnP. *Biotechnology and Bioengineering,* 75, 362-368.

Mylranjan, M., Valli, K., Wariishi, H., Gold, M.H. and Loehr, T.M. (1990). Resonance Raman Spectroscopic characterization of compound III of lignin peroxidase. *Biochemistry,* 29, 9617 – 9623.

Parkinson, N., Smith, I., Weaver, R. and Edwards, J.P. (2001). A new form of arthropod phenoloxidase is abundant in venom of the parasitoid wasp *Pimpla hypochondrica. Insect Biochemistry Molecular Biology,* 31, 57-63.

Peisch, M.C. (1996). Promod and Swis-Model: Internat-based tools for automated comparative protein modelling. *Biochemical Society Transaction,* 24, 274-279.

Pelaez, F., Martinez, J., Martinez, A.T. (1995). Screening of 68 species of basidiomycetes for enzymes involved in lignin degradation. *Mycology Research,* 99, 37-42.

Perez-Boada, M., Ruiz-Duenas, F.J., Pogni, R., Basosi, R., Choinowski, T., Martinez, M.J., Piontek, K. and Martinez, A.T. (2005). Versatile peroxidase: Oxidation of high redox potential aromatic compounds: Site directed mutagenesis spectroscopic and crystallographic investigation of three long-range electron transfer pathways. *Journal of Molecular Biology,* 354, 385-402.

Piontek, K., Glumoff, T., and Winterhatter, K. (1993). Low pH crystal structure of glycosylated lignin peroxidase from *Phanerochaete chrysosporium* at 2.5Å resolution. *FEBS Letters,* 315 119 – 124.

Piotek, K., Antorini, M. and choinowsi, T. (2002). Crystal structure of a laccase from the fungus Trametes versicolor at 1.90A resolution containing a full complement of coppers. *Journal of Biological Chemistry,* 277, 37663-37669.

Pogni, R., Baratto, M.C., Giansanti, S., Tentloff, C., Verdin, J., Valderrama, B., Lendzian, F., Lubitz-Wolfgang, Rafad Vazquez-Duhalt, R. and Basosi, R. (2005).Tryptophane-based rsdical in the catalytic mechanism of versatile peroxidase from *Bjerkandera adusta. Biochemistry,* 44, 4267-4274.

Pointek, K., Smith, A. T., and Blodig, W. (2001). Lignin peroxidase structure and function. *Biochemical Society Transaction,* 29, 111–116.

Poulos T.L., Edwards, S.L., Wariishi, H., and Gold, M. H. (1993). Crystallographic refinement of lignin peroxidase at 2Å. *Journal of Biological Chemistry,* 268, 4429 – 4440.

Quintanar, L., Yoo, J., Aznar, C.P., Palmar, A.E., Andersson, K.K., Britt, R.D. and Solomon, E.I. (2005). Spectroscopic and electronic structure studies of the trinuclear Cu cluster active site of the multicopper oxidase laccase: native of its coordination unsaturation. *Journal of American Chemical Society,* 127, 13832-13845.

Reinhammer, B. and Malstrom, B.G. (1984). Blue copper containing oxidases. In copper proteins and copper enzymes. *(R. Lonticed) CRC Press Inc. Boca Raton, Fla,* 3, 1-35.

Renganathan, V. and Gold, M.H. (1986). Spectral characterisation of the oxidized states of lignin peroxidase, an extracellular heme enzyme from white rot basidiomycete *Phanerochaete chrysosporium. Biochemistry,* 25, 1626 – 1631.

Ruiz-Duenas, F.J., Guillen, F., Camarero, S., Perez-Boada, M., Martinez, M.J., (1999). Regulation of peroxidase transcript levels in liquid cultures of the lignolytic fungus *Pleurotus eryngii. Applied Environmental Microbiology,* 65, 4458-4463.

Ruiz-Duenas, F.J., Martinez, M.J. and Martinez, A.T. (1999). Heterologous expression of *Pleurotus eryngii* peroxidase: Ability to oxidize Mn(II) and different aromatic substrate. *Applied Environmental Microbiology,* 65, 4705-4707.

Ruiz-Duenas, F.J., Martinez, M.J. and Martinez, A.T. (1999). Molecular characterisation of a novel peroxidase isolated from the lignolytic fungus, *Pleurotus eryngii. Molecular Microbiology,* 31, 223-236.

Ruiz-Duenas, F.J., Camarero, S., Perez-Boda, M., Matinez, M.J. and Matinez, A.T. (2001). A new versatile peroxidase from *Pleurotus. Biochemical Society Transaction,* 29, 116-122.

Sarkar, S., Martinez, A.T. and Martinez, M.J. (1997). Biochemical and molecular characterization of a manganese peroxidase isozyme from *Pleurotus ostreatus. Biochemica et Biophysica Acta* 1339, 23-30.

Schultz, A., Jonas, U., Hammer, E. and Schauer, F. (2001). Dehalogenation of chlorinated Hydroxyl biphenyls by fungal laccase. *Applied Environmental Microbiology,* 67, 4377-4381.

Scott Reading, N. and Steven D.A. (2001). Role of disulphide bonds in the stability of recombinant MnP. *Biochemistry,* 40, 8161-8168.

Shumakovich, G.P., Shleev, S.V., Morozova, O.V., Khohlov, P.S., gazaryan, I.G. and Yaropolov, A.I. (2005). Electrochemistry and kinetics of fungal laccase mediators. *Bioelectrochemistry,* 10, (In press).

Sollewijn Gelpke, M.D., Moenne Loccoz, P.and Gold, M.H. (1999). Arginine 177 is involved in Mn(II) binding by MnP. *Biochemistry,* 38, 11482-11489.

Solomon, E.I., Sundaram, U.M. and Machonkin, T.E. (1996). Multicopper oxidases and oxygenases. *Chemistry Review,* 96, 32563-32605.

Souza, D., Merritt, C.S., and Reddy, C.A. (1999). Lignin-modifying Enzymes of the White Rot Basidiomycete *Ganoderma Lucidum. Applied Environmental Microbiology,* 65, 5307-5313.

Steffen, K.T., Hatakka, A. and Hofrichter, M. (2003). Degradation of Benzo[a] pyrene by the litter decomposing basidiomycete *Stropharia coronilla*: Role of MnP. *Applied and Environmental Microbiology,* 69, 3957-3964.

Su, G., An, Z., Zhang, W. and Liu, Y. (2005). Light promotes the synthetisis of lignin through the production of $H_2O_2$ mediated by Diamine oxidase in soyabean hypocotyls. *Journal of Plant Physiology,* 162, 1297-1303.

Sundaramoorthy, M., Kishi, K., Gold, M.H. and Poulos, T.L. (1997). Crystal structures of substrate binding site mutants of MnP. *Journal of Biological Chemistry,* 272, 17574-17580.

Sundaramoorthy, M., Kishni, K., Gold, M.H. and Poulos, T.L. (1994).The crystal structure of MnP from *P.chrysosporium* at 2.06 – A resolution. *Journal of Biological Chemistry,* 269, 32759-32767.

Sundaramoorthy, M., Youngs, H.L., Gold, M.H. and Poulos, T.L.(2005). High-resolution crystal structure of manganese peroxidase: Substrate and inhibitor complexes. *Biochemistry,* 44, 6463-6470.

Tein, M. and Kirk, T..K. (1983). Lignin degrading enzyme from the Hymenomycete *Phanerochaete chrysosporium* burds. *Science,* 221, 661-663.

Tein, M., Kirk, T.K., Bull, C., and Fee, J.A. (1986). Steady state and transient state kinetic studies on the oxidation of 3, 4 – dimethoxy benzyle alcohol catalysed by the ligninase of *Phanerochaete chrysosporium.*Burds. *Journal of Biological Chemistry,* 261, 1687-1693.

Thomas, B. R., Yonakura, M., Morgan, T. D., Czapla, T. H., Hopkins, T.L. and Kramer, K.J. (1989).A trypsin solubilised laccase from phrate pupal integument of the tobacco hornworm, Manduca sixta. *Insect Biochemistry,* 19, 611-622.

Thurston, C. (1994). The structure and function of fungal laccases. *Microbiology,* 140, 19-26.

Tien, M. and Kirk, T.K. (1984). Lignin degrading enzyme from *Phanerochaete chrysosporium*: Purification, characterization and catalytic properties of a unique $H_2O_2$- requiring oxygenase *Biochemistry,* 81, 2280-2284.

Timofeevski, S.L., Nie, G.J., Reading, N.S. and Aust, S.D. (2000). Substrate apecificity of lignin peroxidase and a 5168W varient of manganese peroxidase. *Archives of Biochemistry Biophysics,* 373, 147-153.

Tuor, U., Wariishi, H., Schoemaker, H.E. and Gold, M.H. (1992). Oxidation of phenolic arylglycerol β-aryl ether lignin model compounds by manganese peroxidase from *Phanerochaete chrysosporium*: oxidative cleavage of an α-carbonyl model compound. *Biochemistry,* 31, 4986-4995.

Wariishi, H. and Gold M.H. (1990). Lignin peroxidase compound III: Mechanism of formation and decomposition. *Journal of Biological Chemistry,* 265, 2070 –2077.

Wariishi, H., Akileswaran, L. and Gold, M.H. (1988). Manganese peroxidase from the basidiomycete *Phanerochaete chrysosporium*: Spectral characterization of the oxidized states and the catalytic cycle. *Biochemistry,* 27, 5365-5370.

Wariishi, H., and Gold, M.H. (1989). Lignin peroxidase compound III: Formation, inactivation and conversion to the native enzyme. *FEBS Letters,* 243, 165 – 168.

Wariishi, H., Huang, J., Dunford, H. B., and Gold M. H. (1999). Reaction of lignin peroxidase compounds I and II with veratryl alcohol. Transient state Kinetic characterization. *Journal of Biological Chemistry,* 266, 20694-20699.

Wariishi, H., Valli K. and Gold, M.H. (1992). Manganese(II) oxidation of manganese peroxidase from the basidiomycete *Phanerochaete chrysosporium*: Kinetic mechanism and role of chelators. *Journal of Biological Chemistry,* 267, 23688-23695.

Wariishi, H., valli, K. and Gold M.H. (1992). Manganese (II) oxidation by MnP from the basidiomycete *Phanerochaete chrysosporium.* Kinetic mechanism and role of chelators. *Journal of Biological Chemistry,* 267, 23688-23695.

Wariishi, H., Valli, K., Renganathan, V. and Gold, M.H. (1989). Thiol mediated oxidation of nonphenolic lignin model compounds by MnP of *Phanerochaete chrysosporium. Journal of Biological Chemistry,* 264, 14185-14191.

Wariishi,H., Valli, K. and Gold, M.H. (1991). In vitro depolymerization of lignin by MnP of *Phanerochaete chrysosporium. Biochemical and Biophysical Research Communication,* 176, 269-275.

# 5

# Effect of Phenol and 4-Chlorophenol on Mycelial Biomass and Laccase Activity in Some Fleshy Fungi

**R.C. Upadhyay[1] and W. Fritsche[2]**

[1]*National Research Centre for Mushroom, Solan H.P.- 173213 (India).*
[2]*FSU, Technical Microbiology, Institute of Microbiology, Philosophen Weg-12, 07743- Jena (Germany).*

Phenol and 4-chlorophenol (1 and 0.5mM) were incorporated at the time of inoculation in the CZAPEK-DOX culture medium to see their effect on mycelial biomass and extracellular laccase activity under stationary and shake culture conditions. Mycelial biomass was always more under shaking conditions than in stationary cultures. Phenol and 4-chlorophenol both inhibited mycelial production but inhibition due to later was much higher. Mycelial growth due to 4-chlorophenol was severely inhibited in *Pleurotus sapidus*, *P.*cfr. *florida*, *P. citrionopileatus*, *P. pulmonarius*, *P. ostreatus*, *P. djamore* var *roseus*, *Coprinus comatus*, *Agrocybe aegerita* and *Agaricus bisporus* under stationary conditions.

The laccase activity in general under stationary condition was higher than in the flasks kept under shaking condition. Due to phenol and 4-chlorphenol in the culture medium there was more laccase production by *P. citrionopileatus*, *P. flabellatus*, *P.*cfr. *florida*, *P. sapidus*, *Calocybe indica*,and *A. aegerita*, while 4-chlorophenol slightly inhibited laccase production in *P. ostreatus* Z-15, *P. pulmonarius* and *P. sajor-caju.* Laccase production was completely inhibited by 4-chlorophenol in *P. djamore var. roseus*, *Kuehneromyces mutabilis* and *Lentinus edodes.* Due to phenol and 4-chlorophenol there was less laccase production in *A. bisporus* than in the unincorporated control flasks. Culture of *C. comatus* produced more laccase due to 4-chlorophenol.

## INTRODUCTION

Lignin is one of the most abundant complex aromatic renewable biopolymer present on the earth. White-rot basidiomycetes are one of the most efficient organisms capable of lignin degradation (Kirk and Farell, 1987) due to ligninolytic enzymes like lignin peroxidase (LiP) (Tien and Kirk, 1983, Glenn *et al.,* 1983), manganese peroxidase (MnP) (Kuwahara *et al.,* 1984, Glenn and Gold, 1985) and laccase (Bollag and Leonowicz, 1984). During mineralization of lignin various monomeric, dimeric and polymeric phenolic substances like vanillyl, syringyl and veratryl compounds are produced.

Phenol and its various substituted derivatives are one of the major industrial pollutants from coke ovens, oil refineries and plastic manufacturing industries. Today in USA alone the production of phenol is estimated around 1.7 million tonnes (Annonymous, 1993). Phenols are easily water soluble and cause of concern for polluting drinking water. Many industrially important phenols are halogenated and the halogen substitution is the main cause for environmental persistence and toxicity. The total world production of chlorophenols is estimated around 200,000 tonnes per year (Koch, 1989). These compounds have various industrial and agricultural applications as solvents, lubricants, insulators, fungicides (pentachlorophenol) and herbicides as 2,4-D (2,4-dichlorophenoxyacetic acid) and 2,4,5-T(2,4,5-trichlorophenoxyacetic acid). A number of wood and forest litter degrading basidiomycetous fungi are also known to produce *de-novo* around 1,500 chlorinated aromatic compounds (de Jong, *et al.,*1994).

Various workers have reported antifungal activities of phenolic substances against several fungi inhibiting mycelial growth, sporulation and enzyme production (Mukherjee and Kundu, 1973, Charya and Reddy, 1984). Phenol and substituted phenols can be metabolized by *Candida tropicalis, Rhodotorula glutinis, Fusarium* spp. and *Mycena* spp. (Leibnitz *et al.,* 1960; Harris and Ricketts, 1962; Boominathan and Mahadevan, 1989; and Kremer *et al.,* 1992).

In the present studies effect of phenol and 4-chlorophenol on the mycelial biomass and laccase production under stationary and shaking culture conditions with some white- and brown-rot fungi were investigated.

## MATERIALS AND METHODS

### Organisms

Cultures of *Agrocybe aegerita* (Brig.) Sing., *Pleurotus ostreatus* (Jacq. ex Fr.) Kummer strain Z-15, *P. flabellatus* (Berk and Br.) Sacc. and *P. sajor-caju* (Fr.) Singer. were obtained from Dr. F. Zadrazil, Institute of Soil Biology, Braunschweig (Germany). *Coprinus comatus* (Müller in Fl. Dan. ex. Fr.) S. F. Gray was obtained from Mushroom culture collection, Weimar (Germany). Cultures of *P. sapidus* (Schulzer and Kalchbr) Sacc., *P. ostreatus* cfr. *florida* nom. prov. Eger, *P. citrinopileatus* Sing., *P. pulmonarius* (Fr.) Quel., *P. djamore* Boedjm. var. *roseus*, *Lentinus edodes* (Berg.) Sing., and *Calocybe indica* P. and C. were from N.C.M.R.T. Solan (India). Cultures of *Kuehneromyces mutabilis* (Schaeff. ex. Fr.) Sing and Smith, *Agaricus bisporus* (Lange) Imbach (brown strain), and *Tricholomopsis rutilans* (Schaeff. ex. Fr.) Sing., were from fresh specimens collected in nature.

### Growth Medium

**Malt extract medium** : All the cultures were subcultured on malt extract (2.5%) agar medium plates and incubated (28°C) for 8 days before inoculation in the flasks.

## Modified CZAPEK-DOX medium :

| | |
|---|---|
| Glucose | 10.0 g. |
| $KNO_3$ | 3.0 g. |
| $K_2HPO_4$ | 1.0 g. |
| KCL | 0.5 g. |
| $MgSO_4x7H_2O$ | 0.5 g. |
| $FeSO_4x7H_2O$ | 0.01 g. |
| Yeast extract | 0.2 g |
| Distilled water | 1.0 l. |

## INOCULATION AND INCUBATION

Conical flasks with 50 ml sterilized modified CZAPEK-DOX medium were incorporated with 1mM phenol (94.1 mg/l) and 0.5 mM 4-chlorophenol (64.3 mg/l). Flasks without phenolic substances served as control. All the flasks were inoculated with actively growing 5 mycelial disks (5 mm size) under aseptic conditions. Inoculated flasks were incubated at 28°C in dark for 30 days under stationary and shaking conditions (40 rpm).

## BIOMASS ESTIMATION

For the estimation of mycelial biomass the culture filtrate was filtered under vacuum on pre-weighed filter paper. The mycelial mat was washed twice with distilled water. The filter papers with mycelial mats were oven dried at 80°C to a constant weight.

## ESTIMATION OF ENZYME ACTIVITY

All the enzyme activities were assayed after 10 days of incubation using UV/VIS double beam spectrophotometer (Perkin-Elmer Lambda-2). Enzyme samples were drawn in microcentrifuge tubes and centrifuged. All the estimations were done at room temperature (20-22°C). One International Unit(U) of enzyme activity was defined as the activity that would change 1 μMol of substrate per minute. Enzyme activity divided by mycelial biomass (mg/ml) gave the enzyme activity per mg of biomass.

## LACCASE (p-diphenol oxidase, EC 1.10.3.2.)

The estimation of laccase followed the method of Prillinger (1975) modified by Haars *et al.,* (1981). The enzyme substrate 2,6-dimethoxyphenol is oxidized to 3,5,3'5' tetramethoxydiphenoquinone (a dimer product) (Bocks, 1967), which is a straw coloured product having maxium absorption at 468 nm. 20 μM of 2,6-dimethoxyphenol dissolved in Mc. Ilvaine buffer of pH 4.5 (0.1 M citric acid and 0.2 M $Na_2HPO_4$).

## RESULTS

The effects of phenol and 4-chlorophenol on mycelial biomass are shown in Table 1. Phenol and 4-chlorophenol incorporation in the culture medium inhibited the mycelial biomass production but inhibition due to later was much higher. Maximum biomass was observed under shaking condition in case of *P. citrinopileatus, P. cfr. florida, P. sapidus* and *C. indica* while *P. sajor-caju, C. comatus, K. mutabilis, L. edodes* and *T. rutilans* produced maximum biomass under stationary conditions in the control flasks. Phenol was less inhibitory for the growth of *P. cfr. florida, P. pulmonarius, P. flabellatus, P. sapidus, A. bisporus, K. mutabilis* and *T. rutilans* under stationary and shaking conditions. Cultures of *P. flabellatus, C. indica, L. edodes* and *T. rutilans* were least affected due to 4-chlorophenol while the growth of *P. sapidus, P. cfr. florida* and *P. pulmonarius* was maximum inhibited (70 to 84% inhibition).

**Table 1. Effect of phenol and 4-chlorophenol on mycelial growth after 0 days in stationary and shake conditions (40 rpm) {CZAPEK-DOX 10 g/l glucose, 28°C, 1 mM phenol}**

| | Biomass (mg/ml) stationary | | | Biomass (mg/ml) shake culture | | |
|---|---|---|---|---|---|---|
| *Culture/strain* | *Phenol* | *4-CP* | *Control* | *Phenol* | *4-CP* | *Control* |
| *Pleurotus citrinopileatus* | 4.2 | 1.9 | 5.7 | 5.8 | 3.8 | 7.0 |
| *P. djamore* var. *roseus* | 2.0 | 1.4 | 2.8 | 2.1 | 1.5 | 3.3 |
| *P. flabellatus* | 3.1 | 2.3 | 3.3 | 3.5 | 2.9 | 3.8 |
| *P. cfr. florida* | 4.5 | 1.0 | 4.5 | 6.7 | 1.5 | 7.0 |
| *P. ostreatus* Z-15 | 2.2 | 1.2 | 2.6 | 2.5 | 1.9 | 2.6 |
| *P. pulmonarius* | 2.8 | 0.8 | 2.8 | 3.1 | 0.9 | 3.6 |
| *P. sajor-caju* | 3.1 | 3.7 | 5.0 | 3-5 | 4.1 | 4.3 |
| *P. sapidus* | 4.7 | 0.8 | 4.9 | 6.2 | 1.1 | 6.2 |
| *Agarius bisporus* | 1.1 | 0.4 | 1.2 | 1.2 | 0.5 | 1.2 |
| *Agrocybe aegerita* | 2.1 | 1.0 | 2.4 | 3.8 | 1.6 | 4.4 |
| *Calocybe indica* | 5.8 | 4.0 | 6.0 | 6.0 | 4.4 | 6.8 |
| *Coprinus comatus* | 2.1 | 0.9 | 2.8 | 2.6 | 1.2 | 2.8 |
| *Kuehneromyces mutabilis* | 2.1 | 1.0 | 2.1 | 1.3 | 1.0 | 1.3 |
| *Lentinus edodes* | 1.0 | 1.4 | 2.0 | 1.0 | 1.5 | 1.9 |
| *Tricholomopsis rutilans* | 1.4 | 1.0 | 1.6 | 1.3 | 1.0 | 1.3 |

Laccase production under stationary condition was higher than in shake flasks (Table 2). Maximum laccase activity was observed with *C. indica, P. sajor-caju, P. citrinopileatus, P. flabellatus* , *K. mutabilis* and *A. bisporus* under stationary in the control flasks. However cultural conditions had no influence on laccase activity in *A. bisporus.* Phenol and 4-chlorophenol induced more laccase production in *P. flabellatus, P. citrinopileatus, P. cfr. florida, P.o Z-15, P. sapidus, C. indica* and *A. aegerita* under stationary conditions while phenol and 4-chlorophenol both inhibited laccase production in *A. bisporus.* 4-chlorophenol slightly inhibited laccase production in *P. sajor-caju* and *P. pulmonarius.* It completely inhibited laccase production in *K. mutabilis, L. edodes* and *P. djamore.* Culture of *C. comatus* induced laccase due to phenol and 4-chlorophenol in the culture medium under stationary and shaking condition while it was not detected in control flasks. It shows that fungi react differently to different chemicals.

**Table 2. Effect of phenol (1 mM) and 4-chlorophenol (0.5 mM) on laccase activity (stationary and shake, CZAPEK-DOX 10g/l glucose, estimated after 10 days of growth)**

| | Laccase (mU/ml) stationary | | | Laccase (mU/ml) shake (40 rpm) | | |
|---|---|---|---|---|---|---|
| *Culture/strain* | *Phenol* | *4-CP* | *Control* | *Phenol* | *4-CP* | *Control* |
| *Pleurotus citrinopileatus* | 22.1 | 16.8 | 12.6 | 15.8 | 6.3 | 3.1 |
| *P. djamore* var. *roseus* | 2.1 | 0.0 | 2.1 | 0.0 | 0.0 | 0.0 |
| *P. flabellatus* | 33.6 | 20.0 | 12.6 | 8.4 | 6.3 | 2.1 |
| *P. cfr. florida* | 9.5 | 4.2 | 3.2 | 6.3 | 2.1 | 2.1 |
| *P. ostreatus* Z-15 | 8.4 | 5.3 | 2.1 | 5.3 | 4.2 | 2.1 |
| *P. pulmonarius* | 7.3 | 3.2 | 5.3 | 4.2 | 4.2 | 3.2 |
| *P. sajor-caju* | 16.8 | 6.3 | 14.7 | 8.4 | 6.3 | 4.2 |
| *P. sapidus* | 10.5 | 6.3 | 3.2 | 4.4 | 2.1 | 2.1 |
| *Agarius bisporus* | 5.3 | 2.1 | 12.6 | 5.3 | 0.0 | 12.6 |
| *Agrocybe aegerita* | 4.2 | 4.2 | 1.1 | 0.0 | 0.0 | 0.0 |
| *Calocybe indica* | 24.0 | 21.2 | 18.6 | 6.3 | 8.4 | 6.3 |
| *Coprinus comatus* | 0.0 | 3.2 | 0.0 | 0.0 | 2.1 | 0.0 |
| *Kuehneromyces mutabilis* | 11.6 | 0.0 | 10.5 | 6.3 | 0.0 | 2.1 |
| *Lentinus edodes* | 3.2 | 0.0 | 2.1 | 0.0 | 0.0 | 0.0 |
| *Tricholomopsis rutilans* | 2.1 | 2.1 | 2.1 | 2.1 | 0.0 | 2.1 |

## DISCUSSION

Laccase is a copper containing blue phenol oxidase and is widespread among various groups of organisms including plants (Meyer, 1987), basidiomycetes (Bollag and Leonowicz, 1984, Orth, *et al.,* 1993, Ratcliffe *et al.,* 1994), ascomycetes, *Neurospora crassa* (Frohner and Eriksson, 1974) *Podospora anserina* (Esser, 1968), *Sclerotinia sclerotiorum* (Wong and Willetts, 1974) and also in slime moulds, *Physarum polycephalum* (Daniel, 1963).

Laccases are nonspecific to their electron donating substrates and differently substituted mono- di-, or polyphenolic compounds are preferred. The main role of laccase is considered to be polymerization of phenolics but it can also function in the opposite direction to degrade phenolic substances from lignin model compounds. Laccase is produced by various *Pleurotus* spp. including *P. sapidus, P. ostreatus, A. bisporus* and *L. edodes* (Bollag and Leonowicz, 1984; Wood *et al.,* 1991; Ratcliffe, 1994; Orth *et al.,* 1993). Leonard (1971) reported laccase activity in *Schizophyllum commune* related to the ability of the culture to form a basidiocarp. Fertile mycelial culture produces high laccase activity while sterile culture did not show any laccase activity. Similar changes in laccase activity in saw dust culture have been reported from *P. ostreatus* (Iwahara *et al.,* 1981) and in *L. edodes* (Ishikawa *et al.,* 1983). Paice *et al.,* (1993) reported that laccase and manganese peroxidase from *T. versicolor* are closely associated with pulp bleaching and delignification of craft pulp. Vasdev and Kuhad (1994) reported higher laccase activity in *Cyathus bulleri* under static condition than shaking condition. They also observed suppression of laccase. The results are in agreement with our findings, while Shuttelworth *et al.,* (1986) found higher laccase production in agitated culture than stationary culture of *Rhizoctonia praticola.* It indicates that different groups of organisms react differently to the culture conditions. Laccase production is constitutive or its regulation can occur via inducers such as aromatic compounds (Wood, 1980). Gall and Neujahr (1981) found induction of phenol metabolizing enzymes in *Trichosporon cutaneum* by adding phenol, catechol, cresols and fluorophenols. The induction was inhibited when the phenol concentration was higher than 1mM. Haars and Huttermann (1980) have suggested that phenols are non-toxic to fungal growth but the quinones which are oxidized products of phenols are toxic. Mikulasova *et al.* (1990) observed salylic acid, cinnamic acid, benzoic acid and O-vanillin as the most growth inhibiting substances for *Candida utilis* and *C. albicans* while vanillyl alcohol, coniferyl alcohol and syringic acids were least toxic. The present concentrations of phenol and 4-chlorophenol were inhibitory for mycelial growth but phenol induced more lacasse enzyme than 4-chlorophenol.

Various aromatic substances like gallic acid, cycloheximide, 2,5-xylidine, wheat straw (10%), tobacco juice (10%) (Bollag and Leonowicz, 1984) and microorganisms like *Trichoderma harzianum* are known to induce laccase production. Lobarzewski (1981) also reported induction of peroxidases in *Trametes versicolor* by simple phenolic compounds as ferulic, syringic, vanillic acids and soluble preparation of lignin.

Stimulation of laccase production by incorporating phenol in the spawn substrate can be useful for early and rapid mycelial growth on spawn substrate and quick colonization on the lignocellulosic substrates. Phenol treated grains can minimize the chances of mould contamination in the substrate due to its antifungal activities.

## ACKNOWLEDGEMENTS

The senior author would like to thank Dr. F. Zadrazil, Institute of Soil Biology, Braunschweig (Germany) for giving some cultures of white rot fungi, D.A.A.D., Bonn (Germany), for giving financial support and to the Director N.C.M.R.T., Solan (India) for granting study leave.

## REFERENCES

Anonymous (1993). Facts and figures for the chemical industry. *Chemical and Engineering News,* 71, 38-82.

Bollag, J.M., and Leonowicz, A. (1984). Comparative studies of extracellular fungal laccases. *Applied Environmental and Microbiology,* 48, 849-854.

Boominathan, K. and Mahadevan, A. (1989). Dissimilation of aromatic substances by fungi. *Zentralblatt für Mikrobiologie,* 144, 37-45.

Bonnen, A.M., Anton, L.M. and Orth, A.B. (1994). Lignin-degrading enzymes of the commercial button mushroom, *Agaricus bisporus Applied Environmental and Microbiology,* 60, 960-965.

Buswell, J.A. and Eriksson, K.E.L. (1994). Effect of lignin related phenols and their methylated derivatives on the growth of eight white-rot fingi. *World Journal of Microbiology and Biotechnology,* 10, 169-174.

Charya, M.A.S. and Reddy, S.M. (1984). Inhibition of production of pectic enzymes of *Phoma exigna* seed rot of *Vigna radiata. Indian Phytopathology,* 37, 537-539.

Daniel, J.W. (1963). Metabolism of phenolic precursor of a myxomycete spore pigment. *Journal of Cell Biology,* 19, 18-19.

De Jong, E., Field, J.A., Spinner, H.E., Wijnberg, J.P.B.A. and De-Bont, J.A.M. (1994). Significant biogenesis of chlorinated aromatics by fungi in natural environment. *Applied Environmental and Microbiology,* 60, 264-270.

Esser, K. (1968). Phenol oxidases and morphogenesis in *Podospora anserina. Genetics,* 60, 281-288.

Gaal, A. and Neujahr, H.Y. (1981). Induction of phenol metabolizing enzymes in *Trichosporon cutaneum. Archives of Microbiology,* 130, 54-58.

Glenn, J.K., Morgan, M.A., Mayfield, M.B., Kuwahara M. and Gold, M.H. (1983). An extracellular $H_2O_2$ requiring enzyme preparation involved in lignin biodegradation by the white rot basidiomycete *Phanerochaete chrysosporium. Biochemical and Biophysical Research Communication,* 114, 1077-1083.

Glenn, J.K. and Gold, M.H. (1985). Purification and characterization of an extracellular Mn dependent peroxidase from the lignin degrading basidiomycete *Phanerochaete chrysosporium. Archives of Biochemistry and Biophysics,* 251, 688-796.

Haars, A. and Huttermann, A. (1980). Function of laccase in the white rot fungus *Fomes annosus. Archives of Microbiology,* 125, 233-237.

Harris, G. and Ricketts, R. W. (1962). Metabolism of phenolic compounds by yeasts. *Nature,* 195, 473-474.

Kirk, T.K. and Farrell, R.L. (1987). Enzymatic combustion : the microbial degradation of lignin. *Annual Review of Microbiology,* 41, 465-505.

Koch, R. (1989). *Umwelt Chemikalien,* VCH Weinheim, Deutschland.

Kremer, S., Sterner, O. and Anke, H. (1992). Degradation of pentachlorophenol by *Mycena avenacea* TA 8480-identification of initial dechlorinated metabolites. *Zeitschrift fur Naturforschung,* 47C, 561-566.

Kuwahara, M., Glenn, J.K., Morgan, M.A. and Gold, M.H. (1984). Separation and characterization of two extracellular $H_2O_2$ dependent oxidases from ligninolytic cultures of *Phanerochaete chrysosporium. FEBS Letters,* 169, 247-250.

Leibnitz, E., Beherens, U. and Seifert, H. (1960). Abbau von Phenolen durch Hefen. *Zeitschrift Allegemeine Mikrobiologica,* 1, 13-17.

Leonard, T.J. (1971). Phenol oxidase activity and fruiting body formation in *Schizophyllum commune. Journal of Bacteriology,* 106, 162-167.

Lobarzewski, J. (1981). A comperasion of inductive influence of syringic acid and wastes containing lignosulfonic acids on peroxidase activities and its synthesis in *Trametes versicolor. Acta Microbiologica Polonica,* 30, 145-150.

Meyer, A. M. (1987). Polyphenol oxidases in plants: Recent Progress. *Phytochemistry,* 26, 11-20.

Mikulasova, M., Vodmy, S. and Pekarovicova, A. (1990). Influence of phenolics on biomass priduction by *Candida utilis* and *Candida albicans. Biomass,* 23, 149-154.

Mukherjee, N. and Kundu, B. (1973). Antifungal activities of some phenolics and related compounds to three fungal pathogens. *Phytopathologische Zeitschrift,* 78, 89-92.

Orth, A.B., Royse, D.J. and Tien, M. (1993). Ubiquity of lignin-degrading peroxidases among various wood-degrading fungi. *Applied Environmental and Microbiology,* 59, 4117-4023.

Paise, M.G., Bourbonnais, R., Archibald, F.S. and Jurasek, L. (1993). Manganese peroxidase produced by *Trametes versicolor* during pulp bleaching demethylates and delignifies kraft pulp. *Applied Environmental and Microbiology,* 59, 260-265.

Ratcliffe, B., Flurkey, W.H., Kuglin, J. and Dawley, R. (1994). Tyrosinase, laccase and peroxidase in mushrooms (*Agaricus*, crimini, oyster and shiitake). *Journal of Food Science,* 59, 824-827.

Shuttleworth, K.L., Postie, L. and Bollag, J.M. (1986). Production of induced laccase by the fungus *Rhizoctonia praticola. Canadian Journal of Microbiology,* 32, 867-870.

Siuda, J.F. and Fernandis, J.F. (1973). Naturally occurring halogenated organic compounds. *Lloydia,* 36, 107-143.

Tien, M. and Kirk, T.K. (1983). Lignin degrading enzyme from the hymenomycete *Phanerochaete chrysosporium* Burds. *Science,* 221, 661-663.

Upadhyay, R.C. and Hoffrichter, M. (1993). Effect of phenol on the mycelial growth and fructification of some basidiomycetous fungi. *Journal of Basic Microbiology,* 33, 343-347.

Vasdev, K. and Kuhad, R.C. (1994). Induction of laccase production in *Cyathus bulleri* under shaking and static culture conditions. *Folia Microbiologica.* 39, 326-330.

Wong, A.G. and Willets, H.G. (1974). Polyacryamide-gel elecrophorasis of enzymes during morphogenesis of sclerotia of *Sclerotinia sclerotiorum. Journal of General Microbiology,* 81, 101-109.

# 6

# Molecular Biology of Fungal Biodegradation: Cellulolytic Activity Over Agricultural Wastes

**Maria Sílvia Rossi[1], Marcio José Poças-Fonseca[2*] and Maristella de Oliveira Azevedo[1+]**

*[1]Department of Cellular Biology,*
*[2]Department of Genetics and Morphology, University of Brasilia, 70.910-900, Brazil*

In this work, we have analysed the cellulase production profile of *Humicola grisea* var. *thermoidea* and *Chrysonilia sitophila*, concerning their ability to convert the most abundant agricultural residues in Brazil (ball-milled straw—BMS—and sugar cane bagasse –SCB) to fermentable materials. In comparison to *Trichoderma reesei*, an industrial strain which efficiently degrades crystalline cellulose, we concluded that both fungi have a potent cellulolytic complex, early induced by natural lignocellulosic substrates. In this context, these microrganisms are potential cellulase producers at the industrial level. Alternatively, our data support their use as gene donors for the genetic improvement of microorganisms traditionally used in bioconversion processes.

## INTRODUCTION

Plant biomass is a promising renewable substrate to replace fossil fuels and to generate chemicals and electric power, reducing greenhouse gas emissions and the amount of solid wastes, therefore improving the economy, the energy security and the environmental conditions. The initial steps for processing lignocellulosic wastes include the high temperature and chemical pretreatment of these materials with acids, resulting in a significant industrial cost. Enzymatic hydrolysis is a rational alternative, since cellulolytic enzymes can efficiently access these complex substrates and break down the crystalline fibres, significantly reducing the utilization of organic acids (Wyman, 2003).

Since the rate-limiting step of the bioconversion process by fungi is the cellulose hydrolysis, degradation efficiency depends on two factors: (i) the profile of fungal secreted enzymes and (ii) the nature and complexity of the substrate: the higher the crystallinity and/or structural complexity, the lower the hydrolysis rate (Knowles *et al.,* 1987). An ideal "fungus × substrate" pair would provide an efficient alternative to chemical conversion.

Cellulolytic fungi play a major role in biodegradation of cellulose in natural ecosystems and consequently, present a considerable biotechnological potential, as recently reviewed by Poças-Fonseca & Maranhão (2005). Among these microorganisms, *Trichoderma reesei* has been extensively studied, and its cellulolytic system is considered a model for the hydrolysis of the crystalline cellulose. Conversion of cellulose to reducing sugars involves a cooperative action of the secreted enzymes (endoglucanases, cellobiohydrolases and beta-glucosidases) in a synergistic manner (Henrissat *et al.,* 1985). Initially, studies were mainly performed with commercial cellulose, nonetheless, the economic and environmental importance of cellulosic biomass has drawn the attention to the natural substrates (Wyman, 2003).

Fungal enzymes have been employed both for the production of enzymatic pools and/or purified enzymes. Several research groups have been interested in altering cellulase production by established *T. reesei* and *Aspergillus* strains, aiming the obtention of genetically engineered strains displaying optimized enzymatic profiles. Alternatively, yeast or bacteria recombinant strains, expressing fungal cellulases, were constructed (Nevalainen *et al.,* 1990). Nowadays, efforts also focus on the cellulase production, cellulose hydrolysis and byproducts fermentation in one step (consolidated bioprocessing-CPB) (Lynd *et al.,* 2005).

Developing countries, such as Brazil, generate a huge quantity of sugarcane bagasse and straw, derived from the sugar and/or bioethanol industry and from agricultural activities. In this sense, a special attention should be given to the conversion of these substrates in useful products such as natural fibres, pulps, sweeteners, oils, proteins, acids, microbial enzymes, chelating agents, composites and electric power (Pessoa-Jr. *et al.,* 2005; Gonçalves *et al.,* 2005).

This paper describes the ability of *Humicola grisea* var. *thermoidea* and *Chrysonilia sitophila* to produce and secrete enzymes to degrade specific agricultural wastes, aiming to determine the best "fungus × substrate" pairs. We report the characteristics of the cellulase production by these fungi in the presence of sugarcane bagasse (SCB) and ball-milled straw (BMS) in comparison to a *T. reesei* industrial strain.

## MATERIAL AND METHODS

### Fungal Strains and Growth Conditions

*Humicola grisea* var. *thermoidea* was isolated from compost by Dr. Dayson Olsany, University of Viçosa (UFV), Viçosa-MG, Brazil (Araújo *et al.,* 1983). *Chrysonilia sitophila* was isolated from wheat by Dr. Nelson Duran, University of Campinas (UNICAMP), Campinas-SP, Brazil (Rodriguez *et al.,* 1997). *Trichoderma reesei* strain QM9414 was obtained from ATCC (no. 26921). Spores ($3.0 \times 10^6$) were transfered to 20 mL of minimal medium –MM- (Pontecorvo *et al.,* 1953) supplemented with one of the following carbon sources: 0.1% (w/v) glucose, 0.2% (w/v) ball-milled straw (BMS) or 2.0% (w/v) washed, dried and steam-exploded (17 kgf/cm$^2$ at 265°C) sugarcane bagasse (SCB). *H. grisea* cultures were incubated at 42°C, and *T. reesei* and *C. sitophila* cultures at 37°C, with agitation (100 rpm). Culture supernatants were collected by centrifugation (3,000 g / 5 min) and stored at 20°C.

## CELLULASE ACTIVITY DETERMINATION

Total cellulase activity was based on the filter paper assay (FPA) and determined by the liberation of reducing sugars (RS), essentially as described by Mandels *et al.* (1976): samples of 0.25 mL of culture supernatants were added to 0.5 mL of 50 mM sodium citrate buffer, pH 4.8, containing a strip of Whatman No. 1 filter paper (3.0 × 0.5 cm). Reaction mixtures were incubated at 50°C for 60 min and assayed for RS liberation by the dinitrosalicylic acid (DNS) method, using a glucose reference curve as standard. Assays were performed in triplicate. One enzyme unit was defined as the amount of protein necessary to produce 1 mg RS/h.

## SDS-PAGE ANALYSES

Samples (1.2 mL) of culture supernatants were centrifuged for 3 min at 10,000 g to remove cellular debris. Trichloroacetic acid (TCA) was added to a final concentration of 20% (v/v) and the mixture incubated at 4°C for 15 min. TCA-precipitated material was harvested, washed twice with 1.0 mL of frozen acetone and resuspended in 30 μL of sample buffer. One-third of each sample was subjected to electrophoresis on 13% acrylamide gels under denaturing conditions. Gels were stained with Coomassie Brilliant Blue R-250 and/or silver nitrate.

## WESTERN BLOT ANALYSIS

The procedure was carried out according to the protocol described by Sambrook & Russel (2001), with minor modifications: proteins were transferred to nitrocellulose membranes at 30 V/h (LKB Bromina, 2117 Multiphor II), incubated overnight at 4°C in 7 mM phosphate-buffered saline (PBS: NaCl 137.0 mM; KCl 2.7 mM) containing 5% (w/v) skimmed milk. Membranes were incubated overnight at 4°C with polyclonal anti-*T. reesei* cellobiohydrolase antibody (generously provided by Dr. H. El-Dorry, Institute of Chemistry, University of São Paulo, Brazil), diluted 1:1,000 in PBS/ 1% (w/v) skimmed milk. Two washes with PBS were performed and the second antibody (anti-rabbit IgG-peroxidase conjugate) was added. Membranes were incubated for 6 h and developed with the chloronaphthol reagent.

## PROTEIN DETERMINATION

Protein concentration was determined by the method of Löwry *et al.* (1951), using bovine serum albumin as standard.

## RESULTS AND DISCUSSION

### Fungal Growth

Since fungal growth in the presence of lignocellulosic material resulted in a mixture of fibres and hyphae, impairing the growth ratio determination by conventional parameters, we have adopted two criteria to evaluate mycelial mass increase: (i) the observation of spore germination and mycelial expansion by light microscopy and (ii) the measurement of the culture supernatants pH. In both cases, the growth profile on complex substrates (BMS and SCB) was independent of the available carbon source (Fig. 1 D-I) and was compatible to the expected mycelial growth.

In all the *T. reesei* and *C. sitophila* cultures, cell mass production was associated with an increase in the supernatant pH (Fig. 1A, C, D, F, G and I) and with glucose consumption (Fig. 2). Nevertheless, *H. grisea* showed a very low growth rate on glucose and the maintenance of the pH of the culture filtrate (Fig. 1B), which suggests the need of additional nutrients. The difficulty of *H. grisea* to grow on glucose renders the level of this sugar in the medium essentially unchanged (Fig. 2). Glucose as the sole carbon source did not promote a good hyphal growth and net formation (Fig. 1B), contrasting to cultures supplied with the lignocellulosic materials (Fig. 1E and H).

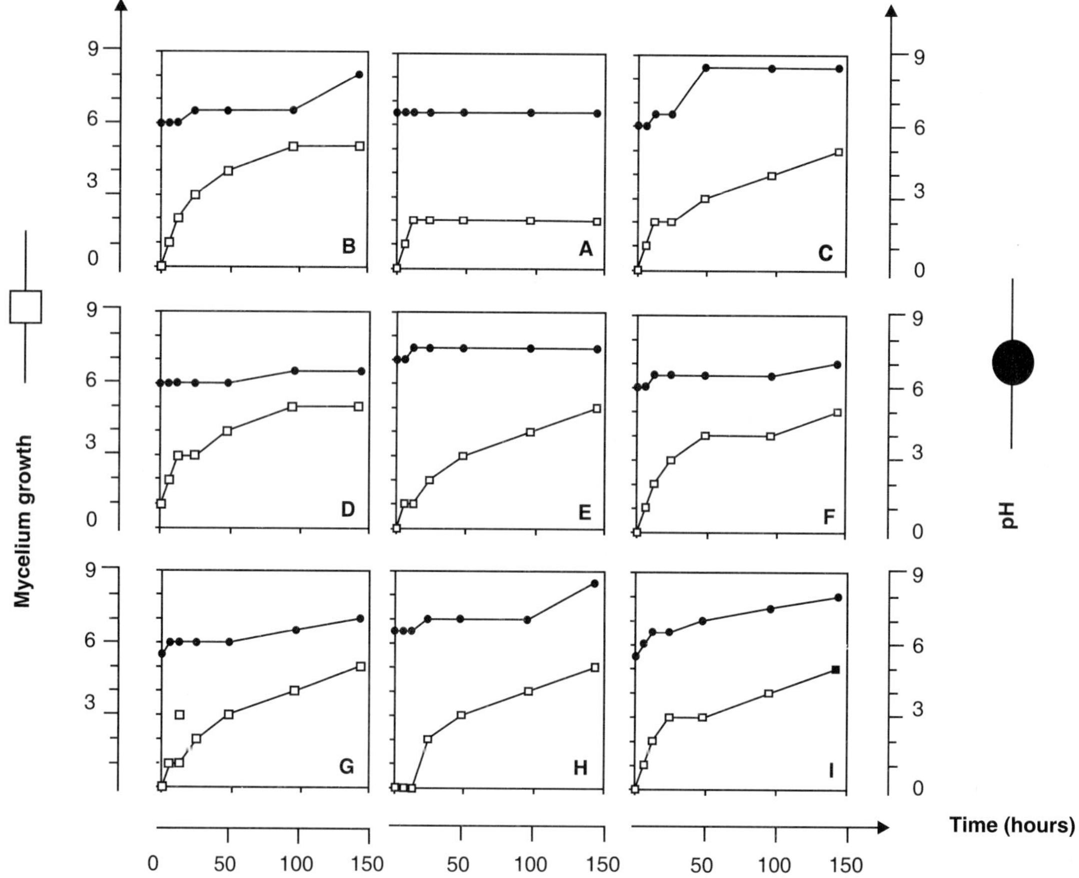


**Fig. 1:** Mycelial growth (open squares) and pH (black diamonds) during incubation of *T. reesei*, *H. grisea* and *C. sitophila* cultures in different substrates. Glucose (GLU), ball-milled straw (BMS) and sugar cane bagasse (SCB). For the estimation of fungal growth, mycelial areas of 0.01; 0.08; 0.12; 1.50 and 5.00 cm$^2$ were determined in a scale ranging from 0 to 5, respectively. Time in hours: 0, 6, 12, 24, 48, 96 and 144.

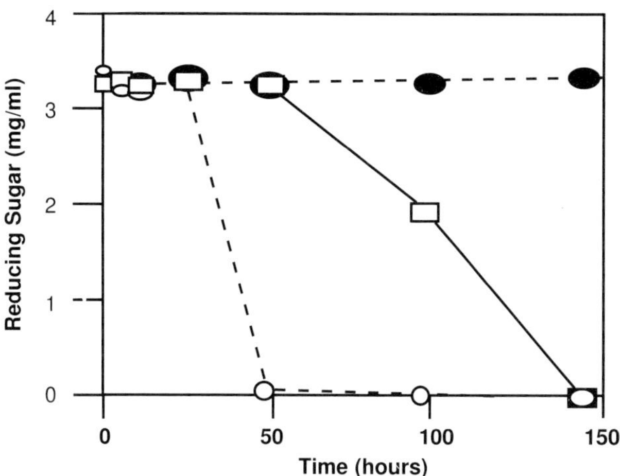


**Fig. 2:** Glucose levels over the period of 144 hours of fungal growth, measured as reducing sugar (RS) by the FPA protocol. Open squares represent *T. reesei*; black diamonds, *H. grisea* and open circles, *C. sitophila*.

## CELLULASES INDUCTION

In the present study, *T. reesei* QM9414 strain grew in BMS and SCB (Fig. 1D and G), but reducing sugar production was poorly detected in the culture supernatants (Fig. 3), according to the FPA. This observation may be related to the low beta-glucosidase and xylanase activities, described for some strains of this fungus, which are essential for the complete hydrolysis of lignocellulose (Sternberg *et al.,* 1977; Berquist *et al.,* 2002). In this view, our results corroborate the assumption that *T. reesei* has limitations in accessing and degrading natural lignocellulosic materials (Mishra *et al.,* 1979), despite its efficiency to act over Avicel or CMCellulose (Nevalainen *et al.,* 1990).

The FPase activity detected in the supernatants of the *H. grisea* cultures on SCB reached 0.36 IU/mL within 48 h and 0.62 IU/mL after 96 h of growth (Fig. 3B). In the BMS cultures, the elicited cellulolytic profile showed an early peak of activity at 12 h (0.33 IU/ml). A higher activity (0.67 IU/ml) was detected after 96 h (Fig. 3A). These data demonstrate the *H. grisea* ability to utilise both natural lignocellulosic substrates, pointing out to an improved adaptation of its hydrolytic complex when compared to *T. reesei.* Taken together with previous results from our group (De-Paula *et al.,* 1999; 2003), the analyses described above suggest a cooperative action of different types of hydrolytic enzyme pools for *H. grisea.* Consistently, Poças-Fonseca *et al.,* (1997) described a multigenic family for *H. grisea* cellobiohydrolases, while distinct xylanase activities have been reported; one of them is already being employed for pulp bleaching (Moriya *et al.,* 2005).

In the *C. sitophila* BMS cultures, the more expressive FPase production (0.36 IU/mL) occurred within 48 h, while in the SCB cultures higher values appeared after 48 h (0.58 IU/ml) and 96 h (1.2 IU/ml) of incubation (Fig. 3A and B). Comparatively, this fungus reveals a better capacity to hydrolyze SCB, apparently the most appropriate natural substrate at the conditions assayed, leading to an ideal pair *C. sitophila* × SCB.

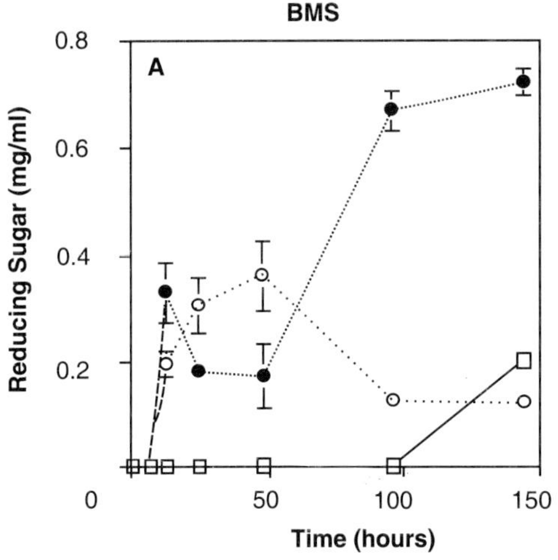


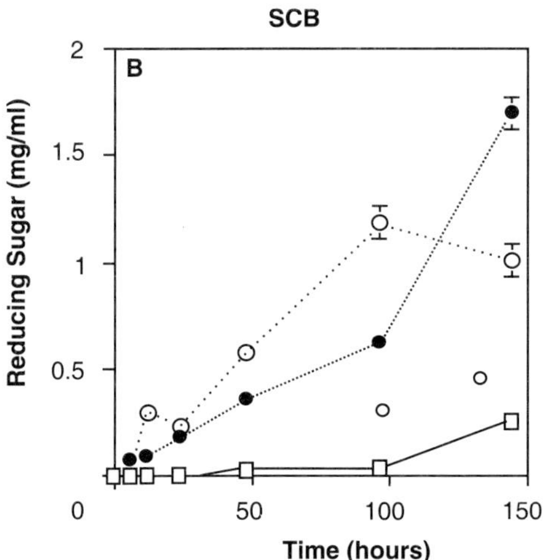


**Fig. 3.** Cellulolytic activity of supernatants from fungal cultures on different carbon sources, determined by the FPA assay. (**A**) BMS cultures; (**B**) SCB cultures. Open squares represent *T. reesei*; black diamonds, *H. grisea* and open circles, *C. sitophila*.

Summarizing, a considerable hydrolytic activity over BMS and SCB was detected in early cultures of *H. grisea* and *C. sitophila*, revealing the prompt response of both fungi cellulase genes to substrate induction.

## PROTEIN SECRETION

Secretion of proteins was investigated by measuring their concentration in the culture supernatants and by SDS-PAGE. Results obtained by both methods correlated well. No proteins were detected in the glucose culture supernatants of all the studied fungi. *T. reesei* cultures grown on BMS and SCB also failed in protein secretion. These results were compatible to the absence of FPase activity in the *T. reesei* BMS and SCB supernatants before 96 h of growth (Fig. 3A and B). However, very low amounts of protein were revealed by silver staining the gels (data not shown), what can reflect the low efficiency of *T. reesei* hydrolytic enzymes genes to respond to the inducers present in these specific culture media.

With respect to *C. sitophila*, low amounts of protein were also detected in BMS and SCB, as revealed by either the colourimetric or the electrophoretic assay. The main SDS-PAGE band corresponds to a protein with a molecular mass of about 45 kDa (data not shown).

A more significant protein secretion was obtained from *H. grisea* cultures. On BMS, a complex pattern of proteins was detected at 12 h of incubation (Fig. 4A, lane 3); while on SCB, the protein secretion profile gradually increased from 12 h to the end of the incubation period (Fig. 4B, lanes 3-6). No proteins were detected in glucose cultures within this period, except by silver-stained SDS gels (data not shown). The presence of proteins detected only by this more sensitive method is in accordance to De-Paula *et al.* (2003) and may represent a low level constitutive expression of

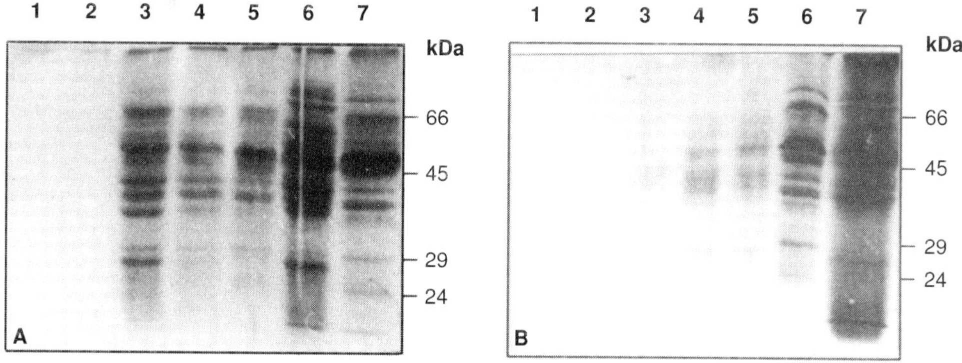


**Fig. 4:** SDS-PAGE analysis showing protein secretion profiles of *H. grisea* var. *thermoidea* grown on (**A**) BMS and (**B**) SCB. Lanes 1 to 7: 0, 6, 12, 24, 48, 96 and 144 h of incubation. Numbers on the right side of the panels refer to protein molecular mass.

*H. grisea* cellulase genes which are not subject to glucose repression, as we have demonstrated before for the *cbh*1.2 gene (Poças-Fonseca *et al.,* 2000).

## IMMUNOSCREENING OF THE SECRETED PROTEINS

The proteins secreted by the fungi were identified by western blot using polyclonal antibodies against *T. reesei* cellulases (Fig. 5). In the conditions assayed, no proteins were recognized in the *T. reesei* (Fig. 5A, lanes 1-2) and *C. sitophila* (Fig. 5C, lanes 1-2) cultures grown on glucose, even at the end of the incubation time. Interestingly, a weak band of about 45 kDa was detected from the 12 h *H. grisea* culture with 0.1% glucose (Fig. 5B, lanes 1-2).

*H. grisea* and *C. sitophila* analyses showed that both natural substrates provided efficient inducers for cellulases expression, since one strong reacting band, with a molecular mass of about 45 kDa, was detected (Fig. 5B and C, lanes 3-6).

A very weak signal, corresponding to the same molecules population (> 45 kDa), was identified in the *T. reesei* BMS and SCB supernatants, only at 144 h of incubation (data not shown), when reducing sugars were detected in the medium (Fig. 3A).

We concluded that *C. sitophila* and *H. grisea* var. *thermoidea* have a potent cellulolytic complex, which is efficiently induced by the natural substrates SCB and BMS. These fungi have potential as cellulase producers or gene donors for the manipulation of industrial microorganisms. Further studies can identify which cellulase genes are expressed and clarify their mode of action during the process of cellulose hydrolysis.

The data presented in this paper instigated us to study the *H. grisea* var. *thermoidea* hydrolytic complex regulation. For an improved exploitation of *H. grisea* genes, a complete programme on genetic manipulation of this fungus is in progress at the Molecular Biology Laboratory of the University of Brasilia, Brazil. Recent data point out to a coordinated regulatory mechanism of cellulases and xylanases, with some genes being expressed constitutively and others subject to carbon source and ambient pH regulation (Poças-Fonseca and Maranhâo, 2005).

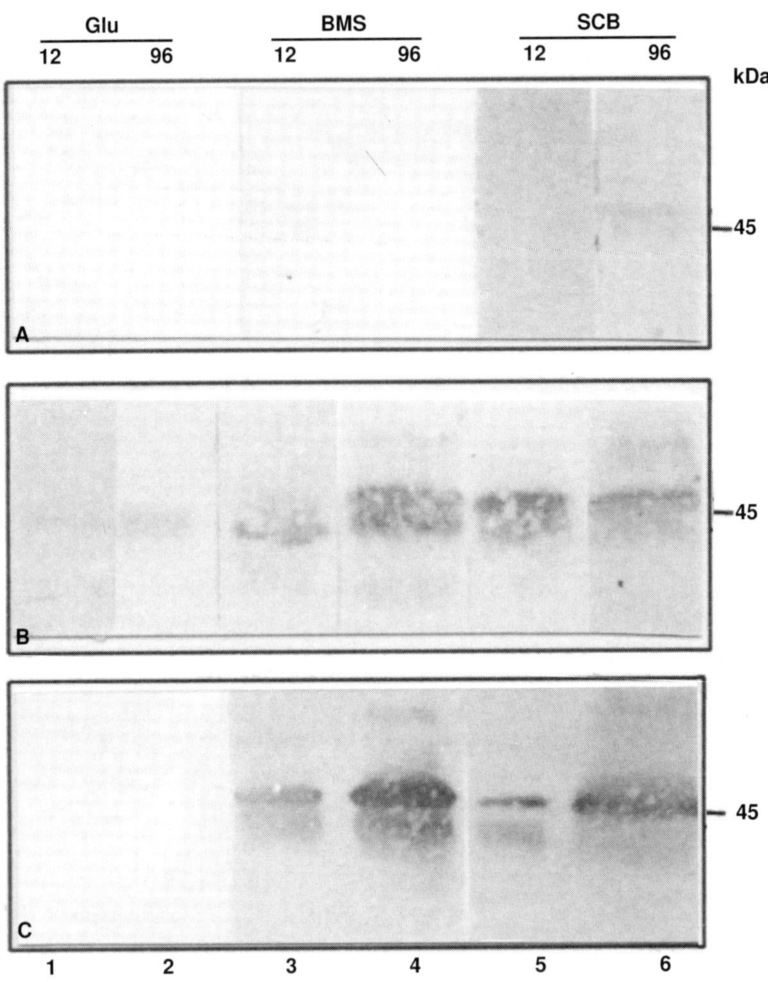


**Fig. 5:** Western blot analysis of secreted cellulases on glucose (GLU), ball-milled straw (BMS) and sugarcane bagasse (SCB) supplemented cultures. (**A**) *T. reesei*; (**B**) *H. grisea*; (**C**) *C. sitophila*. Lanes 1, 3 and 5 correspond to 12 h of incubation and lanes 2, 4 and 6 to 96 hours.

## ACKNOWLEDGEMENTS

The authors thank M. Fátima. L. Cesário for technical assistance. This work was supported by a grant from FAP-DF and by FUB, CAPES and CNPq (Brazil).

## REFERENCES

Araújo, D. F., Barros, E. G., Caldas, R. A. and Silva, D. O. (1983). β–glucosidase activity of thermophilic cellulolytic fungus *Humicola* sp. *Biotechnology Letters,* 5, 781-784.

Bergquist, P., T´eo, V., Gibbs, M., Cziferszky, A., de Faria, F. P., Azevedo, M. O. and Nevalainen, H. (2002). Expression of xylanase enzymes from thermophilic microorganisms in fungal hosts. *Extremophiles,* 6(3), 177-184.

De-Paula, E. H., Poças-Fonseca, M. J. and Azevedo, M. O. (2003). The product of *Humicola grisea* var. *thermoidea cbh*1.2 gene is the major expressed protein under induction by lignocellulosic residues. *World Journal of Microbiology & Biotechnology,* 19, 631-635.

De-Paula, E.H., Ramos, L. and Azevedo, M.O. (1999). The potential of *Humicola grisea* var. *thermoidea* for bioconversion of sugar cane bagasse. *Bioresource Technolology,* 68, 35-41.

Gonçalves, A.R., Benar, P., Costa, S.M., Ruzende, D.S., Moriya, R.Y., Luz, S.M. and Ferretti, L.P. (2005). Integrated processes for use of pulps and lignins obtained from sugarcane bagasse and straw: a review of recent efforts in Brazil. *Applied Biochemistry and Biotechnology,* 121-124, 821-826.

Henrissat, B., Driguez, H., Viet, C. and Schülein, M. (1985). Synergism of cellulases from *Trichoderma reesei* in the degradation of cellulose. *Biotechnology,* 3, 722-726.

Knowles, J.L., Lehtoväara, P., Teeri, T., Pentillä, M., Salowori, J. and André, L. (1987). The application of recombinant-DNA technology to cellulases and lignocellulosic wastes. *Philosophy Transactions of the Royal Society of London,* A-321, 449-454.

Löwry, O., Rosenbrough, N., Farr, A., Randall, R. (1951). Protein measurement with the folin-phenol reagent. *Journal of Biological Chemistry,* 193, 256.

Lynd, L.R., van Zyl, W.H., McBride, J.E. and Laser, M. (2005). Consolidated bioprocessing of cellulosic biomass: an update. *Current Opinion in Biotechnology,* 16, 577-583.

Mandels, M., Andreotti, R. and Roche, C. (1976). Measurement of saccharifying cellulase. *Biotechnology Bioengineering Symposium,* 6, 17-34.

Mishra, M.M., Singh, C.P., Kapoor, K. and Jain, M.K. (1979). Degradation of lignocellulosic material and humus formation by fungi. *Annual Review of Microbiology,* 130A(4), 481-486.

Moriya, R.Y., Gonçalves, A.R. and Faria, F.P. (2005). Enzymatic bleaching of organosolv sugarcane bagasse pulps with recombinant xylanase of the fungus *Humicola grisea* and with commercial Cartazyme HS xylanase. *Applied Biochemistry and Biotechnology,* 121-124, 195-203.

Nevalainen, K.M.H., Pentillä, M.E., Harkki, A., Teeri, T.T. and Knowles, J.(1990). The molecular biology of *Trichoderma* and its application to the expression of both homologous and heterologous genes. In: *Molecular Industrial Mycology.* Eds Leong, S.A. and Berka, R.M. Marcel Dekker Inc., NY, chapter 6, 1990.

Pessoa-Jr., A., Roberto, I.C., Menossi, M., dos Santos, R.R., Filho, S.O. and Penna, T.C. (2005). Perspectives on bioenergy and biotechnology in Brazil. *Applied Biochemistry and Biotechnology,* 121-124, 59-70.

Poças-Fonseca, M.J. and Maranhão, A.Q. (2005). Diversity among cellulolytic fungi. In: *Microbial Diversity: Current Perspectives and Potential Applications.* Eds. Satyanarayana, T. and Johri, B.N., I.K. International Pvt. Ltd., India. Pp 165-180.

Poças-Fonseca, M.J., Lima, B.D., Brígido, M.M., Felipe, M.S.S., Radford, A. and Azevedo, M.O. (1997). *Humicola grisea* var. *thermoidea cbh*1. 2: a new gene in the family of cellobiohydrolases is expressed and encodes a cellulose-binding domain-less enzyme. *Journal of General and Applied Microbiology,* 43(2), 115-120.

Poças-Fonseca, M.J., Silva-Pereira, I., Rocha, B.B. and Azevedo, M.O. (2000). Substrate-dependent differential expression of *Humicola grisea* var. *thermoidea* cellobiohydrolase genes. *Canadian Journal of Microbiology,* 46, 749-752.

Pontecorvo, G., Roper, J.A., Hemmons, L.H., McDonald, K.D. and Bufton, W.J. (1953). The genetics of *Aspergillus nidulans. Advances in Genetics,* 5, 141-238.

Rodriguez, J., Ferraz, A., Nogueira, R.F., Ferrer, I., Esposito, E. and Duran, N. (1997). Lignin biodegradation by the ascomycete *Chrysonilia sitophila. Applied Biochemistry and Biotechnology,* 62, 233-242.

Sambrook, J. and Russel, R.W. (2001). *Molecular Cloning - a laboratory manual.* 3rd ed. Cold Spring Harbor Laboratory Press, Cold Spring Harbor, N.Y-USA.

Sternberg, D, Vijayakumar, P. and Reese, E.T. (1977). β–glucosidase: microbial production and effect on enzymatic hydrolysis of cellulose. *Canadian Journal of Microbiology,* 23(2), 139-147.

Wyman, C.H. (2003). Potential synergies and challenges in refining cellulosic biomass to fuels, chemicals and power. *Biotechnology Progress,* 19, 254-262.

# 7

# Isolation and Characterization of Lignocellulose Biodegrading Fungi from Soil Polluted with Pulp and Paper Mill Effluent

**Piyush Malaviya[1, 2] and V.S. Rathore[1]**

[1]*Department of Environmental Sciences, College of Basic Sciences and Humanities G.B. Pant University of Agriculture and Technology, Pantnagar-263 145*

[2] *Department of Environmental Science, University of Jammu*

Soil inhabitating ligninolytic fungi have been isolated following step-wise enrichment with Indulin AT (an industrial polymeric lignin almost free of polysaccharides). Sixty-six pure fungal isolates obtained were screened for lignin transformation potential by nine chromogenic plate tests, along with *P. chrysosporium* used as a positive control. Four isolates gave 6-7 positive tests as against eight given by *P. chrysosporium*. These four isolates included two hymenomycetous basidiomycete [*Merulius* (Syn. *Phlebia*) *aureus* and an unidentified basidiomycete; one ascomycete (*Libertella betulina* Desm.) and one species belonging to deuteromycetes (*Fusarium sambucinum* Fuckel). The ligninolytic activity of the four isolates was also tested by radiorespirometric assay, with rice $^{14}$C-[lignin]-lignocellulose as substrate. All the four isolates were found capable of solubilization and mineralization of $^{14}$C-[lignin]-lignocellulose. However, the two basidiomycetes were found to posses higher mineralization rates ($^{14}CO_2$ evolution) compared to the ascomycetous isolate (*Libertella betulina*) and deuteromycetous isolate (*Fusarium sambucinum*). The results suggest that enrichment with Indulin AT followed by plate tests could be a simple and useful methodology for isolation of lignocellulose transforming fungi from different types of soils.

## INTRODUCTION

It is estimated that 10 to 100 billion of lignocelluloses biomass is produced annually by plants from photosynthesis (Bassham, 1975; Kuhad and Singh, 1993). Almost 50% of wastes originating from agro-industrial processes, forestry and municipal sources comprise lignocelluloses (Ishaq and Chahal, 1991). There is a great potential for utilization of this plentiful and renewable resource for diverse applications such as production of energy, liquid fuels, pulp and paper, upgraded ruminant feed, single cell protein, novel chemicals and solvents, adhesives, asphalt, and a host of other value-added substances, by development of appropriate technologies (Wood, 1985; Fermor, 1993;

Kamra and Zadrazil, 1998; Pointing, 2001)). However, the presence of lignin in the lignocelluloses is the major barrier in achieving these goals. Lignin—an aromatic heterogeneous polymer consisting of three types of phenylpropanoid units—is highly recalcitrant. It is intimately interspersed and covalently bonded with the hemicelluloses, forming a 3-dimensional matrix surrounding the cellulose fibrils of the cell wall. It thus imposes a great restriction on the hydrolysis of polysaccharide components (cellulose and hemicelluloses) of the lignocelluloses by the microbial enzymes (Kirk and Farrell, 1987; Jeffries, 1994). Chemical de-lignifications are highly energy intensive and produce environmentally undesirable byproducts. Attempts have, therefore, been focused on finding microbial species, which would delignify the lignocelluloses and make it amenable for diverse biotechnological applications (Boominathan and Reddy, 1993; Buswell, 1991; Kuhad, *et al.*, 1997).

Extensive research carried out during the last two decades has revealed that wood-degrading white-rot fungi—a heterogeneous collection of hundreds of basidiomycetous fungi – are most efficient lignin degraders. These species, under restricted nutrient supply (e.g. limited N in culture medium) produce extra-cellular enzymes (ligninolytic system), capable of efficient lignin solubilization and mineralization (Buswell, 1991). The ligninolytic system of white-rot fungi has the additional capability of transforming and mineralization of several types of recalcitrant environmental pollutant like synthetic dyes, dioxins, biphenyls, polyaromatic hydrocarbons, polychlorinated phenols and pesticides (Bumpus and Aust, 1987). However, most of our knowledge on ligninolytic microorganisms has been derived from studies on wood-inhabitating species. There is paucity of information on the role of various microorganisms, particularly basidiomycetes, on transformation of lignin and other organic matter in soil ecosystem. Isolation of such species from soils would be useful in their application in utilization of lignocelluloses wastes from agro-industrial sources, site remediation, treatment of waste waters from food processing and agro-industries, composting, cattle feeds etc. Considering these points, the present research was undertaken to isolate and characterize lignin biodegrading fungi from soil which had been exposed to lignin-rich waste waters of a pulp and paper mill for several years, thus providing substrate for natural enrichment of ligninolytic microorganisms.

## MATERIAL AND METHODS

### Collection and Enrichment of Soil Samples

Soil sample (top 2") were collected in plastic bags from the bank of the effluent discharge channel of Century Pulp and Paper Mill, Lalkuan, Distt. Nainital (Uttaranchal), brought to laboratory and stored at 4° C. After pulverizing and sieving (40 mesh) 10 gm of each sample (in duplicate) was transferred to sterilized Erlenmeyer flasks to which was added 1 gm Indulin AT (Sigma Chemical Co - a commercial polymeric Kraft lignin) or black liquid (1%w/w) – lignin-rich liquid obtained during Kraft pulping of wood chips. The contents were thoroughly mixed with a sterilized glass rod and the flasks were stoppered with cotton plugs and kept for incubation at 30°C in an incubator for 100 days for enrichment. After the enrichment period, 1 gm soil (in duplicate) was weighed out from each flask, suspended in 10 ml of 85% saline and mixed on a vortex. The soil suspension was serially diluted with sterile water ($10^{-3}$- $10^{-5}$ fold) followed by spread plating on specific media in petri plates. For isolation of fungal species, medium of Janshekar *et al.,* (1982) was used supplemented with 1 % Indulin AT as sole carbon source. The plates were incubated at 30°C in inverted position in a BOD incubator for 7-10 days. Distinct colonies were identified and replated on the same medium to obtain pure cultures, which were maintained on slants.

## PLATE TESTS

Each isolated strain was subjected to different colorimetric plate tests to evaluate its lignolytic potential (Falcon *et al.,* 1995). Laccase, Bavendum, Peroxidase and Tyrosinase tests were performed according to the procedure suggested by Brauns and Brauns (1960). Nobles' test was done as described by Nobles (1958). Ferulic acid hydrolysis and poly-R dye decolorization were performed as per the procedures described by Rayner and Boddy (1988) and Falcon *et al.,* (1995), respectively.

## TAXONOMIC IDENTIFICATION OF SELECTED STRAINS

Only those microbial strains that gave sufficiently high number of positive plate tests, and thus possible candidate of being ligninolytic, were selected for taxonomic identification. Basidiomycetous fungal strains were identified according to the identification keys of Bakshi (1971) and Gilbertson and Ryvarden (1986). Deuteromycetes and ascomycetous species were identified in collaboration with Indian Institute of Microbial Technology, Chandigarh, India.

## $^{14}$C-(LIGNIN)- LIGNOCELLULOSE BIODEGRADATION ASSAY

Preparation of $^{14}$C-lignocellulose and its biodegradation by selected microbial strains were undertaken according the procedure described by Crawford and Crawford (1978) and Crawford (1981).

### Preparation of $^{14}$C-(lignin)-Lignocellulose

Lignocellulose specifically labelled with carbon-14 in the lignin fraction was prepared by feeding DL- (U-$^{14}$C) phenylalanine (specific activity= 450 mCi$^{-\text{m mole}}$, obtained from BARC, Trombay) to 4-week-old wheat seedlings through transpiration stream. The dried plant tissues were finely powdered (40 mesh), and extracted with hot water (twice), benzene: ethanol (1:1,v/v, twice), ethanol (twice) and hot water, to remove the radioactivity from non-lignin fractions. The "extractive-free lignocellulose", thus obtained was dried and thoroughly ground. It was then treated successively with the enzymes proteinase K and cellulase (Sigma Chemical CO. USA), to remove residual radioactivity from protein and cellulose fractions (McCarthy *et. al.,* 1984; Reid, 1979). After drying to constant weight, the radiolabelled lignocellulose preparation was stored in a desiccator at 4°C.

## Biodegradation Assay

Assays for biodegradation of the $^{14}C$-(lignin)-lignocellulose by the selected microbial strains were done by the method described by Kirk *et al.,* (1978) with slight modification. The basal medium used contained (gm/l): D-glucose 10; $KH_2PO_4$ 2.0; magnesium sulfate 0.05; ammonium nitrate 0.79; calcium chloride 0.1; L-asparagine 0.048; trace elements and vitamins. Final pH of the basal medium was adjusted to 5.10 ml of sterilized medium was taken per 125 ml Erlenmeyer flask followed by addition of $^{14}C$-(lignin)- lignocellulose powder (Total radioactivity added /flask= 80,000 dpm). Each flask was aseptically inoculated with 5 disks of the desired microbial strain from the active growing zone of the plate. The inoculated flask was immediately closed with a rubber stopper having provision of inlet and outlet tubes for gas flow. Each flask was then secured in gyratory water bath shaker, and inlet and outlet tubes were closed tightly with clips. The flasks were then incubated under shaking in water bath at 30°C. The $^{14}CO_2$ liberated during [$^{14}C$-(lignin)]-lignocellulose biodegradation by the growing cultures was flushed (intervals of 3 days) with sterile air passed from the inlet tube for 25 minutes and trapped in a scintillation vial containing a cocktail of Bray's scintillation fluid and ethanolamine (9:1, v/v) to absorb liberated $^{14}CO_2$. Radioactivity of the samples was measured with a liquid scintillation counter (LKB Wallace, Finland). External standard ratio method was used for quench correction. Liquid samples were measured using Bray's scintillation fluid (Bray, 1960). Powder samples were radioassayed in jelly form according to Degenhart (1974). All assays and radioactivity measurements were done in duplicates.

# RESULTS

## Plate Tests

A total of 66 fungal isolates were obtained in pure culture by enrichment with Indulin AT or black liquor in the laboratory for one hundred days. All these isolates were screened for their ligninolytic potential using 9 different plate tests. Results are summarized in Tables 1 and 2. Out of the 66 fungal isolates, 18 failed to give even a single positive test. Of the remaining, 21 isolates gave only one positive test, six isolates gave 2, four isolates gave 3, twelve isolates gave 4, one isolate gave 5, one isolate gave 6, and three isolates gave 7 positive plate tests, respectively, out of nine plate tests applied. *Phanerorochaete chrysosporium,* used as a positive control, gave 8 positive tests. Some selected positive plate tests are presented in Plate –I and Plate –II.

**Isolate FT 1** (Plate I, F 1) gave 7 positive plate tests and only 2 negative tests viz. tyrosinase and poly-R dye decolorization tests.

**Isolate Ft 43** (Plate I, F 2 and Plate II, F 2) also gave similar results in respect of plate tests as in case of isolate FT 1.

**Isolate FT 65** (F 3, Plate I) also gave seven positive plate tests as in case of earlier mentioned two isolates.

**Isolate FT 66** (Plate II, F4) gave 6 positive plate tests and 3 negative plate tests viz., tyrosinase, poly-R dye decolorization and ferulic acid hydrolysis tests.

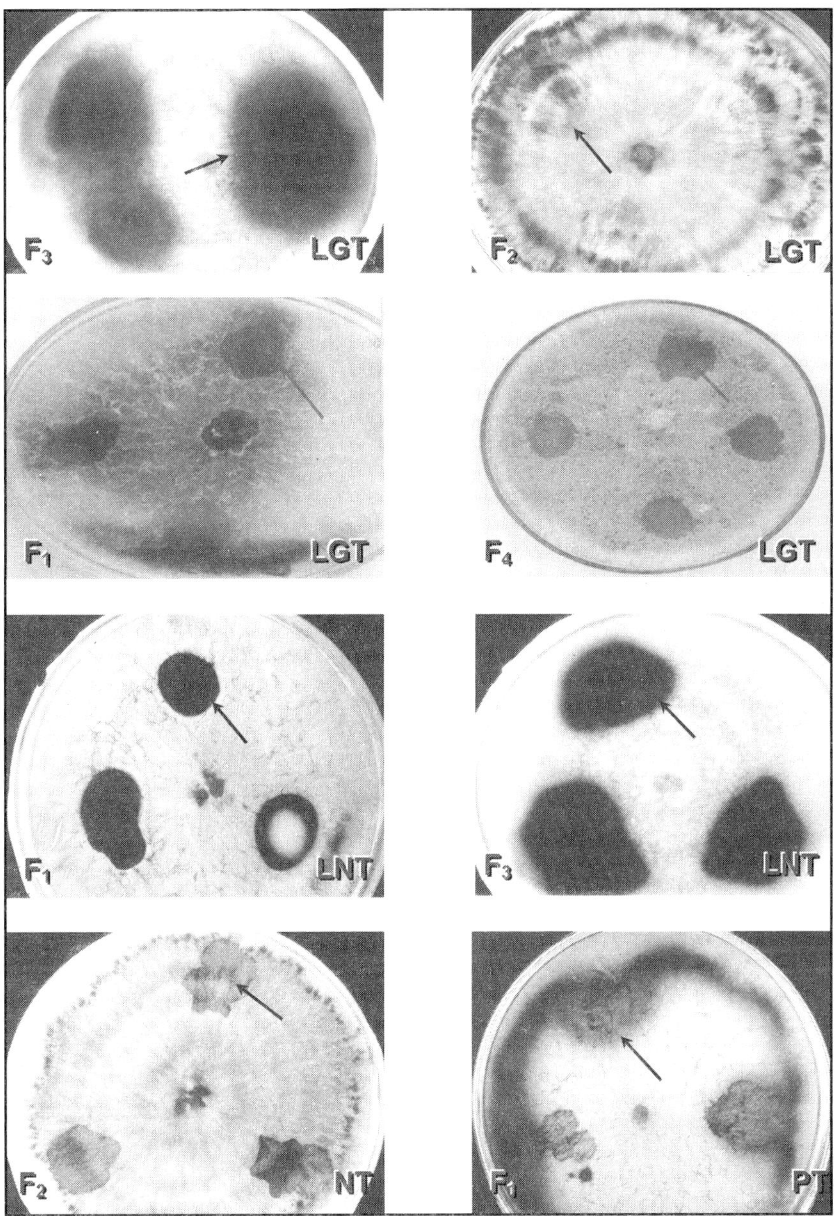


**Plate I: Representative plate tests by four selected fungal isolates from soil**

Row 1: F3 LGT: Laccase test with guaiacol by unidentified basidiomycete (isolate FT 65); F2 LGT: Laccase test with guaiacol by *Merulius (Syn. Phlebia) aureus* (isolate FT 43) ;

Row 2: F1 LGT: Laccase test with guaiacol by *Libertella betulina* (isolate FT 1); F4 LGT Laccase test with guaiacol by Fusarium sambucinum (isolate FT 66)

Row 3: F1 LNT: Laccase test with alpha-naphthol by *L. betulina* (isolate FT 1); F3 LNT: Laccase test with alpha-naphthol by unidentified basidiomycete

Row 4: F2 NT: Nobles test by *Merulius aureus* (isolate FT 43); F1 PT: peroxidase test by *L. betulina* (isolate FT 1)

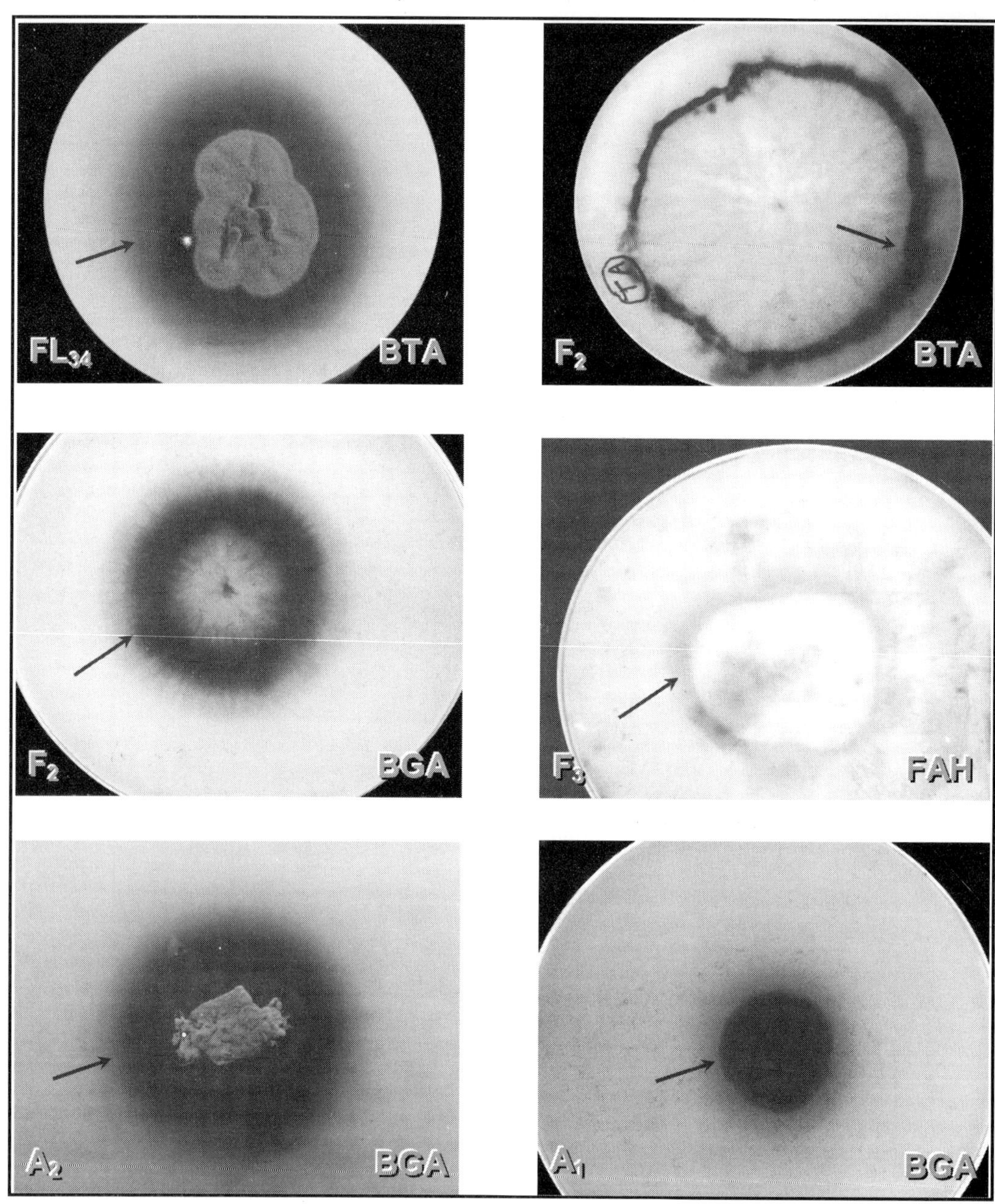


**Plate II: Plate tests with selected fungal isolate (contd)**

Row 1: Fl 34 BTA: Bavendum test by fungal isolate Fl 34 (the isolate was not selected for detailed study as it gave only 3 + plate tests (see Table 2); F2 BTA: Bavendum test with tannic acid by *M.aureus aureus* (isolate FT 43)

Row 2: F2 BGA: Bavendum test with gallic acid by *M. aureus* (isolate FT 43); F3 FAH: ferulic acid hydrolysis test by unidentified basidiomycete (isolate FT 65)

**Table 1: Number of positive plate tests given by various microorganisms isolated from the soils collected from the bank of Century Pulp and Paper Mill effluent discharge channel**

| *Microbial Group* | *Number of Species* | *Number of positive plate tests* |
|---|---|---|
| Fungi | 3 | 7 |
| | 1 | 6 |
| | 1 | 5 |
| | 12 | 4 |
| | 4 | 3 |
| | 6 | 2 |
| | 21 | 1 |
| | 18 | 0 |

**Note :** A total of 66 fungal as per the break-up given in the Table were subjected to the following plate tests.
**Laccase test :** (a) α-naphthol, (b) Guaiacol.
**Bavendamm test:** (c) Gallic acid, (d) Tannic acid,
(e) Peroxidase test, (f) Tyrosinase test, (g) Nobles test,
(h) Ferulic acid hydrolysis test, (i) Poly-R dye decolourization test.

**Table 2: Results of plate tests given by various fungal isolates from soils contaminated by Century Pulp and Paper Mill effluents**

| Sl. | | Plate tests | | | | | | | | |
|---|---|---|---|---|---|---|---|---|---|---|
| *No.* | *Fungal isolates* | *Laccase* | | *Peroxidase* | *Tyrosinase* | *Nobles* | *Ferulic acid hydrolysis* | *Poly-R dye decolourization* | *Bavendamm* | |
| | | **α-Naphthol** | **Guaiacol** | | | | | | **Gallic acid** | **Tannic acid** |
| 1. | FT 1 | + | + | + | - | + | + | - | + | + |
| 2. | FT 43 | + | + | + | - | + | + | - | + | + |
| 3. | FT 65 | + | + | + | - | + | + | - | + | + |
| 4. | FT 66 | + | + | + | - | + | - | - | + | + |
| 5. | FL 33 | + | + | + | + | - | - | - | - | + |
| 6. | FT 8 | + | + | - | + | + | - | - | - | - |
| 7. | FT 12 | + | + | + | - | - | - | - | + | - |
| 8. | FT 17 | + | + | + | + | - | - | - | - | - |
| 9. | FL 21 | + | + | + | + | - | - | - | - | - |
| 10. | FL 22 | + | + | + | - | + | - | - | - | - |
| 11. | FL 23 | + | + | + | + | - | - | - | - | - |
| 12. | FL 27 | + | + | + | + | - | - | - | - | - |
| 13. | FL 28 | + | + | + | - | + | - | - | - | - |
| 14. | FL 35 | + | + | + | + | - | - | - | - | - |
| 15. | FL 38 | + | + | + | - | - | - | - | - | + |
| 16. | FL 46 | + | + | - | + | - | - | - | - | + |
| 17. | FL 57 | + | + | - | + | + | - | - | - | - |
| 18. | FL 19 | + | + | - | + | - | - | - | - | - |
| 19. | FL 24 | + | + | - | + | - | - | - | - | - |
| 20. | FL 34 | - | - | + | - | - | - | - | + | + |
| 21. | FT 47 | - | - | + | - | + | - | - | - | + |
| 22. | *P. chrysosporium* | + | + | + | - | + | + | + | + | + |
| | | | | | | | | | + | + |
| | | | | | | | | | + | + |

+ Positive - Negative
* Only those isolates, which gave three or more positive tests, have been included.

## *Phanerochaete chrysosporium*

This species, in view of its well -established ligninolytic activity and positive response to most plate tests, was used as a positive control in all tests conducted. It gave the same 7 positive plate tests as in case of isolate FT 1, FT 43 and FT 65, and an additional poly-R dye deolorization test. Considering the results of the plate tests, the four fungal isolates (FT 1, FT 43, FT 65, and Ft 66, (labeled F 1, F 2, F3 and F4 in plates I to III) were further evaluated for their ligninolytic potential by radiorespirometric assay, as described below:

## $^{14}C$-(LIGNIN)-LIGNOCELLULOSE BIOTRANSFORMATION BY FUNGAL ISOLATES

Further characterization of lignin biodegradation of the 4 selected fungal isolates, which gave 6 or more + plate tests, was undertaken by testing their capacity to solubilize and mineralize $^{14}C$-(lignin)-lignocellulose prepared from wheat seedlings, and the results are presented in Fig. 1. Expressed as % of the radioactivity added per culture, maximum solubilization was caused by isolate FT 43 (13.47%), followed in decreasing order by *P. chrysoporium* (8.28), FT 66 (8.12%), FT 65 (7.88%) and, FT 1 (7.28%). Mineralization ($^{14}CO_2$ evolved, expressed as % of radioactivity added per culture) was highest for *P. chrysosporium* (3.35%), followed in decreasing order for FT 65 (2.99 %), FT 43 (2.02%), FT 1 (1.81%) and FT 66 (1.01 %), respectively. A more or less similar trend was observed when $^{14}CO_2$ evolution was expressed as per cent of radioactivity solubilized per culture at the end of incubation period of 24 days (Fig. 1).

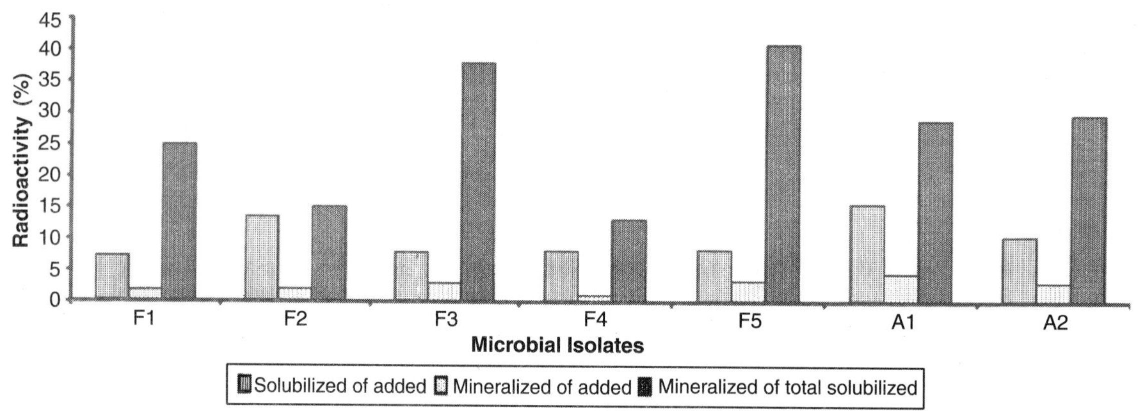


**Fig. 1:** 14C:-lignin solubilization and mineralization by agitated cultures of selected fungi

## TAXONOMIC IDENTIFICATION OF THE SELECTED FUNGAL ISOLATES

The four selected fungal isolates described above in respect of plate tests and $^{14}C$-(lignin)-lignocellulose solubilization and mineralization were subjected to their taxonomic status and the results in this respect are described below:

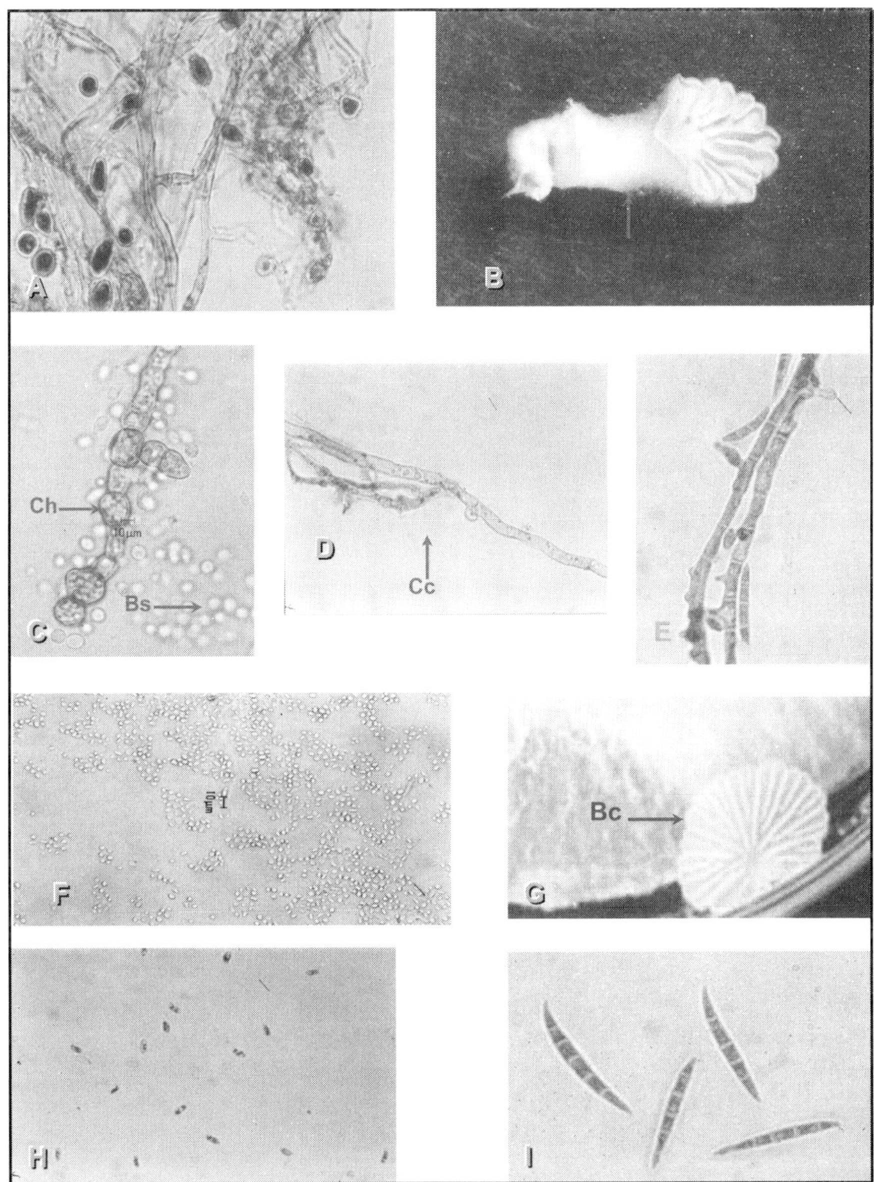


**Plate III: Microscopic study of selected fungal isolates**

**Row 1 (left to right):** A- *Merulius (syn. Phlebia) aureus* (FT 43), mycelium with thick-walled chlamydiosporesthe (B) same isolate showing basidiocarp (sporophore) with longitudinal folds (gills), and stalk (st)

**Row 2 (left to right):** Unidentified basidiomycere (isolate FT 65) – chlamydiopores (Ch) and basidiospores (Bs); (D) *Merulius aureus* (isolate FT 43) – mycelium with clamp connection (Cc); (E) *Fusarium sambucinum* (isolate FT66)-mycelium with fusiform conidiospores

**Row 3 (left to right):** (F) unidentified basidiomycete (isolate FT 65)—a bunch of oval to spherical basidiospores; (G) *Merulius aureus* (isolate FT 43)- a developing basidiocarp in petri plate culture

**Row 4 (left to right):** (H) Basidiospores of *M. aureus* (isolate FT 43) , and (I) *Fusarium sambucinum* ( isolate FT 66)-fusiform macro conidia, multi-celled, curved, apical cells narrowly-beaked and of variable size.

## Isolate FT1 (re-designated F1)

White coelomycetous fungus with acerbular conidiomata producing conidia in tendrils; conidiogenous cells holoblastic, sympodial, condia hyaline, aseptate, thin-walled, smooth, and straight to curved.

**Identification:** Culture identified by Institute of Microbial Technology, Chandigarh as *Libertella betulina* Desm. — An ascomycetous species (Family Xylariaceae (Accession No. MTCC 3787).

## Isolate FT 43 (re-designated F2)

Culture white to light grayish, hyphae branched, septate, multinucleate; clamp connections present; developed fruiting bodies (sporophores) on rice straw, sporophore buff-yellow, size of sporophores (cm) 0.5-4.0 cm, 0.5-3.0 cm, 0.03-0.06 cm; longitudinal folds more prominent than transverse folds giving appearance of gills; basidium clavate, size 10-18 × 2.5-4 microns; basidiospores oval to cylindrical, size 2.5-4.6 × 1.5-1.8 microns. Plate III, B, D, G. H).

**Identification:** *Merulius* (Syn. *Phlebia*) *aureus* (A hymenomyceotus Basidiomycete, family Polyporaceae (according to identification keys of (Bakshi1971; Gilbertson and Ryvarden, 1986).

## Isolate 65 (re-designated FT3)

Culture white, hyphae hyaline, branched, septate, multinucleate, clamp connections present; intercalary, thick-walled chlamydospores; the isolate when inoculated aseptically on rice straw developed profuse mycelia growth but failed to fructify (Plate III).

**Identification:** A basidiomycetous fungus, genus and species not identified.

## Isolate FT 66 (re-designated F4)

Hyphomycetous fungus with purple to slightly reddish culture, macro-conidia fusiform, curved, 3-6 celled, with apical cell narrowly beaked and of variable size.

**Identification:** Identified by IMTECH, Chandigarh as *Fusarium sambucinum* Fuckel, and culture deposited (accession no. MTCC 3788).

## DISCUSSION

Lignocellulose biotransformation in the soil plays an important role in the formation of humus that maintains the soil fertility and release of organically bound carbon, which is an important step in global carbon cycling in nature. However, our knowledge regarding microorganisms involved in these processes has remained largely static chiefly because of difficulties encountered in isolating ligninolytic fungi, particularly basidiomycetous species from soils. The traditional soil dilution and plating method is largely ineffective in isolation of lignilolytic species of basidiomycetes, because the spores of more abundant moulds and ascomycetes present in soils outnumber their propagules.

Several techniques have been applied in the past to overcome this problem, such as soil plate method (Warcup, 1990), particle washing method (Chesters, 1949; Gams and Domsch, 1967, Bääth, 1988), and selective indicator medium method (Thorn *et al.*, 1996). Falcon *et al.,* (1995) used Indulin AT (a commercial lignin containing >99.5 % lignin) for pre-incubation of soil samples followed by preliminary screening with chromogenic plate tests for isolation of ligninolytic microorganisms. In the present study, a modification of this method was used to isolate ligninolytic fungi. The procedure used in the present study took advantage of the fact that soils exposed to lignin rich substrates (paper mill effluents in this case) are likely to get preferentially enriched with species using lignin or its degradation products as carbon and energy source. Additionally, we incubated the effluent polluted soil with Indulin AT for over 3 months for allowing further enrichment. In the next step, the soil dilutes were plated on the medium provided with Indulin AT as sole carbon source. Thus, a step-wise enrichment of species capable of using lignin as carbon source was achieved. The pure isolates were further screened for their ligninolytic potential using specific plate tests such as Bavendum and Nobles' tests. These tests are based on the reaction of specific substrates with the extracellular peroxidases produced only by the ligninolytic organisms, to produce visually identifiable colour products.

Confirmation of the ligninolytic activity of the four isolates was achieved by capability of the isolates to solubilize and mineralize $^{14}$C-(lignin)-lignocellulose, specifically labelled in lignin moiety. To ensure the specificity of $^{14}$C-labelling in lignin moiety, the $^{14}$C-lignocellulose preparation was treated with proteinase enzyme followed by treatment with cellulase enzyme, to remove any $^{14}$C-label incorporated in protein and cellulose fractions. There was a good agreement between the number of positive plate tests and $^{14}$C-(lignin)-lignocellulose mineralization [($^{14}CO_2$ produced during 24-day incubation) (Fig. 1)]. The two basidiomycetous isolates (*Merulius* and the unidentified species) showed 13.47 and 7.88% solubilization (radioactivity as $^{14}CO_{2}$ + radioactivity in the culture filtrate) in 24 days compared to the 8.28% by *P. chrysosporium* during the same period. Mineralization ($^{14}CO_2$ evolution from $^{14}$C-lignin in 24 days) rates by these two basidiomycetous isolates were more or less similar to that in case of *P. chrysosporium. Libertella betulina* (the ascomycetous isolate) and *Fusarium sambucinum* (The deuteromycetous isolate) showed lower rates of mineralization (1.81 and 1.0 2%, respectively) compared to the basidiomycetous isolates (2.02 and 2.99%, respectively) or *P. chrysosporium* (3.35%) The latter, used as a positive control, showed highest rates of solubilization and mineralization in most cases. The relative rates of solubilization and mineralization observed herein for the four fungal isolates, belonging to three different classes of fungi, *vis a vis*, *P. chrysosporium*, are in conformity with their reported ligninolytic activity (Kirk and Farrell, 1987; Kuhad *et al.,* 1997). It is also noteworthy that the four fungal species found to possess higher number of + plate tests and lignocelluloses degradation were derived from soil samples enriched with Indulin AT, suggesting the latter to be a good lignin substrate for enrichment of soils for ligninolytic microorganism. Black liquor, on the other hand, was found to be relatively poor substrate for this purpose (Falcon, *et al.,* 1995).

It may also be mentioned that fungal isolates from soil reported herein have been found earlier also to possess ligninolytic activity in studies involving wood decay or polymeric lignin model compounds. Species of *Merulius* (Syn. *Phlebia*) have been found to cause white-rot decay of wood (Bursdall, 1998), and produce manganese-dependent peroxidase and lignin peroxidase in liquid cultures and solid-state fermentation of lignocelluloses (Jhao *et al.,* 1996; Lundell and Hatakka, 1994). *Libertella betulina* has been reported amongst one of the few ascomycetous fungi (along with *Xylaria* and *Hypoxylon*), which cause white-rot decay of wood (Kirk and Farrell, 1987). Species of *Fusarium* have also been reported to mineralize $^{14}$C-DHP and $^{14}$C-aromatic aromatic acids to $^{14}CO_2$, though to limited extent (Norris, 1980; Sutherland *et al.,* 1983). The presence of two basidiomyctes, one ascomycte and one deuteromycete in the soil as found in our work suggests their involvement in lignocellulose biotransformation in soil ecosystems. Further study is required to identify the sequence and mechanism of their involvement. However, from our knowledge of their relative ligninolytic potential, it may be summarised that a sequential biodegradation in which basidiomycetous fungi first cause native or high molecular weight lignin followed by attack of ascomycetous species and *Fusarium* on low molecular weight lignols, as possible mechanism.

India, being a predominantly agricultural country with varied climatic regimes, has many soils, which are likely to harbour potential lignocelluloses transforming microbial species. In previous studies we reported isolation of potential ligninolytic bacteria and actinomycetes (Kumar *et al.,* 2001; Shah, 1998), using a similar approach as reported herein. Further screening of ligninolytic fungi and other microorganisms, using the approach described in this study, may lead to isolation of new microbial biota from Indian soils, which could have potential applications in lignocelluloses biotechnology.

## ACKNOWLEDGEMENTS

The award of Dr. K.S. Krishnan Fellowship to the first author by B.R.N.S., Department of Atomic Energy, Government of India, during the course of this study, is gratefully acknowledged. The authors are also grateful to G.B. Pant University of Agriculture & Technology, Pantnagar for providing facilities and encouragement, and to Institute of Microbial Technology, Chandigarh, for help in taxonomic identification of some isolates.

## REFERENCES

Bääth, E. (1988). Critical examination of soil washing technique with special reference to size of soil particles. *Canadian Journal of Botany,* 66, 1566-1569.

Bakshi, B.K. (1971). Indian Polyporaceae. *Indian Council of Agricultural Research,* New Delhi, India, Pp 246.

Bassham, A. (1975). Cellulose as a chemical and energy resource. *Biotechnology Bioengineering Symposium,* 5, 9-19.

Boominathan, K. and Reddy, C.A. (1992). Fungal degradation of lignin: Biotechnological applications. In: *Handbook of applied mycology.* Eds. Arora, D.K., Elander, R.P. and Mukerji, K.G., Marcel Decker Inc. New York, 4, Pp 763-822.

Brauns, F. E. and Brauns, D.A. (1960). *The chemistry of lignin-* Supplementary volume, Academic Press, New York.

Bray, G.A. (1960). A simplified efficient liquid scintillator for counting aqueous solutions in a liquid scintillator counter. *Ana. Biochem,* 1, 279-285.

Bumpus, J.A. and Aust, S.D. (1987). Bioremediation of environmental pollutants by the white-rot fungus *Phanerochaete chrysosporium*: A ligninolytic fungus. *Enzyme and Microbial Technology,* 7, 194-196.

Bursdell, Jr., H.H. (1998). Taxonomy of industrially important white-rot fungi. In: *Environmentally friendly technologies for the pulp and paper industry.* Eds Young, R.A. and Akhtar, M. John Wiley & Sons Inc., Pp 250-272.

Buswell, J.A. (1991). Fungal degradation of lignin. In: *Handbook of applied mycology.* Eds Arora, D.K., Rai, B., Mukerji, K.G. and Knudsen, G.R., Marcel Decker Inc., New York, 1, Pp 425-480.

Chesters, C.G.C. (1949). Concerning fungi isolated from soil. *Trans. Br. Myco. Soc.* 32, 197-216.

Crawford, D.L. (1981). *Lignin biodegradation and transformation.* Wiley Inter Science Publications, New York, Pp.154.

Crawford, R. L and Crawford, D.L. (1978). Radio-isotopic methods for the study of lignin biodegradation. *Dev. Indus. Microbio,* 19, 55-65.

Degenhart, H.J. (1974). Isotope techniques- Use of radioactive nucleides in the steroid assays. In. *Clinical Biochemistry.* Eds Curtius, H.C. and Roth, M., Walter de Gruyter Publishers, 1, Pp. 686-697.

Falcon, M.A., Rodriguez, A., Carnicero, A., Regaldo, V., Prestelo, F., Milstein, O. and Fuente, G.D.L. (1995). Isolation of microorganisms with lignin transformation potential from soils of Tenerife Island. *Soil Biol Biochem,* 27, 721-726.

Fermor, T.R. (1993). Applied aspect of composting and bioconversion of lignocellulosic materials. *Internatl. Biodet.Biodegrad,* 31, 87-106.

Gilbertson, R.L. and Ryvarden, L. (1986). *Fungi flora.* Oslo, iv+1.

Gams, W. and Domsch, K.H. (1967). Contributions to the soil washing technique for the isolation of soil fungi. *Arch. Microbiol,* 58, 134-144.

Ishaq, M. and Chahal, D.S. (1991). In: *Feed, food, and fuels from biomass,* Ed Chahal, D.S. Oxford, IBH Publish. Co., New Delhi, India, Pp 15-26.

Janshekar, H., Haltmeir, T. and Brown, C. (1982). Fungal degradation of pine and straw alkali lignins. *Eur. J. Appl. Micro. Biotech,* 14, 174-181.

Jeffries, T.W. (1994). Biodegradation of lignin and hemicelluloses. In: Biochemistry of microbial degradations. Eds Retledge, C., Kluwer Academy Press, The Netherlands, Pp 231-277.

Jhao, J., de Koker, T.H. and Janse, B.J.H. (1996). Comparative studies of lignin peroxidase and manganese-dependent peroxidases produced by selected white-rot fungi in solid media. *FEMS Microbiology Letters,* 145, 393-399.

Kamra, D. and Zadrazil, F. (1988). Microbial improvement of lignocellulose in animal feed production- a review. In: *Treatment of lignocelluloses with white-rot fungi.* Eds Zadrazil, F. and Reiniger, P. Elsevier Applied Science Publishers, London, Pp 56-63.

Kirk, T.K. and Farrell, R.L. (1987). Enzymatic "combustion": The microbial degradation of lignin. *Ann. Rev. Microbiol,* 41, 465-505.

Kirk, T.K., Schultz, E., Connor, W.J., Lorenz, L.F.and Zeikus, G.J. (1978). Influence of culture parameters on lignin metabolism by *Phanerochaete chrysosporium. Arch. Microbiol,* 117, 277-285.

Kuhad, R.C. and Singh, A. (1993) Lignocellulose biotechnology. *Critical Review of Biotechnology,* 13, 151-172.

Kuhad, R.C., Singh, A., and Eriksson, K.-E. L. (1997). Microorganisms and enzymes involved in the degradation of plant fiber cell walls. In K. E. Eriksson (ed) *Advances in Biochemical Engineering Biotechnology,* Springer-Verlag, Germany. 46-125.

Kumar, L, Rathore, V.S. and Srivastava, H.S. (2001). $^{14}$C-[lignin]-lignocellulose biodegradation by bacteria isolated from polluted soil. *Indian Journal of Experimental Biology,* 39, 584-589.

Lundell, T. and Hatakka, A. (1994). Participation of Mn (II) in catalysis of Manganese peroxidase and lignin peroxidase from *Phlebia radiata. FEBS Letters,* 61, 3515-3520.

McCarthy, A.J., MacDonald, M.J., Peterson, A. and Broda, P. (1984). Degradation of $^{14}$C- lignin labeled wheat lignocellulose by white-rot fungi. *Journal of General Microbiology,* 130, 2050-2058.

Nobles, M.K. (1958). A rapid test for extracellular oxidase in cultures of wood- inhabitating hymenomycetes. *Canadian Journal of Botany,* 36, 91-99.

Norris, D.M. (1980). Degradation of $^{14}$C-labelled lignin and $^{14}$C-labelled aromatic acids by *Fusarium solani. Applied Environmental Micobiology,* 40, 376-380.

Pointing, S.B. (2001). Feasibility of bioremediation by white- rot fungi. *Applied Microbiology Biotechnology,* 57, 20-33.

Rayner, A.D.M., and Boddy, L. (1988). *Fungal decomposition of wood: its biology and ecology.* Wiley Publications, New York.

Reid, I.D. (1979). The influence of nutrient balance on lignin degradation by white-rot fungus *Phanerochaete chrysosporium. Canadian Journal of Botany,* 57, 2050-2058.

Shah, S. (1998). Studies on factors affecting lignin biodegradation by actinomycetes (*Streptomyces* species). *M.Sc.Thesis,* G.B.Pant Univ. Agric. Tech., Pantnagar, India, Pp119.

Sutherland, J.B., Pometto III, A.L.and Crawford, D.L. (1983). Lignocellulose degradation by *Fusarium* species. *Canadian Journal of Botany,* 61, 1194-1198.

Thorn, R.G., Reddy, A.C., Harris, D.and Paul, E.A. (1996). Isolation of saprophytic basidiomycetes from soil. *Applied Environmental and Microbiology,* 62, 4288-4292.

Warcup, J.H. (1990). The soil plate method for isolation of fungi from soil. *Nature* (London) 146, Pp 70.

Wood, D.A. (1985). Useful biodegradation of lignocellulose. *Ann. Proc. Phytochem. Soc. Euro.* , Clerendon Press, Oxford, U.K. 26, 1295-1309.

# Biodiversity of Hemicellulose Degrading Microorganisms and their Enzymes

**Bindu Battan[1], Sarika Kuhar[2], Jitender Sharma[1], R.C. Kuhad[1,2]**

[1]*Department of Biotechnology, Kurukshetra University, Kurukshetra-136119, Haryana, India*
[2]*Department of Microbiology, University of Delhi, South Campus, New Delhi-110021, India*

Hemicellulose is the most abundant renewable resource which in most of the cases remains unutilized. Many microorganisms exhibit a complex system of enzymes involved in the degradation of hemicellulose. The heterogenous structure of xylan and mannan, the main components of hemicellulose, is responsible for diversity of these enzymes. Hemicellulases have been reported mainly from bacteria, fungi, actinomycetes and yeast. Hydrolysis of xylan to its monomers requires synergistic action of different enzymes: endo-1,4-β-xylanase (endoxylanase), β-xylosidase, α-arabinofuranosidase, α-glucuronidase and acetyl esterases. Multiple xylanases have been reported in numerous microorganisms. Likewise, for complete hydrolysis of O-acetylgalactoglucomannan, endomannanases, β-mannosidase, β-glucosidase, α-galactosidase and acetylmannan esterases are required. This article reviews diversity of hemicellulose degrading microorganisms and their hemicellulolytic enzymes with special emphasis on the multiplicity of the respective enzymes. In the end, it also briefs the future prospects of hemicellulases.

## INTRODUCTION

Hemicellulose is second to cellulose in abundance on earth and represents a major source of renewable organic matter. Hemicellulose is structurally more complex than cellulose, which contains only 1,4-β–glycosidic linkages. In plants, hemicelluloses are situated between lignin and the collection of cellulose fibres underneath, thus the interactions between these components increases the rigidity and flexibility of fibre cell walls. Hemicelluloses are a group of homo- and heteropolymers consisting largely of anhydro-β–(1→4)-D-xylopyranose, mannopyranose, glucopyranose and galactopyranose main chains with a number of substituents. The major hemicellulose components in softwood are mannan-based, and those in hardwood xylan-based. Hemicelluloses constitute on an average about 26% of hardwood, 22% of softwood, and 30% of various agricultural residues (Brasch, 1956; Rydholm, 1965; Knull *et al.,*1980; Gordon *et al.,*1983 ). Most of the hemicelluloses are built up by β–1,4-linkages between their backbone sugars apart from the galactose-based hemicelluloses, which are characterized by β–1,3-linkages. They usually have degree of polymerization of 100 to 200. Like cellulose they are stereoregular but they differ from cellulose in that they have side groups consisting of sugars, sugar acids, and acetyl esters. Acetyl ester groups are bound to the linear polymer chains of various plant hemicellulose polysaccharides, and the acetyl content of cell walls of higher plants can attain upto 2% of the plant dry mass (Bacon *et al.,* 1975). The degree of substitution with acetyl groups in xylan of hardwoods from temperate zones was found to vary slightly from 1:0.5 to 1:0.6 (xylosyl:acetyl molecules), and in some instances, both the C2 and the C3 positions can be acetylated. Ferulic acid (4-hydroxy-3-methoxy-cinnamic acid) and p-coumaric acid (4-hydroxy-cinnamic acid) are the oxidized forms of the corresponding coniferyl and p-coumaryl alcohols. They are common substituents in the cell walls of several plant families (Mueller-Harvey *et al.,* 1986; Kato *et al.,* 1987; Molgaard *et al.,* 1988) representing upto 2.5% of the graminaceous cell walls (Jung *et al.,* 1989). Diferulic acid may crosslink hemicelluloses (Markwalder and Newkom, 1976) and the formation of diferuloyl units possibly occurs by oxidative coupling of feruloyl groups attached to adjacent xylan chains in the presence of peroxidase and hydrogen peroxide (Hartley and Jones, 1977; Markwalder and Newkom, 1976; Hartley *et al.,* 1990). Christov and Prior (1993) suggested that ferulic acid ethers might form crosslinks between lignin and hemicelluloses by simultaneous esterification of their carboxyl groups to arabinose substituents of wheat straw arabinoglucuronoxylans. These substituents of hemicelluloses render their structure non-crystalline or only poorly crystalline, so that they exist more as a gel than as oriented fibres. It has been suggested that the hemicelluloses in association with the cellulose influence the organization of the lignin (Atalla, 1995; personal communication). O-acetylglucuronoxylan is the major hemicellulose in hardwood making up 15-30% of its weight while O-acetyl galactoglucomannan is the major hemicellulose in gymnosperm wood or softwood comprising about 20% of the wood weight. Hardwoods contain about 3-5% of a β–1,4-linked glucomannan whereas conifer wood also contains 5-10% of a nonacetylated arabinoglucuronoxylan. The structures of these are well described in detail by Timmel (1967). Besides wood, hemicelluloses are also found in grasses, cereals, and some very primitive plants (Whistler *et al.,* 1970). As there is more complexity in the structure of hemicellulose than cellulose, the enzymes that degrade hemicellulose are similarly complex.

The hemicellulolytic enzyme systems of a variety of microorganisms have been extensively investigated and several exhaustive reviews have appeared (Biely, 1985; Dekker, 1989; Wong *et al.,* 1988; Senior *et al.,* 1989; Eriksson *et al.,*1990; Bastawde, 1992; Kuhad *et al.,* 1997). Hemicellulases have been reported mainly from bacteria (Gilbert and Hazlewood, 1993; Sunna and Antranikian, 1997), fungi (Sunna and Antranikian, 1997), actinomycetes (Ball and McCarthy, 1989; Beg *et al.,* 2000a), and yeast (Hrmova *et al.,* 1984; Liu *et al.,* 1999).

## HEMICELLULOSE COMPONENTS

Xylan is a major component found in the hemicellulose fractions of hardwood plant tissues after cellulose (Whistler and Richards, 1970; Biely, 1985) and is located predominantly in the secondary cell walls of angiosperms and gymnosperms (Timmel, 1967; Blanchette *et al.,* 1988; Blanchette *et al.,* 1989). Aside from terrestrial plants, in which xylans are based on a β–1,4-linked

D-xylosyl backbone, marine algae also synthesize xylans of different chemical structures based on a β–1,3-linked D-xylosyl backbone. In some species of Chlorophyceae and Rhodophyceae when cellulose is absent, xylans form a highly crystalline fibrillar material (Bajpai, 1997). The composition and structure of xylan is complicated and can vary quantitatively and qualitatively in various woody plant species (Timell, 1964; Whistler and Richards, 1970), grasses and cereals. It consists of a homopolymeric backbone of 1,4-linked β–D-xylopyranose units. The side chain groups in xylan differ depending on the plant origin. The backbone can be substituted at the C2 and C3 positions with acetic acid (at the C2 and C3 positions in hardwoods and annual plants), 4-O-methylglucuronic acid (at the C2 position in hardwoods, softwoods and annual plants), and arabinose (at the C3 position in softwoods and annual plants). The arabinosyl residues may be further esterified by feruloyl and coumaroyl residues, crosslinking hemicellulose to lignin in the cell wall matrix (Smith and Hartley, 1983; Kato and Nevins, 1985). The degree of branching varies depending on the source (Bajpai, 1997). Thus, complete hydrolysis of xylan requires the interaction of a number of xylanolytic enzymes, produced by many microorganisms that have the ability to cleave main chains and side chains.

The mannan hemicelluloses, galactoglucomannans and glucomannans in softwoods and hardwoods, are both branched heteropolysaccharides. Their backbones are built up of 1,4-linked β–D-glucopyranose and β–D-mannopyranose units. The mannose and glucose units in the backbone are partially substituted in C-2 and C-3 positions by acetyl groups, approximately 1 per 3-4 hexose units (Eriksson *et al.,* 1990; Coughlan and Hazlewood, 1993). The glucose:mannose ratio is reported to vary from 1:1 to 1:2 depending on the wood species (Eriksson *et al.,*1990; Coughlan and Hazlewood, 1993).

## MICROBIAL HEMICELLULOLYTIC SYSTEM

Numerous articles related to the production and various properties of hemicellulases, especially on xylanases, have been published (Biely, 1985; Viikari *et al.,* 1993; Wong *et al.,* 1988). Recent reviews regarding the types, structure, function, classification of microbial hemicellulolytic enzymes have been presented by Rabinovich *et al.* (2002a) and Shallom and Shoham (2003) [Table 1]. Hemicellulases are produced by fungi, bacteria from terrestrial and marine environments, rumen microorganisms (protozoa, fungi and bacteria), yeasts, marine algae, snails, crustaceans, insects and seeds of terrestrial plants (Sunna and Antranikian, 1997; Kuhad *et al.,* 1997). Also xylanase production by plants and higher animals has also been reported (Yamura, 1997). Fungi, yeasts, and some bacteria secrete hemicellulases extracellularly, although cell wall-bound and intracellular hemicellulolytic enzymes have also been reported (Dekker, 1985; Dekker, 1976). For the complete degradation of the hemicellulosic structure, a complement of some 24 enzymes is required (Srinivasan, 1992). The hemicellulolytic enzymes are typically produced as multiple isoenzymes (enzymes with apparently the same specific catalytic function existing in more than one physical form that differ from each other in some properties as optimum pH, temperature); some have cellulose-binding domains, some are part of a multi-domain protein in which other domains degrade cellulose while, in some organisms, xylanase production is more strongly increased by the presence of cellulose than of xylan. More information is available regarding the enzymes degrading xylan, (which constitutes the major proportion of the hemicellulose) than of mannanase (Moracci *et al.,* 2000). Based on the amino acid or nucleic acid sequence of their catalytic modules, they are classified into two groups, glycosidic hydrolases (GHs) which hydrolyze glycosidic bonds, or carbohydrate esterases (CEs) which hydrolyze ester linkages of acetate or ferulic acid side groups and according to their primary sequence homology, they have been grouped into various families (Henrissat and Bairoch, 1996; Rabinovich *et al.,* 2002a,b).

**Table 1. The major hemicellulases and their classification (Shallom and Shoham, 2003)**

| *Enzymes* | *Substrates* | *EC number* | *Family* |
|---|---|---|---|
| Endo-β-1,4 xylanase | β-1,4 xylan | 3.2.1.8 | GH 5<br>8<br>10<br>11<br>43 |
| Exo-β-1,4 xylosidase | β-1,4-Xylooligomers xylobiose | 3.2.1.37 | GH 3<br>39<br>43<br>52<br>54 |
| Endo-β-1,4-mannanase | β-1,4 mannan | 3.2.1.78 | GH 5<br>26 |
| Exo-β-1,4-mannosidase | β-1,4-mannooligomers mannobiose | 3.2.1.25 | GH 1<br>2<br>5 |
| Endo-α-1,5 arabinanase | α-1,5-arabinan | 3.2.1.99 | GH 43 |
| α-L-arabinofuranosidases | α-1,2-arabinofuranosyl (1→2) or (1→3) xylooligomers α-1,5 arabinan | 3.2.1.55 | GH 3<br>43<br>51<br>54<br>62 |
| α-glucuronidase | 4-O-methyl-α-glucuronic acid (1→2) xylooligomers | 3.2.1.139 | GH 67 |
| α-galactosidase | α-galactopyranose (1→6) mannooligomers | 3.2.1.22 | GH 4<br>27<br>36<br>57 |
| Endo-galactanase | β-1,4-galactan | 3.2.1.89 | GH 53 |
| β-glucosidase | β-glucopyranose (1→4) mannopyranose | 3.2.1.21 | GH 1<br>3 |
| Acetyl xylan esterases | 2-or 3-O-acetyl xylan | 3.1.1.72 | CE 1<br>2<br>3<br>4<br>5<br>6 |
| Acetyl mannan esterase | 2-or 3-O-acetyl mannan | 3.1.1.6 | CE 1 |
| Ferulic and *p*-coumaric acid esterases | 2-or 3-O-acetyl mannan | 3.2.1.73 | CE 1 |

## XYLAN HYDROLYTIC ENZYMES

Hydrolysis of O-acetylglucuronoxylan requires four different enzymes: endo-1,4-β–xylanase (endoxylanase), β–xylosidase, α–glucuronidase and acetyl esterase. Softwood xylan degradation requires a fifth enzyme, α–arabinofuranosidase, but not the esterase. Through the cooperative interactions of the debranching enzymes with the xylanases, the constituent sugars of xylan will be produced. Because of this heterogeneity of xylan, there is synergistic action by multiple xylanases. Some hydrolyze the backbone xylan chain near substituents much more efficiently than others. The accessibility of some linkages changes during the course of hydrolysis which necessitate the production of an array of enzymes with specialized functions. Also, α–glucuronidases, α–arabinosidases and acetyl esterases differ in specificity in respect to neighbouring substituents and xylan chain length. The efficient hydrolysis of the xylan substrate seems to involve not only these five types of xylanase but also probably multiple isoenzymes of several of them (Tenkanen *et al.,* 1995). Thus, the production of multiple xylanases is probably one strategy that a microorganism uses to achieve efficient substrate hydrolysis (Wong *et al.,* 1988). Multiple xylanases have been reported in numerous microorganisms (Wong and Saddler, 1988; Kuhad *et al.,* 1997). An array of endoxylanases with different specificities have been found in single fungi (Coughlan *et al.,* 1993) clearly reflecting the complexity of the substrate xylan (Wong *et al.,* 1988). Several factors could be responsible for the multiplicity of xylanases. These include differential RNA processing, post-secretional modification such as glycosylation and autoaggregation (Biely, 1985; Coughlan *et al.,* 1993). Multiple xylanases can also be the product of different alleles of the same gene (Wong *et al.,* 1988). However, some of the multiple xylanases are the result of independent genes (Hazlewood and Gilbert, 1993).

## ENDOXYLANASES (β–1,4-D-Xylanohydrolase, EC 3.2.1.8)

Several enzymes with different functions participate in the hydrolysis of xylan to its monomers, of which the most important are the endo-β–1,4-xylanases (Sunna and Antranikian, 1997). These enzymes degrade xylan to short-chain xylo-oligosaccharides of varying lengths. They can be grouped into two families F and G based on hydrophobic cluster analysis and sequence homology, corresponding to families 10 and 11 in the numerical classification of glycosyl hydrolases. These are the only hemicellulases that have been well characterized (Kuhad *et al.,* 1997). Endoxylanases of fungi apparently vary widely in molecular weight from 16 to 75 kDa and also vary widely in pI (Coughlan *et al.,* 1993) and they do not bind strongly to xylan, mannan or cellulose (Tenkanen, 1995). Some bacterial endoxylanases possess cellulose-binding domain, the function of which is not known, the binding domain does not bind to xylan, nor does it enhance xylan hydrolysis (Coughlan *et al.,* 1993). Endoxylanases are principally of two types, non-debranching or arabinose nonliberating and debranching or arabinose liberating (Dekker and Richards, 1976). Non-arabinose liberating endoxylanases fall into two categories, one releasing xylose and xylobiose as end products and the other giving xylo-oligosaccharides as end products. The former cannot act on L-arabinosyl initiated branch points as β–(1→4) linkages and produce only xylobiose and xylose as the major end products. The latter cannot cleave branch points at α–(1→2) and α–(1→3) and produce mainly xylo-oligosaccharides larger than xylobiose. These endoxylanases have no action on xylotriose and xylobiose. Likewise, the arabinose liberating endoxylanases can be characterized into the one, that can hydrolyze branch points and produce intermediate-sized xylo-oligosaccharides and arabinose as the end products and the other that can cleave the xylan chain and the branch points, releasing mainly xylobiose, xylose and arabinose. Endoxylanases act with diverse modes of action in the presence or absence of number of substituents and location of active sites, which play an important role in the liberating xylo-oligosaccharides. In general, polymeric xylans are cleaved only at unsubstituted regions. In such cases, however it is not clear whether the inhibition of xylanase action by the substituents results from gross stearic hindrance or mere failure to accommodate at the active sites (Coughlan and Hazlewood, 1993). However Frederick *et al.* (1985) reported that some xylanases cleave the main chain only in the vicinity of the substituents.

## β–XYLOSIDASE (β–Xyloside xylohydrolase EC.3.2.1.37)

Other debranching enzymes of xylan which belong to the family hydrolases include β–xylosidase, that hydrolyze short xylo-oligosaccharides and xylobiose from the non-reducing end in an exo-manner to liberate xylose (Wong *et al.,* 1988). β–xylosidases are distributed in families 3,39,43,52 and 54. Bifunctional xylosidase-arabinosidase (Lee *et al.,* 2003) enzymes are found mainly in families 3,43 and 54. These are larger enzymes with molecular weight between 60 and 360 kDa and may be monomeric or dimeric proteins. The affinity of the enzyme towards xylo-oligosaccharides decrease with increasing degree of polymerization. Many β–xylosidases are coupled with transferase activity, helpful for the organism to regulate the xylanolytic systems (Conrad and Noethen, 1984). True β–xylosidases are able to cleave artificial substrate like p-nitrophenyl β–D-xyloside (Coughlan and Hazlewood, 1993).

## α–L-ARABINOFURANOSIDASE (EC 3.2.1.55)

α–L-Arabinofuranosidases, cleave the side chains of xylan and also fall into the group hydrolases. α–L-arabinofuranosidases hydrolyze the terminal, non-reducing α–L-arabinofuranosyl groups of arabinans, arabinoxylans and arabinogalactans. There are two types of arabinases, the enacting α–L-arabinofuranosidase (EC 3.2.1.55), which is active against p-nitrophenyl-arabinofuranosides and on branched arabinans, and the endo-1,5 α–L-arabinan (EC 3.2.1.99), which is active only towards linear arabinans (Vander Veen *et al.,* 1991). Most of the arabinan degrading enzymes studied so far are of the exo-type (Dekker and Richards, 1976). Arabinosidases and α–L-arabinans are distributed in GH families 3,43,51,54 and 62. Some of these enzymes exhibit broad substrate specificity, acting on arabinofuranoside moieties at O-5, O-2 or O-3 bonds as a single substituent, as well as from O-2 and O-3 doubly substituted xylans, xylooligomers and arabinans (Saha, 2000). The size of native arabinofuranosidases may reach upto 495 kDa and found in mono-, di-, tetra-, hexa- and octameric forms (Coughlan and Hazlewood, 1993). They have been studied because of their use in dehazing fruit juices and improving animal feed digestibility. These enzymes vary in their ability to hydrolyze off the arabinose residues in polymeric versus low molecular weight substrates that are produced by endoxylanases.

## GLUCURONIDASES (α–4-O-methyl-glucuronidase; EC 3.2.1.1)

It acts synergistically with endoxylanase and hydrolyze the α–1,2-glycosidic bond of the 4-O-methyl-D-glucuronic acid sidechains of xylan (Coughlan and Hazlewood, 1993; Kuhad *et al.,* 1997). They are found in family 67. Their molecular weight is less than 100 kDa and their substrate specificity differs with its source. 4-O-methyl-glucuronic acid linkage forms a barrier in wood degradation similar to the carbohydrate-lignin complex. Due to the lack of glucuronidase activity in many fungal hemicellulale preparations (Puls and Poutanen, 1989), this enzyme was not described until 1986 (Puls *et al.,* 1987).

## ESTERASES (EC 3.1.1.6)

The esterases in general are low molecular weight enzymes, which enables these enzymes to penetrate the lignocellulose more effectively. Esterases are required to remove the bound acetic and phenolic acids for the complete hydrolysis of natural glucuronoxylans. These enzymes belong to the second group, that rupture the ester linkages between xylose units of the xylan polymer and acetic acid (acetyl xylan esterase, EC3.1.1.6) or between arabinose side chain residues and phenolic acids such as ferulic acid (feruloyl esterase) and *p*-coumaric acid (*p*-coumaryl esterase). The existence of acetyl-esterases in unpurified enzyme preparations was demonstrated by Frohwein *et al.* (1963). The presence of ferulic acid esterase activity in hemicellulase preparations and its coproduction with other xylanolytic enzymes was first described by MacKenzie *et al.* (1987). They suggested that phenolic acid esterases such as ferulic acid esterase and p-coumaric acid esterase perform a similar function to alkali in the deesterification of plant tissues. They aid the release of hemicellulose from lignin and renders the free polysaccharide product more amenable to degradation by the other hemicellulases (Prates *et al.,* 2001). Most of the esterases possess highest activities and isoelectric points in the neutral or low-acidic pH range. Most of them also show maximum activity and good thermostability between 40°C and 50°C. The occurrence of isoenzymes of esterases has been reported indicating the multiplicity of these enzymes. The esterases if used along with the other xylan-degrading enzymes in biopulping and biobleaching of pulps can help in partial disruption and loosening of cell wall structure (Christov and Prior, 1993).

Acetyl esterases and acetyl xylan esterases have been extensively reported in many fungi and bacteria and are apparently involved in the degradation of xylan in rumen (Williams and Withars, 1981; Williams and Withars, 1982; Williams and Withars, 1992; Hespell and Shah, 1988; Borneman *et al.,* 1990; McDermid *et al.,* 1990) and lignocellulose residues in nature (Johnson *et al.,* 1989; Poutanen and Puls, 1989; Poutanen *et al.,* 1990; Ross *et al.,* 1992).

## MANNAN DEGRADING ENZYMES

As in the case of xylan, for the complete hydrolysis of O-acetylglactoglucomannan structure, five types of enzymes are involved. They are endomannanases, β–mannosidase, β–glucosidase, α–galactosidase and acetylmannan esterase. These enzymes bring about efficient hydrolysis of mannan in a synergistic manner.

## 1,4-β–D-MANNANASE (1,4-β–D-mannan mannanohydrolase, EC 3.2.1.78)

1,4-β–D-Mannanase attacks the main backbone polymer and hydrolyzes the 1,4-β–mannopyranosyl linkages of D-mannans and D-galacto-D-mannans releasing oligosaccharide fragments which can be further hydrolyzed to mannose by β–mannosidases. Endomannanases from fungi vary in molecular weight between 42 and 61 kDa and are reported to have acidic pH optima and pI values. About 50 mannanases are found in GH families 5 and 26. Various microorganisms including intestinal and rumen bacteria and fungi are reported to produce mannanases. Microorganisms produce both endo and exo types of D-mannanases which have been well characterized. Microbial endomannanases have been reported to be both inducive and constitutive and are generally secreted extracellularly (Reese and Shibata, 1965). Several fungi have been reported to produce multiple endomannanases, but this is not yet known whether this is the result of multiple structural genes. Endomannanases hydrolyze β–D-mannans to D-mannose and a series of mannose oligosaccharides of degree of polymerization 2-6. Their action on larch glucomannan (mannose : glucose ratio 3:1) also yield D-glucose in addition to mannose (Ishihara and Shimizu, 1980). The endomannases are glycoproteins with carbohydrate contents ranging from 6 to 37%. As reported, most fungal endomannanases require four or more mannosyl residues for efficient hydrolysis of mannan due to a binding requirement near or at the active site (McCleary, 1988). Some mannanases are able to hydrolyze mannotriose but at a slow rate. Like multiple xylanases, a multiplicity of mannanases have been repoted in many fungal species (Johns-on, 1990; Ishihara and Shimizu, 1980; Araujo and Ward, 1990).

## 1,4-β–MANNOSIDASES (β–D-1,4-mannoside mannohydrolase, EC 3.2.1.25)

β–Mannosidases hydrolyze 1,4-linked β–D-mannosyl groups from the non-reducing end of their substrates, mannobiose, mannosylglucose, and higher oligomers, but with decreasing activity with increasing chain length. However, in a few fungal species, the enzyme activity against 1,4-β–mannotriose is higher than the activity against mannobiose. Whereas, the rate of hydrolysis of 1,4-β–linked D-manno-oligosaccharides by β–mannosidase from other fungal species has been found to be markedly reduced with increasing degrees of polymerization (Kuhad *et al.,* 1997). β–Mannosidases are found in GH families 1,2 and 5.

## α–GALACTOSIDASE (α–galactoside galactohydrolase, EC 3.2.1.22)

α–Galactosidase catalyzes the hydrolysis of melibiose, methyl-, ethyl-, phenyl-, and o-nitrophenyl-α–D-galactosides. The presence of any of these α–galactosides induces the production of α–galactosidase in the culture media of some growing fungal species. The enzyme can remove the α–D-galactosyl units found as branched substituents on softwood O-acetylgalactoglucomannans. α–Galactosidases occurs widely in microbes, plants, and animals (Kuhad *et al.,* 1997). α–Galactosidases from fungi have apparently received little attention and the production of this enzyme by white-rot fungi has not been studied.

## ACETYLGLUCOMANNAN ESTERASE (Acetyl mannan esterase, EC 3.1.1.6)

This enzyme liberates acetic acid from O-acetyl-galactoglucomannan, O-acetyl-4-O-methylglucuronoxylan and from α–naphthyl acetate. The enzyme activity is ten times higher on the mannan than on the xylan. Also enzyme's activity gets enhanced in the presence of endomannanase and α–galactosidase. The production of this enzyme by white-rot fungi has not been reported till now.

## β–GLUCOSIDASES (EC 3.2.1.21)

These enzymes are active on various glucosides. These are larger enzymes with higher molecular weight than other polymer degrading enzymes and their molecular weight varies in between 165 and 182 kDa. (Eriksson *et al.,* 1990). Both intracellular and extracellular β–glucosidases have been reported, the extracellular enzyme have an acidic pH optima, whereas the intracellular enzyme is observed to be most active at neutral pH (Smith and Gold, 1979). They are found in GH families 1 and 3.

## MICROORGANISMS PRODUCING HEMICELLULOLYTIC ENZYMES

Hemicellulases are widespread, since they are produced by a variety of microorganisms and also higher plants and animals. A diverse spectrum of hemicellulolytic microorganisms, mainly fungi (Baldrian and Gabriel, 2003; Falcon *et al.,* 1995) and bacteria (McCarthy, 1987; Zimmermann, 1990; Vicuna, 1988) have been isolated and identified over the years and this list still continues to grow rapidly. Table lists some important microorganisms producing hemicellulases. They are produced both constitutively and inductively. The degradation of hemicellulose is associated with a hydrolytic enzyme complex consisting of different enzymes which effectively hydrolyze the hemicellulose backbone chain by splitting off various substituents from the main chain (Eriksson *et al.,* 1990; Biely, 1985; Biely *et al.,* 1980).

The hemicellulases have been reported mainly from bacteria (Gilbert and Hazlewood, 1993; Sunna and Antranikian, 1997), fungi (Sunna and Antranikian, 1997), actinomycetes (Ball and McCarthy, 1989; Beg *et al.,* 2000a), and yeast (Hrmova *et al.,* 1984; Liu *et al.,* 1998, 1999).

## BACTERIA

Hemicellulases of bacterial origin are less studied because fungi are known to be better producers of these enzymes. Most of the hemicellulolytic enzymes are reported to be produced by the species of *Bacillus, Streptomyces, Cellulomonas, Thermomonospora* and *Chainia* sp. (Kuhad *et al.,* 1997). The anaerobic bacteria have also been reported to be producers of hemicellulose degrading enzymes. The major bacteria are of genera *Clostridium, Thermoanaerobacter, Acetovibrio, and Bacteroides* (Eriksson *et al.,* 1990; Lee *et al.,* 1987; Wiegel *et al.,* 1985; Khan *et al.,* 1987). Hemicellulolytic activity is generally associated with an enzyme complex consisting of a variety of enzymes hydrolyzing its complete structure (Biely, 1985; Eriksson *et al.,* 1990; Kuhad *et al.,*1997). Recently, xylan-hydrolyzing enzyme system consisting of β–xylanase, β–xylosidase, β–glucosidase, β–mannanase, arabinosidase, α–glucuronidase, α–galactosidase from *Bacillus pumilus* CBMAI 0008 has been reported having application in pulp bleaching (Teixeira-Duarte *et al.,* 2003). Khan *et al.* (1987) reported the production of endoxylanases, xylosidases, and esterases from *C.thermocellum, A.cellulolyticus, A.cellulosolvens*, and *B.cellulosolvens.* The purification and characterization of two acetyl-xylan esterases, two β–xylosidases, an α–glucuronidase and a cell-associated endoxylanase from the *Thermoanaerobacterium* sp. Strain JW/SL-YS485 has been reported (Shao, 1993; Shao and Wiegel, 1995; Shao *et al.,* 1995).

Among the all xylanolytic enzymes, endo-xylanases are the best characterized and the most widely studied enzymes. Among the recent studies, Gupta *et al.* (2000), Beg *et al.* (2000a), Lappalainen *et al.* (2000), Taneja *et al.* (2002) and Kapoor and Kuhad (2002) observed that *Staphylococcus* sp. SG-13, *Streptomyces* sp. and *Bacillus* sp. are the rich sources of endo-xylanases. Endo-xylanase from *B.pumilus* having application in pulp and paper industry has been reported by Texeira-Duarte *et al.* (2003). *Bacillus pumilus, Bacillus subtilis, Clostridium acetobutylicum, Cellulomonas uda* and *Strepyomyces xylophagus* are reported to be important microbial sources of β–1,4-endoxylanases (Kuhad *et al.,* 1997).

There are reports regarding *Bacillus* sp. (Panbangred *et al.,* 1984) producing intracellular β–xylosidases. The important microbial sources of β–1,4-xylosidases are reported to be *Chlostridium thermocellum, Bacillus pumilus* and *Acetovibrio cellulolyticus* (Kuhad *et al.,* 1997). Xylosidases have also been reported from *Trichoderma koningii* (Li *et al.,* 2000), *Aspergillus phoenicis* (Rizzatti *et al.,* 2001), *Aureobasidium* sp., *Pyrodictium abyssi* (Andrade *et al.,* 2001) and *Bacillus pumilus* (Teixeira-Duarte *et al.,* 2003).

A number of microorganisms including fungi, actinomycetes and bacteria have been studied for α–arabinosidases production (Kaji, 1984; Andrade *et al.,* 2001; Kosugi *et al.,* 2002; Teixeira-Duarte *et al.,* 2003). Arabinofuranosidases from *Bacillus polymyxa* (Morales *et al.,* 1995), *Streptomyces lividans* (Vincent *et al.,* 1997) and *Cytophaga xylanolytica* (Kim *et al.,* 1998) have been studied at molecular level. The main producers of α–arabinfuranosidases are reported to be *Streptomyces purpurascens, Bacillus subtilis, Ruminococcus albus* (Kuhad *et al.,* 1997).

α–glucuronidases are also reported to be produced by a number of microorganisms (Puls,1992). Presence of glucuronidase in the multienzyme xylanolytic system of *B.pumilus* CBMAI 0008 has been reported by Teixeira-Duarte *et al.*(2003).

A number of bacterial species have also been observed to produce acetyl xylan esterase (Johnson *et al.,* 1988). In some *Streptomyces* sp. such as *S. olivochromogenes* and *S. flavogriseus* growing on wheat bran produced significant levels of acetyl xylan esterase. Feruloyl esterase has been reported to be produced by *Bacillus spp.,* and *Lactobacillus fermentum. Fibrobacter succinogenes, Bacteroides cellulosolvens* and *Chlostridium thermocellum* are reported to be good producers of esterases enzymes (Kuhad *et al.,* 1997).

β–1,4-Mannases are reported to be produced mainly by *Aerobacter aerogenes, Bacillus subtilis, Caldocellum saccharolyticum* and *Streptomyces lividans*. Highly purified enzyme preparations from *B. subtilis* and fungal species *A.niger* are also capable of hydrolyzing the D-gluco-D-mannans of konjac and arum root, providing D-glucose, D-mannose and a series of manno- and glucomanno-oligosachharides (Kuhad *et al.,* 1997). In some bacteria such as *Sporocytophaga myxococcoides, Aerobacter mannanolyticus*, and *Xanthomonas campastris*, intracellular or membrane-bound endomannanases have been reported (Eriksson *et al.,* 1990). The major bacterial producer of β–1,4-mannosidase is *Bacillus subtilis* and that of α–galactosidase are *Bacillus subtilis* and *Cellulomonas* sp.

**Table 2. List of bacteria with the highest specific activity (μmol.min$^{-1}$.mg$^{-1}$) for hemicellulases**

| *Enzyme* | *Organism* | *Substrate* | *Specific activity* | *Opt. Temp. (°C)* | *Opt. pH* |
|---|---|---|---|---|---|
| Endo-1,4-β–xylanase | *Bacillus pumilus* | β–1,4-D-xylan | 1780 | 40 | 6.5 |
| β–1,4-xylosidase | *Thermoanaerobacter ethanolicus* | o-nitrophenyl-β–D-xylopyranoside | 1073 | 93 | 6 |
| α–Glucuronidase | *Thermoanaerobacterium saccharolyticum* | 4-O-methyl-glucuronosyl-xylotriose | 9.6 | 50 | 6 |
| α–L-arabinofuranosidase | *Clostridium stercorarium* | alkyl-α–arabinofuranoside/aryl-α–arabinofuranoside/L-arabinogalactan/L-arabinoxylan/methylumbelliferyl-α–L-arabinofuranoside | 883 | 70 | 5 |
| Endo-α–1,5-arabinanase | *Bacillus subtilis* | 1,5-α–L-arabinan | 429 | 60 | 6/8 |
| Feruloyl esterase | *Clostridium stercorarium* | Ethyl ferulate | 88 | 65 | 8 |
| Acetyl xylan esterase | *Fibrobacetr succinogenes* | Acetylxylan/α–naphthyl acetate | 2933 | 47 | 7 |
| Endo-β–1,4-mannanase | *Bacillus subtilis* | Galactoglucomannan/glucomannans/mannan | 514 | 50 | 5/7 |
| Exo-β–1,4-mannosidase | *Pyrococcus furiosus* | p-nitrophenyl-β–D-galactoside | 31.1 | 105 | 7.4 |
| α–Galactosidase | *Escherichia coli* | Raffinose | 27350 | 60 | 6.8 |
| Endo-galactanase | *Bacillus subtilis* | arabinogalactan | 1790 | 48 | 6 |
| β–Glucosidase | *Bacillus polymyxa* | 4-nitrophenyl-β–D-glucopyranoside | 2417 | NA | NA |

In addition to these rumen bacteria have also shown to be producers of hemicellulolytic enzymes that can degrade structural components of cell walls (Akin, 1993). These bacteria are able to adhere to fibres and have all the enzymes essential for the degradation of the polysaccharide components. Among these *Fibrobacter succinogenes, Ruminococcus albus,* and *R. flavifaciens* are most extensively studied (Akin, 1993). Acetyl esterase activities were also found in rumen anaerobic bacteria (Williams and Withars, 1981; Williams and Withars, 1982; Williams and Withars, 1992; Hespell and O'Brayan-Shah,1988). Hespell and O'Brayan-Shah (1988) reported acetyl esterase activities in *Butyrivibrio fibrisolvens, Bacteroides ruminicola, Ruminobacter amylophilus,* and *Streptococcus bovis.*

Recently, xylanolytic activity has been reported from halophilic bacteria, *Halorhabdus utahensis* (Waino and Ingvorsen, 2003). Also, two halotolerant xylanases have been isolated and characterized from a novel halobacterium strain CL8 (Wejse *et al.,* 2003).

## FUNGI

Hemicellulases of fungal origin are well characterized and have been studied in detail (Eriksson *et al.,* 1990; Coughlan *et al.,* 1993). Of the fungal hemicellulases, those of *A. niger* have been best characterized (Eriksson *et al.,* 1990). Both brown- and white-rot fungi are actively involved in the hemicellulolytic degradation. Brown-rot fungi utilize cell wall hemicelluloses and cellulose, leaving the lignin essentially undigested. Brown- and white-rot fungi degrade wood by distinctly different mechanisms. A few studies have been made of the hemicellulases of white-rot fungi and relatively little is known about them. But it is clear from their ability to deplete all the structural components of wood that white-rot fungi have effective hemicellulase systems (Tankanen *et al.,* 1996). Most white rot fungi utilize cellulose and hemicelluloses at approximately the same rate whereas lignin is degraded at relatively faster rate. Whereas few white-rot fungi show preference for removing lignin and hemicelluloses but ultimately they degrade all wood cell wall components. White-rot fungi cause the wood to become pale, eventually reducing it to a fibrous, whitish mass. Degradation of hemicelluloses by white-rot fungi proceeds in a manner roughly analogous to that of cellulose. In the degradation of both the hemicellulosic polymers, glucuronoxylan and galactoglucomannan chains are attacked first by endoxylanases and endomannanases respectively that produce progressively shorter chains, which are hydrolyzed to simple sugars by glycosidases such as mannosidases, xylosidases, glucosidases. The enzymes involved in the removal of side-chain substituents (arabinose, uronic acids, acetyls) have received little attention (Kirk and Cowling, 1984). Simple sugars repress the production of most hemicellulose-degrading enzymes by white-rot fungi. Cellulose is the only carbon source to induce the formation of hemicellulolytic enzymes by these fungi.

Endoxylanases of white-rot fungi have not been investigated to the extent of those of other fungi. Culture filtrates of *T. versicolor* have been shown to be active on a number of polymeric xylan substrates (Highley, 1976). Two endoxylanases from *Irpex lacteus* were shown to be glycoproteins after purification with molecular weights of 38 and 62 kDa, and to have acidic pH optima, they partially depolymerised larch xylan (Kanda *et al.,* 1985; Hoebler and Brillouet, 1984). β–Xylosidases have also been isolated from several different fungi (Eriksson *et al.,* 1990). α–Arabinofuranosidases have also been reported from a number of filamentous fungi. *A. terreus* secretes three α–arabinofuranosidases having molecular masses of 39, 59 and 69 kDa, acidic pH optima and alkaline pIs (Luonteri *et al.,* 1995). *Aspergillus awamori* produces this enzyme of molecular weight 64 kDa, acidic pH optima and it existed as two isoenzymes with acidic pI values (Wood and McCrae, 1996). Two α–arabinofuranosidases have been purified from ascomycete *Penicillium capsulatum* of 64.5 and 62.7 kDa, with acidic pIvalues (Filho *et al.,* 1996). Roche *et al.* (1996) have also reported this enzyme from *T. reesei.* α–Glucuronidases have been reported to be produced by non-wood decaying ascomycetous fungi *Thermoascus aurantiacus* (Khandke *et al.,* 1989), *T. reesei* (Siika-aho *et al.,* 1994), and from *P. chrysosporium.* The enzyme from *Thermoascus* had a molecular weight of 118 kDa and was active on xylan and xylo-oligosaccharides whereas that from *T. reesei* was of 91 kDa molecular weight, was having acidic pH optima and pI value and preferred low-molecular weight xylo-oligosaccharides. *P. chrysosporium* secreted α–glucuronidase having molecular mass of 112 kDa and acidic pH optima. Acetyl xylan esterases have been reported to be produced by a number of filamentous fungi such as *T. reesei, A. niger* and *Schizophyllum commune.* These enzymes from *T. reesei* were glycoproteins having molecular mass of 34 kDa with neutral pI values and acidic pH optima (Sundberg and Poutanen, 1991). Evaluation of different fungal (*Trichoderma reesei, Aspergillus awamori,* and soft-rot causing fungus *Fusarium oxysporum*) xylanolytic systems indicated that *T. reesei* culture filtrate contained relatively higher acetyl xylan esterase activities (Poutanen, 1991). Biely *et al.* also studied the production of acetyl xylan esterase by *T.reesei* and *Schizophyllum commune* (Biely *et al.,* 1988). *Penicillium purpurogenum* also produces two acetyl-xylan esterases of molecular masses, 23 and 48 kDa with neutral pIs and acidic pH optima (Egana *et al.,* 1996). However, the highest specific activities of acetyl xylan esterase reported to be produced in fungi to date were by *Aspergillus niger* (Smith *et al.,* 1991).

The mechanism of hemicellulose degradation in brown-rot fungi is similar to that for white-rot fungi (King, 1966; Keilich *et al.,* 1970; Highley, 1976). Herr *et al.,* (1978) purified β–glucosidase from brown-rot fungus *G.trabeum.* The purified xylanase from *Tyromyces palustris* also had endoglucanase but no glycosidase activities (Ishihara *et al.,* 1978). *Polyporus schwenitzii* secreted endoglucanase enzyme which was accompanied with xylanase and mannanase activities (Keilich *et al.,* 1970). A multiple glycan and glycoside hydrolase were isolated from the brown-rot fungus *Poria placenta* (Highley *et al.,* 1981).

Production and characterization of xylanases from several *Trichoderma, Aspergillus* and *Fusarium* strains have also been reported by many other workers (Dekker, 1985, Singh *et al.,* 1995). Multienzyme complex of xylanolytic system has also been reported from *Curvularia inaequalis. Trichoderma reesei,* and *Aspergillus nidulans* have been found to be rich sources of endo-xylanases (Gupta *et al.,* 2000; Beg *et al.,* 2000a; Lappalainen *et al.,*2000; Taneja *et al.,*2002; Kapoor and Kuhad, 2002). A rare cellulase-free xylanase from alkali-tolerant *Aspergillus fumigatus* AR1 has been isolated by Anthony *et al.* (2003). Xylosidases have been reported from *Trichoderma koningii* (Li *et al.,* 2000), *Aspergillus phoenicis* (Rizzatti *et al.,* 2001), *Aureobasidium* sp. and *Pyrodictium abyssi* (Andrade *et al.,* 2001). Arabinofuranosidases from *Cytophaga xylanolytica* (Kim *et al.,* 1998) have been studied at molecular level. Acetyl xylan esterase has been reported to be produced by a number of microorganisms including *Trichoderma reesei* (Biely *et al.,* 1986), *Candida guillietmondi, Schizophyllum commune* (Biely *et al.,* 1988) and *Orpinomyces* sp. Strain PC-2. The purified p-coumaryl esterase is a dimmer with a molecular weight of 11kDa and feruloyl esterase (two in number) are monomeric with mol. wt. of 68 and 24 kDa, respectively (Mackenzie *et al.,* 1987; Boenman *et al.,* 1993) and by *Aspergillus awamorii* (McCrae *et al.,* 1994). Synergistic interaction of ferulic acid esterase and xylanase from *Trichoderma* sp. has been observed for removal of ferulic acid from oat hulls for improving rumen biodegradability. Production of acetyl xylan esterase enzyme by *Thermonospora fusca* has also been reported by Bachmann and McCarthy (1991).

Endomannanases from *A. niger* degraded galactoglucomannans to mannobiose, mannose, glucose, and mannose oligosaccharides (Sinner, 1979). Four β–D-mannanases and one β–mannosidase components were isolated and characterized from *Thielavia terrestris* (Araujo and Ward, 1990). 1,4-β–D-mannosidase from *Polyporus sulphurius, A.niger,* and *Tremella fuciformis* have also been purified and characterized which liberated mannopyranosyl residues from various natural substrates (Kuhad *et al.,*1997). α–galactosidases have also been reported to be produced by many fungi such as *A. aerogenes, A. niger, Mortierella vinacea, Aspergillus tamarii* and *T. reesei.* The enzymes produced by these fungi have been purified and crystallized (Kuhad *et al.,* 1997).

**Table 3. Major hemicellulases producing fungi**

| *Soft-rot fungi* | *Brown-rot fungi* | *White-rot fungi* |
|---|---|---|
| *Chaetomium cellulolyticum* | *Poria placenta* | *Phanerochaete chrysosporium* |
| *Aspergillus niger* | *Tyromyces balsemeus* | *Trametes versicolor* |
| *Aspergillus fumigatus* | *Gloeophyllum trabeum* | *Dichomitus squalens* |
| *Aspergillus nidulans* | *Lentinus lepidius* | *Phlebia radiata* |
| *Trichoderma viride* | *Lenzites trabeum* | *Heterobasidium annosum* |
| *Trichoderma reesei* | *Coniophora puteana* | *Phellinus pini* |
| *Fusarium oxysporum* | | *Pleurotus ostreatus* |
| *Thielavia terrestris* | | *Polyporus anceps* |
| *Penicillium jenthillenum* | | *Irpex lacteus* |
| *Dactylomyces crustaceous* | | |

**Table 4. List of fungi with the highest specific activity (μ mol.min$^{-1}$.mg$^{-1}$) for hemicellulases**

| *Enzyme* | *Organism* | *Substrate* | *Specific activity* | *Opt. Temp. (°C)* | *Opt. pH* |
|---|---|---|---|---|---|
| Endo-1,4-β–xylanase | *Trichoderma longibrachiatum* | 1,4-β–D-xylan | 6630 | 45 | 5 |
| β–1,4-xylosidase | *Aspergillus nidulans* | p-nitrophenyl-β–D-xylopyranoside | 107.1 | 50 | 5 |
| α–Glucuronidase | *Phanerochaete chrysposporium* | 4-O-methyl-glucuronosyl-xylobiose | 4.5 | 50 | 3.5 |
| α–L-arabinofuranosidase | *Aspergillus niger* | 1,5-α–L-arabinofuranohexose/ 1,5-α–L-arabinotriose/ 1,5-L-arabinan/ α–L-arabinofuranotriose | 396.6 | 50-60 | 3.4-4.5 |
| Endo-α–1,5-arabinanase | *Aspergillus niger* | 1,5-α–L-arabinan | 90.2 | 50-55 | 4.5-5.0 |
| Feruloyl esterase | *Aspergillus niger* | Methyl sinapinate | 156 | 55 | 5 |
| Acetyl xylan esterase | *Schizophyllum commune* | 4-methylumbelliferyl acetate/ 4-nitrophenyl acetate | 227 | 30 | 7.7 |
| Endo-β–1, 4-mannanase | *Sclerotium rolfsii* | Galactoglucomannan/mannans/ galactomannans/glucomannans | 380 | 72-74 | 2.9/ 3.3 |
| Exo-β–1, 4-mannosidase | *Aspergillus niger* | β–D-Man-(1-4)-β–D-GlcNAc-(1-4)-β–D-GlcNAc-Asn-Lys | 188 | 55 | 3.5 |
| α–Galactosidase | *Mortierella vinacea* | melibiose | 2000 | 60 | 4 |
| Endo-galactanase | *Aspergillus niger* | NA | 6593 | 50-55 | 3.5 |
| β–Glucosidase | *Humicola insolvens* | (2-hydroxymethylphenyl)-β–D-glucopyranoside | 266.9 | 50 | 5 |

In addition to aerobic fungi, anaerobic fungi, inhabitants of the alimentary canal of herbivorous animals also have the ability to degrade plant hemicellulose because they can produce an array of all the hemicellulolytic enzymes. Anaerobic fungi can only degrade the structural polysaccharides, but cannot utilize the lignin moieties. Among anaerobic fungi, the most studied are *Neocallimastix frontalis, N.Patriciarum, Piromyces (Piromonas) communis*, and *Caecomyces* (*Sphaeromonas*) *communis* (Wubah *et al.,* 1993; Mountfort, 1987).

Yeasts such as *Aureobasidium, Cryptococcus*, and *Trichosporon* have been reported to produce xylanases. Among these genera, xylanolytic activity from *T. cutaneum* which produces a xylanase of mol. wt. 45 kDa and *C. albidus* has been observed (Eriksson, 1990).

## MULTIPLICITY OF HEMICELLULASES

Occurrence of multiple xylanases have been reported and reviewed by many workers (Kuhad *et al.,* 1997; Sa-Pereira *et al.,* 2002). The extent of xylanase multiplicity is by and large unanswered. It has been found that culture filtrates of *Aspergillus niger* was composed of 15 xylanases while 13 xylanases have been reported from *Trichoderma viride* (Biely *et al.,* 1985). Further, 11 xylanases have been reported from *Butyrivibrio fibrisolvens* (Lin and Thomson, 1991). The most outstanding case regarding multiple forms of xylanases was recorded when more than 30 different protein bands were separated by analytical electro-focussing from *Phanerochaete chrysosporium* grown on avicel (Dobozi *et al.,* 1992). *Aspergillus fumigatus* has been studied to produce multiple xylanases when grown at different temperatures (Lenartovicz *et al.,* 2002). Liu *et al.* (2003) isolated a multiple-xylanase producing bacterium, *Aeromonas caviae* ME-1. This bacterium produced at least nine xylanases. Five xylanases from the culture broth of *A.caviae* ME-1, were purified. Xylanases I, II and III act in an endo-type manner while xylanases IV and V act in an exo-type manner. Several other reports regarding presence of multiple forms of xylanases in *Aspergillus awamori, A.kawachii, A.nidulans, A.sojae, Bacillus circulans, B.polymyxa*, many other species of genus *Bacillus, Cellulomonas fimi, Cellulomonas sp.* N.C.I.M 2353, *Clostridium acetobutylicum, C.stercorarium, C.thermolacticum, Fibrobacter succinogenes, Fusarium oxysporum F3, Humicola grisea var. thermoidea, Melanocarpus albomyces IIS 68, Neocallimastix frontalis, Neurospora crassa, Robillarda sp., Streptomyces cyaneus, S.exfoliates, S.thermoviolaceus, Talaromyces byssochlamydoides, T.emersonii, Thermomonospora curvata, Thermotoga maritima, Thermotoga thermarum, Trichoderma harzianum, T.koningii, T.lignorum,* and *T.reesei* have appeared (Wong *et al.,* 1988; Sunna and Antranikian, 1997; Kimura *et al.,* 1995 Saraswat and Bisaria, 1997; Chaudhary and Deobagkar, 1997; Segura *et al.,* 1998; Tenhanen *et al.,* 1992; Lappalainen *et al.,* 2000; Eleiger *et al.,* 1995; Christakopoulos *et al.,* 1996).

In xylan hydrolysis, co-operation or synergism has been observed between enzymes acting on 1, 4-β–D-xylan backbone (β–1, 4-endo-xylanase) and side chains cleaving enzymes (α–L-arabinofuranosidase, acetylxylan esterase and α–glucuronidase). The synergistic action between acetylxylan esterase and endo-xylanases results in the efficient degradation of acetylated xylan (Biely, 1985; Biely *et al.,* 1988). The thermophilic actinomycete *T.fusca* possesses a complex multi-component xylan-degrading enzyme system that consists of endoxylanase, β–xylosidase, α–arabinofuanosidase and acetyl esterase activites. The cooperative action between these enzymes has been demonstrated (Bachmann and McCarthy, 1991). Xylanolytic enzymes produced by *Penicillium* sp. AHT-1 interacted cooperatively and sequentially in the hydrolysis of oat-spelt xylan in the following order : α–L-arabinofuranosidase→xylanase→β–xylosidase. All three enzymes exhibited optimal activity under the same conditions (temperature, pH, cultivation), thus they alone are sufficient to completely depolymerize the xylan-structure (Rahman *et al.,* 2003). Synergistic action of two types of well-characterized endoxylanases from *Streptomyces* sp. TUBB–12-2 in prebleaching of kraft pulps has been reported (Elegir *et al.,* 1994). The synergistic effect of xylanase in combination with enzymes from other groups is also reported.

A multiplicity of mannanases have been observed in several fungal species including *Tyromyces palustris* (Ishihara and Shiizu, 1980), *Polyporus versicolor* (Johnson and Ross, 1990), and *T.terrestris* (Araujo and Ward, 1990).

## STRAIN IMPROVEMENT OF THE EXISTING MICROBIAL SOURCES OF HEMICELLULOLYTIC ENZYMES

Due to various industrial applications of hemicellulases, it is important to get their higher production level so as to make hemicellulase-based processes economically viable. Successful attempts have been made to increase their production by using physical and chemical mutations. The xylanase A gene from *Streptomyces lividans* was modified by site-directed mutagenesis, selecting for mutations that improved the catalytic activity and the thermostability of the enzyme. The activity half life of the mutant enzyme at 60°C was twice that of the xylanase A (Moreau *et al.,* 1994). And now recent advances in the field of molecular biology have led to significant improvement in increasing the production of hemicellulolytic enzymes, in increasing their specific activity, their pH and temperature optima, their pH and thermal stability. Numerous literature is available dealing with various hemicellulolytic genes identified, their organization, classification, cloning, expression and regulation (Beguin, 1990; Knowles *et al.,* 1987; Gold and Alic, 1993). Recombinant DNA technology has helped to construct genetically modified microorganisms with desired enzyme characteristics. Several attempts have ben made to clone and express xylanase gene from various bacterial, fungal and actinomycete species. The early studies on cloning of xylanase gene include the work on *Bacillus* sp. (Biely, 1985) followed by several other reports. The xylanase gene of *Bacillus circulans* Teri-42 was cloned in both *B.subtilis* and *E.coli*. The recombinant xylanase in *B.subtilis* was more stable at higher temperatures than the parent *B.circulans* Teri-42 (Qureshy *et al.,* 2000). Xylanase I and II of alkalophilic thermophilic *Bacillus* sp. NCIM59 have been cloned into *E. coli* (Shendye and Rao, 1993). Cloned xylanases were different from the parent ones in terms of lower thermostability, higher $K_m$ values, lower molecular weight of xylanase-II and absence of methionine in xylanase-II (Kulkarni *et al.,* 1995). Recombinant xylanases I and II of NCIM59 have been expressed in xylanase negative *B. subtilis* A8. In *B. subtilis* A8, the enzyme activity was 5-fold higher than that in *E. coli* but was lower than that obtained in parent strain (Shendye and Rao, 1993). When expressed in xylanase proficient host *B. subtilis* MI 111, the xylanase activity was 3-fold higher than that in *B. subtilis* A8 and was comparable to that obtained from the parent strain (Shendye *et al.,* 1994). The gene *xya*A encoding alkaline xylanase from alkalophilic *Bacillus* sp. NI37 has been cloned and expressed in *E. coli* (Tabernero *et al.,* 1995). Xylanase J gene from alkalophilic *Bacillus* sp. 41 has been cloned and expressed in *E. coli.*

Highly efficient production of a *Thermomyces lanuginosus* IOC-4145 endoxylanase was achieved in *Pichia pastoris* under the control of the AOX1 promoter (Damaso *et al.,* 2003). Endoxylanase encoding genes of *Trichoderma reesei* (xyn2) and *Aspergillus kawachii* (xynC) were cloned as cDNA copies under transcriptional control of the inducible *Pichia stipitis* xylose reductase gene (XYL1) promoter. The strains expressing the *A.kawachii* xynC gene reached the highest maximum levels of β–xylanase activity and the activity was sustained for a longer period. The recombinant strains expressed higher levels of xylanolytic activity than the controls (Haan and Zyl, 2003).

Gene fusions have been successfully used in the filamentous fungi for increasing the yields of nonfungal heterologous proteins. High yield production of xylanase was obtained when xylanase gene from bacteria Nonomuraea *flexuosa* xyn11A was expressed in the filamentous fungus *Trichoderma reesei* from the strong cellobiohydrolase 1 promoter as fusions to a variety of carrier polypeptides (Paloheimo *et al.,* 2003).

It is being investigated whether transgenic plants expressing bacterial cellulase and xylanase might be useful for forage crops. For these studies, transgenic tobacco was chosen as the model plant (Kawazu *et al.,* 1996, 1999; Sun *et al.,* 1997). Other workers have also shown that microbial xylanase genes can be successfully expressed in tobacco plants and barley seed (Patel *et al.,* 2000). Expression of bacterial xylanase gene in rice plants has also been reported, since rice is agriculturally important in Asia. The gene encoding the catalytic domain of thermostable xylanase from *Clostridium thermocellum* F1 was expressed in rice plants under the control of a constitutive promoter (Kimura *et al.,* 2003).

Significant progress towards isolating the genes encoding the full complement of enzymes involved in xylan hydrolysis from a single organism has been made in studies with *P. fluorescens* subsp. *cellulosa, B. fibrisolvens* and *Caldocellum saccharolyticum, Trichoderma reesei, Aspergilus kawachii* (Hazlewood and Gilbert, 1992; Haan and Zyl, 2003).

There are reports regarding the cloning of xylanases from *Actinomadura* sp. strain FC7, *S. lividans* and *S. halstedii* JMS (Adham *et al.,* 2001). Okada *et al.* (1999) were successful in cloning two xylanase genes (*xyn1* and *xyn2)* from *T. reesei* using the fission yeast *Schizosaccharomyces pombe* as the host. A xylanase gene of *S. halstedii* JM8 was used to isolate a xylanase gene from *Streptomyces chattanoogensis* (Fernandez *et al.,* 2001). A gene encoding a novel extremely thermostable 1, 4-$\beta$ xylanase belonging to family 10 isolated directly from an environmental DNA sample has been cloned in *E. coli* (Sunna and Bergquist, 2003). Various molecular and biotechnological aspects such as gene cloning, expression and regulation for enhanced production of hemicellulases have been described by Kulkarni *et al.,* (1999).

## FUTURE PROSPECTS IN HEMICELLULOSE BIOTECHNOLOGY

Agricultural residues often regarded as waste contain appreciable amounts of hemicellulose. The energy crisis which the world is facing presently is forcing us to realize the value of renewable resources which are abundant in nature. Hemicellulose which is most abundant renewable resource is often burnt in many developing countries whereas it can be efficiently utilized for many industrial purposes. By hemicellulose bioconversion to many useful products the overall economics of the processing of the lignocellulosics biomass can be strengthened and stress should be laid to develop new ways of energy production from renewable resources. Hemicellulose biotechnology from a capital costs investment perspective is an attractive technology as it is suitable for farms and small industrial plants without the need for large engineering infrastructure.

In order for hemicellulases to have significant effect on an industrial scale, they should be stable and effective under all the operating conditions. Isolation of hemicellulases which remain stable over large range of pH and temperature with high specific activity is required. Molecular techniques including recombinant DNA technology, protein engineering and site-directed mutagenesis play a vital role in this regard. To make the overall process result-oriented, resources should be utilized in an efficient way towards modern technologies for identifying novel hemicellulolytic enzymes and engineering enzymes with desired characteristics for industrial applications.

**Table 5. List of different hemicellulase producing microorganisms**

| *Microorganisms* | *Molecular weight (kDa)* | *Optimum* | | *pI* |
|---|---|---|---|---|
| | | *pH* | *Temp. (°C)* | |
| *Acidobacterium capsulatum* | 41 | 5 | 65 | 7.3 |
| *Acrophialophora nainiana* | 17 | 6 | 50 | — |
| *Aeromonas caviae ME 1* | 20 | 7 | 50 | 7.1 |
| *Aspergillus awamori* | 39 | 5.5-6 | 55 | 5.7-6.7 |
| | 23 | 5.0 | 50 | 3.7 |
| | 26 | 4.0 | 45-50 | 3.3-3.5 |
| *Aspergillus fischeri* | 31 | 6 | 60 | — |
| *Aspergillus fumigatus AR1* | >212 | 6.5 | 65 | — |
| *Aspergillus kawachii* | 26-35 | 2-5.5 | 50-60 | 3.5-6.7 |
| *Aspergillus nidulans* | 34 | 6 | 56 | 3.4 |
| *Aspergillus niger* | 24 | 5.5 | 65 | — |
| *Aspergillus sojae* | 32.7 | 5.5 | 60 | 3.5 |
| | 35 | 5.5 | 50 | — |
| *Aspergillus syndowii* | 30 | 5.5 | 60 | — |
| *Aureobasidium pullulans ATCC 42023* | 21 | 3-4.5 | 35 | — |
| *Aureobasidium pullulans Y-2311-1* | 25 | 4.8 | 54°C | 9.4 |
| *Bacillus amyloliquefaciens* | 18.5-19.6 | 6.8-7.0 | 80 | 10.1 |
| *Bacillus sp.strain AR009* | 23 | 9 | 60 | — |
| | 48 | 9-10 | 75 | — |
| *Bacillus circulans WL-12* | 85 | 5.5-7 | — | 4.5 |
| | 15 | 5.5-7 | — | 9.1 |
| *Bacillus sp.CCMI 966* | 340 | 6.0 | 60 | — |
| *Bacillus sp. NG-27* | — | 7,8.4 | 70 | — |
| *Bacillus sp. NCIM 59* | 15.8, 35 | 6 | 50-60 | 4, 8 |
| *Bacillus polymyxa* | 34 | 6.0-7.0 | 45 | 9.3 |
| | 34 | 4.0-6.0 | 50-62 | >9.3 |
| | 22 | 6.0-7.0 | 55 | 9.0 |
| *Bacillus pumilus* | 40 | 8.0 | 55 | 4.8 |
| *Bacillus polymyxa CECT 153* | 61 | 6.5 | 50 | 4.7 |
| *Bacillus sp. 11-1S* | — | 4 | 80 | — |
| *Bacillus sp. strain 41-1(36)* | 36 | 9 | 50 | 5.3 |
| *Bacillus sp.strain BP-23* | 32 | 5.5 | 50 | 9.3 |
| *Bacillus sp. strain BP-7* | 22-120 | 6 | 55 | 7-9 |
| *Bacillus sp. strain K-1* | 23 | 5.5 | 60 | — |
| *Bacillus sp. strain TAR-1* | 40 | 6 | 75 | 4.1 |
| *Bacillus sp. W1* | 21.5 | 6.0 | 65 | 8.5 |
| | 49.5 | 7.0-9.0 | 70 | 3.6 |
| *Bacillus sp.strain SPS-0* | 99 | 6.0 | 75 | — |
| *Bacillus sp. W1 (JCM 2888)* | 21.5 | 6 | 65 | 8.5 |
| | 49.5 | 7.9 | 70 | 3.7 |

| | | | | |
|---|---|---|---|---|
| *Bacillus sp. W2* | 22.5 | 6.0 | 65 | 8.3 |
| | 50 | 7.0-9.5 | 70 | 3.7 |
| *Bacillus stearothermophilus T-6* | 43 | 6.5 | 75 | 9.0 |
| *Bacillus strain XE* | 22 | 6 | 75 | 7.7 |
| *Cellulomonas flavigena PN-120* | 56 | 9.0 | 55 | 6.14 |
| *Cellulomonas fimi* | 14 | 5.0 | 45 | — |
| | 22 | 6.0 | 40 | — |
| | 150 | 5.5-6.5 | 40 | — |
| *Cellulomonas sp. N.C.I.M.2353* | 22,33,53 | 6.5 | 55 | 8 |
| *Cephalosporium sp. strain RYM-202* | 35 | 7.5-8.0 | 50 | 6.3 |
| | 24 | 7.5-8.0 | 50 | 4.4 |
| *Chaetomium cellulolyticum* | 25 | 6.0-7.0 | — | 8.9 |
| | 47 | 6.0-7.0 | — | 8.4 |
| | 57 | 6.0-7.0 | — | 5.0 |
| *Clostridium thermolacticum* | 39 | 6.5 | 80 | 4.45 |
| | 55 | 6.5 | 80 | 4.55 |
| | 65 | 6.5 | 80 | 4.65 |
| *Cryptococcus flavous* | 31 | 4.5 | 55 | 10.0 |
| *Cryptococcus albidus* | 48 | 5 | 25 | — |
| *Erwinia chrysanthemi* | 42 | 5.5 | 55 | 8.8 |
| *Fibrobacter succinogenes* | 537 | 7.0 | 39 | — |
| | 66 | 6.3 | 55 | |
| *Fusarium oxysporum F3* | 20.8 | 6.0 | 60 | 9.5 |
| | 23.5 | 6.0 | 55 | 8.45-8.7 |
| *Humicola grisea var. thermoidea* | 22.55 | 5.5 | 70 | — |
| *Humicola insolens* | 6.0 | 6-6.65 | 55-60 | 9.0 |
| | 21 | 6-6.5 | 55-60 | 7.7 |
| *Humicola lanuginsoa* | 22.5 | 6.0 | 65 | 4.1 |
| *Halophilic bacterium strain CL8* | 43 | 6.0 | 60 | 5.0 |
| | 62 | 6.0 | 65 | 3.4 |
| *Geotrichum candidum* | 60-67 | 4 | 50 | 3.4 |
| *Irpex lacteus* | 38 | 4.6-5.2 | — | 7.6-8 |
| *Lentinula edodes* | 41 | 4.5-5 | 60 | 3.6 |
| *Micrococcus sp. AR-135* | 56 | 7.5-9 | 55 | — |
| *Microtetraspora flexuosa S IIX* | 26.3 | 6 | 80 | 8.4 |
| | 16.8 | 6 | 80 | 9.45 |
| *Myrothecium verrucaria* | 15.9 | 5.5 | 45 | 4.35 |
| *Neocallimastix frontalis* | 45 | 5.5 | 55 | — |
| | 70 | 6.0 | 55 | — |
| *Neurospora crassa* | 33 | 4.8 | — | 4.8 |
| | 30 | 4.8 | — | 4.5 |
| *Paecilomyces varioti* | 25 | 5.5-7.0 | 65 | 3.9 |
| *Pencillium purpurogenum* | 33 | 7.0 | — | 4.5-5.5 |
| | 23 | 3.5 | — | 4.5-5.5 |
| *Piromyces sp. E2* | 12.5 | 6.0 | 50 | 9.06 |
| | 12.5 | 6.0 | 50 | 9.06 |
| *Recombinant E.coli PATBX4.5* | 35 | 6.0 | 60 | 4 |
| | 14.5 | 6.0 | 60 | 8 |
| *Robillarda sp. Y-20* | 17.6 | 4.5- 6.0 | 50 | 9.7 |
| | 59 | 4.5- 6.0 | 50 | 3.5 |
| *Schizophyllum commune* | 21 | 5 | — | 4.5 |
| *Schizophyllum commune* | 33 | 5.0 | 45-50 | — |
| *Staphylococcus sp. SG-13* | 60 | 7.5, 9.2 | 50 | — |
| *Streptomyces lividans 66* | 22 | 6 | 57 | >10.2 |
| *Streptomyces sp. No 3137* | 50 | 5.5-6.5 | 60-65 | 7.1 |
| | 25 | 5.0-6.0 | 60-65 | 10.06 |
| | 25 | 5.0-6.0 | 60-65 | 10.26 |
| *Streptomyces sp. Strain C5-A13* | 19.2 | 8.5-9.5 | 55 | 5 |

| | | | | |
|---|---|---|---|---|
| *Streptomyces sp.B–12-2* | 26.4 | 6.0 | 55 | 7.5 |
| | 23.8 | 6.0 | 60 | 8.3 |
| | 36.2 | 7.0 | 60 | 5.4 |
| | 26.2 | 7.0 | 60 | 5.0 |
| | 40.5 | 6.0 | 60 | 4.8 |
| *Streptomyces sp.strain S38* | 24.5 | 6.0-6.5 | 60 | 9.8 |
| | 37.5 | 6.0-7.5 | 60 | 5.2 |
| | 38 | 6.0-8.0 | 60 | 4.7 |
| *Streptomyces sp.A451* | 22.8 | 5.6 | 50 | 8.6 |
| | 33.1 | 5.4 | 50 | 8.9 |
| *Streptomyces viridosporus T7A* | 59 | 7.0-8.0 | 65-70 | 10.2-10.5 |
| *Streptomyces thermoviolaceus OPC-520* | 33,54 | 7 | 60-70 | 4.2, 8 |
| *Streptomyces chattanogensis CECT 3336* | 48 | 6 | 50 | 9 |
| *Streptomyces sp. QG-11-3* | — | 8.6 | 60 | — |
| *Streptomyces T-7* | 20.643 | 4.5-5.5 | 60 | 7.8 |
| *Thermoanaerobacterium sp. Strain JW/SL-YS485* | 350 | 80 | 6.5 | — |
| *Thermoascus aurantiacus* | 32 | 5.1 | — | 7.1 |
| *Thermomyces lanuginosus DSM 5826* | 25.5 | 7.0 | 60-70 | 4.1 |
| *Thermomyces lanuginosus-SSBP* | 23.6 | 6.5 | 70-75 | 3.8 |
| *Thermomonospora curvata* | 15-36 | 6.8-7.8 | 75 | 4.2-8.4 |
| *Thermotoga thermarum* | 266 | 6 | 80 | — |
| | 35 | 7 | 90-100 | |
| *Thermotoga maritime MSB8* | 40,120 | 5.4, 6.2 | 92-105 | 5.6 |
| *Trichoderma harzianun* | 20 | 5.0 | 50 | — |
| *Trichoderma koningii* | 18 | 4.9-5.5 | — | 7.3 |
| | 29 | 4.9-5.8 | — | 7.2 |
| *Trichoderma longibrachiatum* | 37.7 | 5-6 | 45 | — |
| *Trichoderma reesei* | 20,19 | 5-5.5<br>4-4.5 | 45, 40 | 9, 5.5 |
| *Trichoderma viride* | 22 | 5 | 53 | 9.3 |

## ACKNOWLEDGEMENT

The authors wish to acknowledge financial assistance from Council of Scientific and Industrial Research, New Delhi to Bindu Battan in form of Junior Research Fellowship during the course of investigation.

## REFERENCES

Adham, S.A.I., Camelo, A.B., Ramos, A. and Gil, J.A. (2001). Construction of a Xylanase-Producing Strain of *Brevibacterium lactofermentum* by Stable Integration of an Engineered xysA Gene from *Streptomyces halstedii* JM8. *Applied and Environmental Microbiology,* 67, 5425-5430.

Akin, D.E., Gordon, G.L.R. and Hogan, J.P. (1983). Rumen bacterial and fungal degradation of *Digitaria pentzii* grown with or without sulfur. *Applied and Environmental Microbiology,* 46, 738-748.

Akin, D.E. (1993). In: *Genetics, biochemistry and ecology of lignocellulose degradation.* Eds Shimada,K., Ohmiya, K., Kobayashi, Y., Hoshino, S., Sakka, K., Karita, S, UNI, Tokyo, p95.

Andrade, C.M., Aguiar, W.B. and Antranikian, G. (2001). Physiological aspects involved in production of xylanolytic enzymes by deep-sea hyperthermophilic archaeon *Pyrodictium abyssi. Applied Biochemistry and Biotechnology,* 91-92, 655-669.

Anthony, T., Raj, K.C., Rajendran, A. and Gunasekaran, P. (2003). High molecular weight cellulase free xylanase from alkali-tolerant *Aspergillus fumigatus* AR1. *Enzyme and Microbial Technology,* 32, 647-654.

Anthony, T., Raj, K.C., Rajendran, A. and Gunasekaran, P. (2003). Inhibition of Proteases during Fermentation Improves Xylanase Production by Alkali Tolerant *Aspergillus fumigatus* AR1. *Journal of Bioscience and Bioengineering,* 96, 394-396.

Araujo, A. and Ward, O.P. (1990). Hemicellulases of the Bacillus spp, Preliminary comparative studies on production and properties of mannanases and galactanases. *Journal of Applied Bacteriology,* 68, 253-267.

Atalla, R. 1995. In proceedings Eighth International Symposium Wood and Pulping Chemistry. KCl, Helsinki. P77.

Bachmann, S.L., McCarthy, A.J. (1991). Purification and cooperative activity of enzymes constituting the xylan-degrading system of *Thermomonospora fusca. Applied and Environmental Microbiology,* 57, 2121-2130.

Bacon, J.S.D., Gordon, A.H. and Morris, E.J. (1975). Acetyl groups in cell wall preparations from higher plants. *Biochemistry Journal,* 149, 485-487.

Bajpai, P. (1997). Microbial xylanolytic enzyme system properties and applications. *Advances in Applied and Environmental Microbiology,* 43, 141-189.

Baldrian, T. and Gabriel, J. (2003). Lignocellulose degradation by *Pleurotus ostreatus* in the presence of cadmium. *FEMS Microbiology Letters,* 220,235-240.

Ball, A.S., McCarthy, A.J. (1989). Production and purification of xylanase from actinomycetes. *Journal of General Bacteriology,* 66,439-444.

Bastawde, K.B. (1992). Xylan structure, microbial xylanases and their mode of action. *World Journal of Microbiology and Biotechnology,* 8,353-368.

Beg, Q.K., Bhushan, B., Kapoor, M. and Hoondal, G.S. (2000a). Production and characterization of thermostable xylanase and pectinase from a Streptomyces sp. QG-11-3. *Journal of Indian Microbiology and Biotechnology,* 24,396-402.

Beg, Q.K., Kapoor, M., Mahajan, L. and Hoondal, G.S. (2001). Microbial xylanases and their industrial applications, A review. *Applied Microbiology and Biotechnology,* 56, 326-338.

Beguin, P. (1990). Molecular biology of cellulose degradation. *Annual Reviews in Microbiology,* 44, 219-248.

Biely, P., Kratky, Z., Vrsanska, M and Urmanicove, D. (1980). Induction and inducers of endo-1,4-β–xylanase in the yeast Cryptococcus albidus. *European Journal of Biochemistry,* 108, 323-329.

Biely, P. (1985). Microbial Xylanolytic systems. *Trends in Biotechnology,* 3, 286-290.

Biely, P., Mackenzie, C.R., Puls, J. and Schneider,H. (1986). Cooperativity of esterases and xylanases in the enzymatic degradation of acetyl xylan. *Biotechnology,* 4,731-733.

Biely, P., MacKenzie, C.R. and Schneider, H. (1988). In Methods Enzymology (Wood WA and Kellog ST,eds) Academic Press, San Diego, CA. 160, 700-707.

Biely, P., MacKenzie, C.R. and Schneider, H. (1988). Production of acetyl xylan esterase by *Trichoderma reesei* and *Schizophyllum commune. Canadian Journal of Microbiology,* 34, 767-772.

Blanchette, R.A., Burnes,T.A., Gary, F.L. and Etand, M.J. (1988). Selection of white-rot fungi for biopulping. *Biomass,* 15,93-101.

Blanchette, R.A., Abad, A.R., Farrell, R.L. and Leathers,T.D. (1989). Detection of lignin peroxidase and xylanases by immunocytochemical labeling in wood decayed by basidiomycetes. *Applied and Environmental Microbiology,* 55,1457-1465.

Borneman, W.S., Hartley, R.D., Himmelsbach, D.S. and Ljungdahl, L.G. (1990). Assay for trans-p-coumaroyl esterase using a specific substrate from a plant cell walls. *Analytical Biochemistry,* 190,129-133.

Borneman, W.S., Ljungdahl, L.G., Hartley, R.D. and Akin, D.E. (1993). In,Coughlan, M.P., Hazlewood, G.P. (eds.). Hemicellulose and Hemicellulases. Portland, Chapel Hill, NC, p85.

Brasch, D.J. and Wise, L.E. 1956. *Tappi,* 39,581.

Chaudhary, P. and Deobagkar, D. (1997). Purification and characterization of xylanase from *Cellulomonas* sp. N.C.I.M. 2352. *Biotechnology and Applied Biochemistry,* 25,127-133.

Christakopoulas, P., Nerinckx, W., Kekos, D., Marcis, B. and Claeyssens, M. (1996). Purification and characterization of two low molecular mass alkaline xylanases from *Fusarium oxysporum* F3. *Journal of Biotechnology,* 51,181-189.

Christov, L.P. and Prior, B.A. 1993. Esterases of xylan-degrading microorganisms, production, properties, and significance. *Enzyme and Microbial Technology,* 15,460-475.

Coughlan, M.P. and Hazlewood, G.P. (1993). β–1, 4-D-Xylan degrading systems, biochemistry, molecular biology and applications. *Biotechnology and Applied Biochemistry,* 17,259-289.

Coughlan, M.P., Tuohy, M.G., Filho, E.X.F., Puls, J. and Hughs, M.H. (1993). Enzymological aspects of microbial hemicellulases with emphasis on fungal systems. In, Hemicellulose and Hemicellulases. Coughlan, M.P. and Hazlewood, G.P.(eds.) Portland Press, London, pp.53-84.

Damaso, M.C.T., Almeida, M.S., Kurtenbach, E., Martins, O.B., Junior, N.P., Andrade, C.M.M.C. and Albano, R.M. (2003). Optimized Expression of a Thermostable Xylanse from *Thermomyces lanuginosus* in *Pichia pastoris. Applied and Environmental Microbiology,* 69,6064-6072.

Dekker, R.F.H. and Richards, G.N. (1976). Hemicellulases, their occurrence, purification, properties and mode of action. *Advances in Carbohydrate Chemistry and Biochemistry,* 32,277-352.

Dekker, R.F.H. (1985). In Biosynthesis and Biodegradation of Wood Components (Higuchi T, eds) Academic press, Orlando FL, pp 505-533.

Dekker, R.F.H. (1989). In Plant Cell Wall Polymers, Biogenesis and Biodegradation (Lewis NG and Paice MG,eds) ACS Symposium Series. 399, American Chemical Society, Washington, pp619-629.

Dobozi, M.S., Szakacs, G. and Bruschi, C. (1992). Xylanase activity of *Phanerochaete chrysosporium. Applied and Environmental Microbiology,* 58,3466.

Egana, L., Gutierrez, R., Caputo, V. and Peirano, A., Steiner, J. and Eyzaguirre, J. (1996). Purification and characterization of two acetyl xylan esterases from *Penicillium purpurogenum. Biotechnology and Applied Biochemistry,* 24,33.

Elegir, G., Szakaes, G. and Jeffries, T.W. (1994). Purification, characterization and substrate specificities of multiple xylanases from *Streptomyces* sp. strain B–12-2. *Applied and Environmental Microbiology,* 60,2609-2615.

Elegir, G., Sykes, M. and Jeffries,T.W. (1995). Differential and synergistic action of *Streptomyces* endoxylanases in prebleaching of kraft pulp. *Enzyme and Microbial Technology,* 17,954-959.

Eriksson, K.E.L. (1990). Biotechnology in the pulp and paper industry. *Wood Science and Technology,* 24,79-101.

Eriksson, K-EL., Blanchette, R.A. and Ander, P. (1990). In Microbial and Enzymatic Degradation of Wood and Wood Components (Timell TE,ed) Springer-Verlag, Berlin, pp 181-213.

Falcon, M.A., Rodriguez, A., Carnicero, A., *et al.,* (1995). Isolation of microorganisms with lignin transformation potential from soil of Tenerife Island. *Soil Biology and Biochemistry,* 27(2),121-126.

Fengel, P. and Wegener, G. (1984). Wood, Chemistry, ultrastructure, reactions. Walter de gruyter, New York.

Fernandez-Bolanos, J., Felizon, B., Heredia, A., Rodriguez, R., Guillen, R., Jimenez, A. (2001). Steam-explosion of olive stones, hemicellulose solubilization and enhancement of enzymatic hydrolysis of cellulose. *Bioresource Technology,* 79,53-61.

Filho, E.X., Puls, J. and Coughlan, M.P. (1996). Purification and characterization of two arabinofuranosides from solid state cultures of the fungus *Penicillium capsulatum. Applied and Environmental Microbiology,* 62,168.

Frederick, M.M., Kiang, C., Frederick, J.R. and Reilly, P.J. (1985). Purification and characterization of endo-xylanases from *Aspergillus niger.*I. Two isoezymes active on xylan backbones near branch points. *Biotechnology and Bioengineering,* 27,525-532.

Frowein, Y.Z., Zori, U. and Leibowitz, J. (1963). *Enzymologia,* 26,193-200.

Gilbert, H.J., Hazlewood, G.P. (1993). Bacterial cellulases and xylanases. *Journal of General Microbiology,* 139, 187-194.

Gold, M.H. and Alic, M. 1993. Molecular biology of the lignin-degrading basidiomycetes *Phanerochaete chrysosporium. Microbiological Reviews,* 57(3), 605-622.

Gordon, A.H., Lomax, J.A. and Chesson, A. (1983). Glycosidic linkages of legume, grass and cereal straw cell walls before and after extensive degradation by rumen microorganisms. *Journal of Science and Food Agriculture,* 34,1341.

Gupta, S., Bhushan, B. and Hoondal, G.S. (2000). Isolation, purification and characterization of xylanase from *Staphylococcus* sp. SG-13 and its application in biobleaching of kraft pulp. *Journal of Applied Microbiology,* 88,325-334.

Haan, R.D. and Zyl, W.H.V. (2003). Enhanced xylan degradation and utilization by *Pichia stipitis* overproducing fungal xylanolytic enzymes. *Enzyme and Microbial Technology,* 33,620-628.

Hartley, R.D. and Jones, E.C. (1976). Diferulic acid as a component of cell walls of Lolium multiflorum. *Phytochemistry,* 15,1157-1160.

Hartley, R.D., Morrison, W.H., Balza, F. and Towers, G.H.N. (1990). Substituted truxillic and truxinic acids in cell-walls of cynodon-dactylon. *Phytochemistry,* 29,3699-3703.

Hazlewood, G.P. and Gilbert, H.J. (1992). The molecular architecture of xylanases from Pseudomonas fluorescens subsp. Cellulose, In, Xylans and Xylanases (Visser, J., Beldman, G., Kusters-van Someren, M.A. and Voragen, A.G.J., eds.). Elsevier Science Publishers, Amsterdam. 259-273.

Hazlewood, G.P. and Gilbert, H.J. (1993). Molecular biology of hemicellulases,p.103-126. In, M.P. Coughlan and G.P. Hazlewood (ed.). Hemicellulose and Hemicellulases. Portland Press,London, United Kingdom.

Henrissat, B. and Bairoch, A. (1996). Updating the sequence-based classification of glycosyl hydrolases. *Biochemistry Journal,* 316, 695-696.

Herr, D., Baumer, F. and Dellweg, H. (1978). Purification and properties of an extracellular endo-1,4-β–glucanase from Lenzites trabea. *Archives in Microbiology,* 117,287-292.

Hespell, R.B. and O'Brayan-Shah, P.J. (1988). Esterase activities in *Butyrivibrio* fibrisolvens strains. *Applied and Environmental Microbiology,* 54,1917-1922.

Highley, T.L. (1976). Hemicellulose of white- and brown-rot fungi in relation to host preferences. *Mat u Org.,* 11,33-46.

Highley, T.L., Wolter, K.E. and Evans, F. (1981). Polysaccharide-degrading complex produced in wood and in liquid media by the brown-rot fungus, *Poria placenta. Wood Fibre,* 13,265-274.

Hoebler, C. and Brillouet, J.M. (1984). Purification and properties of an endo- (1→4)-β-D-xylakase from Irpex lacteus (Polyporus tulipiferae). *Carbon hydrate Resource,* 128,141.

Howard, R.L., Abotsi, E., Jansen van Rensburg, E.L. and Howard,S. (2003). Lignocellulose biotechnology, issues of bioconversion and enzyme production. *African Journal of Biotechnology,* 2(12), 602-619.

Hrmova, H., Biely, P., Vrsanka, M. and Petrakova, E. (1984). Induction of cellulose- and xylan-degrading enzyme complex in yeast *Trichosporon cutaneum. Archieves in Microbiology,* 161,371-376.

Ishihara, M., Shimizu, K. and Ishihara, T. (1978). Hemicellulases of brown rotting fungus, Tyromyces palustris, III. Partial purification and mode of action of an extracellular xylanase. *Mokuzai Gakkaishi,* 24,108-115.

Ishihara, M. and Shimizu, K. (1980). Hemicellulases of brown rotting fungus *Tyromyces palustris* IV. Purification and some properties of an extracellular mannanase. *Mokuzai Gakkaishi,* 26,811-818.

Johnson, K.G., Fontana, J.D. and MacKenzie, C.R. (1988). In methods Enzymology (Wood WA and Kellog ST, eds) Academic Press, San Diego, CA. 160,551-560.

Johnson, K.G., Harrison, B.A., Schneider, H., MacKenzie, C.R. and Fontana, J.D. (1988). Xylan-hydrolysing enzymes from *Streptomyces* spp. *Enzyme and Microbial Technology,* 10,403-409.

Johnson, K.G., Silva, M.C., Mackenzie, C.R., Schneider, H. and Fontana, J.D. (1989). *Applied Biochemistry and Biotechnology,* 20/21,245-258.

Johnson, K.G. (1990). Enzymic properties of β-mannanase from *Polyporus versicolor. World Journal of Microbiology and Biotechnology,* 6,209.

Johnson, K.G. and Ross, N.W. (1990). Enzymic properties of β-mannanase from *Polyporus versicolor. Enzyme and Microbial Technology,* 12,960-964.

Jung, H.J.G. and Himmelsbach, D.S. (1989). Isolation and characterization of wheat straw lignin. *Journal of Agriculture and Food Chemistry,* 37,81-87.

Kato, Y. and Nevins, D.J. (1985). Isolation and identification of O-(5-O-feruloyl-α-L-arabinofuranosyl)-1(-3)-O-β-D-xylopyranosyl-(1-4)-D-xylopyranose as a component of *Zea* shoot cell wall. *Carbohydrate Research,* 137,139-150.

Kato, A., Azima, J. and Koshijima, T. (1987). *Agriculture and Biological Chemistry,* 51,1691-1694.

Kaji,A. (1984). In Advances Carbohydrate Chemistry and Biochemistry (Tipson RS and Horton D, eds) Academic Press, Orlando, FL. 42,383-394.

Kanda, T. Amano,Y. and Nisizawa, K. (1985). Purification and properties of Two Endo-1,4-β-Xylanase from *Irpex lacteus* (*Polyporus tulipiferae*). *Journal of Biochemistry,* 98,1545.

Kapoor, M. and Kuhad, R.C. (2002). Improved polygalacturonase production from *Bacillus* sp. MG-cp-2 under submerged (SmF) and solid state (SSF) fermentation. *Letters in Applied Microbiology,* 34,317-322.

Kawazu, T., Ohta, T., Ito, K., Shibata, M., Kimura, T., Sakka, K. and Ohmiya, K. (1996). Expression of a *Ruminococcus albus* cellulase gene in tobacco suspension cells. *Journal of Fermentation Bioengineering,* 82,205-209.

Kawazu, T., Sun, J.L., Shibata, M., Kimura, T., Sakka, K. and Ohmiya, K. (1999). Expression of a bacterial endoglucanase gene in tobacco increased digestibility of its cell wall fibres. *Journal of Bioscience and Bioengineering,* 88,421-425.

Keilich, G., Bailey, P. and Liese, M. (1970). Enzymatic degradation of cellulose, cellulose derivatives and hemicelluloses in relation to the fungal decay of wood. *Wood Science and Technology,* 4,273-283.

Khan, A.E., Lamb, K.A. and Forgie, M.A. (1987). Formation of esterase, xylanase and xylosidase by cellulolytic anaerobes. *Biomass,* 13,135.

Khandke, K. Vithayathil, P. and Murthy, S. (1989). Archieves in Biochemistry and Biophysics. 274,511.

Kim, H., Goto, M., Jeong, H.J., *et al.,* (1998). Functional analysis of a hybrid endoglucanase of bacterial origin having a cellulose binding domain from a fungal exoglucanase. *Applied Biochemistry and Biotechnology,* 75, 193-204.

Kim, S.W., Kang, S.W. and Lee, J.S. (1997). Cellulase and xylanase production by Aspergillus niger KKS in various bioreactors. *Bioresource Technology,* 59,63-67.

Kimura, I., Sasahar, H. and Tajima, S. (1995.) Purification and characterization of two xylanases and an arabinofuranosidase from *Aspergillus sojae. Journal of Fermentation Bioengineering,* 80,334-339.

Kimura, T., Mizutani, T., Tanaka, T., Koyama, T., Sakka, K. and Ohmiya, K. (2003). Molecular breeding of transgenic rice expressing a xylanase domain of the xynA gene from *Clostridium thermocellum. Applied Microbiology and Biotechnology,* 62,374-379.

King, N.J. (1966). The extracellular enzymes of *Coniophora cerebella. Biochemical Journal,* 100,784-792.

Kirk, T.K. and Cowling, E.B. (1984). Biological decomposition of solid wood. In, Rowell RM, ed, Chemistry of Solid Wood. *Advances in Chemistry Series,* 207. Washington DC, American Chemical Society Press. Pp 455-487.

Knowles, J., Lehtovaara, P. and Teeri, T. (1987). Cellulase families and their genes. *Trends in Biotechnology,* 5,255-261.

Knull, L.H. and Inglett, G.E. (1980). Analysis of neutral carbohydrates in agricultural residues by gas-liquid chromatography. *Journal of Agriculture and Food Chemistry,* 28,917-919.

Kuhad, R.C., Singh, A. and Eriksson,K.E.L. (1997). Microorganisms and enzymes involved in the degradation of plant fibre cell walls. Special issue on 'Biotechnology in pulp and paper industry'. *Advances in Biochemical Engineering,* 57,45-127.

Kuhad, R.C. and Singh, A. (1993). Lignocellulose biotechnology, current and future prospecta. *Critical Reviews in Biotechnology,* 13,152-172.

Kulkarni, N., Shendye, A and Rao, M. (1999). Molecular and biotechnological aspects of xylanases. *FEMS Microbiological Reviews,* 23,411-456.

Lapaleinen, A., Siika-aho, M., Kalkkimen, N., Fogerstorm, R. and Tenkanen, M. (2002). Endoxylanase II from *Trichoderma reesei* has several isoforms with different isoelectric points. *Biotechnology and Applied Biochemistry,* 31,61-68.

Lee, R.C., Hrmova, M., Burton, R.A., *et al.,* (2003). Bifunctional family 3 glycoside hydrolases from Barley with alpha-L-Arabinofuranosidase and beta-D-Xylosidase activity characterization, primary structure and COOH-terminal processing. *Journal of Biological Chemistry,* 278,5377-5388.

Lenartovicz, V., Marques, C.G., Moreira, F.G. and Peralta, R.M. (2002). Temperature effect in the production of multiple xylanases by *Aspergillus fumigatus. Journal of Basic Microbiology,* 42(b),388-389.

Li,K., Azadi, P., Collins, R., Tolan, J., Kim, J.S. and Eriksson, K.E.L. (2000). Relationships between activities of xylanases and xylan structures. *Enzyme and Microbial Technology,* 27, 89-94.

Lin, L.L. and Thompson,J.A. (1991). An analysis of the extracellular xylanases and cellulases of *Butryrivibrio fibrisolens* H 17C. *FEMS Microbiology Letters,* 84, 197-204.

Liu,W., Lu,Y. and Ma, G. (1999). Induction and glucose repression of endo-β–xylanase in the yeast *Trichosporon cutaneum* SL 409. *Process Biochemistry,* 34,67-72.

Liu, C.J., Suzuki, T., Hirata, S. and Kawai, K. (2003). Processing of XynE (110-kDa) of *Aeromonas caviae* ME-1 to 72-kDa Xylanase in *Escherichia coli* Transformant. *Journal of Bioscience and Bioengineering,* 96,406-408.

Luonteri, E., Tenkanen, M., Siika-Aho, M., Buchert, J. and Viikari, L. (1995). L-arabinosidases of *Aspergillus terreus* and their potentials in pulp and paper applications. Proceedings in International Conference on Biotechnology Pulp and Paper applications, 6[th], Recent Advances in Applied and Fundamental Research. 11-15 June, Vienna, Austria. p.119.

Maat, J., Roza, M., Verbakel, J., Stam, H., DaSilra, M.J.S., Egmond, M.R., Hagemans, M.L.D., VanGarcom, R.F.M., Hessing, J.G.M., VanDerhondel, C.A.M.J.J., VanRotterdam, C. (1992). Xylanases and their application in bakery. In, Visser, J., Beldman, G., VanSomeren, M.A.K., Voragen, A.G.J. (eds). Xylans and xylanases. Elsevier, Amsterdam, pp 349-360.

MacKenzie, C.R., Bilous, D., Schneider, H. and Johnson, K.G. (1987). Induction of cellulolytic and xylauolytic enzymes systems in *Streptomyces* spp. *Applied and Environmental Microbiology,* 53,2835-2839.

Markwalder, H.U. and Newkom, H. (1976). Diferulic acid as a possible crosslink in hemicelluloses from wheat germ. *Phytochemistry,* 15,836-837.

McCarthy, A.J. (1987). Lignocellulose-degrading actinomycetes. *FEMS Microbiology Letters,* 46(2),145-163.

McCleary, B.V. (1988). *Methods in Enzymology,* 160,596.

McCrae, S.I., Leith, K.M., Gordon, A.H.and Wood, T.M. (1994). Xylan-degrading enzymes system produced by the fungus *Aspergillus awarmon*: isolation and characterization of a feruloye esterase and a p-coumaroyl esterase. *Enzyme and Microbial Technology,* 16,826.

McDermid, K.P., MacKenzie, C.R. and Forsberg, C.W. (1990). Esterase activities of *Fibrobacter succinogenes* sub sp succinogenes 585. *Applied and Environmental Microbiology,* 56,127-132.

Molgaard, P. and Ravn, H. 1988. Evolutionary aspects of caffeoyl ester distribution in Dicotyledons. *Phytochemistry,* 27,2411-2421.

Moracci, M., Ponzano, B.C., Trinkone, A., Fusco,S., De Rosa, M., Vander Oost, J., Sensen, C.W., Charlebols, R.L. and Rossi, M (2000). Identification and Molecular characterization of the first alpha-xylosidase from an archaeon. *Journal of Biological Chemistry,* 275, 22082-22089.

Morales, P., Madrarro, A., Flors, A., Sendra, J.M. and Gonzalez, J.A.P. (1995). Purification and characterization of a xylanase and arabinofuranosidase from *Bacillus polymyxa. Enzyme and Microbial Technology,* 17,424-429.

Moreau, R.A., Powell, M.J., Whitaker, B.D., Bailey, B.A. and Anderson, J.D. (1994). Xylanase treatment of plant cells induces glycosylation and fatty acylation of photosterols. *Physiology of Plants,* 91,575-580.

Mountford, D.O. (1987). The rumen anaerobic fungi. *FEMS Microbiology Reviews,* 46,401.

Mueller-Harvey, I., Hartley, R.D., Harris, P.J. and Curzon, E.H. (1986). Linkage of *p*-coumaroyl and feruloyl groups to cell-wall polysaccharides of barley straw. *Carbohydrate Resources,* 148,71-85.

Okada, H., Mori, K., Tada, K., *et al.,* (2000). Identification of the active site carboxylic residues in *Trichoderma reesei* endoglucanase Cel12A by site-directed mutagenesis. *Journal of Molecular Catalysis B, Enzymatic,* 10,249-255.

Paloheimo, M., Mantyla, A., Kallio, J. and Suominen, P. (2003). High-Yield Production of a Bacterial Xylanase in the Filamentous Fungus *Trichoderma reesei* Requires a Carrier Polypeptide with an Intact Domain Structure. *Applied and Environmental Microbiology,* 69,7073-7082.

Panbangred, W., Shinmyo, A., Kinoshito, S. and Okada, H. (1983). Purification and properties of endoxylanase produced by *Bacillus pumilus. Agriculture Biology and Chemistry,* 47,957-963.

Panbangered, W., Kawaguchi, O., Tomita,T., Sinmyo, O. and Okada, H. (1984). *European Journal of Biochemistry,* 138,267.

Park, G.G., Kusakabe,I., Komatso, Y., Kobayashi, Y.,Yasui,T. and Murakami,K. 1987. *Agricultural Biology and Chemistry,* 51,2709.

Patel, N., Johnson, J.S., Brettell, R.I.S., Jacobsen, J. and Xue,G.P. (2000). Transgenic barley expressing a fungal xylanase gene in the endosperm of the developing grains. *Molecular Breeding,* 6,113-123.

Poutanen, K. and Puls, J. (1989). In Plant Cell Wall Polymers, Biogenesis and Biodegradation (Lewis NG and Paice MG,eds) ACS Symposium Series. 399, American Chemical Society, Washington, pp 630-640.

Poutanen, K., Sundberg, M., Korte, H. and Puls J. (1990). Deacetylation of xylans by acetyl esterases of *Trichoderme reesei. Journal of Applied Microbiology and Biotechnology,* 33,506-510.

Poutanen, K., Tenkanen, M., Korte, H. and Puls, J. (1991). Accessory enzymes involved in the hydrolysis of xylans. In, Enzymes in Biomass Conversion (Leatham, G.F. and Himmel, M.E., eds.) ACS Symp. Ser. 460, American Chemical Society, Washington DC, pp.426-436.

Prates, J.A.M., Tarbouriech, N., Charnock, S.J., *et al.,* (2001). The structure of the feruloyl esterase module of xylanases 10B from *Clostridium thermocellum* provides insight into substrate recognition. *Structure,* 9,1183-1190.

Puls, J., Schmidt, O. and Granzow, C. (1987). α-glucuronidase in two microbial xylanolytic systems. *Enzyme and Microbial Technology,* 9,83-88.

Puls, J., Tenkanen, M., Korte, H. and Poutanen, K. (1991). Products of hydrolysis of beechwood acetylacetyl-4-O-methylglucuronoxylanby a xylanase and an acetyl xylan esterase. *Enzyme and Microbial Technology,* 13,483-486.

Puls, J., Schorn, B. and Schuseil, J. (1992). In Biotechnology in pulp and paper industry. Kuwahara, M. and Shimada, M., editors. Uni Publishers, Tokyo,p357.

Qureshy, A.F., Khan, L.A. and Khanna, S. (2000). Expression of *Bacillus circulans* Teri-42 xylanase gene in *Bacillus subtilis. Enzyme and Microbial Technology,* 27,227-233.

Rabinovich, M.L., Melnik, M.S. and Bolobova, A.V. (2002a). Microbial cellulases, A review. *Applied Biochemistry and Microbiology,* 38(4), 305-321.

Rabinovich, M.L., Melnik, M.S. and Bolobova, A.V. (2002b). The structure and mechanism of action of cellulolytic enzymes. *Biochemistry (Moscow),* 67(8),850-871.

Rahman, A.K.M.S., Sugitani, N., Hatsu, M. and Takamizawa, M. (2003). A role of xylanase, α–L-arabinofuranosidase, and xylosidase in xylan degradation. *Canadian Journal of Microbiology,* 49,58-64.

Reese, E.T. and Shibata, Y. (1965). *Canadian Journal of Microbiology,* 11,167.

Rizzati, A.C.S., Jorge, J.A., Terenzi, H.F., Rechia, C.G.V. and Polizeli, M.L.T.M. (2001). purification and properties of a thermostable β–D-xylosidase produced by a thermotolerant *Aspergillus phonicis. Journal of Indian Microbiology and Biotechnology,* 25,156-160.

Roche, N., Berna, P., Desgranges, P. and Dutand, A. (1995). Substrate use and production of α–L-arabinofuranosidase during solid-state culture of *Trichoderma reesei* on sugar beet pulp. *Enzyme and Microbial Technology,* 17,935.

Ross, N.W., Johnson, K.G., Braun, C., MacKenzie, C.R. and Schneider, H. (1992). Enzyme hydrolysis of water-soluble lignus-carbohydrate complexs from populus deltoides. Effects of combinations of β-mannosidases xylanase and acetylaylan esserase. *Enzyme and Microbial Technology,* 14,90-95.

Rydholm, S.A. (1965). Pulping processes. Wiley, New York, p 95.

Saha, B.C. (2000). α–L-arabinofuranosidases, biochemistry, molecular biology and application in biotechnology. *Biotechnological Advances,* 18,403-423.

Sa-Peireira, P., Costa-Feirreira, M. and Aires-Barros, M.R. (2002). Enzymatic properties of a neutral endo-1,3(4)-β–xylanase Xyl 11 from *Bacillus subtilis. Journal of Biotechnology,* 94,265-275.

Saraswat, V., Bisaria, V.S. (1997). Biosynthesis of xylanolytic and xylan-debranching enzymes in *Melanocarpus albomyces* IIS 68. *Journal of Fermentation Bioengineering,* 83,352-357.

Segura, B.G.D., Durand, R. and Fevre, M. (1998). Multiplicity and expression of xylanases in the rumen fungus *Neocallimastix frontalis. FEMS Microbiology Letters,* 164,47-53.

Senior, D.J., Mayers, P.R. and Saddler, J.N. (1989). In Plant Cell Wall Polymers, Biogenesis and Biodegradation (Lewis NG and Paice MG, eds) ACS Symposium Series. 399, American Chemical Society, Washington, pp 641-654.

Shallom, D. and Shoham, Y. (2003). Purification and characterization of two thermotable acetyle oxylan esscrases from *Thermoanaerobacter* sp. strain JW/GL-YS485. Microbial hemicellulases. *Current Opinion in Microbiology,* 6,219-228.

Shao, W. 1993. Ph.D. Dissertation, University of Georgia, Athens, Georgia.

Shao, W. and Weigel,J. (1995). Purification and characterization of two thermostable acetyl xylan esterases from *Thermoanaerobacterium* sp. strain JW/SL-YS485. *Applied and Environmental Microbiology,* 61, 729.

Shao, W., DeBlois, S. and Wiegel, J. (1995). A high molecular weight, cell-associated xylanase isolated from exponentially growing *Thermoanaerobacterium* sp. strain JW/SL-YS485. *Applied and Environmental Microbiology,* 61,937-940.

Shao, W., Obe, S.K.C., Puls, J. and Weigel, J. (1995). Purification and characterization of α–glucuronidase from *Thermoanaerobacterium sp.* strain JW/SL-Y485, an important enzyme for the utilization of substituted xylans. *Applied and Environmental Microbiology,* 61, 1077-1081.

Shendye, A. and Rao, M. (1993). Molecular cloning and expression of xylanase from an alkaliphilic thermophilic *Bacillus* (NCIM 59) in *Bacillus subtilis* A8. *Enzyme and Microbial Technology,* 15,343-347.

Shendye, A., Gaikaiwari, R. and Rao, M. (1994). Expression of the cloned xylanases from an alkalophilic, thermophilic *Bacillus* in *Bacillus subtilis. World Journal of Microbiology and Biotechnology,* 10,414-416.

Shrinivasan, M.C. (1992). Lignocellulose Biotechnology, Recent Advances and Technology prospects. In, New Trends in Biotechnology. Subba,R.N.S., Balagopalan,C. and Ramakrishna,S.V.eds. Oxford and IBH Publishing Co. Pvt. Ltd., New Delhi, India. P 315-320.

Singh, A., Kuhad, R.C. and Kumar, M. (1995). Xylanase production by a hyperxylanolytic mutant of *Fusarium oxysporium. Enzyme and Microbial Technology,* 17, 551-553.

Sinner, M., Parameswaran, N., Yamazaki, W., Liese,W. and Dietrichs,H.H. (1979). *Advances in Chemistry Series,* 181, 303.

Siika-aho, M., Tenkanen, M., Buchert, J., Puls, J. and Viikari, L.(1994). An α–glucuronidase from *Trichoderma reesei. Enzyme and Microbial Technology,* 16,813.

Smith, M.M. and Hartley, R.D. (1983). *Carbohydrate Resources,* 118,65-80.

Smith, D.C., Bhat, K.M. and Wood, T.M. (1991). *Phanerochaete chrysosporium* β-glucuronidase. Induction, cellular localiration end physiclal characterization. *World Journal of Microbiology and Biotechnology,* 7,475-484.

Smith, M.H. and Gold, M.H. (1979). *Phanerochaete chrysosporium* β-glucosidases: Induction, Cellular localization and Physical characterization. *Applied and Environmental Microbiology,* 37,938.

Subramaniyam, S. and Prema, P. (2002). Biotechnology of microbial xylanases, enzymology, molecular biology and application. *Critical Reviews in Biotechnology,* 22,33-64.

Sundberg, M. and Poutanen, K. (1991). *Biotechnology and Applied Biochemistry,* 13,1-11.

Sun, J.L., Kawazu, T., Kimura, T., Karita, S., Sakka, K. and Ohmiya, K. (1997). High expression of the xylanase B gene from *Clostridium stercorarium* in tobacco cells. *Journal of Fermentation Bioengineering,* 84,219-223.

Sunna, A. and Antranikian, G. (1997). Xylanolytic enzymes from fungi and bacteria. *Critical Reviews in Biotechnology,* 7,425-430.

Sunna, A. and Bergquist, P.L. (2003). A gene coding a novel extremely thermostable 1,4-β–xylanase isolated directly from an environmental sample. *Extremophiles,* 7,63-70.

Tabernero, C., Sanchez-Torres, J., Perez, P. and Santamaria, R.I. (1995). Cloning and DNA sequencing of xyaA, a gene encoding an endo-β–1,4-xylanase from an alkalophilic *Bacillus* strain (N137). *Applied and Environmental Microbiology,* 61,2420-2424.

Taneja, K., Gupta, S. and Kuhad, R.C. (2002). Properties and applications of a partially alkaline xylanase from an alkalophilic fungus *Aspergillus nidulans* KK-99. *Bioresource Technology,* 85, 39-42.

Teixiera-Duarte, M.C., Cristina de S.E., Gomes, I.L., Nunes-Ponezi,A. and Davanzo,E. (2003). Xylan hydrolyzing enzyme system from *Bacillus pumilus* CBMAI 0008 and its effect on Eucalyptus grandis kraft pulp for bleaching improvement. *Bioresource Technology,* 88,9-15.

Tenkanen, H., Puls, J. and Poutanen, K. (1992). Two major xylanases of *Trichoderma reesei. Enzyme and Microbial Technology,* 14, 566-574.

Tenkanen, M., Buchert, J. and Viikari, L. (1995). Binding of hemicellulases on isolated polysaccharide substrates. *Enzyme and Microbial Technology,* 17,499-505.

Tenkanen, M., Thornton, J. and Viikari, L. (1995). An acetylglucomannan esterase of *Aspergillus oryzae*; purification, characterization and role in the hydrolysis of *O*-acetyl-galactoglucomannan. *Journal of Biotechnology,* 42,197.

Tenkanen, M., Siika-aho, M., Hausalo, T., Puls, J. and Viikari, L. (1996). In Biotechnology in the pulp and paper industry. Sribotnik, E. and Messner, K., editors. Facultas-Universitalsverlag, Vienna, p503.

Timmel, T.E. (1964). In Advances Carbohydrate Chemistry (Wolfrom ML and Tipson RS, eds) Academic press, New York 19,247-302.

Timmel, T.E. (1967). Recent progress in the chemistry of wood hemicellulases. *Wood Science and Technology,* 1,45.

Vander veen, P., Flipphi, M.J.A., Voragen, A.G.J. and Visser, J. (1991). Induction, purification and characterization of arabinases produced by *Aspergillus niger. Archieves in Microbiology,* 159,66-71.

Viikari, L., Tenkanen, M., Buchert, J., Ratto, M., Bailey, M., Siika-aho, M. and Linko, M. (1993). In, Saddler, J. (ed.). Bioconversion of forest and agricultural wastes. CAB International, Oxford, p131.

Vincent, P., Shareck, F., Dupont, C., Morosoli, R. and Kluepfel, D. (1997). New α–L-arabinofuranosidase produced by *Streptomyces lividans,* cloning and DNA sequence of the abfB gene and characterization of the enzyme. *Biochemistry Journal,* 322,845-852.

Waino, M. and Ingvorsen, K. (2003). Production of β–xylanase and β–xylosidase by the extremely halophilic archaeon *Halorhabdus utahensis. Extremophiles,* 7,87-93.

Weigel, J., Mothersed, C.P. and Puls, J. (1985). Differences in xylan degradation by various *noncellulolyticus* thermopeutic anaerobes and clostridium hemocellum. amerobes and clostridium thermocellum. *Applied and Environmental Microbiology,* 49,656.

Wejse, P.L., Ingvorsen, K. and Mortensen, K.K. (2003). Purification and characterization of two extremely halotolerant xylanases from a novel halophilic bacterium. *Extremophiles,* 7,423-431.

Whistler, R.L., Richards, E.L. (1970). Hemicelluloses. In, Pigman W, Horton D (eds) The carbohydrates. Academic Press, New York, pp 447-469.

Williams, A.G. and Withars, S.E. (1981). *Journal of Applied Bacteriology,* 51,375-385.

Williams, A.G. and Withars, S.E. (1982). The production of plant cell wall polysaccharide degrading enzymes by hemicellulolytic rumen bacterial isolates grown on a range of carbohydrate substrates. *Journal of Applied Bacteriology,* 52,377-387.

Williams, A.G. and Withars, S.E. (1992). The regulation of xylanolytic enzyme formation by *Butyrivibrio fibrisolvens* NCFB 2249. *Letters in Applied Microbiology,* 14,194-198.

Wong, K.K.Y., Tan, L.U.L. and Saddler, J.N. 1988. Multiplicity of β–1,4-xylanase in microorganisms, functions and applications. *Microbiology Reviews,* 52,305-317.

Wubah, D.A., Akin, D.E. and Borneman, W.S. (1993). *Critical Reviews in Microbiology,* 19,99.

Yamura, I., Koga, T. and Matsumoto, T. (1997). Purification and some properties of endo-1,4-β–xylanase from a fresh water mollusk *Pomacea insularus* (de Oringny). *Bioscience Biotechnology and Biochemistry,* 61,615-620.

# 9

# Xylanases: Structure, Molecular Cloning and Regulation of Expression

**R.P. Singh, Pallavi Dwivedi, Vivekanand and Nidhi Kapur**

*Department of Biotechnology, Indian Institute of Technology Roorkee Roorkee-247 667, India*

Hemicellulolytic enzymes improve the overall economics of bioconversion of abundantly available hemicellulosic biomass into value added products. Hemicellulose constitutes the 20-30% of total plant biomass in association with cellulose. Xylan is the major structural polysaccharide, along with mannans, galactans and arabinans in the complex heteropolysaccharide. Due to complex and branched structure, complete hydrolysis of xylan requires several hydrolytic enzymes including endoxylanases, β-xylosidases, arabinofuranosidases, glucuronidases etc. A repertory of organisms including fungi, bacteria, yeast, marine algae, protozoans, snails, crustaceans, insects and seeds etc are known for the xylanase production ability. Xylanase gene has been isolated, cloned characterized and expressed from several microbes into different homologous as well as heterologous hosts in order to understand its transcriptional and translational regulation. Induction of xylanase biosynthesis in presence of several low molecular weight compounds which are end products of xylan degradation shows inducible nature of enzyme. Effect of these inducer molecules vary from species to species due to structural diversity of polymer. A consensus sequence (TATAAA), which is similar to eukaryotic promoter sequence, is recognized as promoter sequence of xylanase gene in many organisms. The TGT/AAANC/GNTNA/TGA sequence is identified as catabolic repression consensus sequence. The three-dimensional structure of family G / 11 xylanases is β-jelly roll fold while xylanases of family F / 10 show eight fold β / α barrel (TIM barrel) structure. Among all xylanases, two glutamate residues are found to be conserved and involved in the catalysis. Stability of enzyme at higher temperature and pH can be improved by site-directed mutagenesis through specific amino acid replacement. Both natural and genetically engineered xylanases have wide applications in food and feed industry, paper industry, baking industry, biofuel, coffee extraction, clarification of fruit juices and animal feed industry.

## INTRODUCTION

Hemicellulose is an abundantly available renewable organic biomass on the planet with tremendous applications. It constitutes the 20-30% of total plant biomass in association with cellulose (Kulkarni *et al.,* 1999). Hemicellulose is present as a storage polysaccharide in seeds and known as component of plant cell wall (Izydorczyk and Biliaderis 1995; Polizeli *et al.,* 2005). Hemicelluloses are a complex heteropolymer made up of glucuronoxylans, arabinoglucuronoxylans, glucomannans, arabinogalactans and galactomannans (Haltrich *et al.,* 1996; Sunna and Antranikian 1997; Kulkarni *et al.,* 1999; Subramaniyan and Prema 2002). Xylan is a major component of hemicellulose contributing 15-30% of the total dry weight in angiosperm and 7-12 % in gymnosperm (Haltrich *et al.,* 1996; Singh *et al.,* 2003). Other minor polysaccharides are mannans, glucans, arabinans and galactans (Whistler and Richards 1970; Viikari *et al.,* 1994; Uffen 1997; Ebringerova and Heinze 2000). The backbone of xylan consists of β-1, 4-xylopyranosyl residues. Xylanases are glycosidase, which hydrolyze the endo-β-1, 4-xylopyranosyl linkages and degrade xylan into xylooligosaccharides. These enzymes are produced by a repertoire of organisms including fungi, bacteria, yeast, marine algae, protozoans, snails, crustaceans, insects, seeds etc. (Polizeli *et al.,* 2005; Collins *et al.,* 2005). The end product of xylan degradation have considerable industrial applications in bioconversion of lignocellulose to biofuel (ethanol), artificial sweatner (xylitol and furfural), animal feed production, baking and textile industry, clarification of fruit juices and coffee extraction. Besides this there has been an increasing interest in applying xylanases to prebleaching of kraft pulp in the paper industry (Viikari *et al.,* 1994).

Considering the tremendous industrial applications of xylanases it is necessary to understand the regulatory and structural aspects of enzyme, as that would be of immense significance in developing a technology for the economical production of industrially compatible enzyme preparation.

## STRUCTURE OF POLYSACCHARIDE

Xylan is a complex branched heteropolysaccharide. It forms covalent linkage with lignin and noncovalent linkage with cellulose and maintains the structural integrity of plant cell wall. (Kulkarni *et al.,* 1999; Collins *et al.,* 2005; Polizeli *et al.,* 2005). Structure of xylan varies among different plant species. The main chain of xylan consists of endo-β-1,4-xylopyranosyl residues having side group substitution to varying degree with glucuronopyranosyl, 4-O-methyl-D-glucuronopyranosyl, α–L-arabinofuranosyl, acetyl, feruloyl and p-coumaroyl groups (Kulkarni *et al.,* 1999; Li *et al.,* 2000) (Figure 1). Hardwood xylan and softwood xylans are different in terms of their degree of polymerization and side group substitution. The degree of polymerization of hardwood xylan is higher (150-200) as compared to softwood xylan (70-130) (Kulkarni *et al.,* 1999). Hardwood contains acetylated xylans and softwood contains arabinoxylans (Thomson 1993). Linear unsubstituted xylan has also been reported in esparto grass, tobacco and certain marine algae (Collins *et al.,* 2005). Besides this β-1, 3-linked xylopyranosyl units are reported in xylans from certain marine algae.

## XYLANOLYTIC ENZYME SYSTEM

Due to its heterogeneity and complexity, the complete hydrolysis of xylan requires a large variety of cooperatively acting enzymes (Subramaniyan and Prema 2002). Endo-1, 4- β-D-xylanase (EC 3.2.1.8) randomly cleaves the xylan back bone at endo-β-1,4-xylosidic linkage and releases xylooligosaccharides while β-D-xylosidases (EC 3.2.1.37) cleave the terminal β-1,4-xylosidic linkage and releases xylose monomers from the reducing end of xylooligosaccharides and xylobiose. Besides above-mentioned two enzymes, removal of side group is catalyzed by α–L-arabinofuranosidases (EC 3.2.1.55), α–L-glucuronidases (EC 3.2.1.139), acetyl xylan esterases (EC 3.2.1.72) and ferulic acid esterases (EC 3.1.1.73). All these above-mentioned enzymes constitute the xylanase enzyme system (Figure 1).

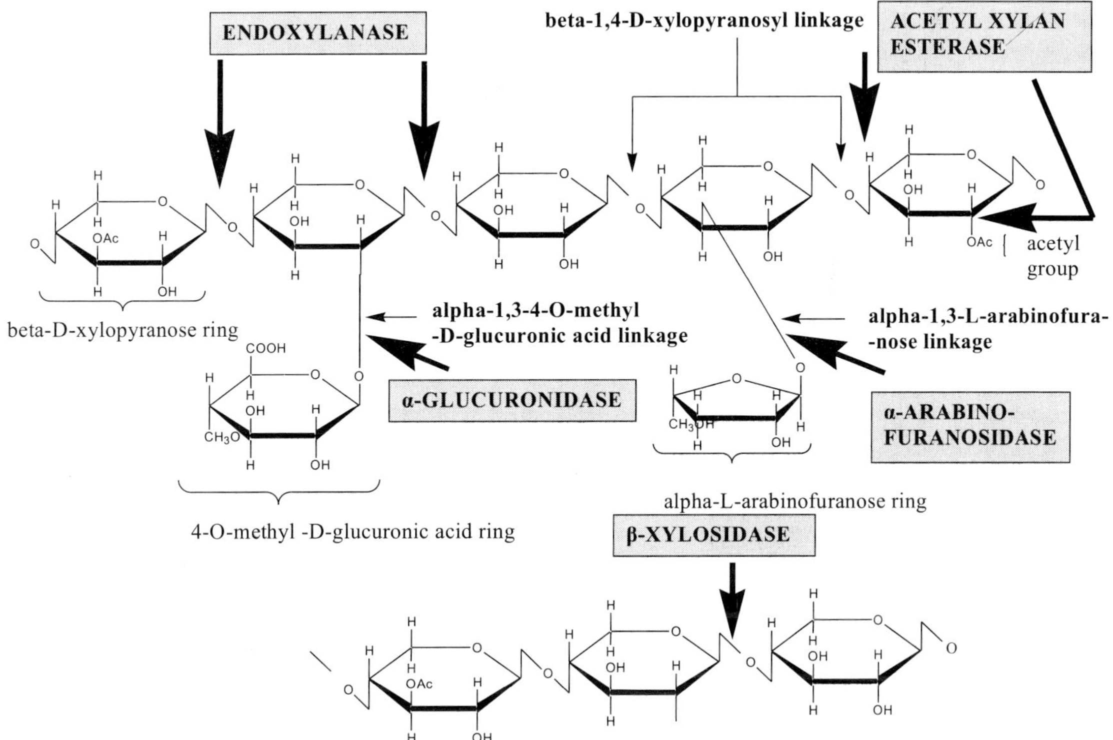


**Fig. 1.** Structure of xylan and various enzymes of xylanolytic system.

## CLASSIFICATION OF XYLANASES

According to the molecular weight and isoelectric point, Wong (1988) classified xylanases into two groups. All high molecular weight xylanases (>30 kDa) with lower isoelectric point were placed in one group while the other group contains xylanases having lower molecular weight (<30 kDa) and with high isoelectric point. The heterogeneity and complex structure of xylan has resulted in diverse xylanases with varying substrate specificities, primary and secondary structures and folds that put a limitation with the classification of these enzymes by substrate specificity alone. Later on Henrissat (1989) classified cellulases and xylanases into six families (A-F) using amino acid sequence similarity and hydrophobic cluster analysis. In 1993, Hanrissat upgraded these families and subdivided xylanases into two families that are family F/10 and family G/11 (Henrissat and Bairoch, 1993) comprising high and low molecular weight xylanase, respectively. Based on primary structure comparisons of the catalytic domains, a more complete classification system was proposed for glycosidases in general including xylanases. Characteristic features of these glycoside hydrolase families are described in Table 1. Currently 96 glycoside hydrolase families have been reported to exist (Coutinho and Henrissat 1999) (http://afmb.cnrs.mrs.fr/~cazy/CAZY/.). Enzymes within a particular family have a similar three-dimensional structure and due to divergent evolution some families have related three-dimensional structures. Thus these families are grouped into higher hierarchical levels known as clans (Bourne and Henrissat 2001). Presently, 14 different clans have been proposed (GH-A to GH-N), with most clans encompassing two to three families, apart from clan GH-A which currently contain 17 families. Within this classification system xylanases mainly confined to family-F/10 and family-G/11 (Table 1). But enzymes with xylanase activity are also found in families 5, 7, 8, 16, 26, 43, 52 and 62. Xylanase in families 5, 7, 8, 10, 11 and 43 contains a distinct catalytic domain with a demonstrated endo-1, 4-β-xylanase activity while xylanases of families 16, 52 and 62 are bifunctional with two catalytic domains.

## CLONING OF XYLANASE GENE

To understand the structure and regulation of xylanase gene at molecular level it needs to be isolated, characterized and cloned into suitable host system. Scaling up production of enzyme with

**Table 1. Characteristic features of glycoside hydrolase families**

| *Glycoside Family* | *Enzyme Activity* | *Structure* | *Conserved Residues* | *Mode of Action* | *Organisms* |
|---|---|---|---|---|---|
| 5 | Cellulose, licheninase, mannosidase, endoglycosylceramidase, glucanase, 1,3-β-glucosidase, endo-1,4-β-xylanase | $(\beta/\alpha)_8$ barrel fold | Glutamate - Glutamate | Retaining | *Aeromonas puncata* ME-1 *Clostridium cellulovorans* *Erwinia chrysanthemi* D1 |
| 7 | Endo glucanase, endo xylanase | β-jelly roll fold | Glutamate - Glutamate | Retaining | *Trichoderma reesei* |
| 8 | Cellulases, Chitosanases, lichenases and endoxylanases | $(\alpha/\alpha)_6$ barrel fold six xylose binding site in substrate binding cleft | Glutamate - Aspartate | Inverting | *Bacillus halodurans* C-125 *Bacillus sp.* KK-1 *Pseudoalteromonas haloplanktis* TAH 3a |
| 10 | Endo-1,4-β-xylanases, endo-1, 3-β-xylanases, cellobiohydrolases | $(\beta/\alpha)_8$ barrel fold four-five substrate binding site | Glutamate - Glutamate | Retaining | *Geobacillus stearothermophilus* T-6 *Streptomyces lividans* *Thermatoga maritima* |
| 11 | Exclusively xylanase activity | β-jelly roll fold | Glutamate - Glutamate | Retaining | *Aspergillus kawachii* *Aspergillus niger* *Bacillus circulans* *Streptomyces* S-38 *Trichoderma harzianum* E-58 |
| 43 | Lichenase, xylanase | 5-blade β-propeller | Glutamate - Aspartate | Inverting | *Caldicellulosiruptor sp.* *Bacillus sp.* |

economic viability and industrial compatibility is possible when attempts are made for its higher level of production. It would require knowledge about regulatory mechanisms and appropriate inducers of xylanase gene. Extremozymes have tremendous industrial applications due to their compatibility with harsh working conditions (Nakamura, 2003). Recombinant DNA technology can be employed as an excellent tool for desired modifications in the enzyme systems. Up till now xylanase gene from various microbes have been cloned and expressed into suitable hosts with various objectives. Structure of xylanase gene of *Streptomyces olivaceoviridis* E-86 was deduced by cloning a GC rich gene (68%) for an F/10 xylanase into suitable vector, using a long range inverse PCR method (Kuno *et al.,* 1998). The open reading frame of the cloned gene, fxyn, contained 1431 bp and encodes 477 amino acid residues. Xylanase gene from an alkalophilic *Bacillus* sp. strain 41 M-1 was cloned and sequenced into *E. coli.* Xylanase J gene of 2.2 kb size contained an open reading frame of 1062 bp and encoded a 27 amino acid leader peptide followed by a 327 amino acid mature enzyme (Nakamura, 2003). Cloning of xylanase gene has been reported in both heterologous and homologous hosts. The xylanase genes have been isolated from different microbial genera and expressed in *E. coli.* A high molecular weight xylanase (110 kDa), which was encoded by xyn-E gene was cloned into *E. coli* (Liu *et al.,* 2003). Upon nucleotide sequence analysis it was found that xyn-E comprises 2823 bp and encodes a protein of 941 amino acid residues, which was similar to endo-β-1, 4-xylanases of family-10 of glycosyl hydrolases. Xylanase gene of *Streptomyces flavogriseus* was cloned into *E. coli* using pUC8 plasmid as vector (Srivastava *et al.,* 1991). The gene was located in the 0.8 kb Hind-III fragment and encoded for 18 kDa molecular weight xylanase. Another example of heterologous cloning was the cloning of foreign xylanase gene (xyn-A) of *Eubacterium ruminatium* in *Butyrivibrio fibriosolvens.* The transformants has 7-14-fold higher xylanolytic activity as compared to parent (Kobayashi *et al.,* 1998). Three genes encoding two types of xylanases (stx-1 and stx-11 and an acetyl xylan esterase (stx-111) from *Streptomyces thermoviolaceus* OPC-520 were cloned into heterologous (*E. coli* JM109) as well as in homologous host (*Streptomyces griseus* PSR2) (Kobayashi *et al.,* 1998). Analysis of the sequence of the pXY 95 revealed the presence of a complete open reading frame, stx-1 (bp 121-1548), coding for a 476 amino acids with a molecular weight of 47 kDa. Similarly nucleotide sequence analysis of pXY 87 showed two open reading frames of 1005 (stx-II) and 993 (stx-III) bp, which are transcribed from the same strand and separated by 37 bp. The open reading frame of stx-II and stx-III encoded a protein of 335 amino acids with a molecular weight of 35 kDa and a protein of 331 amino acids with a molecular weight of 34 kDa respectively. The xylanase gene of *Bacillus circulans* Teri-42 was cloned in both *Bacillus subtilis* and *E. coli* (Qureshy *et al.,* 2000). The enzyme activity was almost 87% higher in *B. subtilis* (pBA7) than in *Eschirichia coli* (pAQ4). The absence of post-translational modifications such as glycosylation in *E. coli* and the intracellular accumulation of the recombinant xylanases have been suggested to be the crucial factors leading to the lower levels of activity (Kulkarni *et al.,* 1999). The characteristic feature of recombinant xylanase in *B. subtilis* was its stability at higher temperatures. Recombinant enzyme was comparatively stable at 60° C whereas the native enzyme was found to be unstable at that temperature (Qureshy *et al.,* 2000). Endoxylanases from a thermophilic bacterium *Dictyoglomus thermophilum* Rt46B, showed maximum activity at 85° C. Further, the endoxylanase gene from this bacterium was cloned and was effectively transcribed in *Trichoderma reesei* (Te'O *et al.,* 2000). There are numerous reports available for cloning of the xylanases of bacterial origin; however, very few reports are available for xylanase gene cloning of fungal origin. dc Faria *et al.,* (2002) described the heterologous expression of the *Humicola grisea* var thermoidea xyn-2 gene (cDNA) in *Trichoderma reesei.* A xyn-I cDNA from *Aureobasidium pullulans* var melanigeneum was cloned and expressed in *Saccharomyces cerevisiae* (Ohta *et al.,* 2001). An open reading frame, consisting of 663 bp, encoded a mature protein of 187 amino acid. For bulk production of enzyme, xylanase gene (xyn-B) from thermophilic bacterium *Dictyoglomus thermophilum* was cloned into two fungal hosts, i e. in *Kluyveromyces lactis* and in the *Trichoderma reesei* (Bergquist *et al.,* 2003). Different cDNAs designated as xyn-3 and xyn-4 were isolated from an expression library of the anaerobic rumen fungus *Neocallimastix frontalis* and cloned into *E. coli.* (Durand *et al.,* 1996).

## REGULATION OF EXPRESSION

Production of xylanolytic enzymes from both bacteria and fungi has been demonstrated to be inducible in nature. Smaller molecules, which are the degradation products of complex polysaccharide xylan, act as inducers. Basal level of these hydrolytic enzymes is always detectable in the cell (Kulkarni *et al.,* 1999). This leads to the generation of small signalling molecules that are generated by the degradation of the substrate molecules. These signalling molecules are able to enter the cell and induce the synthesis of the corresponding enzyme. The potent inducer molecules are xylose, xylobiose, xylooligosaccharides, heteropolysaccharides of xylose and glucose and their positional isomers (Figure 2). In general, the xylanase induction is a complex phenomenon and the level of

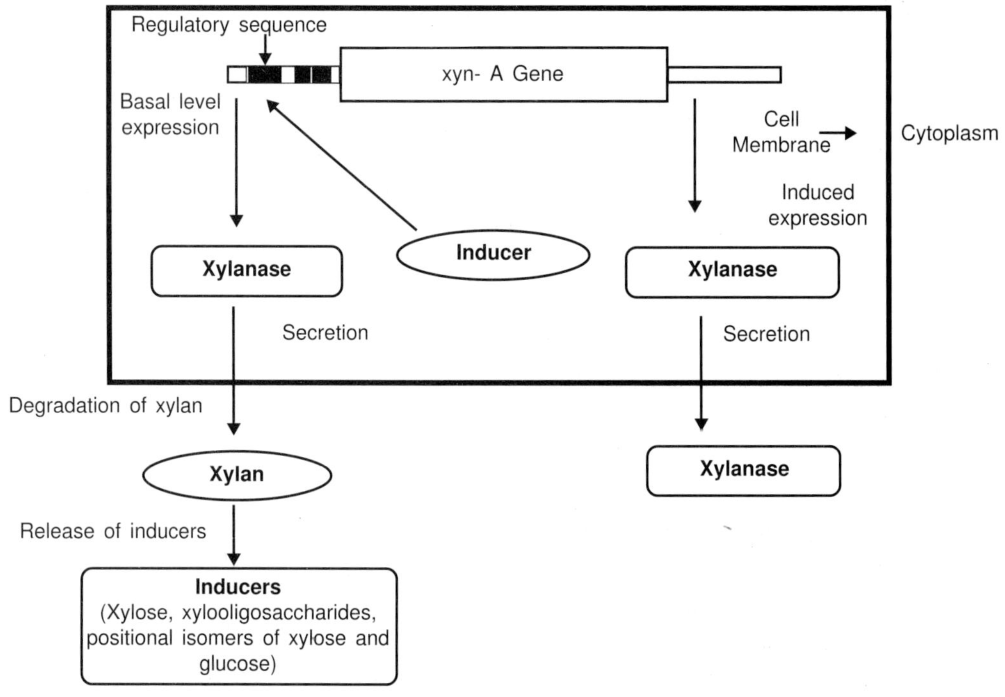


**Fig. 2.** Regulation of xylanase biosynthesis

response against particular inducer varies from species to species. An inducer producing maximum xylanase activity in one species may be the inhibitor of activity in other species (Haltrich *et al.,* 1996; Beg *et al.,* 2001). Xylanase production was examined by Kohli *et al.,* (2001) using different carbon sources. Maximum xylanase production was observed with the xylan-containing medium. Xylanase production was observed to increase in presence of lignocellulosic materials such as wheat bran, rice husk, bagasse, cotton fibres and rice straw (Biely 1985; Kohli *et al.,* 2001; Anthony *et al.,* 2003). It was proposed that xylan or xylan containing carbon sources are necessary for the xylanase production because its hydrolysis products serve both as inducers as well as carbon source for the organism. It is evident that the preference for lignocellulosic substrate depends upon the strain. In *Aspergillus fumigatus* maximum xylanase was produced in the presence of rice straw (Anthony *et al.,* 2003). However, *Aspergillus fischeri* and *Cephalosporium* sp. have been reported to produce high levels of xylanase with wheat bran as compared to the rice straw and bagasse (Bansod *et al.,* 1993; Chandra and Chandra 1995). The effect of different synthetic and natural compounds on xylanase production by *Clostridium absonum* CFR-702 was studied (Swaroopa Rani and Krishna Nand, 2000). Among these, maximum xylanase activity was obtained with birch wood xylan (BWX), larch wood xylan (LWX), and oat spelt xylan (OSX) followed by xylose and xylobiose. However, in *Cryptococcus albidus,* xylose repressed xylanase production (Biely, 1985). The possible reason behind poor inducer response of xylose as compared to xylan is its quick consumption by organism due to easy availability in culture media. The pH dependent induction of xylanase gene by xylan and xylose was reported by Anthony *et al.,* (2003) in *Aspergillus fumigatus.* Cellulose has also been shown to act as an inducer of the xylanase in few strains, namely, *Streptomyces sp.* (Perez-Avalos *et al.,* 1996).

Regulation of xylanase gene at molecular level is comparatively less investigated. Analysis of DNA fragments containing β-xylanase genes from *Bacillus pumilus,* indicated that the xylanase and xylosidase genes are closely associated and linked in a 14.4 kb DNA fragment (Moriyama *et al.,* 1987). In *Caldicellulosiruptor* Rt8B.4 a thermophilic bacterium, xylanase gene (xyn-A) appeared to be an operon-like structure involving genes associated with xylose metabolism. The gene Xyn-A acts as a 1, 4-β-endoxylanase, β-D-xylosidase and α–L-arabinofuranosidases (Dwivedi *et al.,* 1996). Screening of genomic library of *Aspergillus oryzae* KBN616 showed that the structural part is 725 bp long and was interrupted by a single intron of 62 bp. The intron/exon junction followed the GT-AG rule and the size of the introns resembled with introns of other filamentous fungi. The open reading frame consists of 221 amino acid residues. Amino acid sequence comparison showed the presence of a 32 amino acid long signal sequence at the amino terminal of precursor xyn-$G_1$. The 5' non-coding region of the gene contained important consensus sequences. A consensus sequence (TATAAA) which is similar to eukaryotic promoter sequence that is TATA box (TATAAAT) exists 86 bp upstream from the translation initiation site. Furthermore, the CT rich sequence, which is considered as a fungal promoter element, was located just downstream of the TATA box like sequence. A similar promoter sequence (TATAAT) was also detected in *Bacillus* sp. and was located 125 bases upstream of the ribosome-binding site. Promoter like sequence and typical Shine-Dalgarno (SD) sequence were observed upstream from the possible start codon (TTG instead of ATG). A perfect 14-bp inverted repeat, corresponding to a transcriptional terminator, occurred downstream from the TAG stop codon (Nakamura *et al.,* 2003). The nucleotide sequence from *Streptomyces thermoviolaceus* OPC-520 showed that genes stx-II and stx-III were clustered on the genome. GGAG, the potential ribosome binding sites were located upstream of each initiation codon ATG. A characteristic A+T rich promoter sequence was found upstream of the ribosome binding site (Kobayashi *et al.,* 1998). Xylanase gene expression was observed to decrease in presence of easily metabolizable sugars such as glucose. Catabolic repression is the main reason behind this which is a common phenomenon observed in xylanase biosynthesis. The catabolic repression consensus sequence identified was TGT/AAANC/GNTNA/TGA, where underlined letters represented the most critical bases (Kulkarni *et al.,* 1999).

## STRUCTURAL ELEMENTS IN XYLANASES

In order to modify enzymes for their suitability for industrial applications, determination and analysis of structure is an essential prerequisite. X-ray crystallography, NMR and Mass spectrophotometry are potential techniques that have been used extensively for structural determination of the proteins. Chemical modifications using group specific reagents and competitive inhibitors provide important information about active site residues. On the basis of hydrophobic cluster analysis and sequences homology, xylanases can be grouped into two families. Three-dimensional crystal structure of xylanase from *Aspergillus niger* (Krengel and Dijkstra 1996), *Streptomyces sp 38* (de lemos Esteves *et al.,* 2004), *Palcilomyces varioti* (Kumar *et al.,* 2000), *Trichoderma ressei* (Torronen *et al.,* 1994) and *Dictyoglomus thermophilum* have been deciphered and confined to family G/11. Xylanases belonging to this family consist of a single domain with β-jelly roll fold that predominantly consists of β-strands. Two β – sheets are twisted around a deep, long catalytic cleft which is large enough to accommodate atleast four xylose residues. Single α–helix is present in catalytic domain. The catalytic cleft contained two highly conserved glutamate residues but their position may vary from organism to organism. $Glu^{79}$ and $Glu^{170}$ in *A. niger,* $Glu^{86}$ and $Glu^{178}$ in *Palcilomyces varioti,* $Glu^{90}$ and $Glu^{180}$ in *Dictyoglomus* and $Glu^{86}$ and $Glu^{177}$ in *Trichoderma ressei.* Due to conserved nature, it was proposed that they participate in the catalysis where one acts as proton donor and other as nucleophile (Krengel and Dijkstra 1996; Kulkarni *et al.,* 1999; Collins *et al.,* 2005). Chemical modification of amino acids indicated the presence of tryptophan (Nath and Rao 1998) and tyrosine in the substrate-binding site (De lemos Esteves *et al.,* 2004). The family – 10 xylanases have been reported from *Streptomyces olivaceovirdis* E-86 (Fujimoto *et al.,* 2004), *Geobacillus stearothermophilus* (Teplitsky *et al.,* 2004) and *Pseudomonas fluorescens* (Harris *et al.,* 1996). The tertiary structure consists of an eight-fold β/α barrel, denoted as TIM fold. Such xylanases contain a catalytic domain and a substrate-binding domain, which are joined by a Gly/Pro – rich linker. The two essential catalytic residues are $Glu^{159}$ and $Glu^{265}$ in *Geobacillus* (Teplitsky *et al.,* 2004) and $Glu^{127}$ and $Glu^{246}$ in catalytic cleft. A multidomain enzyme xylanase J had been reported from alkalophilic *Bacillus* sp. Strain 41 M – 1 (Nakamura *et al.,* 2003). It consists of two functional domains – a family G/11 catalytic domains and a non-catalytic xylan binding domain.

## PROTEIN ENGINEERING

Many microorganisms have been discovered at unusual environments such as extreme temperature, pH and salinity. Such microbes are known as extremophiles and enzymes produced by them as "extremozymes". Extremozymes have tremendous potential for industrial applications as well as for academic directions (Nakamura *et al.,* 2003). Biotechnological applications of the xylanases require thermostable enzyme preparation with stability at wide pH and temperature range. Since availability of the ideal enzyme preparation is limited, the application of protein engineering has gained importance. Proteins earlier were modified through random replacement of active site residues through chemical modifications but advances in recombinant DNA technology have made it possible for the individual amino acids to be substituted by site directed mutagenesis. The highly conserved amino acid residues located at specific position in xylanase are important in structure function analysis and hence are targeted for site directed mutagenesis (Kulkarni *et al.,* 1999). A shift in the optimal pH for enzyme activity from pH 9.0 to pH 5.0 and 7.0 was observed by Nakamura (2003) by replacing the acidic amino acids (Asp$^{14}$, Glu$^{16}$, Glu$^{55}$ and Glu$^{177}$) in the catalytic domain with alkaline one. This indicated that these amino acids were responsible for maintaining structural integrity. Presence of disulfide linkage, specific ion pairs, buried water molecules and the lower packing value are some factors known to contribute to the thermostability of the xylanase from the thermophilic fungus. Further thermophilic fungus possesses a disulfide linkage between Cys$^{110}$ and Cys$^{154}$ (Kumar *et al.,* 2000) while mesophilic xylanases have Ser and Asn at these positions. Mutational studies carried out on xylanase gene of *B. circulans* showed that the introduction of a disulfide linkage at this position confers a slight increase in thermal stability (Wakarchuk *et al.,* 1994). Similarly, increase in optimal pH for activity by one unit was observed by replacing Asn$^{48}$ to Asp$^{48}$ in *Streptomyces* sp. S-38 xylanase (de lemos Esteves *et al.,* 2004). Site directed mutagenesis is also helpful in increasing the substrate specificity of xylanase. Xylanase having activity for the xylans from various sources are technically more favourable. Any mutation at the conserved Pro$^{126}$ at the tip of the thumb and Trp/Tyr at the end of strand B2 can change geometry of active-site cleft (Kumar *et al.,* 2000). This allows the entry and binding of small as well as large substrate molecules, hence improvement in substrate specificity. In case of xylanase from *Streptomyces lividans* 1326 the thermo stability is increased by replacing Arg$^{156}$ by Glu$^{156}$. The modified enzyme had the optimal temperature 5°C higher than that of the wild type (Moreau *et al.,* 1994).

## APPLICATIONS

In recent years the biotechnological use of xylanase has grown remarkably (Beg *et al.,* 2001; Collins *et al.,* 2005; Polizeli *et al.,* 2005). The end products of xylan degradation are of considerable importance. High specificity, mild reaction conditions, environmental safety and no loss of substrates due to chemical modifications are some advantages of enzymatic hydrolysis as compared to chemical one. The main industrial application of xylanase is for the bleaching of pulp during paper manufacturing. Removal of lignin, which is mainly responsible for the colour of pulp, is known as bleaching. Final brightness of paper mainly depends upon the efficiency of bleaching stage. Up to now paper industry relies mainly upon chemical bleaching for the production of fully bleached white pulp. Effluent from pulp and paper industry contains absorbable organic halides (AOX) as well as total organic halides (TOX) and other chlorine based compounds. These compounds cause air and water pollution. These are toxic, mutagenic, persistent, bioaccumulating in nature and cause numerous harmful disturbances in biological systems. Some of the other alternatives of chlorine bleaching are the use of $O_2$, $O_3$, $H_2O_2$ etc. for the same. These are less harmful as compared to the chlorine compounds but are less preferable due to their non-selective nature. They also degrade cellulose fibres, which are responsible for the strength of paper. Removal of lignin with the help of microbes or microbial enzymes is known as biobleaching and it proved to be the best alternative of chemical bleaching in pulp and paper industry. Efficiency of biobleaching mainly relies on the capability of xylanase enzyme system along with lignolytic enzymes. Xylanase pretreatment of pulp, prior to bleaching reduces the requirement of bleaching chemicals. The decreased chemical consumption results into significant reduction in production cost. Results from laboratory studies and mill trials had shown about 35-41% reduction in chlorine consumption at the chlorination stage for hardwoods and 10-20% for softwoods. Xylanase treatment also results into increased final brightness of pulp. Xylanase pretreatment has led to reduction in effluent AOX and dioxin concentration due to the reduced chlorine requirement to achieve the desired final brightness. Xylanase treated pulp shows improved strength properties and improved viscosity. Thus, biobleaching is the simple, economical and appears to be the best alternative of chemical bleaching. Besides their role in pulp and paper industry xylanases have potential applications in wide range of industrial processes. In food industry they are used for food and vegetable processing, brewing and wine production. They are also used specially for the baking industry for improved quality of bakery products with increased shelf-life. They also have important role in improving the quality of animal food by decreasing the content of non-starch polysaccharides. Xylanases are also used for bioconversion of hemicellulosic raw materials into fermentable sugars e.g. D-xylose. These sugars can further be converted into fuel ethanol and fine chemicals by fermentation. D-xylose can be converted into D- xylitol, which is used as artificial sweetener. Some minor uses of xylanases are in brewing, coffee extraction, in detergents and in the production of pharmaceutically active polysaccharides for use as antimicrobial agents or antioxidants. Commercial xylanases for various applications are industrially produced in Japan, Finland, Germany, Ireland, Denmark, Canada and USA. The microorganisms used to obtain these enzymes are *A. niger, Trichoderma sp* and *Humicola onsolens* (Polizeli *et al.,* 2005). At present, biobleaching is on trail in many countries around the globe.

## CONCLUSION AND FUTURE PROSPECTS

Global markets for industrial enzymes reached from $ 1 billion in 1995 to almost $ 2 billion in 2001 and continue to increase as new enzymes and applications are discovered. Hydrolases constitute approximately 75% of the market for industrial enzymes with the glycosidases including cellulases, amylases and hemicellulases constituting the second largest group after proteases. The application of enzymes for various processes is relatively simpler, safe, produces no harmful byproducts and is an ecofriendly approach. Considering these, governments in many countries are encouraging industries to adopt these safer and ecofriendly approaches rather than the conventional chemical based processes. Thus, in order to make these enzymes a better and economical choice, it is necessary to develop enzyme preparations that are low priced and are more compatible for industrial applications. Recent technological advancements and exploration of novel microflora on the planet are the two approaches that can be used for compatible enzyme preparations. The processes can further be economized by evaluating and choosing the proper fermentation systems and using the cheaper and easily available lignocellulosic residues in the process for production.

Therefore, in future, search should be focused for the development of improved enzyme with suitable properties having better performance achieved 'at the lower cost.

## REFERENCES


Anthony, T., Raj, K.C., Rajendran, A. and Gunasekar, P. (2003). High molecular weight cellulase free xylanase from alkali tolerant *Aspergillus fumigatus* AR-1. *Enzyme and Microbial Technology,* 32, 647-654.

Bansod, S.M., Dutta-Choudary, M., Srinivasan, M.C. and Rele, M.V. (1993). Xylanase active at high pH from an alkalo-tolerant *Cephalosporium* sp. *Biotechnology Letters,* 35, 965-970.

Beg, Q.K., Kapoor, M., Mahajan, L. and Hoondal, G.S. (2001). Microbiol xylanases and their industrial application: a review. *Applied Microbiology and Biotechnology,* 56, 326-338.

Bergquist PL, Te'o V.J.S., Gibbs MD, Curach, N.C. and Nevalainen, K.M.H. (2003). Recombinant enzymes from thermophlic Micro-organisms expressed in fungal hosts. *Thermophiles,* 293-297.

Biely, P. (1985). Microbiol xylanolytic systems. *Trends in Biotechnology,* 11, 286-290.

Bourne, Y. and Henrissat, B. (2001). Glycoside hydrolases and glycosyltransferases: families and functional modules. *Current Opinion in Structural Biology,* 11, 593-600.

Chandra, R.K. and Chandra, T.S. (1995). Production of cellulose-free xylanase from an alkalo-tolerant *Aspergillus fischeri* Fxn 1. *Biotechnology Letters,* 17, 309-314.

Collins, T., Gerday, C. and Feller, G. (2005). Xylanases, xylanase families and extremophilic xylanases. *FEMS Microbiology Review,* 29, 3-23.

Coutinho, P.M. and Henrissat, B. (1999). Carbohydrate- active enzyme server (CAZY) at *URL: http:// afmb.cnrs-mrs.fr/ ~cazy/CAZY/.*

De Faria, F.P., Te'O, V.S.J., Berggquist, P.L., Azevedo, M.O. and Nevalainen, K.M.H. (2002). Expression and processing of a major xylanase (xyn 2) from the thermophlic fungus *Humicola grisea* var. thermoidea in *Trichoderma reesei. Letters in applied Microbiology,* 34, 119-123.

De lemos Esteves, F., Ruelle, V., Lamotte-Brasseur, J., Quinting, B. and Frere, J.M. (2004). Acidophilic adaptation of family 11 endo-b-1,4-xylanases: modeling and mutational analysis. *Protein Science,* 13, 1209-1218.

Durand, R., Rascle, C. and Fevre, M. (1996). Molecular characterization of xyn 3, a member of the endoxylanase multigene family of the rumen anaerobic fungus *Neocallimastix frontalis. Current Genetics,* 30, 531-540.

Dwivedi, P.P., Gibbs, M.D., Saul, D.J. and Bergquist, P.L. (1996). Cloning, sequencing and overexpression in *Eschirichia coli* of a xylanase gene, xyn A from the thermophilic bacterium Rt8B.4 genus *Caldicellulosiruptor. Applied Microbiology and Biotechnology,* 45, 86-93.

Ebringerova, A. and Heinze, T. (2000). Xylan and xylan derivatives- biopolymers with valuble properties. 1. Naturally occurring xylans structures, isolation procedures and properties. *Macromol Rapid Commun,* 21, 542-556.

Fujimoto, Z., Kaneko, S., Kuno, A., Kobayashi, H., Kusakabe, I. and Mizuno, H. (2004). Crystal structures of decorated xylooligosaccharides bound to a family-10 xylanase from *Streptomyces olivaceoviridis* E-86. *Journal of Biology Chemistry,* 279, 9606-9614.

Haltrich, D., Bernd, N., Kulbe, K.D., Steiner, W. and Zupancic, S. (1996). Production of fungal xylanases. *Bioresource Technology,* 58, 137-161.

Harris, G.W., Jen Kins, J.A., Connerton, I. and Pickersgill, R.W. (1996). Refined crystal structure of the catalytic domain of xylanase A from Pseudomonas fluorescens at 1.8 Aº resolution. *Acta Crystallography,* 52, 393-401.

Henrissat, B. and Bairoch, A. (1993). New families in the classification of glycosyl hydrolases based on amino acid sequence similarities. *Biochemical Journal,* 293, 781-788.

Henrissat, B., Claeyssens, M., Tomme, P., Lemesle, L. and Mornon, J.P. (1989). Cellulase families revealed by hydrophobic cluster analysis. *Gene,* 81, 83-95.

Izydorczyk, M.S. and Biliaderis, C.G. (1995). Cereal arabinoxylans: advances in structure and physicochemical properties. *Carbohydrate Polymers,* 28, 33-48.

Kobayashi, Y., Okuda, N., Matsumoto, M., Inoue, K., Wakita, M. and Hoshino, S. (1998). Constitutive expression of a heterologous *Eubacterium ruminantium* xylanase gene (xyn A) in *Butyrivibrio fibriosolvens. FEMS Microbiology Letters,* 163, 11-17.

Kohli, U., Nigam, P., Singh, D. and Chaudhary, K. (2001). Thermostable, alkalophlic and cellulose free xylanase production by *Thermoactinomyces thalophilus* subgroup C. *Enzyme and Microbial Technology,* 28, 606-610.

Krengel, U. and Dijkstra, B.W. (1996). Three dimentional structure of endo-1,4-b-xylanase I from *Aspergillus niger.* Molecular basis for its low pH optimum. *Journal of Molecular Biology,* 263, 70-78.

Kulkarni, N., Shendye, A. and Rao, M. (1999). Molecular and biotechnological aspects of xylanases. *FEMS Microbiology Review,* 23, 411-456.

Kumar, P.R., Eswaramoorthy, S., Vithayathil, P.J. and Viswamitra, M.A. (2000). The tertiary structure at 1.59 A° resolution and the proposed amino acid sequence of a family-11 xylanase from the thermophilic fungus *Paecilomyces varioti* Bainier. *Journal of Molecular Biology,* 295, 581-593.

Kuno, A., Shimizu, D., Kaneko, S., Koyama, Y., Yoshida, S., Kobayashi, H., Hayashi, K., Taria, K. and Kusakabe, I. (1998). PCR cloning and expression of the F/10 family xylanase gene from *Streptomyces olivaceoviridis* E-86. *Journal of Fermentation and Bioengineering,* 86, 434-439.

Li, K., Azadi, P., Collins, R., Tolan, J., Kim, J. and Eriksson, K. (2000). Relationships between activities of xylanases and xylan structures. *Enzyme and Microbial Technology,* 27, 89-94.

Liu, C.J., Suzuki, T., Hirata, S. and Kawai, K. (2003). The processing of high- molecular weight xylanase (Xyn E, 110 kDa) from *Aeromonas caviae* ME-1 to 60 -kDa xylanase (XynE $_{60}$) in *E. coli* and purification and characterization of XynE$_{60}$. *Journal of Bioscience and Bioengineering,* 95, 95-101.

Moreau, A., Shareck, F., Kluepfel, D. and Morosoli, R. (1994). Increase in catalytic activity and thermostability of the xylanase A of *Streptomyces lividans* 1326 by site specific mutagenesis. *Enzyme and Microbial Technology,* 16, 420-424.

Moriyama, S., Fukusaki, E., Cabrera, J.C., Shinmyo, A. and Okada, H. (1987). Structure and expression of genes coding for xylan-degrading enzymes of *Bacillus pumilus. European Journal of Biochemistry,* 166, 539-545.

Nakamura, S. (2003). Structure and function of a multidomain alkaline xylanase from alkalophilic *Bacillus sp.* Strain 41 M-1. *Catalysis Surveys from Asia,* 7, 157-164.

Nath, D. and Rao, M. (1998). Structural and functional role of Tryptophan in xylanase from an extremophilic *Bacillus*: assessment of the active site. *Biochemical and Biophysical Research Communications,* 249, 207-212.

Ohta, K., Moriyama, S., Tanaka, H., Shige, T. and Akimoto, H. (2001). Purification and characterization of an acidophilic xylanase from *Aureobasidium pullulans* var. melanigenum and sequence analysis of the encoding gene. *Journal of Bioscience and Bioengineering,* 92, 262-270.

Perez-Avalos, O.P., Teresa, P., Ignacio, M. and Mayra, D. (1996). Induction of xylanase and b-xylosidase in *Cellulomonas flavigena* growing on different carbon sources. *Applied Microbiology and Biotechnology,* 46, 405-409.

Polizeli, M.LT.M., Rizzatti, A.C.S., Monti, R., Terenzi, H.F., Jorge, J.A. and Amorim, D.S. (2005). Xylanases from fungi: properties and industrial applications. *Applied Microbiology and Biotechnology.*

Qureshy, A.F., Khan, L.A. and Khanna, S. (2000). Expression of *Bacillus circulans* Teri-42 xylanase gene in *Bacillus subtilis. Enzyme and Microbial Technology,* 27, 227-233.

Singh, S., Madlala, A.M. and Prior, B.A. (2003). *Thermomyces lanuginisus*: properties of strains and their hemicellulases. *FEMS Microbiology Review,* 27, 3-16.

Srivastava, R., Ali, S.S. and Srivastava, B.S. (1991). Cloning of xylanase gene of *Streptomyces flavogriseus* in *E. coli* and bacteriophage l-induced lysis for the release of cloned enzyme. *FEMS Microbiology Letters,* 78, 201-206.

Subramaniyan, S. and Prema, P. (2002). Biotechnology of microbial xylanases, enzymology, molecular biology and application. *Critical Review Biotechnology,* 22, 33-64.

Sunna, A., and Antranikian, G. (1997). Xylanolytic enzymes from fungi and bacteria. *Critical Review of Biotechnology,* 17, 39-67.

Swaroopa Rani, D. and Krishna Nand (2000). Production of thermostable cellulose free xylanase by Clostridium absonum CFR-702. *Process Biochemistry,* 36, 355-362.

Te'O, V.S.J., Cziferszky, A.E., Berggquist, P.L. and Nevalainen, K.M.H. (2000). Codon optimization of xylanase gene xyn B from the thermophlic bacterium *Diotyoglomus thermophilum* for expression in the filamentous fungus *Trichoderma reesei. FEMS Microbiology Letters,* 190, 13-19.

Teplitsky, A., Mechaly, A., Stojanoff, V., Sainz, G., Golan, G., Feinberg, H., Gilboa, R., Reiland, V., Zolotnitsky, G., Shallom, D., Thompson, A., Shoham, Y. and Shoham, G. (2004). Structure determination of the Extracellular xylanase from Geobacillus stearothermophilus by Selenomethioxyl MAD phasing. *Acta Crystallogr D Biol Crystallogr,* 60, 836-848.

Thomson, J. A. (1993). Molecular biology of xylan degradation. *FEMS Microbiology Review,* 104, 65-82.

Torronen, A., Harkki, A. and Rouvinen, J. (1994). Three dimentional structure of endo-1,4-b-xylanase II from *Trichoderma reesei*: two conformational states in the active site. *European Molecular Biology Organization,* 13, 2493-2501.

Uffen, R.L. (1997). Xylan degradation: a glimpse at microbial diversity. *Journal of Indian Microbiology Biotechnology,* 19, 1-6.

Viikari, L., Kantelinen, A., Sundguist, J. and Linko, M. (1994). Xylanases in bleaching, from an idea to the industry. *FEMS Microbiology Review,* 13, 335-350.

Wakarchuk, W.W., Sung, W.L., Campbell, R.L., Cunningham, A., Watson, D.C. and Yaguchi, M. (1994). Thermostabilization of the *Bacillus circulans* xylanase by the introduction of disulfide bonds. *Protein Engineering,* 7, 1379-1386.

Whistler, R.L. and Richards, E.L. (1970). Hemicelluloses. In: *The carbohydrates.* Eds Pigman, W. and Horton, D. Academic Press, New York. Pp 447-469.

Wong, K.K.Y., Tan, L.U.L. and Saddler, J.N. (1988). Multiplicity of β-1, 4-xylanases in microorganisms: functions and applications. *Microbiology Review,* 52, 305-317.

# 10

# Methods of Purification and Characterization of Xylanases

Kalika Kuhar[1], Suchita Ninawe[2,3], V.K. Gupta[1], Sarika Kuhar[3], K.K. Tripathi[2] and Ramesh Chander Kuhad[3]

[1]*Department of Biochemistry, Kurukshetra University, Kurukshetra-136 119, India*
[2]*Department of Biotechnology, C.G.O. Complex, Lodhi Road, New Delhi -110003*
[3]*Department of Microbiology, University of Delhi, South Campus, New Delhi-110021, India*

Xylanases, the group of enzymes, which degrade the xylan backbone, are classified on the basis of the mechanism of action. Due to their biotechnological applications in various industrial processes, xylanases from microorganism have attracted increasing attention in the last decade. Therefore, xylanases need to get purified and characterized in order to fulfill the increasing demand of the market. Also, purification and characterization costs are becoming important issues in modern biotechnology as the industry matures and competitive products reach the marketplace. Thus, new paths for successful and efficient xylanase recovery are the need of the day. Also, characterization methods are of utmost importance. Hence this article reviews various purification and characterization methods used for the recovery of microbial xylanases.

## INTRODUCTION

Lignocellulosics is the term used collectively for cellulose (40-60%), hemicellulose (20-30%) and lignin (15-30%) (Kuhad *et al.,* 1997). These are the most abundant renewable natural materials in the biosphere accounting more than 60% of the total biomass produced. Xylan is the second most abundant biopolymer after cellulose (Biely, 1985) and the major hemicellulosic polysaccharide of wood (Timell, 1967). It is a heteropolymer with a backbone of β–1,4-D-xylanopyranosyl residues substituted with acetyl, arabinosyl and uronyl side chains (Joseleau *et al.,* 1992). Several hydrolytic enzymes are required for the complete hydrolysis of xylan (Dekker and Richards, 1976), the most important of which is the xylanases (E.C. 3.2.1.8.) (Reilly, 1981). Xylanases have been classified on the basis of their mechanism of action as: endoxylanases (1,4-β–D-xylan xylanohydrolase; E.C.

3.2.1.8), which catalyze the hydrolysis of xylan to xylo-oligosaccharides and xylose and β–xylosidase (1,4-β–D-xylan xylohydrolase; E.C. 3.2.1.37) that releases xylose residues from the non-reducing ends of xylo-oligosaccharides (Kuhad *et al.,* 1997). Xylanases have been reported to be produced by several genera of marine and terrestrial bacteria, ruminant bacteria, yeast, fungi, marine algae, protozoa, snails, crustaceans, insects and seeds of terrestrial plants (Kuhad *et al.,* 1997; Sunna and Antranikian, 1997).

Xylanases from microorganisms have attracted increasing attention in the last decade due to their biotechnological applications in various industrial processes (Bajpai, 1999; Kuhad and Singh, 1993; Niehaus *et al.,* 1999; Wong and Saddler, 1992). Their applications include biopulping wood and pulp bleaching (Bajpai *et al.,* 1994; Srinivasan and Rele, 1999; Viikari *et al.,* 1994), wastewater treatment (Rani and Nand, 1996), treating animal feed to increase digestibility (Bedford and Classen, 1992; Vanparidon *et al.,* 1992), processing food to increase clarification (Biely, 1985; Dekker, 1985), pretreatment of forage crop converting lignocellulosic substances into feedstocks and fuels (Eriksson, 1985; Jeffries, 1985), improve cell wall maceration for the production of plant protoplasts (Wong *et al.,* 1988) and many more. This puts up the need of purification and characterization of xylanases. Presently, purification and characterization costs are becoming important issues in modern biotechnology as the industry matures and competitive products reach the market place. Purification has become a necessity as it leads to reduction in bulk, concentration enrichment, removal of specific impurities (e.g. toxins from therapeutic products), prevention of catalysis other than the type desired (as with enzymes), prevention of catalyst poisoning (as with enzymes), recommended product specification (e.g. pharmacopoeia requirement), enhancement of protein stability and reduction of protein degradation (e.g. by proteolysis). Thus, new paths for successful and efficient xylanase recovery have to be followed. Also, to improve the process of purification, to quantify and to determine their primary, secondary and tertiary structures, to identify physico-chemical changes (deamidation, racemisation, oxidation, aggregation, fibrillation etc.) and to establish evolutionary relationships between proteins in diverse organisms, characterization methods are desirable. Therefore, reviewing this article various purification and characterization methods would be discussed.

## ISOLATION OF XYLANASES

Xylanases have been isolated from a number of hemicellulosic microorganisms such as bacteria, fungi, actinomycetes, yeast, plants, animals etc.

**(i)** **Bacteria**: *Bacillus circulans* AB16 (Dhillon *et al.,* 2000); *Bacillus* sp. (Okazaki *et al.,* 1984); *Aeromonas caviae* W-61 (Viet *et al.,* 1991); *Thermoanaerobacterium saccharolyticum* (Lee *et al.,* 1993); *Thermotoga thermarum* (Sunna *et al.,* 1996); *Bacillus stearothermophilus* T-6 (Khasin *et al.,* 1993); *Actinomadura* sp strain FC7 (Ethier *et al.,* 1994); *Streptomyces* (Keskar *et al.,* 1989); *Acidobacterium capsulatum* (Inagaki *et al.,* 1998); *Bacillus* sp. (Lopez *et al.,* 1998); *Fibrobacter succinogenes* S85 (Matte and Forsberg, 1992); *Bacillus* sp. Strain BP-23 (Blanco *et al.,* 1995); *Bacillus* sp. Strain K-1 (Ratannakanokchai *et al.,* 1999); *Bacillus* sp. (Bataillon *et al.,* 1998) etc.

**(ii) Fungi**: *Lentinula edodes* (Mishra *et al.*, 1990); *Aureobasidium pullulans* Y-2311-1 (Li *et al.*, 1993); Mutant of *Aspergillus ochraceus* (Biswas *et al.*, 1990); *Humicola grisea* var. thermoidea (Monti *et al.*, 1991); *Fusarium verticillioides* (Saha, 2001); *Neocallimastix frontalis* (De Segura and Fevre, 1993); *Piromyces* sp. Strain E2 (Teunissen *et al.*, 1993); *Pleurotus ostreatus* (Qinnghe *et al.*, 2003); *Aspergillus fischeri* Fxn1 (Raj and Chandra, 1996); *Acrophialophora nainiana* (Ximenes *et al.*, 1999), *Aspergillus sojae* (Kimura *et al.*, 1995); *Aspergillus sydowii* MG 49 (Ghosh and Nanda, 1994); *Fusarium oxysporum* (Christakopolous *et al.*, 1996); *Geotrichum candidum* (Radionova *et al.*, 2000); *Thermomyces lanuginosus* (Gomes *et al.*, 1993); *Penicillium brasilianum* (Jorgensen *et al.*, 2003); *Aspergillus fumigatus* AR1 (Anthony *et al.*, 2003) etc.

**(iii) Actinomycetes**: *Streptomyces* sp. QG-11-3 (Beg *et al.*, 2000); *Streptomyces lividans* (Blanco *et al.*, 1997); *Saccharomonospora viridis* (McCarthy *et al.*, 1985); *Actinomadura* sp. strain FC7 (Ethier *et al.*, 1994); *Streptomyces thermoviolaceus* OPC-520 (Tsujibo *et al.*, 1992); *Streptomyces chattanoogensis* CECT 3336 (Lopez-Fernandez *et al.*, 1998); *Streptomyces viridosporus* (Magnuson and Crawford, 1997); *Thermomonospora curvata* (Stutzenberger and Bodine, 1992); *Streptomyces* T7 (Kesker, 1992) etc.

**(iv) Yeast**: *Aureobasidium pullulans* Y-2311-1 (Li *et al.*, 1993); *Cryptococcus albidus* (Morosoli *et al.*, 1986); *Trichosporon cutaneum* SL409 (Liu *et al.*, 1998); *Trichosporon cutaneum* (Hrmova *et al.*, 1984) etc.

**(v) Plants**: Japanese pear fruit (Yamaki and Kakiuchi, 1979); flour of European wheat (Cleempurt *et al.*, 1997) etc.

**(vi) Animals**: Freshwater mollusk (Yamura *et al.*, 1997) etc.

Different media have been used for culturing xylanase-producing organisms. These include basal medium containing rice straw (Dhillon *et al.*, 2000), Yeast extract-malt extract agar (YMS); Defined xylan agar (DX) and trypticase soy broth (TSB) (Grabski and Jeffries, 1991), TYE medium (Lee *et al.*, 1993), oatmeal baby food Quaker and solid medium (Monti *et al.*, 1991), Humic acid-vitamin agar (Ethier *et al.*, 1994), YM medium (Li *et al.*, 1993), oakwood medium (Mishra *et al.*, 1990), Berg's mineral salt medium (Ratanakhanokchai *et al.*, 1998), MYG medium (Franco *et al.*, 2004) and CYC medium (Blanco *et al.*, 1997).

## ENZYME PURIFICATION

Purity and purification costs are becoming important issues in modern biotechnology as the industry matures and competitive products reach the marketplace. Thus, new paths for successful and efficient xylanase recovery have to be followed. For industry, time and money are an important criterion for selection of technique. Low cost methods are always used first in an industrial purification, even if other methods are more effective. High cost methods are applied last (for clean-up) where volumes and amounts of material tend to be smaller (and therefore cheaper to handle). Most of the purification methods, which are used in laboratory research, can be scaled up to industrial processes. Such methods are filtration, centrifugation, microfiltration, ultrafiltration, diafiltration, precipitation, ion-exchange chromatography and gel filtration. Some of the laboratory methods are not really available or possible on large scale, like HPLC, affinity chromatography and electrophoresis etc. Many of the laboratory reagents are too expensive or toxic for industrial use. The purification of enzymes is easier because the activity of the enzyme can be monitored rapidly, but also more difficult because the activity of the enzyme must be maintained. The total xylanase activity was dropped during purification due to the removal of synergistic enzymes (Mishra *et al.*, 1990).

Xylanases are considered to be mostly extracellular but intracellular xylanases have also been reported in some organisms such as *Bacillus firmus, Bacteroide amylogens, Bacteroide ruminicola, Butyrivibrio fibriosolvens, Ruminicoccus flavefaciens, Bacillus stearothermophilus* T-6, *Cellvibrio mixtus, Penicillium janthinellum* (Fontes *et al.*, 2000; Hespell, 1988; Rodrigues and Tambourgi, 2001 and Teplitsky *et al.*, 2000) and in some protozoans (Dekker and Richards, 1976). Cell wall associated xylanases have been reported in *Thermoanaerobacterium* sp. (Shao *et al.*, 1995). Extracellular xylanases occur as a high molecular-weight complex (molecular weight> $10^6$), which cannot be separated by standard methods (e.g. ammonium sulphate precipitation, anion-exchange chromatography or cation-exchange chromatography using a number of different buffer systems at different pH values).

## CLARIFICATION

Clarification is an important step in the purification of extracellular xylanases. The importance of this step lies in the removal of polyphenols, pigments, polysaccharides and nucleic acids which if not removed, bind to ion-exchange resin and hinder separation of proteins (Mishra *et al.*, 1990). Clarification has been done by precipitation, polymerization, centrifugation or filtration.

Elegir and coworkers (1994) used bioprocessing acid 1000, a cross-linked polymer, for the removal of pigments from the crude extract of *Streptomyces* sp. B-12-2 but Mishra and coworkers (1990) used polyethylenimine for the removal of pigments from the culture filtrate of *Lentinula edodes* which led to an increase in purification up to 2.3-fold. Monti and coworkers (1991) used kaoline to remove undesirable matter (polysaccharide) from the crude extract of *Humicola grisea* by precipitation. Nucleic acids were removed from the crude extract of *Bacillus* sp. with streptomycin sulphate (Okazaki *et al.*, 1985). One commonly applied method for clarification is centrifugation. It is helpful in removal of large cellular debris and unwanted material which may block the column. Some typical centrifugation protocols for xylanase purification are: 5000xg, 5 min (Li *et al.*, 1993); 9000xg, 30 min (Ethier *et al.*, 1994); 10,000xg, 15 min (Dhillon *et al.*, 2000); 10,000xg, 20 min (Li *et al.*, 1993); 10,000xg, 30 min (Monti *et al.*, 1991); 11000xg, 30 min (Grabski and Jeffries, 1991) 16000xg, 30 min (Nakamura *et al.*, 1993); 48000xg, 30 min (Gomez *et al.*, 1993); 100,000xg, 2hr (Matte and Forsberg, 1992). The fermentation broth of xylanase T-6 from *Bacillus stearothermophilus* was partially clarified by continuous centrifugation (Fishman *et al.*, 1995). Bragger *et al.* (1989) removed cells from cultures using hollow fibre filtration but Maheshwari and Kamalam (1985) collected the precipitate after ammonium sulphate precipitation by filtering through a celite bed.

## PRELIMINARY PURIFICATION METHODS

### Concentration

Concentration of the material is important for costs and for later purification stages. This can be crucial in the planning of industrial scale purification since large volumes require large plant and machinery and, therefore, greater costs. Concentration is done using precipitation, ultrafiltration or lyophilization. The chemicals employed for precipitation include ethanol (Keskar *et al.,* 1989; Park *et al.,* 1992), acetone (Anand *et al.,* 1990; Tsujibo *et al.,* 1990), isopropanol (Tuohy *et al.,* 1993), ammonium sulphate (Honda *et al.,* 1985; Okazaki *et al.,* 1985 and Biswas *et al.,* 1990), combination of ammonium sulphate and acetone (Anand *et al.,* 1990) and a combination of ammonium sulphate and NaCl (Monti *et al.,* 1991). Use of ethanol for precipitation of culture filtrate by Keskar *et al.* (1989) led to 4.14-fold purification but ethanol precipitation of culture filtrate of *S. lividans* led to 1.5-fold (xylanaseI) and 1.9-fold (xylanaseII). Honda *et al.* (1985) and Ghosh and Nanda (1994) used polyethyleneglycol for concentration of crude extract. Khasin *et al.* (1993) also used polyethylene glycol 20000 to concentrate the crude extract of *Bacillus* sp. Cardinale and Matta (2001) concentrated the crude filtrate of *Fusarium oxysporum* up to 10-fold using ultrafiltration. Grabski and Jeffries (1991) concentrated the crude extract by ultrafiltration, which led to 2.2-fold purification leading to increase in total activity up to 13%. Dhillon *et al.* (2000) and Raj and Chandra (1996) used lyophillisation process to concentrate the extract, thus facilitating ammonium sulphate precipitation.

### Ammonium sulphate precipitation

Ammonium sulphate precipitation is an efficient method for removal of lower molecular mass xylanases (Nakamura *et al.,* 1993). This is often employed to clean up the material as an initial separation step. The material needed for this method is cheap, and finds much use in industrial purifications. Ammonium sulphate precipitation of the crude concentrated filtrate of *Streptomyces roseiscleroticus* led to 11-fold increase in purity (Grabski and Jeffries, 1991). Ammonium sulphate precipitation led to several fold purification such as 3.6-fold (Li *et al.,* 1993), 2.6 fold (Biswas *et al.,* 1989), 1.2-fold (De Segura and Fevre, 1993) and so on. Kansoh and Nagieb (2004) partially purified xylanase in the culture supernatant of *Streptomyces galbus NR* by salting out at 40–60 and 60–80% ammonium sulphate saturation which led to the purification of 9.63-fold and 68.80% recovery. Ammonium sulphate has to be removed at a later stage by dialysis, ultrafiltration or size-exclusion chromatography (desalting) because salt interferes with some later methods such as ion-exchange chromatography. Also, at very high ammonium sulphate concentrations, the density of the solution becomes greater than that of the insoluble protein. This means that rather than being pelleted, the insoluble protein will "float" on the surface as a scummy layer.

### Dialysis

Proteins can be separated from small molecules by dialysis through a semipermeable membrane, made up of cellulose. Molecules having dimensions significantly greater than the pore diameter are retained inside the dialysis bag, whereas smaller molecules and ions traverse the pores of such a membrane and emerge in the dialysate outside the bag. This technique is useful for removing a salt or other small molecule. Dialysis is often done using different buffers of different pH such as sodium acetate 50mM, pH 4.0 (Grabski and Jeffries, 1991; Mishra *et al.,* 1990), sodium acetate 50mM, pH 5.5 (Lee *et al.,* 1993), citrate phosphate 0.01M, pH 6.0 (De Segura and Fevre, 1993), Tris HCl 10mM pH 8.0 (Black *et al.,* 1995), potassium phosphate 20mM pH 6.0 (Ethier *et al.,* 1994), sodium phosphate 50mM pH 6.8 (Li *et al.,* 1993), Sodium phosphate 10mM pH 7.0 (Nakamura *et al.,* 1993), citrate 0.005M pH 6.0 (Biswas *et al.,* 1990) and acetate buffer 50mM pH 5.0 (Saha, 2001). Some workers have performed dialysis against water (Morosoli *et al.,* 1986; Monti *et al.,* 1991).

### Diafiltration

On industrial scale, a related method of diafiltration is employed. In this case, a rigid nitrocellulose membrane of defined pore size is used. Diafiltration is also employed when there is loss of activity by dialysis (Grabski and Jeffries, 1991). Dialysis and diafiltration have a few drawbacks that lead to losses with small sample sizes, removal of low-molecular weight compounds and tendency to froth especially during diafiltration. Due to above drawbacks, dialysis is replaced by concentration, dilution and reconcentration with the use of hollow-fibre ultrafiltration equipment.

## CHROMATOGRAPHY METHODS

Direct scaling-up of column chromatographic procedures is usually possible. Modifications to column geometry to maximize flow rates are often desirable but purification factors and recoveries comparable with those obtained on the laboratory scale can be achieved relatively easily. In chromatography, the column matrix provides a non-mobile microenvironment into which proteins can penetrate and with which they can interact. At high flow rates, high protein concentrations, and with restricted/ convoluted pores in the beads, equilibrium is often not the case; under these conditions, separation is considerably reduced.

### Ion-exchange chromatography

In ion-exchange chromatography, separation depends on: (i) the binding strength of the protein at a given ionic strength, (ii) the partition constant as it passes down the column after its initial displacement from the start of the column. Both cation exchange and anion exchange chromatography have been used for xylanase purification but anion exchange chromatography is more common method of purification than cation exchange chromatography. Table 1 below shows the utilization of various cation and anion exchange matrices for different organisms.

**Table 1. Ion-exchange matrices employed for purification of xylanases**

| *Name of organism* | *Ion exchange matrix* | *Type of exchanger* | *Purification Fold* | *References* |
|---|---|---|---|---|
| *Fibrobacter succinogens* S85 (Endoxylanase 1) (Endoxylanase 2) | Cation | CM-Sepharose CL-6B | 24<br>48 | Matte and Forsberg, 1992 |
| *Streptomyces rosei scleroticus* | Cation | Carboxymethyl Bio-Gel A<br>Mono-S | 16.3<br>45.1 | Grabski and Jeffries, 1991 |
| *Trichoderma reesei* (Xyl II) | Cation | Carboxymethyl-Sepharose | - | Lappalainen *et al.*, 2000 |
| *Aureobasidium pullulans* Y-2311-1 | Anion | DEAE-Sephadex A-50 | 4.0 | Li *et al.*, 1993 |
| *Bacillus circulans* AB16 | Anion | Sepharose Q<br>Sepharose 6B (Xyl B) | 23<br>125 | Dhillon *et al.*, 2000 |
| *Bacillus* sp. | Anion | DEAE-Toyopearl 650M | - | Nakamura *et al.*, 1993 |
| *Cellulomonas fimi* | Anion | DEAE Trisacryl M | 150 | Black *et al.*, 1995 |
| *Fibrobacter succinogens* S85 (Endoxylanase 2) | Anion | Q-Sepharose | 2.5 | Matte and Forsberg, 1992 |
| *Humicola grisea* var. thermoidea | Anion | DEAE-cellulose | 9.9 | Monti *et al.*, 1991 |
| *Lentinula edodes* | Anion | DEAE Trisacryl M | 1.8 | Mishra *et al.*, 1990 |
| *Neocallimastix frontalis* | Anion | DEAE-Bio-Gel A | 4.1 | De Segura and Fevre, 1993 |
| *Piromyces* sp. strain E2 (xylanase I) | Anion | Protein PAK DEAE-5PW | 112.8 | Teunissen *et al.*, 1993 |
| *Piromyces* sp. strain E2 (xylanase II) | Anion | Superose 12 | 28.5 | Teunissen *et al.*, 1993 |
| *Streptomyces lividans*(xylanase II) | Anion | DEAE Bio-Gel A | 4.6 | Ethier *et al.*, 1994 |
| *Thermotoga thermarum* (endoxylanase 1) | Anion | Mono Q | 22.5 | Sunna *et al.*, 1996 |
| *Trichoderma reesei* | Anion | DEAE Sepharose | - | Lappalainen *et al.*, 2000 |

## Gel filtration chromatography

In maximum cases, gel filtration chromatography has been used in combination with other column chromatography for xylanase production (Lee *et al.*, 1993). There has been report of use of different gel filtration matrices as depicted in the Table 2 below:

**Table 2. Gel filtration matrices for purification of xylanases**

| *Name of organism* | *Gel filtration matrice* | *Purification Fold* | *References* |
|---|---|---|---|
| *Aeromonas caviae* W-61 | Sephadex G-100 | 23 | Viet *et al.*, 1991 |
| *Aureobasidium pullulans* Y-2311-1 | Sephadex G-75 | 5.8 | Li *et al.*, 1993 |
| *Clostridium thermocellum* | Sepharose 4B | - | Morag *et al.*, 1990 |
| *Fusarium verticillioides* | Bio Gel A | 378 | Saha, 2001 |
| *Humicola grisea* var. thermoidea | Sephadex G-200 | 18.7 | Monti *et al.*, 1991 |
| Mutant of *Aspergillus ochraceus* | Sephadex G75 | 3.0 | Biswas *et al.*, 1990 |
| *Neocallimastix frontalis* | Ultrogel AcA 44 | 13.0 | De Segura and Fevre, 1993 |
| *Piromyces* sp. strain E2 (xylanase I) | Sephacryl S-300 HR | 0.9 | Teunissen *et al.*, 1993 |
| *Thermoanaerobacterium saccharolyticum* | Sephacryl S-300 | - | Lee *et al.*, 1993 |
| *Thermotoga thermarum* (endoxylanase 1) | Superdex 200 | 1.9 | Sunna *et al.*, 1996 |
| *Thermotoga thermarum* (endoxylanase 2) | TSK G3000 | 1.9 | Sunna *et al.*, 1996 |
| *Trichoderma reesei* (Xyl II) | Sephacryl S-100 | - | Lappalainen *et al.*, 2000 |

## Affinity chromatography

Classical affinity chromatographic techniques have not proved so amenable to large-scale work, mainly because of the enormous expense and rather short life of the matrices. However, the quasi-affinity chromatography afforded by the triazine dye conjugates has proved of great benefit. The materials are cheap to prepare. Xylanase from *Thermoanaerobacterium saccharolyticum* B6A-RI after purifying through gel filtration and concentrating was further purified by affinity chromatography with xylan-coupled Sepharose CL-4B (Lee *et al.*, 1993). Teunissen *et al.* (1993) purified the sample of *Piromyces* sp. strain E2 on cellobiose-sepharose-6B, which led to 4.3-fold increase in purification. Morag *et al.* (1990) and Black *et al.* (1995) also purified xylanase by affinity chromatography. Saha (2001) purified xylanase by the use of octyl-sepharose chromatography, which led to the 31-fold increase in purification. An affinity based matrix Cellobiose-Sepharose-6B has been reported to increase xylanase purification (Teunissen *et al.*, 1991). Lee *et al.* (1993) purified xylanase from *T. saccharolyticum* B6A-R1 using xylan coupled Sepharose CL-4B. The shortcoming of this technique is that the enzyme is denatured due to the use of harsh conditions because of the strong binding to ligand or antibody. The non-specific binding (in case of antibody) can overcome the "correct" binding sites at very high protein concentrations, leading to poor purification. Thus, there is often a limit to the protein concentration that can be applied to an affinity column.

## Hydroxyapatite Chromatography

Hydroxyapatite chromatography is very useful in protein separation as it gives good resolution. It is mainly useful in concentrating proteins from a very dilute mixture. Biogel HPHT hydroxyapatite chromatography has been reported to bring about 66-fold xylanase purification (Simpson *et al.,* 1991). To remove the contaminating proteins, the sample of *F. succinogens* was further purified by hydroxyapatite chromatography, which led to 20-fold increase in purification (Matte and Forsberg, 1992). Morosoli *et al.* (1986) used hydroxyapatite chromatography to purify the sample of *Cryptococcus albidus* which led to 27-fold increase in purification. Tuohy *et al.,* 1993 purified xylanase II from *T. emersonii* by 11-fold using hydroxyapatite chromatography. Lappalainen *et al.* (2000) purified xylanase II isolated from *Trichoderma reesei* using Phenyl-Sepharose matrice but Franco *et al.* (2004) purified xylanase from *T. harzianum* using same matrix which led to 15.11-fold purification.

## CHROMATOFOCUSING/ ISOELECTRIC FOCUSING (IEF)

In this technique, proteins are separated on the basis of their net charge. Chromatofocusing and isoelectric focusing are related methods involving the production of a stable pH gradient. In chromatofocusing, matrix is a column but in IEF, it is a polyacrylamide or agarose gel. Matte and Forsberg (1992) used polybuffer exchanger 94 chromatofocusing column for purification of xylanase from *Fibrobacter succinogens* S85 (Endoxylanase 1) which led to 0.9-fold purification and 8.5% recovery. Xylanase isolated from *Thermotoga thermarum* (endoxylanase 2) was purified by Sunna *et al.* (1996) using this chromatofocusing technique. Preparative isoelectric focusing yielded a single xylanase activity peak in *Lentinula edodes* (Mishra *et al.,* 1990). Isoelectric focusing was employed for the separation of xylanases in *Phanerochaete chrysosporium* (Dobozi *et al.,* 1992). The main drawback of these techniques is that neither technique is very suitable for industrial scale since capacity is low. The other shortcomings are that the high concentrations of protein lead to the precipitation of material at the isoelectric point which can block the matrix or gel also due to which resistance rises tremendously leading to denaturation of protein.

## HIGH PERFORMANCE LIQUID CHROMATOGRAPHY (HPLC)

The chromatographic methods are sometime being used with HPLC columns and equipment, but with HPLC the resolution is almost always better than with standard chromatography columns because of the packing of HPLC columns with much smaller beads than the standard columns. Rao *et al.* (1996) purified xylanase from *Chainia* sp. using reverse-phase HPLC column but Kluepfel *et al.* (1990) purified the crude filtrate of xylanase from *Streptomyces lividans* using anion-exchange DEAE Protein Pak semi-preparatory HPLC column which led to 19.6% yield. It was further purified by gel chromatography on two Waters PW 300 HPLC columns leading to 4.4% recovery.

## FAST PROTEIN LIQUID CHROMATOGRAPHY (FPLC)

FPLC is used for rapid protein purification. The system currently in place includes a broad selection of columns for ion exchange and gel filtration purification at either analytical or preparative levels. Jorgensen *et al.* (2003) used ion exchange and hydrophobic interaction chromatography on a FPLC system for the purification of xylanase from *Penicillium brasilianum* IBT 20888.

## ELECTROELUTION

Electroelution is a simple, fast, cost-effective and reproducible method for eluting multiple protein bands. It also does not need former knowledge of the protein characteristics. As this consideration varies with the protein under study, it is not possible to predict protein recovery by this method. Due to very small production of extracellular xylanase from *Humicola grisea* var. *thermoidea*, the component form 1 was purified using electroelution method (Monti *et al.,* 1991). Similarly, high molecular weight xylanase excreted by a strain of *Bacillus* sp. CCMI 966 has been purified using electroelution (Sa-Pereira *et al.,* 2000).

## CRYSTALLIZATION

Crystallization is the highest yield step of the whole process of purification and also gives the best purification in a single process step. Margolin *et al.* (2002) got a process yield of 96% for xylanase XYNII when the initial concentration was 50 g/l. A set of empirical phase diagrams for xylanase was generated with the help of microfluidic formulator, which was used to devise a rational crystallization screen that resulted in 72-fold improvement in successful crystallization hits (Hansen *et al.,* 2004). Torronen *et al.* (1994) found that xylanase from *Trichoderma* sp. readily crystallizes in ammonium sulphate and sodium/potassium phosphate. Xylanase can also be crystallized with other salts, polymers and organic solvents.

## CHECKING OF ENZYME PURITY

To determine the success of a protein purification scheme, the procedure is monitored at each step by performing Polyacrylamide gel electrophoresis (Native and SDS) and by determining specific activity. SDS-PAGE is purely an analytical method in which polyacrylamide gel of varying percentage is loaded with enzyme obtained at different steps of purification. Native—PAGE revealing single band confirms the purity of an enzyme. If the purified enzyme shows a single band on Native polyacrylamide gel then SDS-PAGE is also performed to see whether the enzyme consists of identical or non-identical subunits.

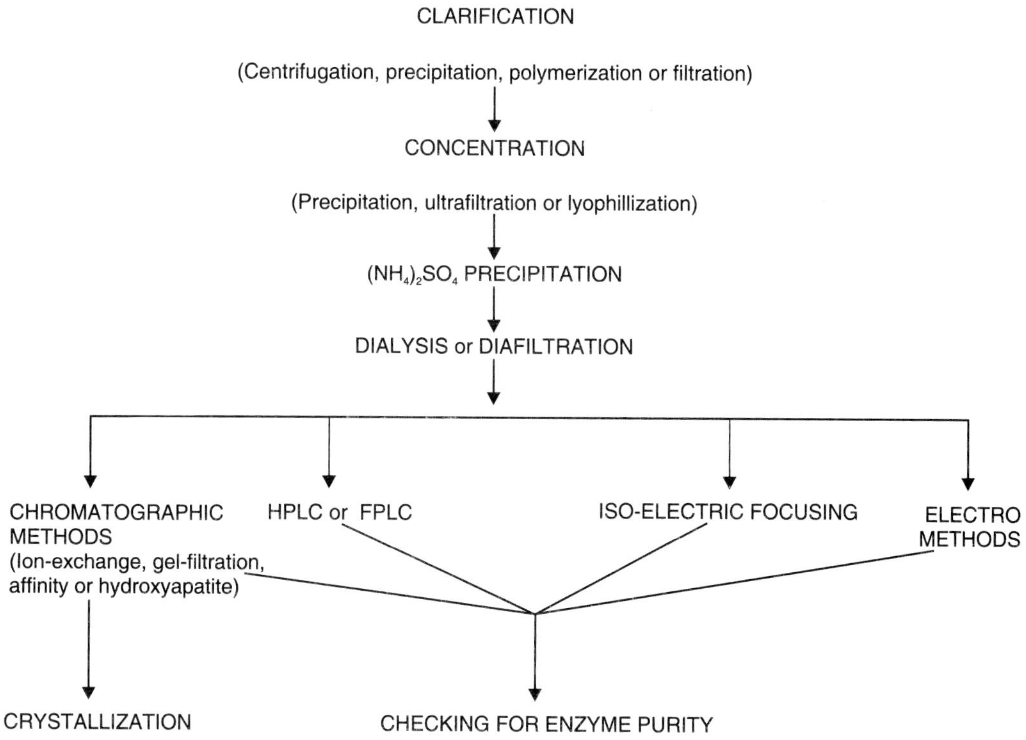

CLARIFICATION

(Centrifugation, precipitation, polymerization or filtration)

↓

CONCENTRATION

(Precipitation, ultrafiltration or lyophillization)

↓

$(NH_4)_2SO_4$ PRECIPITATION

↓

DIALYSIS or DIAFILTRATION

↓

CHROMATOGRAPHIC METHODS (Ion-exchange, gel-filtration, affinity or hydroxyapatite) | HPLC or FPLC | ISO-ELECTRIC FOCUSING | ELECTRO METHODS

CRYSTALLIZATION

CHECKING FOR ENZYME PURITY

**Fig. 1:** Flow-chart representing various steps involved in purification of xylanases

## CHARACTERIZATION

It is considered very important to characterize the purified enzymes for their physico-chemical and biological properties. For xylanase enzyme, various parameters taken into consideration for their characterization are as follows:

### Molecular Weight

Whether or not the apparent native molecular weight is representative of the mass of the molecule, the ability of the molecule to diffuse through small pores would be advantageous to the organism for the penetration of the hemicellulose-cellulose-lignin matrix of wood (Grabski and Jeffries, 1991).

The molecular weight of xylanase can be determined by Gel filtration chromatography: Two conclusions come out in relation of molecular weight determination with gel filtration chromatography technique, firstly if a xylanase was a small molecule then the xylanase would adsorb to the matrix resulting in its elution, secondly, it might be possible that the protein could be elongated or tapered molecule and such shape would allow it to penetrate the matrix resulting in its later elution. Second possibility is more common. Molecular weight determination by gel filtration gave varied results, which was strongly dependent on the ionic strength of the elution buffer. The retention coefficient of the enzyme decreased with the increase of the buffer ionic strength suggesting that the enzyme interacts with the matrix (Khasin *et al.,* 1993).

The subunit molecular weight of xylanase can be determined by SDS-PAGE: Polyacrylamide gels of varying percentage loaded with different standard molecular mass markers are used.

Wong *et al.* (1988) reported that many xylanases belong to the category of low molecular weight (16-22 kDa) with basic pI (8.3-10.0) or high molecular weight (43-50 kDa) with acidic pI (3.6-4.5). *Trichoderma* and *Aspergillus* xylanases belong to the low molecular weight category (Espinar *et al.,* 1993) but several *Aspergillus* xylanases have been found to possess molecular weights, intermediate between those two ranges, deviating from the above dichotomy (Kimura *et al.,* 1995; Khanna *et al.,* 1995). *Aspergillus fumigatus* xylanase deviates from the above classification since its molecular mass is very high (Anthony *et al.,* 2003).

Most xylanases are often found as monomeric protein with a molecular weight ranging from 11 - 80 kDa (Sunna *et al.,* 1996). But xylanases of molecular weight as high as 537 kDa (Matte and Forsberg, 1992), 350 kDa (Shao *et al.,* 1995), 340 kDa (Sa-Pereira *et al.,* 2000), 212-253 kDa (Anthony *et al.,* 2003), 150 kDa (Sunna *et al.,* 1996), 131 kDa (Tuohy *et al.,* 1993), 150 kDa (Khanna and Gauri, 1993), 99 kDa (Bataillon *et al.,* 1999) and as low as 5.5 kDa (Bastawde, 1992) and 6 kDa (Bandivadekar and Deshpande, 1996) have also been reported. Some microorganisms secrete multiple xylanases, which can be attributed mainly to the heterogeneity of xylan, which needs multiple xylanases with different specificity for complete hydrolysis (Morales *et al.,* 1995). In multiple xylanase secreting microorganisms, smallest xylanase is always the predominant as, 18 kDa xylanase from *T. koningii* (Wood and McCrae, 1986), 20 kDa xylanase from *T. harzianum* (Tan *et al.,* 1985) and 21.5 kDa xylanase from *Trichoderma longibrachiatum* (Tenkanen *et al.,* 1992). Some of the xylanases especially belonging to larger molecular weight are found to possess more than one subunit, which were of same molecular weight as in *T. emersonii* (Tuohy *et al.,* 1993) or different molecular weights as in *Thermotoga thermarum* (Sunna *et al.,* 1996).

## pH OPTIMA AND TEMPERATURE OPTIMA

The effect of pH on xylanase activity is tested using different buffers of varying pH under standard assay conditions. Most xylanases known so far are active at neutral (Marui *et al.,* 1985; Okada and Shinmyo, 1988) or acidic pH (Tavobilov *et al.,* 1981). However, several alkaline-tolerant xylanases have also been characterized (Dey *et al.,* 1991; Ratto *et al.,* 1992). The pH optima of fungal xylanases (4.0-7.0) are somewhat lower than those of bacterial xylanase (6.0-10.0) (Poutanen *et al.,* 1987). Xylanases having alkaline pH optima found use in paper and pulp industry (Srinivasan and Rele, 1995). Usefulness of such xylanases is further enhanced if they have broad pH optima. The optimum pH of xylanase A of *Bacillus sp.* was 6-10 (Honda *et al.,* 1985) and the pH of xylanase activity from *Thermomonospora fusca, T. curvata, T. chromogena* and *Saccharomonospora viridis* was optimal in the range pH 5-8 (McCarthy *et al.,* 1985). Xylanase J from alkalophilic *Bacillus* sp. strain 41M-1 showed a broad pH activity profile and was active over a pH range of 4-11 (Nakamura *et al.,* 1993). The highest reported value for pH optima for a xylanase is 11 from *Aeromonas sp.* 212 (Kudo *et al.,* 1985).

To determine the optimal incubation temperature for xylanase activity, the reaction mixture is incubated at different temperatures for varying time intervals and the enzyme activity is assayed under standard conditions. Most xylanases known so far are active between 40 °C-70°C (Wong *et al.,* 1988). It has been observed that microorganisms producing xylanase with temperature optima lower than 40°C are less in number in comparison to those organisms producing xylanases having temperature optima higher than 70°C. Few exceptions to this are of those xylanases produced by *Clostridium* strain which has a low temperature optima of 20°C (Akila and Chandra, 2003) whereas xylanases produced by *Pleurotus ostreatus* had a optimum temperature between 25°C-40°C (Qinnghe *et al.,* 2003). Optimum temperature of 80°C has also been reported (Shao *et al.,* 1995; Tuohy *et al.,* 1993; Sunna *et al.,* 1996). The optimum temperature of 90°C-100°C has been reported for endoxylanase 2 from *T. thermarum* (Sunna *et al.,* 1996). The most thermoactive purified xylanases reported so far are those produced by extreme thermophiles such as *Thermotoga sp.* Fj-SS3-B1 and *Thermotoga maritima* M SB8 with temperature optima between 92°C-105°C (Simpson *et al.,* 1991; Winterhalter *et al.,* 1995).

## ISOELECTRIC FOCUSING POINT

Isoelectric focusing point has been determined by analytical isoelectric focusing using Serva Procote gel (Grabski and Jefferies, 1991), PBE 94 column (Teunissen *et al.,* 1993), PhastGel (Kluepfel *et al.,* 1990); by chromatofocusing on a FPLC Mono P column (Khasin *et al.,* 1993) and by chromatofocusing carried out in a PBE-Mono P anion-exchange resin (Saha, 2001). Gels of different pH including 7-9.2 (Teunissen *et al.,* 1993); 3-9 (Blanco *et al.,* 1997); 3.5-10 (Keskar *et al.,* 1989); 3.5-7.35 (Copa-Patino *et al.,* 1993) are being used for pI determination. The isoeletric point has also been determined under native and denaturing conditions of xylanase T-6 by Khasin *et al.* (1993). It has been found that the enzyme if have a high isoelectric point would facilitate purification by cation-exchange chromatography. In most cases, pI values for xylanases are found in the range of 3.6-10.3 but a pI of 10.2-10.5 has been reported in *Streptomyces viridosporus* T7A (Magnuson *et al.,* 1997). Wong *et al.* (1988) have put forward a classification according to which most of the xylanases belong to two categories of low molecular weight (16-22 kDa) with basic pI (8.3-10.0) and high molecular weight (43-50 kDa) with acidic pI (3.6-4.5). Wong *et al.* (1988) have reported xylanase multiplicity in *Bacillus* sp. suggesting that these bacteria produce two types of xylanases; one is basic (pI 8.3-10.0) with low molecular weight (16-22 kDa), and the other is acidic (pI 3.6-4.5) with high molecular weight (43-50 kDa). *Fibrobacter succinogens* S85 also produces multiple endoxylanases both acidic and basic in nature (Matte and Forsberg, 1992). This dichotomous pattern has also been found in *Clostridium* and *Streptomyces* sp. (Wong *et al.,* 1988), *T. harzianum* (Tan *et al.,* 1985), *Trichoderma pseudokoningii* (Baker *et al.,* 1977), *A. caviae* (Viet *et al.,* 1991) and the recombinant xylanase from *Bacillus* sp. NCIM 59 (Kulkarni *et al.,* 1995). But this dichotomous pattern was deviated in *A. niger* (Shei *et al.,* 1985), *H. lanuginosa* (Anand *et al.,* 1990), *A. fumigatus* (Anthony *et al.,* 2003) and *Nocardiopsis dassonvillei* (Tsujibo *et al.,* 1990).

## KINETIC PARAMETERS

Common kinetic parameters, which have been calculated for xylanases, are Michaelis Menten constant ($K_m$), maximum velocity ($V_{max}$) and enzyme turnover number ($k_{cat}$). The values for xylanases have been calculated using Lineweaver-Burk plot or Eadie-Hofstee plot. The unit of $K_m$ representation for xylanases is mg/ml but no uniformity has been observed in the unit for $V_{max}$ representation, which makes their comparison complicated. Many researchers have calculated $k_{cat}$, which is directly proportional to $V_{max}$ for xylanase. The value of $K_m$ indicates the effect of substrate concentration on xylanase activity. If the value of $K_m$ is known then it can be used to improve the assay conditions by changing the substrate concentration so that it should not be limiting. Different substrates have been used for determination of kinetic parameters for example larchwood xylan (Nakamura *et al.,* 1993), birchwood xylan (Raj and Chandra, 1996), oatspelt xylan (Dhillon *et al.,* 2000; De Segura and Fevre, 1993) and aspen glucuronoxylan (Mishra *et al.,* 1990).

The values of $K_m$ for xylanases generally range from 0.1 – 14 mg/ml depending upon the substrate used. The lowest $K_m$ value (0.14 mg/ml) has been reported for the xylanase (xln 2) from *T. reesei* (Torronen *et al.,* 1992). $K_m$ value of 22.3 mg/ml on glucuronoxylan for *T. reesei* xyalanase (Tenkanen *et al.,* 1992) has also been reported. The $k_{cat}$ values of xylanases from *T. emersonii* are reported to be in range of 216-9213 $s^{-1}$ whereas xylanase of *Thermoanaerobacterium* sp. has the $k_{cat}$ value of 26,680 U/µmol (Shao *et al.,* 1995). The kinetic parameters of xylanase from different microorganisms are listed in the Table 3 below:

**Table 3. Kinetic parameters of xylanases from different microorganisms**

| S. No. | Organism | Molecular Weight (kDa) | Optimum pH | Optimum Temp. (°C) | Optimum pI | Km (mg/ml) | Vmax (µM/min/ mg) | References |
|---|---|---|---|---|---|---|---|---|
| 1. | *Acidobacterium capsulatum* | 41 | 5 | 65 | 7.3 | 3.5 | 403 | Inagaki *et al.*, 1998 |
| 2. | *Acrophialophor a nainiana* | 17 | 6 | 50 | - | 0.731, 0.343 | - | Ximenes *et al.*, 1999 |
| 3. | *Aspergillus fischeri* Fxn1 | 31 | 6 | 60 | - | 4.88 | 5.88 | Raj and Chandra, 1996 |
| 4. | *Aspergillus kawachii* IFO 4308 | 26-35 | 2-5.5 | 50-60 | 3.5-6.7 | - | - | Ito *et al.*, 1992 |
| 5. | *Aspergillus nidulans* | 22-34 | 5.4 | 55 | - | - | - | Fernandez-Espinar *et al.*, 1992 |
| 6. | *Aspergillus niger* | 13.5-14.0 | 5.5 | 45 | 9 | - | - | Frederick *et al.*, 1985 |
| 7 | *Aspergillus sydowii* MG 49 | 30 | 5.5 | 60 | - | - | - | Ghosh and Nanda, 1994 |
| 8. | *Aureobasidium pullulans* Y-2311-1 | 25 | 4.4 | 54 | 9.4 | 7.6 | 2650 | Li *et al.*, 1993 |
| 9. | *Bacillus circulans* WL-12 | 15 | 5.5-7 | - | 9.1 | 4 | - | Esteban *et al.*, 1982 |
| 10. | *Bacillus* sp. NCIM 59 | 15.8, 35 | 6 | 50-60 | 4,8 | 1.58, 3.50 | 0.017, 0.742 | Dey *et al.*, 1992 |
| 11. | *Bacillus* sp. NG-27 | - | 7, 8.4 | 70 | - | - | - | Gupta *et al.*, 1992 |
| 12. | *Bacillus* sp. SPS-0 | - | 6 | 75 | - | - | - | Bataillon *et al.*, 1998 |

| | | | | | | | | |
|---|---|---|---|---|---|---|---|---|
| 13. | *Bacillus sp.* *strain* AR-009 | 23, 48 | 9-10 | 60-75 | - | - | - | Gessesse, 1998 |
| 14. | *Bacillus sp.* *strain* BP-23 | 32 | 5.5 | 50 | 9.3 | - | - | Blanco *et al.*, 1995 |
| 15. | *Bacillus sp.* *strain* BP-7 | 22-120 | 6 | 55 | 7-9 | - | - | Lopez *et al.*, 1998 |
| 16. | *Bacillus sp.* *strain* K-1 | 23 | 5.5 | 60 | - | - | - | Ratannakanokchai *et al.*, 1999 |
| 17. | *Cellulomonas sp.* NCIM 2353 | 22, 33, 53 | 6.5 | 55 | 8 | 1.7, 1.5 | 380, 690 | Chaudhary and Deobagkar, 1997 |
| 18. | *Cephalosporium sp.* | 30, 70 | 8 | 40 | - | 0.15 | - | Bansod *et al.*, 1993 |
| 19. | *Cryptococcus albidus* | 48 | 5 | 25 | - | 5.7, 5.3 | - | Morosoli *et al.*, 1986 |
| 20. | *Fusarium oxysporum* | 20.8, 23.5 | 6 | 60, 55 | - | 9.5; 8.45, 8.7 | 0.41, 0.37 | Christakopolous *et al.*, 1996 |
| 21. | *Geotrichum candidum* | 60-67 | 4 | 50 | 3.4 | - | - | Radionova *et al.*, 2000 |
| 22. | *Micrococcus sp.* AR-135 | 56 | 7.5-9 | 55 | - | - | - | Gessesse and Mamo, 1998 |
| 22. | *Paecilomyces varioti* | 20 | 4 | 50 | 5.2 | 49.5 | - | Kelly *et al.*, 1989 |
| 23. | *Penicillium purpurogenum* | 33, 23 | 7, 3.5 | 60, 50 | 8.6, 5.9 | - | - | Belancic *et al.*, 1995 |
| 24. | *Staphylococcus sp.* SG-13 | 60 | 7.5-9.2 | 50 | - | 4 | 90 | Gupta *et al.*, 2000 |
| 25. | *Streptomyces chattanoogensis* CECT 3336 | 48 | 6 | 50 | 9 | 4, 0.3 | 78.2, 19.1 | Lopez-Fernandez *et al.*, 1998 |
| 26. | *Streptomyces sp.* EC 10 | 32 | 7-8 | 60 | 6.8 | 3 | - | Lumba and Pennickx, 1992 |
| 27. | *Streptomyces sp.* QG-11-3 | - | 8.6 | 60 | - | 1.2 | 158.85 | Beg *et al.*, 2000 |
| 28. | *Streptomyces* T7 | 20 | 4.5-5.5 | 60 | 7.8 | 10 | 7610 | Kesker, 1992 |
| 29. | *Streptomyces thermoviolaceus* OPC-520 | 33, 54 | 7 | 60-70 | 4.2, 8 | - | - | Tsujibo *et al.*, 1992 |
| 30. | *Thermomonospora curvata* | 15-36 | 6.8-7.8 | 75 | 4.2-8.4 | 1.4-2.5 | - | Stutzenberger and Bodine, 1992 |
| 31. | *Thermomyces lanuginosus* DSM 5826 | 25.5 | 7 | 60-70 | 4.1 | 7.3 | - | Cesar and Mrsa, 1996 |
| 32. | *Thermomyces lanuginosus*-SSBP | 23.6 | 6.5 | 70-75 | 3.8 | 3.26 | 6300 | Lin *et al.*, 1999 |
| 33. | *Trichosporon cutaneum* SL409 | - | 6.5 | 50 | - | - | - | Liu *et al.*, 1998 |

## pH STABILITY AND THERMAL STABILITY

For pH stability determination, purified samples were incubated in buffers of varying pH at a particular temperature and for a fixed time interval. For thermal stability characterization, various temperatures were used at a particular pH for a definite time interval. The residual xylanase activity was assayed under standard conditions. Table 4 shows the pH and thermal stability of xylanases from different microorganisms:

**Table 4. pH and thermal stability of xylanases from different microorganisms**

| S.No. | Organism | pH stability | Thermal stability | References |
|---|---|---|---|---|
| 1. | *Acidobacterium capsulatum* | 3.0-8.0 | 20 °C-50 °C | Inagaki *et al.*, 1998 |
| 2. | *Acrophialophora nainiana* | 5.0 | 50 °C | Ximenes *et al.*, 1999 |
| 3. | *Aeromonas caviae* W-61 | - | 50 °C for 1 hr | Viet *et al.*, 1991 |
| 4. | *Aspergillus fischeri* Fxn1 | 5-9.5 | 55 °C | Raj and Chandra, 1996 |
| 5. | *Aspergillus fumigatus* AR1 | - | 40 °C | Anthony *et al.*, 2003 |
| 6. | *Aspergillus kawachii* IFO 4308 | 1.0-10.0 | 30 °C-60 °C | Ito *et al.*, 1992 |
| 7. | *Aspergillus nidulans* | 5.4 | 24 °C-40 °C | Fernandez-Espinar *et al.*, 1992 |
| 8. | *Aspergillus niger* | 5.0-6.0 | 60 °C | Frederick *et al.*, 1985 |
| 9. | *Aspergillus sojae* | 5.0-8.0 | 50 °C | Kimura *et al.*, 1995 |
| 10. | *Aspergillus sydowii* MG 49 | | | Ghosh and Nanda, 1994 |
| 11. | *Aureobasidium pullulans* Y-2311-1 | - | 40 | Li *et al.*, 1993 |
| 12. | *Bacillus sp.* NCIM 59 | 7.0 | 50 °C | Dey *et al.*, 1992 |
| 13. | *Bacillus sp.* NG-27 | 6.0-11.0 | 40 °C-90 °C | Gupta *et al.*, 1992 |
| 14. | *Bacillus sp.* SPS-0 | 6.0-9.0 | 85 °C | Bataillon *et al.*, 1998 |
| 15. | *Bacillus sp. strain* 41M-1 | - | 55 °C | Nakamura *et al.*, 1993 |
| 16. | *Bacillus sp. strain* AR-009 | 8.0-9.0 | 60 °C-65 °C | Gessesse, 1998 |
| 17. | *Bacillus sp. strain* BP-23 | 5.0-10.0 | 40 °C for 3 hr | Blanco and Pastor, 1993 |
| 18. | *Bacillus sp. strain* BP-23 | 9.5-11 | 55 °C | Blanco *et al.*, 1995 |
| 19. | *Bacillus sp. strain* BP-7 | 8.0-9.0 | 65 °C | Lopez *et al.*, 1998 |
| 20. | *Bacillus sp. strain* K-1 | 5.0-12.0 | 50 °C-60 °C | Ratannakanokchai *et al.*, 1999 |
| 21. | *Bacillus sp.* W-1 | 4.0-10.0 | 40 °C | Okazaki *et al.*, 1985 |
| 22. | *Bacillus spp.* | 4.5-10.5 at 45 °C for 1 hr | 60 °C | Okazaki *et al.*, 1984 |
| 23. | *Bacillus stearothermophilus* T-6 | 6.5-10.0 | 70 °C | Khasin *et al.*, 1993 |
| 24. | *Cephalosporium sp.* | 8.0-10.0 | | Bansod *et al.*, 1993 |
| 25. | *Fibrobacter succinogenes* S85 *(endoxylanase 2)* | - | 100 °C for 30 min | Matte and Forsberg, 1992 |
| 26. | *Fusarium oxysporum* | 7.0-10.0 | 30 | Christakopolous *et al.*, 1996 |
| 27. | *Fusarium verticillioides* | 5.0 Also 4.0-9.5 (1 hr at 40 °C) | 50°C for 30min | Saha, 2001 |
| 28. | *Geotrichum candidum* | 3-4.5 | 45 | Radionova *et al.*, 2000 |
| 29. | *Humicola grisea* var. thermoidea | - | 50°C for 30 min | Monti *et al.*, 1991 |
| 30. | *Micrococcus sp.* AR-135 | 6.5-10.0 | 40 °C | Gessesse and Mamo, 1998 |
| 31. | Mutant of *Aspergillus ochraceus* | 5.0-10.0 for 30 min at 45 °C | 45 °C for 30 min | Biswas *et al.*, 1990 |
| 32. | *Neocallimastix frontalis* | - | Below 45 °C | De-Segura and Fevre, 1993 |
| 33. | *Penicillium purpurogenum* | 6-7.5, 4.5-7.5 | 40 | Belancic *et al.*, 1995 |
| 34. | *Phanerochaete chrysosporium* | 4.0-5.5 | 37 °C | Dobozi *et al.*, 1992 |
| 35. | *Phanerochaete chrysosporium* | 4.0-9.0 | - | Copa-Patino *et al.*, 1993 |
| 36. | *Piromyces sp.* strain E2 | - | 30°C and 40 °C for 72 hr | Teunissen *et al.*, 1993 |
| 37. | *Pleurotus ostreatus* | 3.0-7.0 | 40 °C | Qinnghe *et al.*, 2003 |
| 38. | *Staphylococcus sp.* SG-13 | 7.5-9.5 | 50 °C | Gupta *et al.*, 2000 |
| 39. | *Streptomyces lividans* 66 | - | 30 °C | Kluepfel *et al.*, 1990 |
| 40. | *Thermomonospora alba* ULJB1 | 4.0-10.0 | 20-70 °C | Blanco *et al.*, 1997 |
| 41. | *Thermomonospora fusca* MT 816 Thermomonospora *Curvata* MT 815 Thermomonospora MT 814 Thermomonospora *viridis* NCIB 9605 | - | 70 °C for 60 min | McCarthy *et al.*, 1985 |
| 42. | *Thermomyces lanuginosus* DSM 5826 | 5.0-9.0 | 60 | Cesar and Mrsa, 1996 |
| 43. | *Thermomyces lanuginosus* SSBP | 5.0-12.0 | 60 | Lin *et al.*, 1999 |

## SUBSTRATE SPECIFICITY

Xylanases differ with respect to their substrate specificity. Table 5 below shows variations in substrate specificity of microbial xylanases:

**Table 5. Variations in substrate specificity of microbial xylanases from different microorganism**

| *Organism* | *Substrate* | *Activity* | *References* |
|---|---|---|---|
| *Aeromonas caviae* W-61 | β–1,4-xylan | ++ | Viet *et al.*, 1991 |
| | β–1,3-xylan, Cellulose, CM-cellulose, Starch | 0 | |
| *Bacillus stearothermophilus* T-6 | Oat spelt xylan | ++ | Khasin *et al.*, 1993 |
| | Guar gum, Locust bean gum,Arabinogalactan CM-cellulose, | 0 | |
| | p-nitrophenyl β–D-xylopyranoside, p-nitrophenyl α-L-arabinofuranoside, pNP-β–galactopyranoside, p-nitrophenyl β–glucopyranoside, p-nitrophenyl β–cellobioside, | + | |
| *Fibrobacter succinogenes* S85 | Rye flour arabinoxylan, Oat spelt xylan | ++ | Matte and Forsberg, 1992 |
| | Larchwood xylan, Birchwood xylan | + | |
| *Fusarium verticillioides* | Corn fiber xylan | 0 | Saha, 2001 |
| | Corn cob xylan, Oatspelt xylan | ++ | |
| | Birchwood xylan, Larchwood xylan | + | |
| *Humicola grisea* var. *thermoidea* | CM-cellulose, Starch, p-nitrophenyl β–D-xylopyranoside, Microcrystalline cellulose, Salicin, p-nitrophenyl β–D-galactopyranoside | 0 | Monti *et al.*, 1991 |
| | p-nitrophenyl β–glucopyranoside | + | |
| *Lentinula edodes* | Oatspelt arabinoxylan, Aspen glucuronoxylan | ++ | Mishra *et al.*, 1990 |
| | Laminarin, Pachyman, Pustulan | + | |
| | Acid-swollen cellulose, Salicin, p-nitrophenyl β–D-xylopyranoside, pNP-β–galactopyranoside, p-nitrophenyl β–glucopyranoside, p-nitrophenyl β–cellobioside | 0 | |
| Mutant of *Aspergillus ochraceus* | CM-cellulose, Araban | + | Biswas *et al.*, 1990 |
| | Starch, Laminarin,Cellobiose | 0 | |
| *Neocallimastix frontalis* | CM-cellulose | ++ | De Segura and Fevre, 1993 |
| | pNP-β–glucoside, pNP-β–xyloside, pNP-β–fucoside, pNP- β–lactoside, pNP-β–cellobioside, pNP-β–galactoside, Laminarin | 0 | |
| *Streptomyces lividans* | Soluble xylan, Insoluble xylan, Birchwood xylan | ++ | Blanco *et al.*, 1997 |
| | Lichenan, CM-cellulose | 0 | |
| *Thermotoga thermarum* | Oat spelt xylan, Larchwood xylan, Birchwood xylan, CM-cellulose, Starch, Pullulan, Galactomannan | ++ | Sunna *et al.*, 1996 |
| | Beechwood xylan | + | |
| | p-nitrophenyl β–D-xylopyranoside, p-nitrophenyl α-L-arabinofuranoside | 0 | |

(++) High activity; (+) Low activity; (0) No activity

## EFFECT OF METAL IONS AND CHEMICAL AGENTS

The xylanase activity is measured under the standard assay conditions in the presence of metal ions and other agents.

Monovalent cations such as $Na^+$ and $K^+$ have been reported to increase (Biswas *et al.,* 1990; Khasin *et al.,* 1993), inhibit (Shei *et al.,* 1985) and have no effect on xylanase activity (Anthony *et al.,* 2003; Raj and Chandra, 1996). $Ca^{2+}$ have been found to stimulate xylanase activity (Ghanem *et al.,* 2000; Qureshy *et al.,* 2002) but in some cases have no influence (Anthony *et al.,* 2003; Nakamura *et al.,* 1996; Raj and Chandra, 1996; Saha, 2001). $Fe^{2+}$ partially inhibited xylanase activity (Dhillon *et al.,* 2000; Khasin *et al.,* 1993; Sunna *et al.,* 1996). $MnCl_2$ was found to increase (Gupta *et al.,* 2000; Bataillon *et al.,* 2000) or inhibit the activity (Kang *et al.,*1996; Araki *et al.,* 1999). $Hg^{2+}$significantly increased the xylanase activity in *Aspergillus fischeri* Fxn 1 (Raj and Chandra, 1996) but in most cases this ion has been reported to completely inhibit the activity (Bataillon *et al.,* 2000; Gupta *et al.,* 2000; Kang *et al.,* 1996; Qureshy *et al.,* 2002).

Xylanase activities are found to be inhibited by chemicals such as 4-[2-aminoethyl] benzylsulphonyl fluoride (Sunna *et al.,* 1996), p-hydroxymercuribenzoate (Bataillon *et al.,* 2000; Sunna *et al.,* 1996), N-ethyl maleimide and monoiodoacetic acid (Gupta *et al.,* 2000; Qureshy *et al.,* 2002). In most cases, EDTA does not cause any change in the activity of xylanase (Khasin *et al.,* 1993; Nakamura *et al.,* 1993; Raj and Chandra, 1996; Saha, 2001) but in few inhibits the xylanase activity (Ghanem *et al.,* 2000; Gupta *et al.,* 2000; Okazaki *et al.,* 1985; Sunna *et al.,* 1996). Deactivation of xylanase by detergents such as SDS has also been reported (Liu *et al.,* 1998; Gupta *et al.,* 2000). N-bromosuccinimide has reported to significantly inhibit xylanase activity (Araki *et al.,* 1999; Bataillon *et al.,* 2000; Khasin *et al.,* 1993; Nakamura *et al.,* 1993; Sunna *et al.,* 1996). β-mercaptoethanol and dithiothreitol were found to have stimulatory effect on xylanase activity (Bataillon *et al.,* 2000; Gupta *et al.,* 2000; Cesar and Mrsa, 1996) but on the other hand, no enhancement in activity was reported in the presence of dithiothreitol (Saha, 2001).

## BINDING STUDIES

Jorgensen *et al.* (2003) found that xylanase isolated from *Penicillium brasilianum* IBT 20888 showed only limited adsorption (4-9%) onto cellulose whereas Blanco *et al.* (1997) showed that XylA isolated from *Streptomyces lividans* was able to bind to Avicel (microcrystalline cellulose) but not to insoluble xylan.

## ANALYSIS OF HYDROLYSIS PRODUCTS

The degradation of various xylans by purified xylanase was followed by analyzing the reaction products by High-Performance liquid Chromatography (Dhillon *et al.,* 2000; Jorgensen *et al.,* 2003; Kluepfel *et al.,* 1990; Saha, 2001; Teunissen *et al.,* 1993), Thin-layer chromatography (Blanco *et al.,* 1997; De Segura and Fevre, 1993; Ethier *et al.,* 1994; Khasin *et al.,* 1993; Lee *et al.,* 1993; Li *et al.,* 1993; Morosoli *et al.,* 1986; McCarthy *et al.,* 1985; Mishra *et al.,* 1990; Nakamura *et al.,* 1993; Okazaki *et al.,* 1984; Raj and Chandra, 1996) or paper chromatography (Keskar *et al.,* 1989; Mishra *et al.,* 1990; Viet *et al.,* 1991) as given in the Table 6:

**Table 6. Degradation of various xylans by xylanase from different microorganisms**

| *S. No.* | *Organism producing xylanase* | *Assay Substrate* | *End-Products* | *References* |
|---|---|---|---|---|
| 1. | *Actinomadura* sp. strain FC7 | Oat spelt xylan | | Ethier *et al.,* 1994 |
| | Xyl I | | Traces of xylose, xylobiose, xylotriose | |
| | Xyl II | | xylobiose, xylotriose | |
| 2. | *Aeromonas caviae* W-61 | Oat spelt xylan | Xylobiose, xylotriose, xylotetroase, xylopentaose and xylose | Viet *et al.,* 1991 |
| 3. | *Aspergillus fischeri* Fxn 1 | Oat spelt xylan | Xylose, arabinose, glucose | Raj and Chandra, 1996 |
| 4. | *Aureobasidium pullulans* Y-2311-1 | Oat spelt xylan | Xylobiose, Xylose, xylopentaose | Li *et al.,* 1993 |
| 5. | *Bacillus circulans* AB16 | Oat spelt xylan | | Dhillon *et al.,* 2000 |
| | Xyl A | | Xylobiose, xylose, higher oligosaccharides | |
| | Xyl B | | Higher oligosaccharides, xylobiose, xylose | |
| 6. | *Bacillus* sp. Strain 41M-1 | Larchwood xylan | Xylose, xylobiose, xylotriose, xylotetraose, xylopentaose | Nakamura *et al.,* 1993 |
| 7. | *Bacillus* sp. strain BP-23 | Oat spelt xylan | Xylobiose, xylotriose, xylose | Viet *et al.,* 1991 |
| 8. | *Bacillus spp.* | Larchwood xylan | Xylose and xylobiose | Okazaki *et al.,* 1984 |
| 9. | *Bacillus stearothermophilus* T-6 | Oat spelt xylan | Xylose, xylobiose | Khasin *et al.,* 1993 |
| 10. | *Cryptococcus albidus* | Oat spelt xylan | Xylose and xylobiose | Morosoli *et al.,* 1986 |
| 11. | *Fibrobacter succinogenes* S85 | Oat spelt xylan or rye flour arabinoxylan | Glucose, arabinose, xylooligosaccharides with chain lengths less than or equal to that of xylotetrose | Smith and Forsberg, 1991 |
| 12. | *Fusarium verticillioides* | Corn fibre xylan | Arabinose, xylose, xylobiose | Saha, 2001 |
| 13. | *Lentinula edodes* | Aspen glucuronoxylan | Xylobiose, xylo-oligosaccharides (including $X_3$ & $X_4$), xylose | Mishra *et al.,* 1990 |
| 14. | *Neocallimastix frontalis* | Oat spelt xylan | Xylose and xylobiose | De Segura and Fevre, 1993 |
| 15. | *Piromyces* sp. strain E2 | Oat spelt xylan | Xylo-oligosaccharides ($>X_3$), xylose, arabinose | Teunissen *et al.,* 1993 |
| 16. | *Saccharomonospora viridis* | Oat spelt xylan | Xylobiose, xylose | McCarthy *et al.,* 1985 |
| 17. | *Streptomyces* | Larchwood xylan | Xylobiose, xylo-oligosaccharides ($Xyl_3$-$Xyl_6$), xylose | Keskar *et al.,* 1989 |
| 18. | *Streptomyces lividans* 66 | Oat spelt xylan | | |
| | Xylanase A | | Xylotriose, xylobiose Oligo- | Kluepfel *et al.,* 1990 |
| | Xylanase B | | xylosides ranging in size from 1 to 8 xylose molecules | |
| 19. | *Thermoanaerobacterium saccharolyticum* B6A-RI | Oat spelt xylan | Xylobiose, xylotriose | Lee *et al.,* 1993 |
| 20. | *Thermomonospora alba* JB1 | Oat spelt xylan | Xylobiose and traces of xylose | Blanco *et al.,* 1997 |
| 21. | *Thermotoga thermarum* Endoxylanase 1 | Beechwood xylan | Xylobiose, xylo-oligomers upto a degree of polymerization of 10 | |

## CONCLUSION

The specific activity of microbial xylanases in industrial applications is 2 to 3 times greater than the hydrolases of other polymers like crystalline cellulose (Gilbert and Hazlewood, 1993); thus making them industrially important. In view of the varied applications of xylanases, it is very important to get their higher production level to make xylanases based processes economically viable. In order to have significant impact on an industrial scale, they must need to be consistently effective under various operating conditions. Standardization of various purification and characterization methods of xylanases is necessary so as to evaluate the suitability of different xylanolytic systems for different applications. Programmes involving microbiologists, biochemists and process engineers in collaboration with the paper industries should be established to view the overall perspective and to create effective and functional networks that are progressive and result oriented.

## REFERENCES


Akila, G. and Chandra, T.S. (2003). A novel cold-tolerant *Clostridium* strain PXYL1 isolated from a psychrophilic cattle manure digester that secretes thermolabile xylanase and cellullase. *FEMS Microbiology Letters,* 219(1), 63-67.

Anand, L., Krishanamurthy, S. and Vithayathil, P.J. (1990). Purification and properties of xylanase from the thermophilic fungus, *Humicola lanuginosa. Archives of Biochemistry and Biophysics,* 276, 546-553.

Anthony, T., Chandra R.K., Rajendran, A. and Gunasekaran, P. (2003). High molecular weight cellulase-free xylanase fro alkali-tolerant *Aspergillus fumigatus* AR1. *Enzyme and Microbial Technology,* 32, 647-654.

Bajpai, P. (1999). Application of enzymes in the pulp and paper industry. *Biotechnology Progress,* 15, 147-157.

Bajpai, P., Bhardwaj, N.K., Bajpai, P.K and Jauhari, M.B . (1994). The impact of xylanase on bleaching of eucalyptus kraft pulp. *Journal of Biotechnology,* 38, 1-6.

Baker, C.J., Whalen, C.H. and Bateman, D.F. (1977). Xylanase from *Trichoderma pseudokoningii:* purification characterization and effects on isolated plant cell walls. *Phytopathology,* 67, 1250-1258.

Ball, A.S. and McCarthy, A.J. (1989). Production and purification of xylanase from actinomycetes. *Journal of Applied Bacteriology,* 66, 439-444.

Bandivadekar, K.R. and Deshpande, V.V. (1996). Structure-function relationship of xylanase: Fluorimetric analysis of the tryptophan environment. *Biochemical Journal,* 315, 583-587.

Bansod, S.M., Chaudhary, M.D., Srinivasan, M.D. and Rele, M.V. (1993). Xylanase active at high pH from an alkalotolerant *Cephalosporium* species. *Biotechnology Letters,* 15, 965-970.

Bastawde, K.B. (1992). Xylan structure, microbial xylanases, and their mode of action. *World Journal of Microbiology and Biotechnology,* 8, 353-368.

Bataillon, M., Cardinali, A.P.N. and Duchiron, F. (1998). Production of xylanases from a newly isolated alkalophilic thermophilic *Bacillus* sp. *Biotechnology Letters,* 20, 1067-1071.

Bedford, M.R. and Classon, H.L. (1992). The influence of dietary xylanase on intestinal viscosity and molecular weight distribution of carbohydrates in rye-fed broiler chick. In: *Xylans and Xylanases.* Eds Visser, J., Beldman, G., Vansomenon, A.K. and Voragen, A.G.J. Elsevier Science Publishers, Amsterdam. Pp 361-370.

Beg, A.Q., Bhushan, B., Kapoor, M. and Hoondal, G.S. (2000). Production and characterization of thermostable xylanase and pectinase from a *Streptomyces* sp. QG-11-3. *Journal of Industrial Microbiology and Biotechnology,* 24, 396-402.

Belancic, A., Scarpa, J., Peirano, A., Diaz, R., Stemer, J., and Eyzayuine, J. (1995). *Penicillium purpurogenum* produces several xylanases: purification and properties of two of the enzymes. *Journal of Biotechnology,* 41: 71-79.

Biely, P. (1985). Microbial xylanolytic systems. *Trends in Biotechnology,* 3(11), 286-290.

Biswas, S.R., Jana, S.C., Mishra, A.K. and Nanda, G. (1990). Production, Purification and Characterization of Xylanase from a Hyperxylanolytic Mutant of *Aspergillus ochraceus. Biotechnology and Bioengineering,* 35, 244-251.

Black, G.W., Hazlewood, G.P., Millaward-Saddler, S.J., Laurie, J.I. and Gilbert, H. J. (1995). A modular xylanase containing a novel noncatalytic xylan specific binding domains. *Biochemical Journal,* 307, 191-195.

Blanco, A. and Pastor, F.I.J. (1993). Characterization of cellulose free xylanases from the newly isolated *Bacillus* sp. strain BP-23. *Canadian Journal of Microbiology,* 39, 1162-1166.

Blanco, A., Vidal, T., Colon, J.F. and Pastor, F.I.J. (1995). Purification and properties of xylanase from alkali-tolerant *Bacillus* sp. strain BP-23. *Applied Environmental and Microbiology,* 61, 4468-4470.

Blanco, J., Coque, J.J.R., Velasco, J. and Martin, J.F. (1997). Cloning, expression in *Streptomycin lividans* and biochemical characterization of a thermostable endo-β–1,4-xylanase of *Thermomonospora alba* ULJB1 with cellulose-binding ability. *Applied Microbiology and Biotechnology,* 48, 208-217.

Bragger, J.M., Daniel, R.M., Coolbear, T. and Morgan, H.W. (1989). Very stable enzymes from extremely thermophilic archaebaçteria and eubacteria. *Applied Microbiology and Biotechnology,* 31, 556-561.

Cardinale, F. and Matta, A. (2001). Basic xylanases from the fungal tomato pathogen *Fusariun oxysporum* F. Sp. lycopersici. *Journal of Plant Pathology,* 83(1): 27-36.

Cesar, T. and Mrsa, V. (1996). Purification and properties of xylanase produced by *Thermomyces lanuginosus. Enzyme and Microbial Technology,* 19: 289-296.

Chaudhary, P. and Debunker, D. (1997). Purification and characterization of xylanase from *Cellulomonas* sp. N.C.I.M. 2353. *Biotechnology and Applied Biochemistry,* 25, 127-133.

Christakopolous, P., Nerinckx, W., Kekos, D., Marcis, B. and Claeyssens, M. (1996). Purification and characterization of two low molecular mass alkaline xylanases from *Fusarium oxysporum* F3. *Journal of Biotechnology* 51, 181-189.

Christakopoulos, P., Nerinckx, W., Kekos, D., Marcis, B. and Claeyssens, M. (1996). Purification and characterization of two low molecular mass alkaline xylanases from *Fusarium oxysporum* F3. *Journal Biotechnology* 51, 181-189.

Cleemput, G., Hessing, M., Van-Oort, M., Deconynck, M. and Delcour, J.A. (1997). Purification and characterization of β–D-xylosidase and an endo-xylanase from wheat flour. *Plant Physiology,* 113, 377-386.

Copa-Patino, J.L., Kim, Y.G. and Broda, P. (1993). Production and initial characterization of the xylan-degrading system of *Phanerochaete chrysosporium. Applied Microbiology and Biotechnology,* 40, 69-70.

De Segura, B.G. and Fevre, M. (1993). Purification and characterization of two 1,4-β–xylan endohydrolases from the rumen fungus *Neocallimastrix frontalis. Applied and Environmental Microbiology,* 59(11), 3654-3660.

Dekker, R.F.H. (1985). Biodegradation of hemicellulose In: *Biosynthesis and Biodegradation of Wood Components.* Ed. Higuchi, T. Academic Press, Tokyo. Pp 505-531.

Dekker, R.F.H. and Richards, G.N. (1976). Hemicellulases, their occurrence, purification, properties and mode of action. *Advances in Carbohydrate Chemistry and Biochemistry,* 32, 277-352.

De-Segura, B. G. and Fevre, M. (1993). Purification and characterization of two 1,4-beta-xylan endohydrolases from the rumen fungus *Neocallimastix frontalis. Applied and Environmental Microbiology,* 59, 3654-3660.

Dey, D., Hinge, J., Shendye, A. and Rao, M. (1992). Purification and properties of extracellular endo-xylanase from alkalophilic thermophilic *Bacillus* sp. *Canadian Journal of Microbiology,* 38, 436-442.

Dhillon, A., Gupta, J.K. and Khanna S. (2000). Enhanced production, purification and characterization of a novel cellulase-poor thermostable, alkali tolerant xylanase from *Bacillus circulans* AB 16. *Process Biochemistry,* 35, 849-856.

Dobozi, M.S., Szakacs, G. and Bruschi, C.V. (1992). Xylanase Activity of *Phanerochaete chrysosporium. Applied and Environmental Microbiology,* 58(11), 3466-3471.

Elegir, G., Sykes, M. and Jeffries, T.W. (1994). Purification, characterization and substrate specificities of multiple xylanases from *Streptomyces* sp. B-12-2. *Applied and Environmental Microbiology,* 60(7), 2609-2615.

Eriksson, K.E.L. (1985). Swedish developments in biotechnology related to the pulp and paper industry. *TAPPI (Tech. Assoc. Pulp Pap. Ind.),* 68, 46-55.

Espinar, M.T.F., Pinaga, F., Sanz, P., Ramon, D. and Valles, S. (1993). Purification and characterization of a neutral endoxylanase from *Aspergillus nidulans. FEMS Microbiology Letters,* 113, 223-228.

Esteban, R., Villanueva, J.R. and Villa, T.G. (1982). β–D-xylanases of *Bacillus circulans* WL-12. *Canadian Journal of Microbiology,* 28, 733-739.

Ethier, J.F., Harpin, S., Girard, C., Beaulieu, C., Dery, C.V. and Brzezinski, R. (1994). Cloning of two xylanase genes from the newly isolated actinomycete *Actinomadure* sp. strain FC7 and characterization of the gene products. *Canadian Journal of Microbiology,* 40, 362-368.

Fernandez-Espinar, M.T., Ramon, D., Pinaga, F. and Valles, S. (1992). Xylanase production by *Aspergillus nidulans. FEMS Microbiology Letters,* 91, 91-96.

Fishman, A., Berk, Z. and Shoham, Y. (1995). Large-scale purification of Xylanase T-6. *Applied Microbiology and Biotechnology,* 44, 88-93.

Fontes, C.M.G.A., Gilbert, H.J., Hazlewood, G.P., Clarke, J.H., Prates, J.A.M., Mckie, V.A., Nagy, T., Fernandes, T.H. and Ferreira, L.A.M. (2000). A novel *Cellvibrio mixtus* family 10 xylanase that is both intracellular and expressed under non-inducing conditions. *Microbiology,* 146, 1959-1967.

Franco, P.F., Ferreira, H.M. and Filho, E.X.F. (2004). Production and characterization of hemicellulase activities from *Trichoderma harzianum* strain T4. *Biotechnology and Applied Biochemistry,* 40, 255-259.

Frederick, M.M., Kiang, C., Frederick, J.R. and Reilly, P.J. (1985). Purification and characterization of endo-xylanases from *Aspergillus niger* I. Two isozymes active on xylan backbones near branch points. *Biotechnology and Bioengineering,* 27, 525-532.

Gessesse, A. (1998). Purification and characterization of two thermostable alkaline xylanases from an alkalophilic *Bacillus* sp. *Applied and Environmental Microbiology,* 64, 3533-3535.

Gessesse, A. and Mamo, G. (1999). High-level xylanase production by an alkalophilic *Bacillus* sp. by using solid-state fermentation. *Enzyme and Microbial Technology,* 25, 68-72.

Ghosh, M., and Nanda, G. (1994). Purification and some properties of xylanase from *Aspergillus sydowii* MG 49. *Applied and Environmental Microbiology,* 60, 4620-4623.

Gilbert, H.J. and Hazlewood, G.P. (1993). Bacterial cellulases and xylanases. *Journal of General Microbiology,* 139, 187-194.

Gomes, J., Purkarthofer, H., Hayn, M., Kappimillen, J., Sinner, M. and Steiner, W. (1993). Production of a high level of cellulose free xylanase by thermophilic fungus *Thermomyces lanuginosus* in laboratory and pilot scales using lignocellulosic material. *Applied Microbiology and Biotechnology,* 39: 700-707.

Grabski, A.C. and Jeffries, T.W. (1991). Production, purification and characterization of β–(1,4)-endoxylanase of *Streptomyces roseiscleroticus. Applied and Environmental Microbiology,* 57(4), 987-992.

Gupta, N., Vohra, R.M. and Hoondal, G.S. (1992). A thermostable extracellular xylanase from alkalophilic *Bacillus* sp. NG27. *Biotechnology Letters,* 14, 1045-1046.

Gupta, S., Bhushan, B. and Hoondal, G.S. (2000). Isolation, purification and characterization of xylanase from *Staphylococcus* sp. SG-13 and its application in biobleaching of kraft pulp. *Journal of Applied Microbiology,* 88, 325-334.

Hansen, C.L., Sommer, M.O.A. and Quake, S.R. (2004). Systematic investigation of protein phase behavior with a microfluidic formulator. *Proceedings of National Academy of Sciences,* 101(40), 14431-14436.

Hespell, R.B. and Bryan-Shah, P.J.O. (1988). Esterase activity in *Butyrivibrio fibrisolvens* strains. *Applied and Environmental Microbiology,* 54, 1917-1922.

Honda, H., Kudo, T., Ikura, Y. and Horikoshi, K. (1985). Two types of xylanases of alkalophilic *Bacillus* sp. no. C-125. *Canadian Journal of Microbiology,* 31, 538-542.

Hori, H. and Elbein, A.D. (1985). The biosynthesis of plant cell wall polysaccharide. *In*: T.Higuchi (ed.). *Biosynthesis and biodegradation of wood components.* Academic Press, Inc., Orlando, Fla. 109-135.

Hrmova, M., Biely, P., Vrsanka, M. and Petrakova, E. (1984). Induction of cellulose- and xylan-degrading enzyme complex in yeast *Trichosporon cutaneum. Archives of Microbiology,* 161: 371-376.

Inagaki, K., Nakashira, K., Mukai, K., Tamura, T. and Tanaka, H. (1998). Gene cloning and characterization of an acidic xylanase from *Acidobacterium capsulatum. Bioscience Biotechnology Biochemistry,* 62: 1061-1067.

Ito, K., Ogasawara, H., Sugimoto, T. and Ishikawa, T. (1992). Purification and properties of acid stable xylanases from *Aspergillus kawachii. Bioscience Biotechnology Biochemistry,* 56: 547-550.

Jeffries, T.W. (1985). Emerging technology for fermenting D-xylose. *Trends in Biotechnology,* 3: 208-212.

Jorgensen, H., Eriksson, T., Borjesson, J., Tjerneld, F. and Olsson, L. (2003). Purification and Characterization of five Cellulase and one Xylanase from *Penicillium brasilianum* IBT 20888. *Enzyme and Microbial Technology,* 32: 851-861.

Joseleau, J.P., Comta, J. and Ruel, K. (1992). Chemical structure of xylans and their interaction in the plant cell walls. In: *Progress in Biotechnology. Xylans and Xylanases.* Eds Visser, J., Beldman, G., Kusters-van Someren, M.A.and Voragen, A.G.J., Amsterdam: Elsevier. Pp 1-16.

Kansoh, A.L. and Nagieb, Z.A. (2004). Xylanase and Mannanase enzymes from *Streptomyces galbus* NR and their use in biobleaching of softwood kraft pulp. *Antonie van Leeuwenhoek* 85(2), 103-114(12).

Kelly, C.T., O'Mahony, M.R. and Fogarty, W.M. (1989). Extracellular xylanolytic enzymes of *Paecilomyces varioti. Biotechnology Letters,* 11, 885-890.

Keskar, S.S., Srinivasan, M.C. and Deshpande, V.V. (1989). Chemical modification of a xylanase from a thermotolerant *Streptomyces.* Evidence for essential tryptophan and cysteine residues at the active site. *Biochemistry Journal,* 261: 49-55.

Kesker, S.S. (1992). High activity xylanase from thermotolerant *Streptomyces* T7, cultural conditions and enzyme properties. *Biotechnology Letters,* 14, 481-486.

Khanna, P., Sundri, S.S. and Kumar, N. J. (1995). Production, isolation and partial purification of xylanase from an *Aspergillus* sp. *World Journal of Microbiology and Biotechnology,* 11, 242-243.

Khanna, S. and Gauri, P. (1993). Regulation, purification and properties of xylanase from *Cellulomonas fimi. Enzyme and Microbial Technology,* 15: 990-995.

Khasin, A., Alchanati, I. and Shoham, Y. (1993). Purification and characterization of a thermostable xylanase from *Bacillus stearothermophilus* T-6. *Applied and Environmental Microbiology,* 59(6), 1725-1730.

Kimura, I., Sasahar, H. and Tajima, S. (1995). Purification and characterization of two xylanases and an arabinofuranosidase from *Aspergillus sojae*. *Journal of Fermentation Bioengineering,* 80, 334-339.

Kluepfel, D., Vats-Mehta, S., Aumont, F., Shareck, F. and Morosoli, R. (1990). Purification and Characterization of a New Xylanase (Xylanase B) Produced By *Streptomyces lividans* 66. *Biochemistry Journal,* 267, 45-50.

Kudo, T., Ohkoshi, A. and Horikoshi, K. (1985). Molecular Cloning and Expression of a Xylanase Gene of Alkalophilic *Aeromonas* sp. no. 212 in *Escherichia coli. Journal of General Microbiology,* 131, 2825-2830.

Kuhad, R.C. and Singh, A. (1993). Lignocellulosic biotechnology: current and future prospectus. *Critical Review of Biotechnology,* 13, 151-172.

Kuhad, R.C., Singh, A. and Eriksson, K.E.L. (1997). Microorganisms and enzymes involved in the degradation of plant fiber cell wall. *Advances in Biochemical Engineering and Biotechnology,* 57, 47-125.

Kulkarni, N., Chanthaiwale, J. and Rao, M. (1995). Characterization of the recombinant xylanase in *Escherichia coli* from an alkalophilic thermophilic *Bacillus* sp. NCIM59. *Enzyme and Microbial Technology,* 17, 972-976.

Lappalainen, A., Sikka-Aho, M., Kalkkinen, N., Fagerstorm, R. and Tenkanen, M. (2000). Endoxylanase II from *Trichoderma reesei* has several isoforms with different isoelectric points. *Biotechnology and Applied Biochemistry,* 31, 61-68.

Lee, Y.E., Lowe, S.E. and Zeikusi, J.G. (1993). Regulation and Characterization of Xylanolytic Enzymes of *Thermoanaerobacterium saccharolyticum* B6A- RI. *Applied and Environmental Microbiology,* 59(3), 763-771.

Li, X.L., Zhang, Z.Q., Dean, J.F.D., Eriksson K.E.L. and Ljunghahl, L.G. (1993). Purification and characterization of a new xylanase (APX-II) from the fungus *Aureobasidium pullulans* Y-2311-1. *Applied and Environmental Microbiology,* 59(10), 3212-3218.

Lin, J., Ndlovu, L.M., Singh, S. and Pillay, B. (1999). Purification and biochemical characteristics of β–D-xylanase from a thermophilic fungus *Thermomyces lanuginosus*-SSBP. *Biotechnology and Applied Biochemistry* 30, 73-79.

Liu, W., Zhu, W., Lu, Y., Kong, Y. and Ma, G. (1998). Production, partial purification and characterization of xylanase from *Trichosporon cutaneum* SL409. *Process Biochemistry,* 33, 331-326.

Lopez, C., Blanco, A. and Pastor, F. I. J. (1998). Xylanase production by a new alkali-tolerant isolate of *Bacillus. Biotechnology Letters,* 20, 243-246.

Lopez-Fernandez, C., Rodriguez, J., Ball, A.S., Lopa-Patino, J.L., Periz-Lebic, M.L. and Arias, M.E. (1998). Application of the affinity binding of xylanases to oat-spelt xylan in the purification of endoxylanase CM-2 from *Streptomyces chattanoogensis* CECT 3336. *Applied Microbiology and Biotechnology,* 50, 284-287.

Lumba, F.L. and Penninckx, M.J. (1992). Characterization of multiple forms of β–xylanase produced by a *Streptomyces* sp. growing on lignocellulose. *Applied Microbiology and Biotechnology,* 36, 733-738.

Magnuson, T.S. and Crawford, D.L. (1997). Purification and characterization of an alkaline xylanase from *Streptomyces viridosporus* T7A. *Enzyme and Microbial Technology,* 21, 160-164.

Maheshwari, R. and Kamalam, P.T. (1985). Isolation and Culture of a Thermophilic Fungus, *Melanocarpus albomyces*, and Factors Influencing the Production and Activity of Xylanase. *Journal of General Microbiology,* 131, 3017-3027.

Margolin, A.L. Govardhan, C.P., Visuri, K. and Uotila, S. (2002). *United States Patent,* No. 6, 359, 118. Carbohydrate Crosslinked Glycoprotein Crystals.

Marui, M., Nakanishi, K. and Ysui, T. (1985). Purification and properties of three types of xylanases induced by methyl β–xyloside from *Streptomyces* sp. *Agricultural and Biological Chemistry,* 49, 3399-3407.

Matte, A. and Forsberg, C.W. (1992). Purification, Characterization, and Mode of Action of Endoxylanases 1 and 2 from *Flbrobacter succinogenes* S85. *Applied and Environmental Microbiology,* 59(3), 157-168.

McCarthy, A.J., Peace, E. and Broda, P. (1985). Studies on the extracellular xylanase activity of some thermophilic actinomycetes. *Applied Microbiology and Biotechnology,* 21, 238-244.

Mishra, C., Forrester, I.T., Kelley, B.D., Burgess, R.R. and Leathem, G.F. (1990). Characterization of a major xylanase purified from *Lentinula edodes* culture grown on a commercial solid lignocellulosic substrate. *Applied Microbiology and Biotechnology,* 33, 226-232.

Monti, R., Terenzi, H.F. and Jorge, J.A. (1991). Purification and Properties of an Extracellular Xylanase from the Thermophilic Fungus *Humicola grisea* var. *thermoidea Canadian Journal of Microbiology,* 37, 675-681.

Morag, E., Bayer, E.A. and Lamed, R. (1990). Relationship of cellulosomal and noncellulosomal xylanases of *Clostridium thermocellum* to cellulose-degrading enzymes. *Journal of Bacteriology,* 172, 6098-6105.

Morales, P., Madrarro, A., Flors, A., Sendra, J.M. and Gonzalez, J.A.P. (1995). Purification and characterization of a xylanase and an arabinofuranosidase from *Bacillus polymyxa. Enzyme and Microbial Technology,* 17, 424-429.

Morosoli, R., Roy, C. and Yaguchi, M. (1986). Isolation and partial primary sequence of a xylanase from the yeast *Cryptococcus albidus. Biochimica et Biophysica Acta,* 870, 473-478.

Nakamura, S., Wakabayashi, K., Nakai, R., Aono, R. and Horikoshi, K. (1993). Purification and some properties of an alkaline xylanase from alkaliphilic *Bacillus* sp. Strain 41M-1. *Applied and Environmental Microbiology,* 59(7), 2311-2316.

Niehaus, F., Bertoldo, C., Kahler, M. and Antranikian, G. (1999). Extremophiles as a source of novel enzymes for industrial applications. *Applied Microbiology and Biotechnology,* 51, 711-729.

Okada, H. and Shinmyo, A. (1988). Xylanase from *Bacillus pumilus. Methods in Enzymology,* 160, 632-637.

Okazaki, W., Akiba, T., Horikoshi, K. and Akahoshi, R. (1984). Production and properties of two types of xylanases from alkalophilic thermophilic *Bacillus* spp. *Applied Microbiology Biotechnology,* 19, 335-340.

Okazaki, W., Akiba, T., Horikoshi, K. and Akahoshi, R. (1985). Purification and characterization of xylanases from alkalophilic thermophilic *Bacillus* spp. *Agricultural Biology Chemistry,* 49, 2033-2039.

Park, Y.K., Yum, D.Y., Bal, D.H. and Yu, J.H. (1992). Xyianase from alkalophilic *Bacillus* sp. YC-335. *Bioscience, Biotechnology and Bioengineering,* 56, 1355-1356.

Poutanen, K., Ratto, M., Puls, J. and Viikari, L. (1987). Evaluation of different microbial xylanolytic systems. *Journal of Biotechnology,* 6, 49-60.

Qinnghe, C., Xiaoyu, Y., Tiangui, N., Cheng, J. and Qiugang, M. (1993). The screening of culture condition and properties of xylanase by white-rot fungus *Pleurotus ostreatus. Process Biochemistry,* 39(11), 1516-1566.

Radionova, N.A., Dubovaya, N.V., Eneiskaya, E.V., Matrinovich, L.I., Gracheva, I.M., and Bezborodov, A.M. (2000). Purification and characterization of endo-1-4-β–xylanase from *Geotrichum candidum* 3C. *Applied Biochemistry and Microbiology,* 36, 460-465.

Raj, K.C. and Chandra, T.S. (1996). Purification and characterization of xylanase from alkali-tolerant *Aspergillus fischeri* Fxn 1. *FEMS Microbiology Letter,* 145, 457-461.

Rani, S. and Nand, K. (1996). Development of cellulose-free xylanase producing anaerobic consortia for the use of lignocellulosic waste. *Enzyme and Microbial Technology,* 18, 23-28.

Rao, M., Khadilkar, S., Bandivadekar, K.R. and Deshpande, V. (1996). Structural environmental of an essential cysteine residue of xylanase from *Chainia* sp. *Biochemistry Journal,* 316, 771-775.

Ratanakhanokchai, K., Kyu, K.L. and Tanticharoen, M. (1999). Purification and Properties of a Xylan-Binding Endoxylanase from Alkalophilic *Bacillus* sp. Strain K-1. *Applied and Environmental Microbiology,* 65(2), 694-697.

Ratto, M., Poutanen, K., and Viikari, L. (1992). Production of xylanolytic enzymes by an alkalitolerant *Bacillus circulans* strain. *Applied Microbiology and Biotechnology,* 37(4), 470-473.

Reilly, P.J. (1981). Xylanases: structure and function. In: *Trends in the Biology of Fermentations for Fuels and Chemicals,* Pp 111-129.

Rodrigues, E.M.G. and Tambourgi, E.B. (2001). Continuous extraction of xylanase from *Penicillium janthinellum* with reversed micelles using experimental design mathematical model. *Biotechnology Letters,* 25(5), 365-367(3).

Saha, B.C. (2001). Xylanase from a newly isolated *Fusarium verticillioides* capable of utilizing corn fiber xylan. *Applied Microbiology and Biotechnology,* 56(5-6), 765-766.

Sa-Pereira, P., Duarte, J. and Costa-Ferreira, M. (2000). Electroelution as a simple and fast protein purification method: isolation of an extracellular xylanase from *Bacillus* sp. CCMI 966. *Enzyme and Microbial Technology,* 27(1-2), 95-99.

Shao, W., Obe, S.K.C., Puls, J. and Wiegel, J. (1995). Purification and characterization of α-glucuronidase from *Thermoanaerobacterium* sp. strain JW/SL-Y485: an important enzyme for the utilization of substituted xylans. *Applied and Environmental Microbiology,* 63(3), 1077-1081.

Shei, J.C., Fretzke, A.R., Fredrik, M.M., Fredrick, J.R. and Reilly, P.J. (1985). Purification and characterization of endo-xylanases from *Aspergillus niger* II: an enzyme of pl 4.5. *Biotechnology and Bioengineering,* 27, 533-538.

Simpson, H.D., Haufler, U.R. and Daniel, R.M. (1991). An extremely thermostable xylanase from the thermophilic eubacterium *Thermotoga. Journal of Biochemistry,* 277, 413-417.

Smith, D.C. and Forsberg, C.W. (1991). α-Glucuronidase and other hemicellulase activities of *Fibrobacter succinogenes* S85 grown on crystalline cellulose a ball-milled barley straw. *Applied and Environmental Microbiology,* 57(7) 3552-3557.

Srinivasan, M.D. and Rele, M.V. (1995). Cellulase-free xylanases from microorganisms and their applications to pulp and paper biotechnology: an overview. *Indian Journal of Microbiology,* 35, 93-101.

Srinivasan, M.D. and Rele, M.V. (1999). Microbial xylanases for paper industry. *Current Science,* 77, 137-142.

Stutzenberger, F.J. and Bodine, A.B. (1992). Xylanase production by *Thermomonospora curvata. Journal of Applied Biotechnology,* 72, 509-511.

Sunna, A. and Antranikian, G. (1997). Xylanolytic enzymes from fungi and bacteria. *Critical Reviews in Biotechnology,* 17, 39-67.

Sunna, A., Puls, J. and Antranikian, G. (1996). Purification and characterization of two thermostable endo-β–1,4-xylanases from *Thermotoga thermarum. Biotechnology and Applied Biochemistry,* 24(2), 177-185.

Tan, L.U.L., Wong, K.K.Y., Yu, E.K.C. and Saddler, J.N. (1985). Purification and characterization of two D-xylanases from *Trichoderma harzianum. Enzyme and Microbial Technology,* 7, 425-430.

Tavobilov, I.M., Gorbacheva, N.A., Rodionova and Bezborodov, A.M. (1981). Purification of endo-1-4-β–xylanase from the fungus *Aspergillus niger* strain 15. *Applied Biochemistry and Microbiology,* 17, 320-324.

Tenkanen, H., Puls, J. and Poutanen, K. (1992). Two major xylanases of *Trichoderma reesei. Enzyme and Microbial Technology,* 14, 566-574.

Teplitsky, A., Shulami, S., Moryles, S., Shoham, Y., and Shoham, G. (2000). Crystallization and preliminary X-ray analysis of an intracellular xylanase from *Bacillus stearothermophilus* T-6. *Acta Crystallogr D Biol Crystallogr.* 56, 181-4.

Teunissen, M.J., Hermans, J.M.H., Huis in't Veld, J.H.J. and Vogels, G. D. (1993). Purification and characterization of a complex bound and a free β–1,4-endoxylanase from the culture fluid of the anaerobic fungus *Piromyces* sp. strain E2. *Archive Fur Mikrobiologie,* 159, 265-271.

Teunissen, M.J., Smits, A.A.M., Opdencamp, H.J.M., Huis, in't Veld, J.H.J. and Vogels, G.D. (1991). Fermentation of cellulose and production of cellulolytic and xylanolytic enzymes by anaerobic fungi from ruminant and non-ruminant herbivores. *Archive Fur Mikrobiologie,* 156, 290-296.

Timell, T.E. (1967). Recent progress in the chemistry of wood hemicellulose. *Wood Science and Technology,* 1, 45-70.

Torronen, A, *et al.* (1994). Three-dimensional structure of endo-1,4-beta-xylanase II from *Trichoderma reesei*: two conformational states in the active site. *The EMBO Journal,* 13(11), 2493-2501.

Torronen, A., Mach, R.L., Messner, R., Gonzalez, R., Kalkkinen, N., Harkil, A. and Kubicek, C.P. (1992). The two major xylanases from *Trichoderma reesei:* Characterization of both enzymes and genes. *Biotechnology* 10(11), 1461-1465.

Tsujibo, H., Miyamoto, K., Kudo, T., Minami, K., Sakamoto, T., Hasegawa, T. and Inamori, Y. (1992). Purification, properties and partial amino acid sequences of thermostable xylanases from *Streptomyces thermoviolaccus* OPC-520. *Applied and Environmental Microbiology* 58, 371-375.

Tuohy, M.G., Puls, J., Claeyssens, M., Vrsanska, M., and Coughlan, M.P. (1993). The xylan-degrading enzyme system of *Talaromyces emersonii*: novel enzymes with activity against aryl beta-D-xylosides and unsubstituted xylans. *Journal of Biochemistry* 290, 515-523.

Vanparidon, P.A., Booman, J.C.P., Selten, G.C.M., Geerse, C., Barug, D., deBot, P.H.M. and Hemke, G. (1992). Application of fungal endoxylanase in poultry diets. In: *Xylans and xylanases.* Eds Visser J. Beldman G, vanSomeran M.A.K, Voragen, A.G.J. Elsevier, Amsterdam. Pp 371-378.

Viet, D.N., Kamio, Y., Abe, N., Kaneko, J. and Izaki, K. (1991). Purification and Properties of β–1,4-Xylanase from *Aeromonas caviae* W -61. *Applied and Environmental Microbiology* 57(2), 445-449.

Viikari, L., Kantelinen, A., Sundquist, J. and Linko, M. (1994). Xylanase in bleaching: from an idea to the industry. *FEMS Microbiology Reviews* 13, 335-350.

Winterhalter, C., Heinrich, P., Candussin, A., Wich, G. and Liebl, W. (1995). Identification of a novel cellulose-binding domain within the multidomain 120 kDa xylanase XynA of the hyperthermophilic bacterium *Thermotoga maritima. Molecular Microbiology,* 15, 431-444.

Wong, K.K.Y. and Saddler, J.N. (1992). *Trichoderma* xylanases, their properties and purification. *Critical Reviews in Biotechnology,* 12, 413-435.

Wong, K.K.Y., Tan, L.U.L. and Saddler, J.N. (1988). Multiplicity of β–1, 4-xylanase in microorganisms: functions and applications. *Microbiological Reviews,* 52, 305-317.

Wood, T.M. and McCrae, S.I. (1986). Studies on two low molecular weight endo-1, 4-beta-D-xylanases constitutively synthesized by the cellulolytic fungus *Trichoderma koningii. Carbohydrate Research,* 149, 321-330.

Ximenes, F.A., Sousa, M.V., Puls, J., SilvaJr, F.G. and Filho E.X.F. (1999). Purification and characterization of a low-molecular weight xylanase produced by *Acrophialophora nainiana. Current Microbiology,* 38, 18-21.

Yamaki, S. and Kakiuchi, S. (1979). Changes in hemicellulose degrading enzymes during development and ripening of Japanese pear fruit. *Plant Cell Physiology,* 20, 301-306.

Yamura, I., Koga, T. and Matsumoto, T. (1997). Purification and some properties of endo-1,4-β–xylanase from a fresh water mollusk *Pomacea insularus* (de Oringny). *Bioscience Biotechnology Biochemistry,* 61, 615-620.

# 11

# Cellulose Degradation in the Rumen : An Anaerobic Eco-system

**Neeta Agarwal**

*Microbiology Section, Centre of Advanced Studies in Animal Nutrition, Indian Veterinary Research Institute, Izatnagar-243122, India*

Lignocellulosics are the most abundant organic compounds in the biosphere, accounting for approximately 50% of the biomass in the world. Biomass in the form of cellulose, hemicellulose and lignin provides a means of collecting and storing solar energy and hence represents an important energy and material resource. The structural component of the plants comprises two major polymers cellulose and hemicellulose. The cellulose is the most abundant biopolymer on earth. The animals including human being have no system by which they can degrade these polymers to extract energy from them. The microbes have a well-developed enzyme system for the degradation of these structural carbohydrates, thus the animals have to depend on these microbes to utilize these carbohydrate polymers as a source of energy. This article reviews the cellulase enzyme complex, compartmentation of enzymes, rumen cellulolytic microbes, mode of action of rumen microbes and factors affecting cellulose degradation in rumen. It became clear to improve the utilization of lignocellulosic feeds by the animals after noting the scenario of animal husbandry.

## INTRODUCTION

Enzymatic degradation of lignocellulosic material has a potential application not only in animal feed but also in waste management, pulp and paper industry, fuel production, oxychemical production etc, hence attracting the attention of scientists to work on enzyme systems of microbes since last three decades. Cellulose is a straight chain polymer of hydrogen bonded β–1,4-linked D-glucans which accounts for 20 to 30% of the dry weight of most primary cell walls (McNeil *et al.,* 1984). In immature plants, cellulose is mainly of amorphous type but as plants grow older, the hydrogen bonds become more and more compact providing the crystalline structure (hydrogen bond-ordered) to cellulose, which is less degradable than amorphous. The crystallinity of cellulose can vary from 20% in primary cell wall to 70% in secondary cell wall. With the maturity of the plants, lignin content also increases, which protects cellulose from the attack of cellulose degrading enzyme system.

To improve the performance of any system, the first step is to analyze the constraints responsible for the prevailing performance. After discussing these factors, strategy can be planned to manipulate the rumen for the improvement of cellulose degradation. The major factors affecting the cellulose degradation in rumen are:

- Chemical and physical nature of diet.
- Rumen digestion kinetics.
- Rumen microbial population dynamics in a prevailing condition.
- Enzyme profile of rumen microbes and cellulose degradation.

## CELLULASE ENZYME COMPLEX

Cellulase enzyme complex is a mixture of various enzymes, which act synergistically on the plant fibre to degrade cellulose completely. The rumen bacterial and fungal cellulases are partially constitutive and partially inducible as it has been observed that the cellulase enzyme production is there even when the microbes are grown on soluble sugars but the production increases when the cellulolytic microbes are grown on cellulosic substrate (Ho and Abdullah, 1999). The enzymes are:

- Endoglucanases (β–1-4-D-glucan glucanhydrolase, endo-β–1,4-glucanase, carboxymethylcellulase (CMCase), $C_x$ cellulase, EC 3.2.1.4)
- Exoglucanases (exo-β–1,4-glucanase, β–1,4-D-glucan cellobiohydrolase, Avicelase, $C_1$ cellulase, cellobiohydrolase EC 3.2.1.91)
- β–glucosidase (β–D-glucoside glucohydrolase, cellobiase, aryl-β–glucosidase, β–1,4-Glucoside, EC 3.2.1.21)
- Cellulodextrinase

Esterases, though not a part of the cellulase enzyme complex, are very important to make cellulose available from the fibrous feed to the microbial digestion by delinking the ester bonds between hemicellulose and lignin. The important phenolic esterases are:

- p-Coumaroyl esterase
- Feruloyl esterase
- Acetyl-xylan esterases

The endoglucanases break the β–1,4-linkages of glucan polymer in the randomized manner, thus cleaving the polymer in small chains with one reducing end releasing cellodextrins, cellobiose and glucose. Endoglucanase is most active against carboxymethyl cellulose and acid swollen cellulose but cannot degrade crystalline cellulose efficiently. The exoglucanses break the β–1,4-linkages from the non-reducing end of the cellulose chain releasing cellobiose unit. Exoglucanse is very active towards crystalline cellulose but degrades CMC inefficiently. The hydrolysis of cellobiose into glucose residues is catalyzed by β–glucosidase. The enzyme is active towards cellobiose, salicin and esculin releasing glucose and aromatic residues. Exoglucanase is an important enzyme for crystalline cellulose degradation, but the exoglucanase activity is very low as compared to endoglucanase (Agarwal *et al.,* 2000). The total endoglucanase activity was 363 to 1183.56 IU and exoglucanase activity was 8.03 to 30.55 in Herford cows fed 100% alfalfa hay (Lee *et al.,* 1999). Cellulodextrinase is an enzyme of different group and is reported from some rumen bacteria. The enzyme degrades small chain soluble glucose polymers releasing cellobiose and triose.

The cellulase enzyme systems of bacteria, fungi and protozoa are quite different. The rumen fungi have extracellular enzymes (release the enzymes in the medium) with an acidic pH (5 to 6) and mesophillic temperature optima. However, in *Neocallimastix frontalis* a small component of cellulases are in the form of cellulosome-like structure which is secreted and adsorbed to cellulose; thus though slowly, it is able to degrade microcrystalline cellulose. In *Piromyces communis*, a highly active cellulase system is available which solubilizes crystalline cotton fibre at a rate faster than any other extracellular enzyme. Unlike rumen fungi, rumen cellulolytic bacteria, secrete the enzymes in a complex form called cellulosomes which cannot be broken without any loss of enzyme activity. The cellulosomes are present on the cell surface and attach with the insoluble substrate. When the substrate is solubilized the enzymes bound in clusters in cellulosome are released in the medium. In protozoa the enzymes are intracellular as they engulf the fibrous material and then degrade them.

## COMPARTMENTATION OF ENZYMES

In our laboratory, the rumen contents were fractionated into three fractions, viz., particulate material (containing microbes adhered to feed particles), cellular (containing microbes free in liquid portion) and extracellular (the liquid portion of rumen contents without cells). About 80 to 90% fibre degrading enzyme activites like microcrystalline cellulose degrading activity, carboxymethylcellulase, filter paper degrading activity, xylanases are associated with the particulate material, 8 to 15% is associated with cellular fraction and rest (1 to 4%) with extracellular fraction indicating that the microbes adhered to the fibrous plant material have the most active cellulase enzyme system (Agarwal *et al.,* 2000). Minato *et al.,* (1993) also reported about 88 to 91% of total endoglucanase, xylanase activity, 70% of amylase activity and 75% protease activity in the fraction of microbes associated to the feed particles. The distribution pattern of enzymes indicated a direct involvement of the microbes attached with the feed particles in cellulose degradation.

## RUMEN CELLULOLYTIC MICROBES

In ruminants, rumen is the main chamber of the gastrointestinal tract where the microbial eco-system thrives. These anaerobic microbes have an excellent enzyme system to degrade cellulose. The fibrous feed taken by the animals is degraded by the rumen microbes and fermented into volatile fatty acids, which are utilized by the host as a source of energy. The rumen microbial ecosystem comprises bacteria, protozoa and fungi. The three groups of microbes differ in their potential and mode of action for cellulose degradation and the cellulolytic potential of the three groups can be described as most active fungi>bacteria>protozoa.

### Bacteria

Bacteria is the major group of the rumen microbial ecosystem with about 200 species and number upto $10^{11}$/ml rumen liquor out of which cellulolytic bacteria number can be as high as $10^{8}$/ml rumen liquor, thus reflecting an important role in degradation of fibrous feed. The major cellulolytic bacteria are

| *Primary cellulolytics* | *Secondary cellulolytics* |
|---|---|
| Fibrobactor succinogenes | Butyrivibrio fibrisolvens |
| Fibrobacter ruminocola | Clostridium longisporum |
| Ruminococcus flavifaciens | Clostridium locheadii |
| Ruminococcus albus | |

*C. longisporum* has been identified as hypercellulolytic but the organism is occasionally present in the rumen and never reaches to abundance (Varet *et al.,* 1995). The authors explained the inability of *C. longisporum* to persist in competition with other cellulolytic bacteria due to the incapability of synthesizing cellulosome-like structure. The specific characteristics of these two categories are:

- Primary cellulolytics—nonmotile, extensive adherence, cellulose predominant substrate, enzymes are on cell surface
- Secondary cellulolytics—motile or nonmotile, minimum adherence to the feed particles, enzymes extracellular, cellulose and sugars as substrate.

## Protozoa

Protozoa constitute a major portion of microbial biomass in the rumen with about 20 species and number varying from $10^{5}$ to $10^{6}$. Though protozoa are very less in number as compared to bacteria but they do have a significant role to play in digestion of cell wall since they secrete the enzymes required for cellulose degradation (Williams and Coleman, 1988; Agarwal *et al.,* 1991). In the rumen, protozoa contribute about 43% of cellulase, 53% protease and 38% amylase activity of the total present in the rumen liquor of buffalo (Agarwal *et al.,* 1991). While Dijkstra and Tamminga, (1995) reported 17-21% contribution of protozoa in fibre degrading activity in animals fed on diet containing low level of concentrate mixture. Previously it was believed that the cellulose degrading capability of protozoa was due to the bacteria engulfed by them but later on, work done with antibiotics established that protozoa also secrete significant amount of cellulose degrading enzymes. Recently cloning of endo-glucanase and xylanase genes of *Entodinium caudatum* confirmed that the cellulases are of protozoal origin (Takenaka and Habashi, 1994). Among Spirotrichs, except small Entodinia, all Ophryoscolecids have the capacity to degrade cellulose (Jouany and Martin, 1997). Holotrichs cannot degrade avicel but can attack carboxymethylcellulose (Prins and Van Hoven, 1977). The major cellulolytic protozoa identified so far are:

- *Polyplastron multivesculatum*
- *Epidinium ecaudatum*
- *Eudiplodinium maggi*
- *Ophryoscolex caudatus*
- *Entodinium caudatum* was classified as non-cellulose utilizer (Lee *et al.,* 1999) but Koenig *et al.,* (2000) found that it has lesser capacity to digest cellulose as compared to other cellulolytic protozoa.

## Fungi

The rumen fungi belong to chytridiomycetes group contributing a very small fraction of rumen microbiota ($10^4$ to $10^5$/ml rumen liquor) with 7 known species. The cellulases of rumen fungi are the most active fibrolyitc enzymes as compared to the cellulases of various other anaerobic and aerobic microbes (Trinci *et al.,* 1994) but its role in fibre degradation in rumen is still debatable. *In vitro* dry matter degradability of wheat straw was reduced by 7% when fungicide was added to the medium whereas the reduction was about 72.48% when antibiotics were added to the medium. When both fungicide and antibiotics were added to the medium, IVDMD of wheat straw was reduced by 77.03% indicating very little contribution of fungal fraction of rumen microbiota in degradation of lignocellulosic feed material (unpublished data). But there are reports showing increased fibre digestibility in rumen by introducing the fungal isolates (Ha *et al.,* 1994) and decreased fibre degradation due to elimination of fungi by chemical means (Gordon and Phillips, 1993), justifying the significance of rumen fungi in fibre degradation. The cellulase produced by *Neocallimastix frontalis* can degrade microcrystalline cellulase at a remarkably higher rate as compared to the cellulase from *Trichoderma reesei* an aerobic fungi and being presently used commercially for enzyme production (Wood, 1992). But recently *Piromuces communis* is supposed to produce more active cellulase than *N. frontalis* (Wood and Wilson, 1994) and both degrade microcrystalline cellulose most rapidly on earth. The growth conditions also influence enzyme production, being three times more cellulase production in stirred fermenter (Dijkerman *et al.,* 1996) and twenty times more cellulase production in continuous flow system (Zhu *et al.,* 1996) as compared to static batch culture. The rumen fungi being rich in phenolic acid esterases (Borneman *et al.,* 1990) de-link lignin from cellulose thus exposing cellulose for microbial cellulolysis and thus improve the utilization of lignocellulosic plant material. Presence of proteolytic activity in rumen cellulotyic fungi leads to its superiority over cellulolytic bacteria, which do not contain proteolytic activity (Ushida *et al.,* 1997). Having both proteolytic and cellulolytic activity in the same organism facilitates the fibre degrading activity since the protease act on the protein structure of plant cell wall together with cellulose. The metallo-protease inhibitor (L-cysteine) does not alter endoglucanase and exoglucanase activities of rumen fungi but increased timothy hay dry matter, NDF and ADF degradation (Asao *et al.,* 1993). The work done in our laboratory on compartmentation of hydrolytic enzymes revealed absence of proteolytic activity in the particulate material (microbes adhered to feed particle) of rumen contents, the fraction of which cellulolytic bacteria are the major component thus indicating absence of proteolytic activity in cellulolytic bacteria (Agarwal *et al.,* 2000). The main cellulose utilizing fungi of the rumen are:

- *Neocallimastix frontallis*
- *Neocallimastix patriciarum*
- *Piromyces communis*
- *Caecomyces communis*
- *Orpinomyces joyonii*

## MODE OF ACTION

Understanding of the mechanism by which the rumen microbes attack the cellulosic plant material is of prime importance to facilitate the cellulose degradation. The three of the components of rumen microbiota though degrade cellulose in the same manner but mode of action is quite different.

## BACTERIA

The process of adhesion is completed in three phases. The first step is the transportation of bacteria to the feed particles. Since the primary cellulolytics are non-motile, it is through physical contact. This is followed by non-specific adhesion including both irreversible as well as reversible process, without any ligands between bacterial cell surface and substrate receptor (Miron *et al.,* 2001). This association starts immediately when the bacteria reach within range (2 to 50 nm) of van der waals forces—hydrophobic, ionic, and electrostatic interaction with the solid substrate. The nonspecific binding is by the formation of glycocalyx, which is a glycoprotein of bacterial origin outside the outer membrane of the gram-negative bacteria (*F. succinogens*) and the peptidoglycan of gram-positive cells (*R. flavefaciens, R. albus*). The components of glycocalx involved in non-specific adhesion are constitutive carbohydrates epitopes (not induced by cellulose substrate), cellulose binding protein (CBP) or other factors (Morrison and Miron, 2000; Ohara *et al.,* 2000). Non-specific adhesion is followed by specific adhesion where adhesins on the bacterial cell surface recognize receptors on the substrate and aggregation of protuberances takes place on the glycocalyx layer and forms adhesins connections between bacteria cell wall and the substrate. These protuberances are cellulosome complexes and have fimbrae structure (Morrison and Miron, 2000; Pegden *et al.,* 1998). Cellulosome contain cellulases secreted by the bacterial cells and are held so firmly that it is not possible to break cellulosome without losing the activity of cellulases. The feed particles are degraded by these cellulosomes. The activity of cellulases liberated outside are about 10 times more active than the cellulases of cellulosome but the liberated enzymes are partly degraded by the protease secreted by proteolytic bacteria (Miwa *et al.,* 2000). The adhesion process is affected by the pH, temperature, presence of cations, soluble substrates and surface area of substrate. Grinding of feed particles results in better adhesion by increasing surface area.

The adhesion of bacteria to the cellulosic substrate provides superiority to this group over the other microbes. By this close association, the cellulases are concentrated around the substrate and expel other microbes from the site of hydrolysis, thus the products of cellulose hydrolysis are first available to adhered microbes. Adhesion also protects these bacteria from the grazing by other microbes. Adhesion mechanism also increases their retention time in the rumen to about three times more than the microbes free in rumen liquor, thus providing them better opportunity of cellulose degradation (Minato *et al.,* 1993; Weimer, 1996).

The hydrolysis of cellulose is accompanied by the absorption of enzyme on the substrate. A yellow carotenoid pigment called yellow affinity substrate (YAS) produced by the bacteria has been found during cellulose degradation. It is supposed to not only stimulate the absorption of the cellulase enzyme on the substrate but also acts as a signal to bacteria for the presence of cellulose (Ljungdahl *et al.,* 1988). The presence of yellow pigment enhances the amount of enzyme absorbed and the rate of absorption. Kopeèný and Hodrová, (1997) have demonstrated better binding of cellobiohydrolase and endoglucanase to cellulose substrate in the presence of YAS isolated from *R. flavefaciens*. The reports indicate a significant role of YAS in regulating cellulose degradation in rumen.

## PROTOZOA

Protozoa attack the feed particles either in active or passive manner. In active association, protozoa attach with the damaged plant material ingesting solid material simultaneously (Epidinium) thus weakening the plant material and breaking it mechanically. In passive association, there is a week joining between protozoa cell wall and plant cell wall without ingesting solid material. Protozoa consume only suspended particles with the help of vestibular lips and cilia. The ingested material is enclosed by the vacuole where enzymatic degradation of feed takes place. However isotrichs do not attach with the feed particles and hence depend on starch particles. Coleman (1992) observed that *Eudiplodinium maggi* engulfed cellulose particles in preference to starch particles.

## FUNGI

Rumen anaerobic fungi have a unique pattern of attack on the feed entering the rumen. The zoosporangium release zoospores, which attach to the freshly ingested lignocellulosic feed due to chemotactic movement. The zoospores of *Neocallimastix* have chemoreceptors for glucose, sucrose, mannose and sorbitol (Orpin and Bountiff, 1978) and those of other fungi like *N. frontalis, Piromyces communis, Orpinomyces joyonii* and *Anaeromyces* sp. have chemoreceptors for p-coumaric acid, ferulic acid and syringic acid, which are essential components of lignocellulosic feeds (Wubah and Kim, 1996). Once attached, they shed flagella, encyst and develop a rhizoidal system, which penetrates the substrate with the help of polysaccharide degrading enzymes (Munn *et al.,* 1981). The rumen fungi prefer lignified tissue like xylem and sclerenchyma. In comparison to bacteria and protozoa, the fungi have a unique capability of penetrating the substrate with the help of filamentous rhizoidal system. The rumen fungi are able to greatly reduce the tensile strength of lignified tissues. The stems of bermuda grass and alfalfa incubated with fungi were weakened by 57% and 55%, while bacteria could weaken only to an extent of 29% and 42%, respectively (Akin *et al.,* 1990). The properties of rumen fungi discussed above about their mode of action on the lignocellulosic feeds, weakening of feed tissue and penetration of feed particles explain their superiority over bacteria and protozoa in the degradation of lignocellulosic feeds in the rumen but its significance in rumen is still debatable because of its very little biomass contributing only 1 to 3% of total microbial nitrogen.

## FACTORS AFFECTING CELLULOSE DEGRADATION IN RUMEN

### Effect of pH

Change in rumen pH also alters the rate of cellulose digestion. The optimum pH for the activity of cellulase from various sources (pure as well as mixed rumen cultures) has been worked out at pH 4.5-5.0 (Agarwal *et al.,* 1991) but in rumen, the cellulose degradation is reduced if the pH is below 6.0. The isolated cellulolytic bacteria when grown on pure cellulose at low pH for a longer period (48h) slowed down the rate of cellulose degradation but when it came to the normal pH to above 6.0, the rate of cellulose degradation was resumed to the normal (Hiltner and Dehority, 1983). The enrichment medium of pH 5.5 with pure cellulose, when used to inoculate the strained rumen fluid, the cellulolytic microbe (*Ruminococcus flavefaciens*) could not be established, indicating that the rumen cellulolytic bacteria could not initiate growth at this pH (Weimer *et al.,* 1991). Cheng *et al.,* (1984) have demonstrated by electron microscopic studies that low ruminal pH prevented the firm attachment of the cellulolytic bacteria to the plant cell wall. The number of microbes associated with the feed particles were reduced by 43% at pH 5.8 as compared to the pH 6.2 through 7.0 resulting in only 15% NDF digestibility as compared to 32.5% at higher pH (Shriver *et al.,* 1986). This indicates that lower rate of cellulose degradation at low pH in the rumen is attributed to the cellulolytic microbes which are very sensitive to low pH and become inactive and cannot be attributed to the cellulase enzyme secreted by these microbes. But the rumen cellulolytic bacteria have excellent capacity to resume to the normal activity rapidly as they get favourable pH.

## Effect of Diet

The quality of forages, composition of diet, level of concentrate, level of grain and maturity of forages are some of the important factors, which affect the cellulase enzyme production of rumen microbiota. Recently in our laboratory studies conducted on buffaloes fed on diets comprised four ratios of concentrate (6-15% grain) and roughage revealed no effect of diet on carboxymethyl cellulase, filter paper degrading activity, xylanase, β–glucosidase, β–xylosidase and amylase, but the activities of micro-crystalline cellulase, acetyl esterase and protease activities increased significantly ($P<0.05$) with increasing level of concentrate in the diet (Kamra *et al.,* (2003). Increase in the number of cellulose degrading bacteria with increase in corn in the diet, but no change in the number of hemicellulose degrading bacteria was also reported by Van der Linden (1984). Martin and Michalet-Doreau, (1995) and Hristov *et al.,* (1999) also demonstrated change in the enzyme profile with the change in the type of the diet. There was a depression in carboxymethylcellulase and xylanase activities and increase in amylase activity on a shift from forage based diet to high grain based diet (91%) indicating the shifting of cellulose degrading microbes to starch degrading microbes. The differences in the results of the two experiments were due to the vast difference in the level of grain in the diets. The forage quality also affects the cellulase production. *In vitro* screening of four tropical grasses revealed that the grass having lower NDF and lignin and higher nitrogen contents, promoted cellulase enzyme activity (Filho *et al.,* 2000).

The effect of composition of diet on cellulase enzyme production is mainly due to change in the generic composition of the rumen microbiota. There was 20- and 10-fold decrease in the numbers of cellulose degrading bacteria e.g. *Fibrobacter succinogenes* and *Ruminococcus flavefaciens* and 14 folds in xylan degrading bacteria *Eubacterium ruminantium* within 3 days of the shift in diet and those of starch utilizing bacteria like *Prevotella ruminicola* and *P. bryantii* increased 7- and 263-folds, respectively, when the diet of Holstein cows was shifted from hay to grain based diet

(Tajima *et al.,* 2001). The work done at Indian Veterinary Research Institute, Izatnagar demonstrated an increase in counts of cellulolytic, hemicellulolytic bacteria and fungi along with the increase in CMCase, xylanase and acetyl esterase when the roughage concentrate ratio in the diet was shifted from 30:70 to 70:30 whereas amylase activity was higher in the high concentrate group indicating the shift of populations and their enzyme profile with the change in diet (Agarwal *et al.,* 2004). It was also observed that the enzymes associated with the particulate material of the rumen contents are most affected by the nature of feed indicating that the particle associated enzymes are most sensitive in detecting change in rumen environment related to change in fibre digestibility (Silva *et al.,* 1987; Huhtanen and Khalili, 1992; Kamra *et al.,* 2003). The protein contents of the diet also affect rumen microbiota.

### Effect of Soluble Sugars

The level of soluble sugars entering with the feed or produced in the rumen by hydrolysis of cellulosic material also affects cellulolysis in rumen. By supplementing sucrose in the diet there was a depression in the particle associated carboxymethylcellulase and xylanase and this depression was postulated to be due to the depression in the adherence of microbes to the substrate or because of end product inhibition in enzyme activity (Huhtanen and Khalili, 1952). In the presence of soluble sugars in the medium, shifting of cellulolytic bacteria from cellulose to soluble sugars is well documented, as they prefer to utilize soluble sugars before starting cell wall degradation. Martin and Michalet-Doreau (1995) demonstrated shifting of solid associated bacteria to free bacteria in liquid by changing the diet from sole hay feeding to hay plus barley feeding and it was due to high starch contents of the diet. Presence of glucose also inhibits adhesion of bacteria to cellulolytic substrate. Presence of monomers like glucose, xylose, arabinose produced by degradation of cellulose and xylan inhibit the activity of cellulase and xylanase activity, respectively (Mountfort, 1994).

## INTERDEPENDENCE OF RUMEN MICROBES

It is not possible to increase the cellulose degradation in rumen simply by manipulating the cellulolytic microbes because for efficient working of this group of microbes other non-cellulolytic microbes are also required. The fibrous plant material entering the rumen is colonized by a variety of microbes and type of consortium is an important factor for degradation of plant fibre. For example the cellulose degraded by the cellulolytic bacteria release sugar and hydrogen, which are utilized by the sugar utilizing microbes and removal of these cellulose-degrading products from the environment, is necessary for the continuous steady state cellulase activity. The mixed rumen protozoa population also affects cellulolysis though indirectly by stimulating the cellulolytic bacterial metabolism. Decreased digestibility of NDF, ADF and cellulase has been reported in lambs by removing the protozoa from the rumen (Santra *et al.,* 2000) whereas, other reports on defaunation showed no change in rumen fermentation pattern, carboxymethyl cellulase activity and *in sacco* digestibility in buffaloes (Chaudhary *et al.,* 2000, Kamra *et al.,* 2000).

Though it is well established that protozoa secrete cellulases but monoculture of protozoa is not able to degrade cellulose, which indicates that, the protozoa depends on bacteria for their cellulolysis. The rumen protozoa are able to engulf only fine particles of cellulose while the average cellulose particle size is 100 to150 mm in diameter (Ushida *et al.,* 1987). So it is necessary that the cellulose particles should be first broken down in small particles by cellulolytic bacteria, which is then engulfed by the protozoa.

The bacterial (B), protozoal (P) and fungal (F) populations separated from the rumen degraded cellulose *in vitro* in a order positive F > B and negative P and when there was coculture of populations than P affects B and F negatively but F and B worked synergistically (Park *et al.,* 2000). But on the other hand increased bacterial population adhered to feed particles, fungal zoospores and carboxymethyl cellulase activity in refaunated animals as compared to defaunated sheep was reported by Williams and Withers (1991). This was attributed to protozoa contribution to the cellulase enzyme activity through own enzymes directly or indirectly by stimulating the fibrolytic bacterial growth resulting in increased population size and also due to the pH stable rumen environment provided by protozoa population.

## COMPETITIVENESS FOR SUBSTRATE

There is no competition among bacteria, fungi and protozoa for the substrate since mode of action of three types of microbes is different. There is no competition between cellulolytic bacteria, protozoa and fungi for adhesion to cellulosic substrate (Miron *et al.,* 2001). However, there are some reports that rumen fungi can inhibit bacterial cellulolytic activity (Joblin, 1997) but studies on the cellulolysis of orchard grass cell wall in monoculture demonstrated that the cellulolysis was high in early stage of incubation in bacterial fraction and it is high in later stage of incubation in fungal fraction, whereas, protozoa could not degrade cellulose alone (Lee *et al.,* 1999). This delayed cellulolysis by fungi may be due to their complicated life cycle. But cellulolytic bacteria compete among themselves for the substrate. When the substrate is in excess, all live in harmony but when there is scarcity of substrate, *F. succinogenes* dominates over *R. albus* and *R. flavefaciens.* The probable reason might be that the former one adheres to the cellulose particle more rapidly and has high affinity towards the cellulose degrading products (Shi *et al.,* 1997).

## WHERE DO WE STAND?

In India the research on rumen microbiology was limited upto the rumen fermentation and counting of protozoa. Molecular rumen microbiology has been started recently and there are only very few laboratories having the facilities to work on anaerobic microbes at molecular level because maintaining the anaerobic cultures is a difficult job. The work has been initiated on isolation and characterization of few rumen isolates in various laboratories but cultures could not be maintained. We at IVRI have isolated a few buffalo, goat and wild animals rumen bacteria and screened them on the basis of their microcrystalline degrading efficiency since we are looking for the improvement of utilization of lignocellulosic agriculture byproduct for which microcrystalline degrading enzyme is of prime importance. Isolation of rumen fungi have also been started which is a pioneer effort in this direction (Agarwal *et al.,* 2003, 2004 and Sahu *et al.,* 2003). The work of rumen protozoa has not yet been started since it is very difficult to maintain (Paul *et al.*, 2004a, 2004b) rumen protozoa, as they are very sensitive to the outer atmosphere.

## FUTURE NEEDS

In the present scenario of animal husbandry there is a great need to improve the utilization of lignocellulosic feeds by the animals. As discussed above, there are a number of factors, which are regulating cellulose degradation in the rumen and for the manipulation of rumen to utilize this cellulose source efficiently, all the factors should be considered collectively to achieve the goal.

- The optimum level of soluble sugars in the roughage-based diet should be worked out to get a suitable rumen environment for cellulose degradation.
- There is need to generate data on rumen population diversity on roughage based diet.
- Enzyme profile of rumen microbes should be worked out on roughage-based diet to pin point the marker enzyme and/or the microbe responsible for cellulose degradation in the rumen.
- A potential strategy is to enhance the population or activity of the desired native ruminal strains.
- Since the secondary cellulolytic bacteria have advantage of being motile and resistant towards low pH, introduction of high cellulose capabilities in these microbes can be beneficial.
- Use of strains termed as hypercellulolytic like *Clostridium longisporum* can be a positive step in this direction.
- While using the non-rumen or genetically modified organism with high cellulose degrading activity, possibilities should be explored to provide conditions selectively beneficial for the desired microbes for their establishment in the rumen.
- Availability of cellulose is also a constraint for limited cellulose degradation hence, increase in the microbes secreting phenolic esterases so that lignin can be delinked from cellulose and thus made cellulose available to the microbes.
- Since the cellulases of fungi are the most active enzyme system and fungi in the rumen are the primary colonizers of lignocellulosic feeds, exploitation of this enzyme system in the rumen may be a positive step in this direction.
- One promising direction may be exploitation of proteins of cellulosomes containing binding domains facilitating the binding of cellulolytic bacteria to the fibrous feed particles.
- Yellow affinity substrate, an important regulatory substance for cellulose degradation should be studied in detail so that it can be used to enhance the cellulose degradation in the rumen.

## REFERENCES

Agarwal, N., Agarwal, I, Kamra, D.N. and Chaudhary, L.C. (2000). Diurnal variations in the activities of hydrolytic enzymes in different fractions of rumen liquor of Murrah buffalo. *Journal of Applied Animal Research,* 18, 73-80.

Agarwal, N., Kewalramani, N., Kamra, D.N., Agrawal, D.K. and Nath, K. (1991). Hydrolytic enzymes of buffalo rumen: Comparison of cell free rumen fluid, bacterial and protozoal fractions. *Buffalo Journal,* 7, 203-207.

Agarwal, N., Saxena, J., Saha, S., Chaudhary, L.C. and Kamra, D.N. (2004). Changes in fermentation characteristics, microbial populations and enzyme profile in the rumen of buffalo as affected by roughage level in the diet. *Bubalus bubali,* 111,81-90.

Agarwal. N., Saxena J., Kamra D.N. and Chaudhary, L.C. (2003). Influence of bacterial isolates from the rumen of buffalo or *in vitro* gas production and hydrolytic enzyme activities in the rumen liquor. Fourth Asian Buffalo Congress, February 25-28, New Delhi.

Agarwal, N., Srivastava, R.K. and Kamra, D.N. (2004). Growth performance and nutrient utilization in buffalo calves inoculated with cellulose degrading bacteria. $x_1$ Animal Nutrition conference, Jabalpur, Jan. 5-7.

Akin, D.E., Rigsby, L.L., Lyon, C.E. and Widham, W.R. (1990). Relationship of tissue digestion to textural strength in bermuda grass and alfalfa stems. *Crop Science,* 30, 990-993.

Asao N., Ushida K. and Kojima, Y. (1993). Proteolytic activity of rumen fungi belonging to the genera *Neocallimastix* and *Piromyces. Letters in Applied Microbiology,* 16, 247-250.

Borneman, W.S., Hartley, R.D., Morrison, W.H., Akin, D.E. and Ljungdahl L.G. (1990). Feruloyl and p-coumaroyl esterase from anaerobic fungi in relation to plant cell wall degradation. *Applied Microbiology and Biotechnology,* 33, 345-351.

Chaudhary, L.C., Kamra, D.N., Singh, R., Agarwal N. and Pathak, N.N. (2000). Effect of feeding bentonite on rumen fermentation, enzyme activity and protozoal population in adult buffaloes (*Bubalis bubalis*). *Buffalo Journal,* 1, 73-79.

Cheng, K.J., Stewart, C.S., Dinsdale, D. and Costerton, J.W. (1984). Electron microscopy of bacteria involved in the digestion of plant cell wall. *Animal Feed Science and Technology,* 10, 93.

Coleman, G.S. (1985). The cellulase content of 15 species of entodiniomorphid protozoa, mixed bacteria and plant debris isolated from the ovine rumen. *Journal of Agriculture Science,* 104, 349-360.

Dijkerman, R., Op Den Camp, H.J.M. and Van Der Drift, C. (1996). Cultivation of anaerobic fungi in a 10-1 fermenter system for the production of (hemi)-cellulolytic enzymes. *Applied Microbiology and Biotechnology,* 46, 85-91.

Dijkstra, J. and Tamminga, S. (1995). Simulation of the effects of diet on the contribution of rumen protozoa to degradation of fibre in the rumen. *British Journal of Nutrition,* 74, 617-634.

Filho, J.C.M., Noguuueira, F.M., Urdaneta, A.B. and Ronquillo, M.G. (2000). In vitro microbial fermentation of tropical grasses at an advanced maturity stage. *Animal Feed Science and Technology,* 83, 145-157.

Gordon, G.L.R. and Phillips, M.W. (1993). Removal of anaerobic fungi from the rumen of sheep by chemical treatment and the effect on feed consumption and in vivo fibre digestion. *Letters in Applied Microbiology,* 17, 220-223.

Ha, J.K., Lee, S.S., Kim, C.H., Choi, Y.J. and Min, H.K. (1994). Effect of fungal inoculation on ruminal fermentation characteristics, enzyme activities and nutrient digestion in sheep. *Proceedings of Society of Nutrition and Physiology,* 3, 197.

Ho, Y.W. and Abdullah N. (1999). The role of rumen fungi in fibre digestion –Review. *Asian Australasian Journal of Animal Science,* 12, 104-112.

Hiltner, P. and Dehority, B.A. (1983). Effects of soluble carbohydrates on digestion of cellulose by pure cultures of rumen bacteria. *Applied and Environmental Microbiology,* 46, 642-648.

Hristov, A.N., Mcallister, T.A. and Cheng, K.J. (1999). Effect of diet, digesta processing, freezing and extraction procedure on some polysaccharide degrading activities of ruminal contents. *Canadian Journal of Animal Science,* 79, 73-81.

Huhtanen, B.P. and Khalili, H. (1992). The effect of sucrose supplements on particle associated carboxymethylcellulase (EC 3.2.1.4) and xylanase (EC 3.2.1.8) activities in cattle given grass-silage-based diet. *British Journal of Nutrition,* 67, 245-255.

Hungate, R.E. (1966). *The Rumen and Its Microbes.* Academic Press, New York, NY.

Joblin, K.N. (1997). Interactions between ruminal fibrolytic bacteria and fungi in Rumen In: *Microbes and Digestive Physiology in Ruminants,* Eds Onodera R., Itabashi H., Ushida K., Yano H. and Sasaki Y., Japan Scientific Societies Press, Tokyo, Japan.

Kabré, P., Martin, C., and Michalet-doreau, B. (1994). Enzyme activities of rumen solid-adherent microorganisms in chronically underfed ewes. *Journal of Science of Food and Agriculture* 65, 423-428.

Kamra, D.N., Singh, R., Chaudhary, L.C., Agarwal, N. and Pathak, N.N. (2000). Soapnut as natural defaunating agent: Its effect on rumen fermentation and *in sacco* degradability of jowar hay in buffaloes. *Buffalo Journal,* 1, 99-104.

Kamra, D.N., Saha, S., Neeru, B. and Chaudhary, L.C. (2003). Effect of diet on enzyme profile, biochemical changes and in sacco degradability of feeds in the rumen of buffalo. Asian - Aust. J. Anim. Sci. 16, 274-379.

Koenig, K.M., Newbold, C.J., Mcintosh, F.M. and Rode, L.M. (2000). Effect of protozoa on bacterial nitrogen recycling in the rumen. *Journal of Animal Science,* 78, 2431-2445.

Kopeèný, J. and Hodrová, B. (1997). The effect of yellow affinity substance on cellulases of *Ruminococcus flavefaciens. Letters in Applied Microbiology,* 25, 191-196.

Lee, S.S., Shin, K.J., Kim, W.Y., Ha, J.K. and Han, K. (1999). The rumen eco-system: As a fountain source of noble enzymes – Review. *Asian-Australasian Journal of Animal Science,* 12, 988-1001.

Ljungdahl, L.G., Petterson, B., Eriksson, K-E.L. and Weigel, J. (1983). A yellow affinity substance involved in cellulolytic system of *Clostridium thermocellum. Current Microbiology,* 9, 195-2000.

Martin, C. and Michalet-doreau, B. (1995). Variations in mass and enzyme activity of rumen microorganisms: Effect of barley and buffer supplements. *Journal of Science of Food and Agriculture,* 67, 407-413.

Mcneil, D.L., Carroll, B.J. and Greehoff, P.M. (1984). The interaction between nitrogen and carbon metabolism in nitrogen fixing soybean bacteroids. *Advances in Agricultural Biotechnology,* 4, 515-519.

Miron, J., Ben-ghedalia, D. and Morrison, M. (2001). Invited review: Adhesion mechanism of rumen cellulolytic bacteria. *Journal of Dairy Science,* 84, 1294-1309.

Miwa, T., Takagi, S. and Hino, T. (2000). Degradation of the cellulase of ruminal bacteria by the protease of coexisting bacteria in vitro. *Animal Science Journal,* 71, 157-163.

Morrison, M. and Miron, J. (2000). Adhesion of cellulose by *Ruminococcus albus*: a combination of cellulosomes and Pil-proteins? *FEMS Microbiology Letters,* 185, 109-115.

Mountfort, D.O. (1994). Regulatory constraints in the degradation and fermentation of carbohydrate by anaerobic fungi. In: *Anaerobic Fungi,* Eds Mountfort D.O. and Orpin C.G. New York: Marcel Dekker. Pp 147-168.

Park, J.G., Lee, S.S., Ha, J.K., Shin, K.J. and Cheng, K.J. (2000). Relative contributions of rumen anaerobic microbial fractions (bacteria, protozoa and fungi) in the degradation of cell wall constituents. *Korean Journal of Animal Science,* 42, 605-618.

Munn, B.A., Orpin, C.G. and Hall, F.J. (1981). Ultrastructural studies of the free zoospore of the rumen phycomycete *Neocallimastix frontalis. Journal of General Microbiology,* 125, 311.

Paul, S.S., Kamra, D.N., Sastry, V.R.B., Sahu, M.P. and Agarwal N. (2004a). Effect of administration of an anaerobic gut fungus isolated from wild blue bull (*Boselaphus tragocamelus*) to buffaloes (*Bubalus bubalis*) or *in vivo* ruminal fermentation and digestion of nutrient. Anim. Feed Sci. Technol 115(1-2): 143-157.

Paul, S.S., Kamra, D.N., Sastry, V.R.B., Sahu N.P. and Agarwal, N. (2004b). Effect of anaerobic fungal isolates from wild and domestic nominants on *in vitro* digestebility, fermentation characteristics and febrolytic enzyme production on lignocellulosic feed by mxied rumen microflora of buffalo. Nutr. Repro. Develop. 44,313-319.

Pegden, R.S., Larson, M.A., Grant, R.J. and Morrison, M. (1998). Adherence of the gram-positive bacterium *Ruminococcus albus* to cellulose and identification of a novel form of cellulose-binding protein, which belongs to the Pil family of proteins. *Journal of Bacteriology,* 180, 5921-5927.

Ohara, H.S., Kimura, Karita, T., Sakka, K. and Ohmiya, K. (2000). Characterization of cellulolytic complex (cellulosome) from *Ruminococcus albus. Bioscience,- Biotechnology and Biochemistry,* 64, 254-260.

Orpin, C.G. and Bountiff, I. (1978). Zoospore chemotaxis in the rumen phycomycete *Neocallimastix frontalis. Journal of General Microbiology,* 104, 113-122.

Sahu, N.P., Kamra D.N. and Paul, S.S. (2003). Effect of cellulose degrading bacteria isolated from wild and domestic ruminants on *in vitro* dry matter digestibility and enzyme production. Asian-Aust. J. Anim. Sci. 17,199-202.

Shi, Y., Odt, C.L., Weimer, P.J. and Shi Y. (1997). Competition for cellulose among the predominant cellulolytic bacteria under substrate excess and substrate limited conditions. *Applied and Environmental Microbiology,* 63, 734-742.

Shriver, B.J., Hoover, W.H., Sargent, J.P., Crawford, R.J. Jr. and Thayne, W. V. (1986). Fermentation of a high concentrate diet as affected by ruminal pH and digesta flow. *Journal of Dairy Science,* 69, 413.

Silva, A.T., Wallace, R.J. and Ørskov, E.R. (1987). Use of particle-bound microbial activity predict the rate and extent of fibre degradation in the rumen. *British Journal of Nutrition,* 57, 407-415.

Tajima, K., Aminov, R.I., Nagamine, T., Matsui, H., Nakamura, M. and Benno, Y. (2001). Diet-dependent shifts in the bacterial population of the rumen revealed with real-time PCR. *Applied and Environmental Microbiology,* 67, 2766-2774.

Trinci, A.P.J., Davies, D.R., Gull, K., Lawrence, M.I., Nielsen, B. B., Rickers, A. and Theodorou, M. K. (1994). Anaerobic fungi in herbivorous animals. *Mycological Research,* 98, 129-152.

Ushida, K., Kaneko, T. and Kojima, Y. (1987). Effect of presence of large Entodiniomorphid protozoa on the rumen bacterial flora, fauna composition of small entodinia and in vitro cellulolysis and xylanolysis. *Japanese Journal of Zootechnology Science,* 58, 893-902.

Ushida K., Matsui, H., Fujino, Yuko and Ha, J.K. (1997). Role and potential of ruminal fungi in fibre digestion-Review. *Asian-Australasian Journal of Animal Science,* 10, 541-550.

Varel, V.H., Yen, J.T. and Kreikemeier, K.K. (1995). Addition of cellulolytic clostridia to the bovine rumen and pig intestinal tract. *Applied and Environmental Microbiology,* 61, 1116-1119.

Weimer, P.J. (1996). Why don't ruminal bacteria digest cellulose faster? *Journal of Dairy Science,* 79, 1496-1502.

Weimer, P.J., Shi, Y. and Odt, C.L. (1991). A segmented gas liquid delivery system for continuous culture microorganism on insoluble substrates and its use for growth of *Ruminococcus flavefaciens* on cellulose. *Applied Microbiology and Biotechnology,* 36, 178-183.

Weimer, P.J., Waghorn, G.C., Odt, C.L. and Mertens, D.R. (1998). Effect of diet on populations of three species of ruminal cellulolytic bacteria in lactating dairy cows. *Journal of Dairy Science* 82, 122-134.

Williams, A.G. and Coleman, G.S. (1988). Rumen ciliate protozoa. In: *Rumen microbial eco-system,* Ed Hobson, P. N., Elsevier Applied Science Publication, Amsterdam. Pp 77-128.

Williams, A.G. and Withers, S.E. (1991). Effect of ciliate protozoa on the activity of polysaccharide-degrading enzymes and fibre breakdown in the rumen eco-system. *Journal of Applied Bacteriology,* 70, 144-155.

Wood T.M. (1992). Fungal cellulases. *Biochemical Society Transanctions,* 20, 46-53.

Wood, T.M. and Wilson C.A. (1995). Studies on the capacity of the cellulase of the anaerobic rumen fungus *P. communis* to degrade hydrogen bond-ordered cellulose. *Applied Microbiology and Biotechnology,* 43, 572-578.

Wubah, D.A. and Kim, S.H. (1996). Chemo-attraction of anaerobic ruminal fungi zoospores to selected phenolic acids. *Microbiology Research,* 151, 257-262.

Zhu, W.Y., Theodorou, M.K., Longland, A.C., Nielson, B.B., Dijkstra, J. and Trinci, A.P.J. (1996). Growth and survival of anaerobic fungi in batch and continuous-flow culture. *Anaerobe,* 2, 29-37.

# PART III

## LIGNOCELLULOSE BIOCONVERSION

# 12

# Bioconversion of Lignocellulosic Feeds for the Feeding of Ruminants

**D.N. Kamra**

*Microbiology Section, Centre of Advanced Studies in Animal Nutrition, Indian Veterinary Research Institute, Izatnagar – 243 122, India*

In India lignocellulosic agricultural byproducts constitute staple ruminant feed supplemented with small quantity of green forage, oil cakes and wheat or rice bran. The lignocellulosic feeds, commonly used contain 60-70% of dry matter in the form of cellulose and hemicellulose, but only 60-65% of this holocellulose is available to the animals as energy source. Therefore the efficiency of utilization of these feeds is very poor. This is due to shielding of the carbohydrates by lignin, which acts as a barrier. Efforts have been going on for the last several decades to improve the availability of nutrients from such feeds, either by chemical, physical or microbiological treatments. In the present article, bioconversion of lignocellulosics with white rot fungi along with factors affecting delignification of such feeds are discussed.

## INTRODUCTION

In India lignocellulosic agricultural byproducts constitute staple ruminant feed supplemented with small quantity of green forage, oil cakes and wheat or rice bran. The lignocellulosic feeds, commonly used contain 60-70% of dry matter in the form of cellulose and hemicellulose, but only 60-65% of this holocellulose is available to the animals as energy source. Therefore the efficiency of utilization of these feeds is very poor. This is due to shielding of the carbohydrates by lignin, which acts as a barrier. Efforts have been going on for the last several decades to improve the availability of nutrients from such feeds, either by chemical, physical or microbiological treatments. In the present lecture, bioconversion of lignocellulosics with white rot fungi along with factors affecting delignification of such feeds will be discussed. The microbiological work on treatment of lignocellulosic feeds for the feeding of ruminants has been extensively reviewed by various authors (Kamra and Singh, 1989, Zadrazil *et al.,* 1996, Flachowsky *et al.,* 1999).

Most of the agricultural byproducts are rich sources of lignocellulose as these are residues of food crops. Lignin acts as a barrier between cellulose and the microbes are able to utilise it. Thus the digestibility of lignocellulosic feeds is generally inversely related to the lignin content of that feed (Kirk and Moore, 1972). In addition to the lignification, other factors responsible for reduced digestibility of straws are silicification and crystallinity of cellulose. Silica is one of the major anti-nutritional factors present in rice straw and rice husk. For the plants silica is as important as lignin for imparting structural strength. The soluble silica is mainly responsible for the reduced digestibility of straw and its removal causes an improvement in the digestibility of organic matter of straw.

The crystallinity index of cellulose is another important factor responsible for the lower digestibility of lignocellulosic feeds. The crystalline cellulose is more resistant to penetration by chemicals, enzymes and water than the amorphous cellulose and is therefore more resistant to microbial degradation. The crystallinity index of cellulose can be reduced by treatment with oxidising agents.

## STRUCTURE OF LIGNOCELLULOSE

Lignin gives the plants structural rigidity and protection against invasion by the pathogenic microbes. Only next to cellulose, lignin is the second most common organic compound recycled on the earth. After the death of plants, the lignocellulose is degraded by the microbial activity and lignin is converted to humus. After many complex biochemical reactions the humus ingredients are again incorporated into the plant tissues. Lignin is a natural polymer of three primary precursors e.g. p-coumaryl, coniferyl and sinapyl alcohols. In lignins from different sources the ratio of these three monomers varies. The lignin from grass has relatively more p-coumaryl alcohol than that present in softwood or hardwood lignins (Kirk, 1971). These alcohols are dehydrated by the activity of phenol oxidases, which give rise to free radicals. These free radicals immediately polymerise to give rise to a highly branched lignin polymer, with phenylpropane units bound covalently to each other.

In addition to lignin, the plant cell walls have cellulose and hemicellulose as the major polysaccharides. Some plants may also have pectic polysaccharides. Cellulose is a simple linear polysaccharide containing about 10,000 molecules of glucopyranose linked with each other by β–1,4-linkage. The cellulose chains are packed together tightly and are surrounded with other cell wall components like lignin and silica. This gives resistance to cellulose against hydrolysis by enzymes and microbes. The hemicelluloses of cereal straws have a backbone of β–1,4-linked xylopyranosyl units, with arabinofuranosyl units linked to some of the C3 atoms of xylose in the main backbone. The chain length of xylose in xylan is 50-200 units, much smaller as compared to that of cellulose molecule. This is perhaps the reason hemicelluloses are extracted with mild alkali while cellulose is not extracted.

The transverse section of wheat straw internode clearly shows that the outermost layer is made of sclerenchyma cells, which are highly lignified, in which small vascular bundles are also present. Inner to this is the lignified parenchyma cells containing large vascular bundles with an outer ring of lignified tissue. The innermost layer of tissue, which lines up the central lumen, is unlignified pith. Thus the sclerenchyma tissue and the cells of vascular bundles are highly lignified tissues of wheat straw cell wall.

## TREATMENT OF LIGNOCELLULOSIC FEEDS

The nutritional quality of lignocellulosic byproducts can be increased by biodelignification, before feeding to animals. A researcher may have three objectives in mind :

- Improvement in palatability of dry roughages.
- Increase in digestibility of lignocellulosic feeds for ruminants.
- Increase in protein content.

The palatability can be enhanced significantly by soaking, ensiling or mixing with other more palatable feeds. It is a common practice in villages of India that wheat or paddy straw is soaked in water and mixed with a small quantity of wheat/corn flour and offered to the animals. Such treatment of lignocellulosic feeds increases the dry matter intake by the animals. As far as the ruminants are concerned, the increase in true protein content of fermented lignocellulosic feeds, is not of great significance, because the rumen microflora is able to utilise non-protein nitrogen sources and convert it into microbial protein. Most of the studies, which have been conducted for bioconversion of lignocellulosics into protein rich animal feed, indicate that the fermented product has poorer digestibility as compared with untreated lignocelluloses, although the treatment is accompanied with an increase in crude protein content from 8 to 20% of dry matter. Therefore, the primary object in the process of delignification should be to increase the digestibility of agricultural byproducts so that the animals get more energy from the treated roughage source.

## BIODELIGNIFICATION

### Microbes: Types and their characteristics

The major objective of the treatment process as discussed above is the selective removal of lignin from lignocellulosic feeds so that the treated product is free of lignin and easily digested in the rumen. In nature the process of delignification has been observed in the forests of Southern Chile. The fungi decompose the forest wood and the end product known as *Palo Podrido* is used as feed for the ruminants. The healthy wood of that forest has an *in vitro* digestibility from 0-3% while the fungus infected wood can be degraded up to 60-77% and has lignin content as low as 1% of total dry matter of partially degraded wood. The fungi responsible for the natural degradation of wood have been identified as *Ganoderma applanatum* and *Armillariella* spp. (Zadrazil *et al.,* 1982).

Similarly the lignocellulosic feeds can be delignified under laboratory or simulated conditions, if proper environment is provided. We must have a microorganism which has the desired characteristics. The wood degrading fungi have been classified into three major groups:

- Soft-rot fungi
- Brown-rot fungi
- White-rot fungi

Soft-rot fungi usually grow on soluble sugars and brown-rot fungi degrade cellulose and hemicellulose. White-rot fungi are the efficient lignin degraders and have been studied extensively.

Although these fungi degrade lignin selectively, but lignin cannot serve as a sole source of energy for the fungus. There is not even a single strain of white-rot fungi which can use lignin as a growth substrate. During lignin degradation cellulose, hemicellulose, sugars and glycerol serve as a source of energy and as a co-metabolism lignin is degraded. As soon as the easily utilizable energy sources are exhausted from the medium, the fungus shifts towards secondary metabolism and lignin degradation starts. Nitrogen, carbon or sulphur limitation in the medium triggers the secondary metabolism of fungus (Jefferies *et al.,* 1981).

Different strains of white-rot fungi (*Pleurotus sajorcaju, Pleurotus ostreatus, Pleurotus florida, Pleurotus cornucopiae, Pleurotus 131, Pleurotus 137, Polyporus hirsutus, Polyporus capratus, Polyporus versicolor, Polyporus hirsutus, Polyporus sanguineous, Polyporus cinnibarinus, Stropharia rugosoannulata, Ganoderma lucidum*) were grown on sugarcane bagasse. During the first phase of growth (0-20 days) hemicellulose was selectively degraded whereas in the second phase (21-40 days) lignin was preferentially degraded in comparison to other substrate components. Among the strains tested *Pleurotus sajorcaju, Pleurotus ostreatus, Pleurotus florida, Ganoderma lucidum, Stropharia rugosoannulata* and *Pleurotus 137* were found to be good lignin degraders and caused a significant increase in IVDMD of sugarcane bagasse (Kewalramani *et al.,* 1988; Kamra *et al.,* 1993; Chaudhary *et al.,* 1994; Pal *et al.,* 1999). *Pleurotus ostreatus* and *P. sajorcaju* degrade about 73% and 63% of total lignin present in wheat straw. Total soluble phenolics increased significantly on fungal treatment in both the strains indicating the release of lignin monomers in the growth medium.

## SOLID STATE FERMENTATION

The submerged culture fermentation of lignocellulosics has many advantages and the contents of fermentor are homogenous and the rate of fermentative activity is maximum under the optimum conditions of the organism used, but there are many disadvantages like it requires high volume of liquid medium per unit weight of lignocellulose treated and dewatering of fermented product is a costly process. Above all the liquid byproduct creates problems of pollution if released in the atmosphere without stabilization. The solid-state fermentation (SSF) of lignocellulosic byproducts overcomes many of the limitations of submerged culture fermentation discussed above. The solid-state fermentation is attractive due to the following:

- Its low cost
- Lower reactor volume per unit substrate converted
- Direct use of bioconverted product as feed
- No processing for dewatering
- No drying of product.

The solid-state fermentation may be defined as a system of microbial growth on the surface of solid particles with no free flowing water. Depending on the substrate utilised the level of moisture differs in the substrate. The moisture content of substrate in SSF has been reported in the range of 30-85% in different systems. In koji process, the moisture level is 35-40% with a relative humidity of 80-90% for maintaining the required moisture level in the substrate. In bioconversion of wheat straw and sugarcane bagasse 75% moisture has been used in the substrate (Zadrazil, 1977; 1985; Kewalramani *et al.,* 1988; Kamra *et al.,* 1993). For the production of secondary metabolites like aflatoxins the moisture content is required at 31% with the need of remoistening the substrate. In spite of many advantages there are a few unique limiting factors in SSF e.g.

- Limited availability of substrate in the absence of free flowing liquid medium.
- Scanty availability of oxygen due to passive air flow through the substrate.
- Occasional drying of substrate due to low moisture content.
- Poor repeatability of results due to non-homogeneity of the substrate.

In spite of many drawbacks of SSF discussed above, it is still considered the most practicable technology due to low level of sophistication and appropriate technology at the farm level e.g. silage making and composting. For the bioconversion of lignocellulosic feeds the process has to be standardised which is operational at the farm level with minimum of mechanisation.

## FACTORS AFFECTING SOLID STATE FERMENTATION

A physical contact of fungus with the substrate is essential for lignin degradation during bioconversion of wheat straw with white rot fungi. The unique feature of the filamentous fungi is that these can grow on a solid surface in the absence of free flowing water, while the single cell microbes like yeast and the bacteria require free water in the medium for optimum growth. The process of lignin degradation is completed in two distinct phases i.e. the primary phase of filamentous growth and the secondary phase of lignin degradation (Zadrazil, 1977). Some of the factors affecting SSF of lignocellulosic substrates are discussed below.

### Moisture Content

As there is no free water available in SSF, the moisture content is very critical in the substrate. The moisture content affects the bioconversion process in two ways i.e. providing the optimum water supply for the growth of filamentous fungi and the level of moisture determines the rate of exchange of gases through the substrate. Moisture content more than 85% may make the conditions almost anaerobic at the bottom of the reactor due to accumulation of free water of the substrate. About 70-80% moisture in the substrate has been found to be optimum for the bioconversion of wheat straw with white-rot fungi (Zadrazil, 1980). Moisture content lower than this results in incomplete fermentation due to drying of the substrate.

The optimum water activity of the substrate is usually adjusted at the beginning of the process. Due to aeration of substrate with gases, moisture may be evaporated until fully water saturated air is passed through the substrate. Water availability also changes during the course of microbial growth due to changed water sorption capacity of substrate due to microbial growth. The moisture content varies mainly due to three reasons e.g. evaporation caused by the metabolic heat evolved during microbial growth, water utilization for polysaccharide hydrolysis and water synthesis by sugar metabolism.

## Effect of aeration and gas phase composition

The aeration of solid state fermentor is essential not only to provide oxygen for the respiration of white-rot fungi, but also is required for dissipation of metabolic heat generated during microbial growth and elimination of carbon dioxide and other volatile metabolic products from the substrate. The rate of aeration of substrate depends upon the fungal culture used and the stage of microbial growth. The primary and secondary metabolism and the process of lignin degradation is affected differently by the gas phase. The first phase of mycelial growth can tolerate higher levels of carbon dioxide (16-22%) in gas phase in *Pleurotus ostreatus* and *Pleurotus eryngii* (Zadrazil, 1975). In the secondary phase, the process of lignin degradation is strictly aerobic and needs the presence of oxygen at a partial pressure either equal to or more than that in the atmosphere (Kamra and Zadrazil, 1985; 1986). Thus the rate of aeration should be regulated in such a way that the above needs of the fungal culture are met out depending upon the stage of growth. The high rate of aeration often leads to drying of substrate and, therefore, efforts should be made to optimize the humidity of input air for the fermentor. The porosity of the substrate is important as far as the exchange of gases from gas phase and the solid substrate is concerned. A better oxygen transfer and free movement of air through the substrate can be achieved by arranging lower thickness of substrate layers, use of perforated trays or wire mesh trays etc.

Lignin is degraded slightly faster than the other wood components in the presence of oxygen by *Phanerochaete chrysosporium*, while in the presence of air, lignin is degraded slightly slower than the decomposition of other wood components (Reid and Seifert, 1982). *Pleurotus eryngii* degrades lignin more selectively in the presence of oxygen than *Pleurotus sajorcaju* or *Stropharia rugosoannulata* (Kamra and Zadrazil, 1986). As the requirements of different fungi are different under various stages of growth, it is better if a suitable gas mixture is passed through/over the substrate for the optimum results.

## Temperature

The control of temperature in solid state fermentor is very difficult as the heat transfer between the solid particles is poor due to low moisture content and the presence of air in the interstitial spaces in the solid substrate. In aerobic microbes like white-rot fungi, the generation of heat is enormous, which is related to oxygen consumption and carbon dioxide generation and is also an index of microbial growth. The metabolic heat produced during SSF of cassava with *Aspergillus niger* is as high as 80 Kcal/h/kg dry matter; which must be dissipated to keep the temperature of fermentor within optimum limits of microbial growth. The use of stirred solid state fermentor can be an answer to the above problem, but in fungal growth the mycelial growth is disrupted which stops the activity of fungus completely. For such a growth the use of forced air current through the stationary solid bed reactors is recommended for better results.

## Pretreatment of substrate

The substrates used in most of the solid state fermentations are coarse, insoluble in water and usually have a good water holding capacity. The grinding of lignocellulosic substrate is usually helpful in increasing the surface area and it may be better approachable by the attacking fungal cultures. The grinding may also increase the water holding capacity of lignocellulosic substrate and may hinder the gaseous exchange through the substrate. A simple crushing of straw may be sufficient for a better growth of white-rot fungi.

The indigenous microbial population present in the substrate determines the fate of added inoculum of white-rot fungi. On getting moisture on wetting, these microbes get favourable conditions for growth, the numbers of these microbes increase considerably and compete with the white-rot fungi for the available nutrients. If the white-rot fungus has good saprophytic competitive ability, it will succeed in colonising the substrate and dominate the process of fermentation. This process of colonisation will be made easier if the microbial load of substrate is decreased with some pretreatment. The pasteurisation of wheat straw at 90°C for 3 hours helps in increasing the *in vitro* dry matter digestibility of wheat straw treated with *Pleurotus sajorcaju* (Kamra and Zadrazil, 1986; Kamra *et al.,* 1989). Bano *et al.,* (1978) got a good yield of sporophores with *P. flabellatus* on wheat and paddy straws only after hot water treatment of substrate. A still milder treatment of straw at 50°C for 48 hours under anaerobic conditions was as effective as sterilised substrate for the growth of white rot fungi (Zadrazil and Peerally, 1986). At Indian Veterinary Research Institute, it was found that fumigation of substrate with formaldehyde could sterilize the substrate. To achieve good results in the increase in digestibility of lignocellulosic feeds, the natural microflora must be suppressed and the added inoculum must be given the maximum opportunity to colonise the substrate as soon as possible after the inoculum is added to the substrate.

### Additives

The chemical additives also affect the process of bioconversion. By combining the process of urea treatment with fungal treatment of straw, lignin degradation is inhibited. Urea treatment of straw followed by treatment with *Coprinus fimetarius* results in reduced digestion of feeds in comparison to urea treated rice straw (Gupta *et al.,* 1988; Singh *et al.,* 1990). Therefore the bioconversion of straw is recommended in nitrogen deficit conditions to avoid inhibition of lignin degradation.

The inclusion of organic source of nitrogen in the form of either wheat bran (10%) or mustard seed cake (5%) in wheat straw stimulated the growth of *P. ostreatus* and caused higher lignin loss and higher *in sacco* degradability in comparison to untreated controls. Inclusion of urea with or without molasses in the growth medium, had an adverse effect on lignin degradation. This might be due to adverse effect of high concentration of non-protein nitrogen on lignin metabolism of white rot fungus (Pal, 1996).

Supplementation of wheat straw with sugarcane molasses at 2.5% level increased lignin degradation and *in sacco* degradability by *P. ostreatus.* It also resulted in reduced dry matter loss in comparison to unsupplemented controls.

## FEEDING OF FUNGAL TREATED LIGNOCELLULOSIC FEEDS TO RUMINANTS

Most of the studies on bioconversion of lignocellulosic feeds have been conducted on laboratory scale as the scaling up of the process involves many difficulties which need further attention of the researchers. Therefore the studies on nutritional evaluation and *in vivo* feeding trials are also very scarce as large quantities of treated material are required for such experiments. The *in vivo* feeding trials conducted in buffaloes with wheat straw treated with *Pleurotus sajorcaju* have shown that the digestibility of wheat straw was reduced on fungal treatment and fruit body harvest (Bakshi *et al.,* 1985), while Calzada *et al.,* (1987) observed a slight increase in digestibility of wheat straw treated with *P. sajocaju* when fed to lambs. When wheat straw treated with *Coprinus fimetarius* was fed as a sole source of nutrients to goats, there was either no change or a decline in the digestibility of wheat straw (Rai *et al.,* 1987). Walli *et al.,* (1993) reported that high protein content in fungal treated straw could be utilized as efficiently as groundnut cake by crossbred calves. In another study by Dhandha *et al.,* (1994), it has been reported that the digestibility of various nutrients was improved by replacing untreated straw with *P. sajorcaju* treated straw at the rate of 50% in buffaloes.

The results of experiments on treatment of rice straw with *Coprinus fimetarius* combined with urea treatment showed that on fungal treatment the dry matter intake in goats was increased significantly, but the digestibility of nutrients was either not affected or adversely affected and there was a significantly lower availability of total digestible nutrients (51.28% vs 38.38% and 50.28% vs 46.70% in two different experiments) from fungal treated straw in comparison to that from urea treated straw (Gupta *et al.,* 1988 and Singh *et al.,* 1990).

The energy availability is reduced on prolonged fermentation which might be due to utilization of energy by the fungal biomass and the resultant treated product has lower content of energy left in the treated product. The treatment with fungi in the second stage does help in decreasing ammonia loss due to evaporation during urea ensiling of straws (Gupta and Singh, 1991).

Wheat straw treated with *Pleurotus florida* and after harvest of fruit bodies resulted in reduced availability of organic matter (62% vs 49%) and TDN intake (44% vs 26.5%) when fed to ruminants (Langar and Bakshi, 1986). But later on the same laboratory reported that spent wheat straw after harvest of fruit bodies of *P. florida* was found to be as good as the untreated wheat straw in terms of DCP and TDN contents and there was no effect on the digestibilities of nutrients when 50% and 100% wheat straw of control diet was replaced with spent wheat straw (Kakkar *et al.,* 1990). The authors recommended the use of spent wheat straw to buffaloes after raising fruit bodies of *P. florida* for human consumption.

In one experiment conducted at the author's laboratory, it was observed that *P. ostreatus* treated wheat straw improved the intake by the animals and the digestibility of various nutrients was significantly improved in comparison to untreated wheat straw (Pal, 1996). In this experiment wheat straw was treated with white-rot fungi in nitrogen deficit conditions and soluble sugars in the form of sugarcane molasses (2.5%) was provided which stimulated the primary growth of fungi and lignin degradation was stimulated in the secondary phase of growth. Based on the limited *in vivo* studies conducted, it appears that bioconversion should be conducted in the absence of excess nitrogen and if needed, nitrogen should be supplemented at the time of feeding the animals.

## LIMITATIONS FOR PRACTICAL APPLICATION

- Selection of proper strains of white-rot fungi for selective lignin degradation.
- Excessive cellulose and hemicellulose degradation.

- Substrate pretreatment methods.
- Scale-up of the treatment process.
- Process repeatability.
- Economics of treatment process and effect on animal productivity.

## FUTURE RESEARCH NEEDS

- Screening of tropical white-rot fungi should be continued so that the treatment is feasible at elevated temperatures in tropical conditions.
- The improvement of fungal strains should be tried for better lignin degradation accompanied with minimum loss of dry matter during the process of treatment.
- A good amount of work has been done on biodelignification in laboratory scale. But studies on treatment of straw at a large scale has not been tried seriously, except a few scattered reports. The efforts should now be aimed at evolving a process of treatment which requires the minimum sophistication and instrumentation. The process must be practicable at the farmers' level.
- The nutritional evaluation of treated lignocellulosic feeds should be done with *in vivo* feeding experiments and long time effect of feeding fungal treated straw on the health of ruminants should be studied.

## REFERENCES


Bakshi, M.P.S., Gupta, V.K. and Langar, P.N. (1985). Acceptability and nutritive evaluation of *Pleurotus* harvested spent wheat straw in buffaloes. *Agricultural Wastes,* 13, 51-57.

Bano, Z., Rajarathnam, S. and Nagaraja, N.L. (1978). Some aspects on the cultivation of *Pleurotus flabellatus* in India. *Mushroom Science,* 10, 597-608.

Calzada, J.F., De Leon, R., De Arriola, M.C. and Rolz, C. (1987). Growth of mushrooms on wheat straw and coffee pulp: Strain selection. *Biological Wastes,* 20, 217-226.

Chaudhary, L.C., Singh, R. and Kamra, D.N. (1994). Biodelignification of sugarcane bagasse by *Pleurotus florida* and *P. cornucopiae. Indian Journal of Microbiology,* 34, 55-57.

Dhandha, S., Kakkar, V.K., Garcha, H.S. and Makkar, G.S. (1994). Biological treatment of paddy straw and its evaluation through ruminant feeding. *Indian Journal of Animal Nutrition,* 11, 73-79.

Flachowsky, G., Kamra, D.N. and Zadrazil, F. (1999). Cereal straws as animal feed – possibilities and limitations. *Journal of Applied Animal Research,* 16, 105-118.

Gupta, B.N. and Singh, G.P. (1991). Bioconversion of ammonia to biological protein by *Coprinus fimetarius* during solid-state fermentation of ammoniated (urea) paddy straw. *Indian Journal of Animal Nutrition,* 8, 163-168.

Gupta, B.N., Rai, S.N., Walli, T.K. and Singh, K. (1988). Evaluation of fungal treated rice straw at different fermentation periods in relation to dry matter intake and nutrient utilization in goats. *Indian Journal of Animal Nutrition,* 5, 273-279.

Jeffries, T.W., Choi, S. and Kirk, T.K. (1981). Nutritional regulations of lignin degradation by Phanerochaete chrysosporium. *Applied Environmental Microbiology,* 42, 290-296.

Kakkar, V.K., Garcha, H.S., Dhanda, S. and Makkar, G.S. (1990). Mushroom harvested spent straw as feed for buffaloes. *Indian Journal of Animal Nutrition,* 7, 267-272.

Kamra, D.N., Kewalramani, N., Lall, D. and Pathak, N.N. (1989). Effect of substrate pretreatment on degradation of sugarcane bagasse by white rot fungi. *Indian Journal of Microbiology,* 29, 391-393.

Kamra, D.N., Kewalramani, N., Lall, D. and Pathak, N.N. (1993). Biodelignification and changes in in vitro digestibility of sugarcane bagasse treated with white rot fungi. *Journal of Applied Animal Research,* 4, 133-140.

Kamra, D.N. and Singh, R. (1989). Genetic manipulation of microbes for efficient utilisation of lignocellulosic feeds by ruminants. In: *Animal Productivity,* Eds P.N.Bhat, Menon, K. K.G. and Srivastava, H. C. Oxford and IBH Publishing Co., Pvt. Ltd., New Delhi. Pp 387-404.

Kamra, D.N. and Zadrazil, F. (1985). Influence of oxygen and carbon dioxide on lignin degradation in solid state fermentation of wheat straw with *Stropharia rugoso-annulata. Biotechnology Letters,* 7, 335-340.

Kamra, D.N. and Zadrazil, F. (1986). Influence of gaseous phase, light and substrate pretreatment on fruit body formation, lignin degradation and *in vitro* digestibility of wheat straw fermented with *Pleurotus* spp. *Agricultural Wastes,* 18, 1-17.

Kewalramani, N., Kamra, D.N., Lall, D. and Pathak, N.N. (1988). Bioconversion of sugarcane bagasse with white rot fungi. *Biotechnology Letters,* 10, 369-372.

Kirk, T.K. (1971). *Annual Review of Phytopathology,* 9, 185-210.

Kirk, T.K. and Moore, W.E. (1972). Removing lignin in wood with white rot fungi and digestibility of resulting wood. *Wood Fiber,* 4, 72-79.

Langar, P.N. and Bakshi, M.P.S. (1986). Evaluation in vitro of *Volvariella* harvested spent rice straw. *Agricultural Wastes,* 15, 309-311.

Pal, D. (1996). Studies on Nutritional Improvement of Wheat Straw with Fungal Treatment. *Ph. D. Thesis.* IVRI Deemed to be University, Izatnagar. Pp 1-125.

Pal, D., Kamra, D.N. and Pathak, N.N. (1999). Changes in chemical composition and *in sacco* degradability of wheat straw treated with *Pleurotus sajorcaju* and *P. ostreatus. International Journal of Animal Science,* 14, 61-67.

Rai, S.N., Walli, T.K. and Gupta, B.N. (1987). The chemical composition and nutritive value of rice straw after treatment with *Coprinus fimetarius* in a solid sate fermentation system. *Animal Feed Science Technology,* 26, 81-92.

Reid, I.D. and Seifert, K.A. (1982). Effect of an atmosphere of oxygen on growth, respiration and lignin degradation by white rot fungi. *Canadian Journal of Botany,* 60, 252-260.

Singh, G.P., Rai, S.N., Walli, T.K. and Singh, K. (1990). Effect of microbial treatment of paddy straw on the chemical composition and nutrient utilization in crossbred goats. *Indian Journal of Animal Nutrition,* 7, 251.

Walli, T.K., Rai, S.N., Singh, G.P. and Gupta, B.N. (1993). Nitrogen status of biologically treated straw and its N-utilization by animals. In: *Feeding of Ruminants on Fibrous Crop residues,* Eds Singh, K. and. Schiere J. B., Proceedings of International Workshop, NDRI, Karnal, Feb. 4-8, 1991, pp 237-247.

Zadrazil, F. (1975). Influence of carbondioxide concentration on the mycelium growth of three *Pleurotus* spp. *European Journal of Applied Microbiology and Biotechnology,* 1, 327-335.

Zadrazil, F. (1977). The conversion of straw into feed by basidiomycetes. European *Journal of Applied Microbiology and Biotechnology,* 4, 291-294.

Zadrazil, F. (1980). Conversion of different plant waste into feed by basidiomycetes. *European Journal of Applied Microbiology and Biotechnology,* 9, 243-248.

Zadrazil, F. (1985). Screening of fungi for lignin decomposition and conversion of straw into feed. *Angewandte Bonanik,* 59, 433-453.

Zadrazil, F., Grinsberg, J. and Gonzales, A. (1982). "Palo-podrido" – decomposed wood, which was used as feed. *European Journal of Applied Microbiology and Biotechnology,* 15, 167-171.

Zadrazil, F., Kamra, D.N., Isikhuemhen, O.S., Schuchardt, F. and Flachowsky, G. (1996). Bioconversion of lignocellulose into ruminant feed with white rot fungi - Review of work done at FAL, Braunschweig. *Journal of Applied Animal Research,* 10, 105-124.

Zadrazil, F. and Peerally, A. (1986). Effects of heat microbiological treatments of wheat straw substrate on the initial phase of solid-state fermentation (SSF) by white rot fungi. *Biotechnology Letters,* 8, 663-666.

# 13

# Effect of Cell Wall Degrading Enzymes on Starch Recovery from Cassava Bagasse

**M. Eyini and V. Rajapandy**

*Research Center in Botany, Thiagarajar College (Autonomous), Madurai-625 009, India.*

The solid waste of sago and starch industry using cassava tubers contains 40-65% starch on dry weight basis. These starch granules are located in the root cells, which remained unruptured during the rasping process and hence are not easily recoverable. In the present investigation, the effect of cellulase and pectinase enzymes on cell wall softening and on the recovery of starch from cassava bagasse was studied.

Cassava bagasse contained 47% starch, 17.5% cellulose and 1.1% free reducing sugars by dry weight. Crude pectinase culture filtrates were obtained from *Aspergillus niger* (a laboratory isolate), *Trichoderma viride* (MTCC 800) and *Trichoderma harzianum* (MTCC 792) growing in Czapek's medium with pectin replacing sucrose as the carbon source. The pectinase activity of *Aspergillus niger* was higher than that of the other two organisms (0.121, 0.076 and 0.107 IU/ml). A significant decrease in cellulose from 17.5% to 10% and 4.9% was observed in cassava bagasse incubated with *Aspergillus niger* pectinase (0.2 ml/g) along with commercial cellulase (0.025 ml/g substrate) in solid state and in submerged state respectively. Increasing the pectinase dose to 0.4 ml/g did not have any significant effect on the parameters studied. But incubation with pectinase at 0.4 ml/g and double the concentration of cellulase at 0.05 ml/g significantly decreased the cellulose to 5.5% and 1.6% in solid state and submerged state fermentation respectively. The release of reducing sugars increased from 1.1% to 11.4% and 16.7% and the starch recovery increased from 47% to 55.9% and 57.6% respectively under these conditions. The possibility of adopting fermentation for starch recovery from cassava tubers will be discussed.

## INTRODUCTION

A number of agro-industrial residues are thrown away as waste or are underutilized. The solid waste of sago industry producing sago and starch from cassava tubers is called 'Thippi'. During the industrial manufacture of starch and sago, about 20 percent of the cassava tubers that are processed form the waste fibrous residue. This solid waste contains 40%-65% starch on a dry weight basis

(Pandey *et al.,* 2001). These starch granules are located in the root cells, which remained unruptured during the rasping process and hence are not easily recoverable (Srikanta *et al.,* 1987).

Recovery of starch from cassava tuber waste may be achieved by the overall action of cell wall degrading enzymes. A wide variety of fungi are able to produce these enzymes, namely, cellulase and pectinase (Puniya and Singh, 1995). Fermentation of cassava tubers by a pure culture or mixed culture inoculum of cellulolytic organisms had been found to enhance starch extraction (George *et al.,* 1991). The present study was carried out to find the effect of cell wall degrading enzymes on the release of starch from cassava bagasse by softening the cell walls of the unruptured tuber cells.

## MATERIALS AND METHODS

A laboratory isolate of *Aspergillus niger* and the strains of *Trichoderma viride* (MTCC 800) and *Trichoderma harzianum* (MTCC 792) were used for the induction and recovery of pectinase enzyme. The organisms were subcultured on PDA slants once a month and were maintained at 4°C. Cassava waste was procured from Varalakshmi sago industry, Namakkal, Tamilnadu. It was sun dried, coarsely ground to uniform size to pass though a sieve (6 mm).

## CRUDE PECTINASE ENZYME PRODUCTION

The inoculum was developed by transferring spores from a ten-day-old fungal culture to Czapek's medium (50 ml) in 250 ml Erlenmeyer flasks containing 1% pectin replacing sucrose as the carbon source. Inoculated flasks were incubated on a rotary shaker at 120 rpm for 48 h at 30 ± 2°C.

For enzyme production, the organisms were grown in 250 ml Erlenmeyer flasks containing 100 ml of Czapek's medium containing $NaNO_3$ - 0.2%, KCl - 0.05%, $MgSO_4$ - 0.05%, $K_2HPO_4$ - 0.05%, $FeSO_4$ - 0.01% and 2% pectin which replaced sucrose as the carbon source. Pectin was dissolved in Tris - HCI buffer containing 0.1 M NaCl (pH 7.5) and added to the medium. The pH was adjusted to 7.5 with 1 M NaOH or 1 M HCI. The two-day-old fungal inocula in Czapek's broth (10% v/v) were added separately and aseptically to individual flasks. The flasks were incubated at room temperature (30 ± 2°C) for four days. Samples were withdrawn every day to assay pectin methyl esterase activity (PME). The medium was filtered through a muslin cloth and centrifuged (800 rpm for 5 min) to remove fungal mycelia. The clear fermentation broth was used for enzyme assay and as pectinolytic enzyme source.

## PECTIN METHYL ESTERASE ACTIVITY

The assay system consisted of 10 ml of 1 % pectin in Tris - HCl buffer containing 0.1 M NaCl (pH 7.5) and 2 ml enzyme source. Incubation was carried out at room temperature (30 ± 2°C) for a day. The ester groups released were quantified by titration against 0.2 N NaOH. Enzyme activity was expressed in International Units (IU) which is defined as the micromoles of ester equivalents released by 1 ml of enzyme in 1 min. (George *et al.,* 1991).

## ENZYME SACCHARIFICATION

For solid-state enzyme saccharification, cassava waste (25 g) was taken in 250 ml Erlenmeyer flasks. Water was added to give 60% moisture content. For submerged state experiments, cassava waste (25 g) made into slurry with 100 ml of water was taken in 250 ml Erlenmeyer flasks. The flasks were sterilized at 121°C for 15 min. Three replicate flasks were maintained for each experiment.

The crude PME enzyme filtrate obtained from 3-day-old cultures of *T. harzianum, T. viride* or *A. niger* was added separately and aseptically to individual flasks in SSF and SmSF experiments at the concentration of 0.2 ml or 0.4 ml/g cassava waste. The substrate was mixed well with the enzyme using a sterile glass rod. The flasks were incubated for a day at room temperature (30 ± 2°C). For the combined treatment, the cassava waste was incubated with the selected concentrations of crude pectinase culture filtrate and incubated for a day at room temperature and at pH 7.5, followed by the addition of commercial cellulase aseptically at the concentrations 0.025 or 0.05 ml/g thippi. The pH was then adjusted to 5.5 with 1 M HCl and the flasks were incubated at 55°C for a day. At the end of incubation period, the samples were analysed for cellulose (Updegraff, 1969) and reducing sugars (Miller, 1959) content. The extractable starch was estimated by the method of George *et al.,* (1991).

**Table 1. Effect of enzyme saccharification of cassava waste in solid state or submerged state on the content of residual cellulose (% dry weight)**

| S. No | Enzyme | Enzyme concentration (ml/g thippi) | *Trichoderma harzianum* | | *Trichoderma viride* | | *Aspergillus niger* | |
|---|---|---|---|---|---|---|---|---|
| | | | SSF | SmSF | SSF | SmSF | SSF | SmSF |
| 1 | Pectinase | 0.2 | 17.6 ± 1.8 | 17.7 ± 1.5 | 17.51 ± 1.5 | 17.6 ± 1.4 | 17.2 ± 1.3 | 17.4 ± 1.0 |
| 2. | Pectinase | 0.4 | 17.2 ± 1.8 | 17.8 ± 1.6 | 17.5 ± 1.2 | 17.8 ± 1.4 | 17.4 ± 1.5 | 17.5 ± 1.2 |
| 3. | Pectinase + Cellulase | 0.2 + 0.025 | 10.2 ± 1.0 | 5.1 ± 0.4 | 11.7 ± 1.0 | 5.2 ± 0.4 | 9.95 ± 0.8 | 4.9 ± 0.5 |
| 4. | Pectinase + Cellulase | 0.2 + 0.05 | 6.9 ± 0.6 | 2.1 ± 0.2 | 6.8 ± 0.5 | 2.2 ± 0.2 | 6.4 ± 0.6 | 2.3 ± 0.1 |
| 5. | Pectinase + Cellulase | 0.4 + 0.025 | 9.6 ± 0.8 | 4.6 ± 0.4 | 9.7 ± 0.9 | 4.7 ± 0.4 | 3.18 ± 0.2 | 4.5 ± 0.5 |
| 6. | Pectinase + Cellulase | 0.4 + 0.05 | 6.0 ± 0.5 | 1.7 ± 0. 2 | 6.2 ± 0.5 | 2.0 ± 0.1 | 5.5 ± 0.6 | 1.6 ± 0.2 |

Values are mean ± SE of 3 samples

**Table 2. Effect of enzyme saccharification of cassava waste in solid state or submerged state on the content of reducing sugars (% dry weight)**

| S. No | Enzyme | Enzyme concentration (ml/g thippi) | *Trichoderma harzianum* SSF | *Trichoderma harzianum* SmSF | *Trichoderma viride* SSF | *Trichoderma viride* SmSF | *Aspergillus niger* SSF | *Aspergillus niger* SmSF |
|---|---|---|---|---|---|---|---|---|
| 1 | Pectinase | 0.2 | 1.15 ± 0.1 | 1.14 ± 0.1 | 1.12 ± 0.1 | 1.12 ± 0.1 | 1.17 ± 0.1 | 1.12 ± 0.1 |
| 2. | Pectinase | 0.40 | 1.13 ± 0.1 | 1.15 ± 0.1 | 1.16 ± 0.1 | 1.18 ± 0.1 | 1.15 ± 0.1 | 1.15 ± 0.1 |
| 3. | Pectinase + Cellulase | 0.2 + 0.025 | 4.47 ± 0.5 | 9.11 ± 0.6 | 4.60 ± 0.5 | 8.90 ± 0.5 | 4.85 ± 0.3 | 9.75 ± 0.7 |
| 4. | Pectinase + Cellulase | 0.2 + 0.05 | 9.29 ± 1.0 | 14.14 ± 1.2 | 9.65 ± 0.5 | 13.40 ± 1.2 | 10.90 ± 1.0 | 14.8 ± 1.5 |
| 5. | Pectinase + Cellulase | 0.4 + 0.025 | 5.0 ± 0.5 | 9.92 ± 0.8 | 5.01 ± 0.3 | 9.12 ± 0.6 | 5.18 ± 0.4 | 10.97 ± 1.0 |
| 6. | Pectinase + Cellulase | 0.4 + 0.05 | 10.75 ± 1.2 | 14.75 ± 1.2 | 10.75 ± 1.1 | 14.50 ± 1.2 | 11.4 ± 1.1 | 16.72 ± 1.5 |

Values are mean ± SE of 3 samples

## RESULTS AND DISCUSSION

Cassava waste contained on a dry weight basis: 47.5% starch, 17.5% cellulose, 6.43% hemicellulose, 12% lignin, 1.14 % free reducing sugars and 0.99 % protein. Pectin methyl esterase (PME) activity of the three selected fungi steadily increased and reached the maximum on the third day of fungal growth in Czapek's medium containing pectin as the carbon source. *Aspergillus niger* isolate showed the highest PME activity (0.121 IU/ml) while the pectinolytic activity shown by *Trichoderma harzianum* was significantly higher than that of *T. viride* on all the days of study (Fig. 1). Commercial pectinase

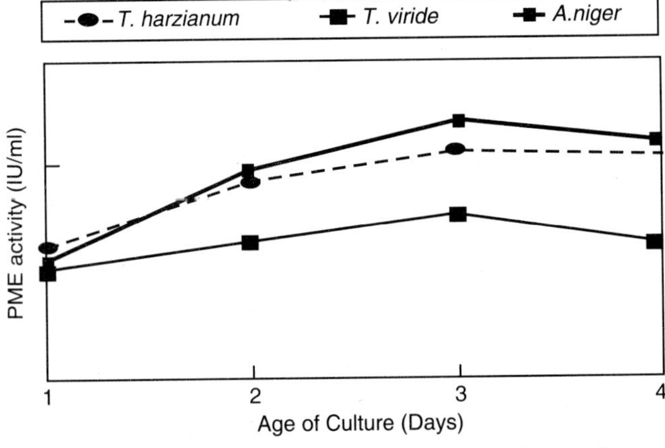


**Fig. 1.** Pectin methyl esterase activity related to fungal growth in Czapek's medium with pectin as carbon source

preparations are obtained mainly from *Aspergillus* and *Penicillium* (Aguilar and Huitron, 1987). *Aspergillus niger* was shown to have high pectinolytic enzyme activity in solid state with citrus pectin as carbon source (Ghosh *et al.,* 2000). Said *et al.,* (1991) observed similar high production and activities of pectin methyl esterase activity in *Penicillium frequentans* after 48 hours at 35°C in agitated cultures supplemented with 0.5% citrus pectin at pH 7 and 8.

The extractability of starch as flour from cassava waste, the extent of cellulose degradation and release of reducing sugars depended on the type and concentration of enzymes used. PME enzyme alone could not cause any significant change in the content of cellulose or reducing sugars in cassava waste. But irrespective of the source, PME enzyme along with commercial cellulase resulted in 90% loss in cellulose, 12-17-fold increase in reducing sugars and 13-18% increase in extractable starch over the control at the concentrations of 0.4 ml and 0.05 ml per g thippi respectively.

While doubling the PME enzyme concentration produced only a slight change in the rate of cellulose degradation, addition of double the quantity of commercial cellulase resulted in a significant decrease in cellulose content and in a dramatic increase in the content of reducing sugars. Significant elevation in the activities of amylase, cellulase and the pectolytic enzymes during the post-harvest deterioration of cassava tubers had been reported by Padmaja and Balagobal (1985) indicating the role of microbial cellulolytic enzymes in cell lysis. This synergistic or enhanced activity may be attributed to the complementary role of pectinolytic extra and cellulolytic enzymes in cell wall degradation.

Saccharification of cassava waste by PME enzyme derived from any of the three fungi in SSF or SmSF resulted in 46.3% to 49.6% starch extraction. The extraction of starch was markedly higher (57.6%) in the SmSF experiment using *A. niger* PME at 0.4 ml/g and commercial cellulase at 0.05 ml/g cassava waste (thippi) concentration respectively. The combined enzyme treatment

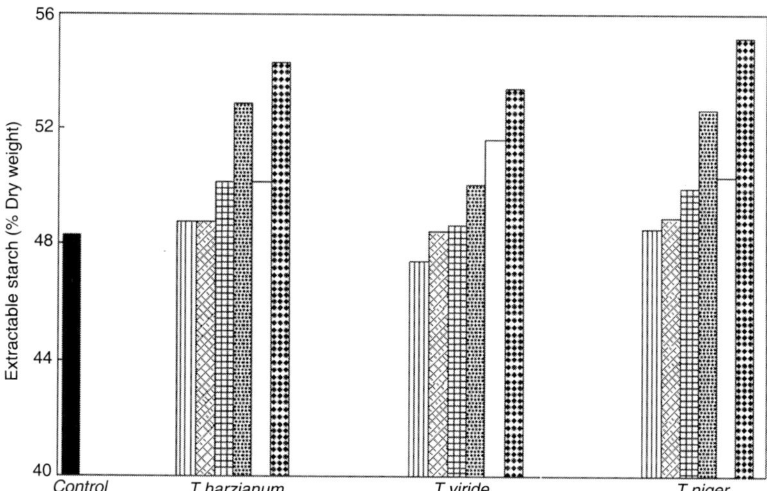


**Fig. 2:** Effect of solid state enzyme saccharification on the content of extractable starch from cassava waste

1. PME 0. 2 ml/g,
2. PME 0. 4 ml/g,
3. PME 0. 2 ml/g + cellulase 0. 025 ml/g,
4. PME 0. 2 ml/g + cellulase 0.05 ml/g,
5. PME 0. 4ml/g + cellulase 0. 025 ml/g,
6. PME 0. 4 ml/g + cellulase 0. 05 ml/g

extracted 15.8% more starch compared to the untreated control (Figs. 2 & 3). The enhanced starch recovery in the combined treatment may be due to the role played by the cell wall degrading enzymes at micro level liberating starch granules from all cells. Even though earlier studies (Pandey *et al.,* 1999) have reported that enzyme activities were higher in SSF rather than in SmSF experiments, we observed that saccharification of cassava waste under submerged state by PME (0.4 ml/g) and commercial cellulase (0.05 ml/g) resulted in 20% higher cellulose degradation, 4.6 fold higher reducing sugars and 3% more extracted starch than under solid state.

Industrial starch finds important applications in processed foods and in the production of many non-food products including cosmetics, adhesives, lubricants and tablets and is also being increasingly used in the paper and pharmaceutical industries. Industrial starch is commercially produced from starchy tuber or cereal crops by the physical extraction process, which requires voluminous quantity of water. Due to the inefficiency of mechanical extraction, a large volume of effluent with high amount of carbohydrates is generated, posing serious threat to the environment. The inefficiency of the conventional process is ascribed to the incomplete rupture of root cells during the rasping process. The results of the present investigation indicate that microbial softening with cellulolytic enzymes or fermentation may be adopted as a viable technology to extract higher amount of starch from cassava and other starchy tuber crops.

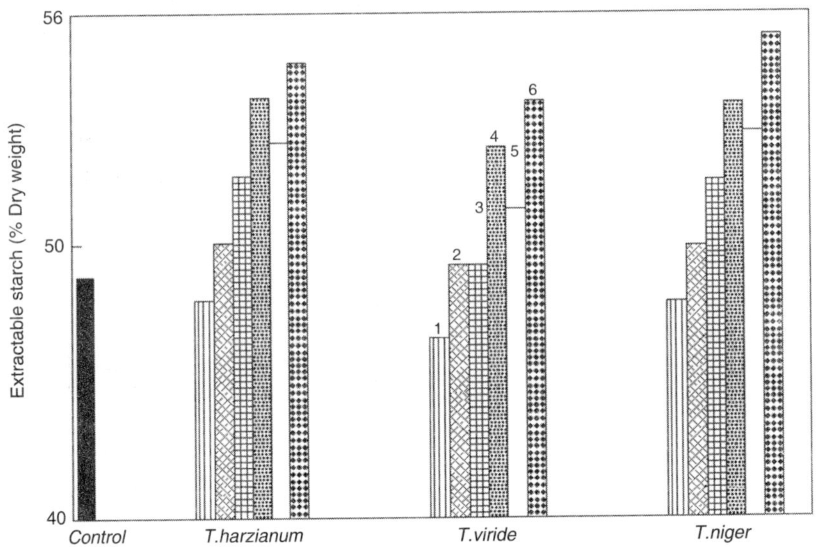


**Fig. 3:** Effect of submerged state enzyme saccharification on the content of extractable starch from cassava waste

1. PME 0. 2 ml/g,
2. PME 0. 4 ml/g,
3. PME 0. 2 ml/g + cellulase 0. 025 ml/g,
4. PME 0. 2 ml/g + cellulase 0.05 ml/g,
5. PME 0. 4ml/g + cellulase 0. 025 ml/g,
6. PME 0. 4 ml/g + cellulase 0. 05 ml/g

## ACKNOWLEDGEMENTS

This work is part of the project funded by UGC, New Delhi under the UGC Research Award Scheme to M.E. Research assistant stipend to V.R is gratefully acknowledged.

## REFERENCES


Aguilar, G and Huitron, C. (1987). Stimulation of the production of extra cellular pectinolytic activities of *Aspergillus* sp. by galacturonic acid and glucose addition. *Enzyme and Microbial Technology,* 9, 690-696.

Ferrer, S., Ramon, D., Salom, J., Vicente, E. and Uruburu, F. (1985). Protoplasts from *Podospora anserine*: Isolation, Purification and Transformation. *Current Microbiology,* 12, 301-306.

George, M., Padmaja, G., and Moorthy, S.N. (1991). Enhancement in starch extractability from cassava (*Manihot esculenta* Crantz) tuber through fermentation with a mixed culture inoculum. *Journal of Root Crops,* 17, 1-9.

Ghosh, S., Kumar, A., Gupta, R. and Bhalla, T. C. (2000). Production of polygalacturonase by *Aspergillus niger* using submerged and solid-state fermentation (Abs). *AMI 41st Annual conference,* Jaipur. Pp 89.

Miller, G.K. (1959). Use of dinitro salicylic reagent for the determination of reducing sugars. *Analytical Chemistry,* 31, 426-428.

Padmaja, G. and Balagobal, C. (1985). Cellular and extracellular enzymes associated with the post-harvest deterioration of cassava tubers. *Journal of Food Science and Technology,* 22, 82-87.

Pandey, A., Selvakumar, P., Soccol, C.R. and Nigam, P. (1999). Solid-state fermentation for the production of industrial enzymes. *Current Science,* 77,149-162

Pandey, A., Soccol, C.R., Rodriguez -Leon, J.A. and Nigam, P. (2001). *Solid - state fermentation in Biotechnology: Fundamentals and Applications,* Asiatech Publishers Inc. Pp 84.

Puniya, A.K., and Singh, K. (1995). Determination of enzyme activities of lignocellulosic strains in solid substrate fermented products. *Indian Journal of Animal Nutrition,* 12, 205-212.

Said, S., Fonseca, M.J.V. and Siessere, V. (1991). Pectinase production by *Penicillium frequentans. World Journal of Microbiology and Biotechnology,* 7, 607-608.

Srikanta, S., Jaleel, S.A., Ghildyal, N.P., Lonsane, B.K. and Karanth, N.G. (1987). Novel technique for saccharification of cassava fibrous waste for alcohol production. *Starke,* 39, 234-237.

Updegraff, D.M. (1969). Semi micro determination of cellulose in biological materials. *Analytical Biochemistry,* 32, 420-424.

# 14

# Combinatorial Approaches to Improve Plant Cell Wall Digestion: Possible Solution for Cattle Feed Problems

**Krishna Kant Sharma[1], Sarika Kuhar[1], Ramesh Chander Kuhad[1] and P.N. Bhat[2]**

*[1]Lignocellulose Biotechnology Laboratory, Department of Microbiology, University of Delhi South Campus, Benito Juarez Road, New Delhi-110021, India.*
*[2]World Buffalo Trust, 205, E-64, C/9, Sector-40, Noida, U.P.-201303, India.*

Ruminants represent a significant proportion of the domesticated animal species worldwide, and among farmed livestock are the best adapted to utilize plant material as feed. Almost half of the global carbon fixed annually by photosynthesis is incorporated into plant cell wall (lignocellulosics) making it most renewable carbon source on earth. Lignocellulose, therefore, will always remain important for ruminant diets. Improvement in the ability of the rumen biota as well as external degrading factors (ligninolytic microorganisms and their enzymes) to modify plant cell wall is highly desirable, which in turn will upgrade the animal feed. The enzymes involved in degradation of fibre holds significance. Here, we attempt to discuss the application of ligninolytic fungi in solid state fermentation and the use of exogenous enzymes in improving animal feed and the use of plant genetic manipulation techniques to bring down lignin content and eventually increasing cellulose fibre in crop residues. Moreover, we propose newer and benign approaches such as genomic and metagenomic techniques to improve our knowledge about the functional genomic framework of plant cell wall degradation, which can be used as basic tool in ruminant feed development.

## INTRODUCTION

World population is expected to reach 8 billion by the year 2020 and maximum population growth will occur in developing countries. Increasing demand for livestock products as a result of rapid growth in the world economics, future hopes of feeding cattle and securing their food security will depend on the better utilization of non-conventional feed resources, which do not compete with human food. The primary constraint of such feeds is the low digestibility of lignocellulosic material, which is maximally effected by lignin content.

At the time scale of the evolution of photosynthetic organisms, lignification is a relatively recent process, which emerged about 430 million years ago. The deposition of lignin in plant cell wall was part of the mechanism, which allowed the development of upright, positively phototropic plants adapted to terrestrial habitat. Lignin present in plants is interspersed with hemicellulose forming a matrix that surrounds the orderly cellulose microfibrils and reduces their digestibility when fed to animals. Lignocellulosic feeds such as grasses and straws are the agricultural by-product with cellulose, hemicellulose and lignin as their major structural components. The cereal straws are good source of energy as it contains about 60-70% carbohydrates (Jackson, 1977). The 65-80% dry matter comprised cellulose, hemicellulose, soluble proteins, sugars and rest 20-35% dry matter comprised recalcitrant substance like lignin, phenolics, and silica (Table 14.1) (Besle *et al.,* 1994). Lignin, phenolics, silica and other antinutritional factors, affect the digestibility of these lignocellulosic feeds from 40-50% and hence only 60% of the energy is available to the animal. The removal of lignin from lignocellulosic residues using physical, chemical or biological treatment enhances the digestibility of these materials. But biological treatment is preferred over other methods because it is specific for lignin, produces minimum harmful byproducts and has low energy requirements.

**Table 1. Composition (%) of various agricultural and wood byproducts**

| *Substrate* | *Hexosans* | *Pentosans* | *Lignin* | *Ash* |
|---|---|---|---|---|
| Bagasse | 33 | 30 | 29 | 4 |
| Barely Straw | 40 | 20 | 15 | 11 |
| Birch Wood | 40 | 33 | 21 | 4 |
| Corn Cobs | 42 | 39 | 14 | 2 |
| Corn Stalk | 35 | 15 | 19 | 5 |
| Groundnut Shell | 38 | 36 | 16 | 5 |
| Oat Straw | 41 | 16 | 11 | 12 |
| Rice Straw | 32 | 24 | 13 | 12 |
| Rice Hulls | 36 | 15 | 19 | 20 |
| Sorghum Straw | 33 | 18 | 15 | 10 |
| Wheat Straw | 30 | 24 | 18 | 10 |
| Pine | 41 | 10 | 27 | 8 |

Biological delignification under solid state fermentation (SSF) processes using selective lignin degraders have applications in improving the cellulose content of plant byproducts, which can be used as animal feed. The ruminating fungi have been reported to contribute in fibre digestion as they colonize in a better way than bacteria and effectively degrade lignified tissues (Barrasa *et al.*, 1998). One of the uses of the SSF of lignocellulosic residues by white-rot fungi is to convert the renewable materials to more digestible feed for ruminants (Kamra *et al.*, 1988; Basu *et al.*, 2002).

The main problem in biological upgrading of poor quality materials is to find suitable microorganisms, which decompose lignin without using much of the hemicellulose and cellulose. The *in-vitro* digestibility of fungal substrate decreases in the beginning of colonization by white-rot fungi and increases afterwards, while the contents of soluble substances increases during the incubation (Zadrazil, 1984; Sangwan *et al.*, 2001).

**Table 2. White-rot fungi and their ligninolytic enzyme profile**

| *White-rot fungi* | | | *Enzymes* | | |
|---|---|---|---|---|---|
| | *LiP-* | *MnP-* | *Laccase* | *AAO* | *VP* |
| *Coriolus versicolor* | + | + | + | | |
| *Cyathus bulleri* | + | + | + | | |
| *Pycnoporus sanguineus* | + | + | + | | |
| *Pleurotus ostreatus* | + | + | + | | |
| *Pleurotus sajo-caju* | + | + | + | + | + |
| *Pleurotus florida* | + | | + | | |
| *Coriolus pruinosum* | + | + | | | |
| *Phanerochaete chrysosporium* | + | + | | | |
| *Ceriporiopsis subvermispora* | | + | + | | |
| *Dichomitus squalens* | | + | + | | |
| *Lentinus edodes* | | + | + | | |
| *Panus tigrinus* | | + | + | | |
| *Rigidoporus lignosus* | | + | + | | |
| *Oudemansiella radicata* | + | | + | | |
| *Polyporus pletenis* | + | | + | | |
| *Polyporus brumalis* | + | | + | | |
| *Phlebia tremellosus* | + | | + | | |
| *Phlebia ochraceofulva* | + | | + | | |
| *Phlebia radiata* | + | + | + | | |
| *Bjerkandera adusta* | + | + | + | + | + |
| *Daedaleopsis confragosa* | + | + | + | | |
| *Polyporus varius* | + | + | + | | |
| *Pycnoporus cinnabarinus* | | | + | | |
| *Flavodon flavus* | + | + | + | | |
| *Trametes pubescens* | | | | | |
| *Coriolus hirsutus* | | | | | |
| *Agaricus bisporus* | | | + | | |
| *Armillaria mellia* | | | + | | |
| *Pleurotus eryngii* | | + | + | + | |

where LiP: Lignin peroxidase; MnP: Manganese peroxidase, AAO: Aryl alcohol oxidase and VP: Versatile peroxidases.

The fungi take very long in degrading lignin, because they produce low amount of lignin modifying enzymes. Therefore, producing fast lignin degrading recombinants and using them for solid state fermentation of plant byproducts would help in improving the animal feed. Another way of improving animal feed could be the production of transgenics with low lignin content. Now, there are several reports on the manipulation of lignin metabolism by suppressing the expression of single gene (Dixon *et al.,* 2001). The potential of such work could be greatly expanded if multiple genes could be coordinately suppressed (Boudet *et al.,* 2001). Application of standard microbiological methods for the recovery of microorganisms from the environment has limited success in providing access to the true extent of microbial biodiversity. Recent developments in

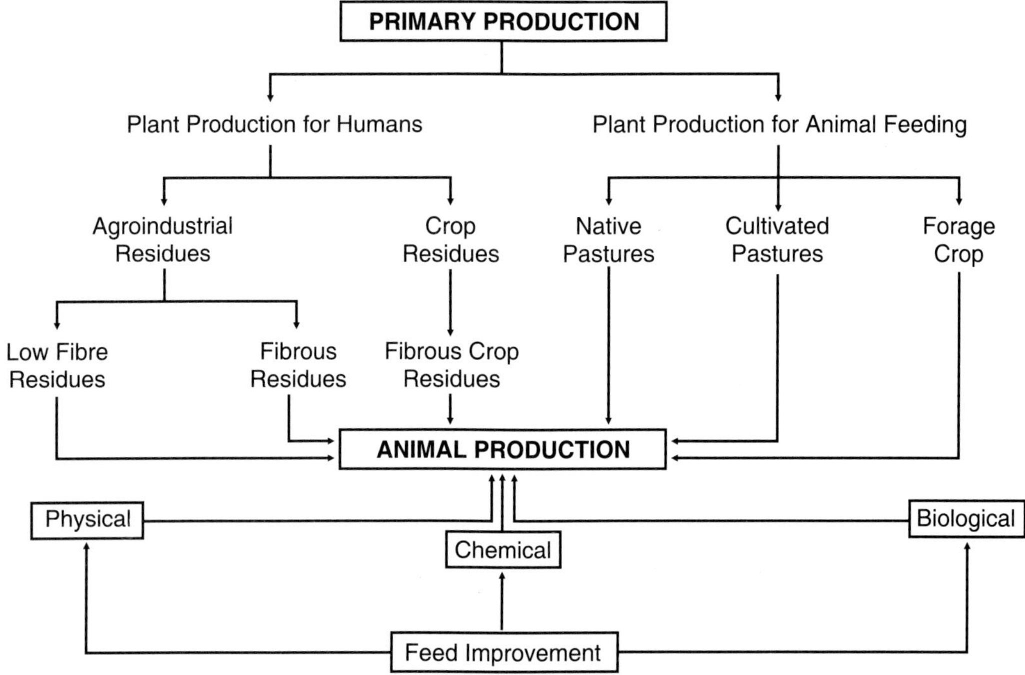


**Fig. 1.** Conventional and non-conventional feed crop

metagenomic technologies have paved the access to the prokaryotic genetic information available in environmental samples, independent of culturability (Cowan, 2000; Ferrer *et al.,* 2005). This article is an attempt to discuss briefly the physico-chemical and biological methods to improve the crop residues as animal feed. Recent approaches will also be highlighted which may contribute in solving the problem of lignin degradation of crop residues so that they can be used in animal feeding.

## IMPROVEMENT OF LIGNOCELLULOSIC FEED

Improvement in the animal feed through value addition of the lignocellulosic waste is possible by accomplishing the following objectives:

(a) Increasing the digestibility of lignocellulosic materials.
(b) Increasing the protein content of the feeds.
(c) Improvement in palatability of the roughages.

Lignin is removed from lignocellulosic feeds so that the treated product is free of amorphous lignin, which can be easily digested in the rumen. Several methods have been developed to convert lignin into simpler forms, which can be utilized by the animals. The different methods can be grouped in two broad classes:

(a) Physico-chemical methods
(b) Biological methods

## PHYSICO-CHEMICAL METHODS

Treatment of crop residues with alkaline reagents, in particular sodium hydroxide and ammonia, have been extensively used in the past.

### Treatment with Sodium Hydroxide

#### *Wet Treatment*

Chemical treatment started way back in early 1900, which involves boiling of straw in NaOH solution under pressure. Thereafter Beckman and Fingerling introduced their modified method, which involves soaking fibrous agro-product in cold NaOH solution for 12 hours followed by washing out with water to neutrality, allowing cellulose to remain in the fodder while about 30% of lignin and 10-15% of the pentosans are solubilized and washed out. The organic matter digestibility of rye straw increased from about 46 to 71% (Sundstol, 1988). The Norwegian farmers from the beginning of World War II routinely used the modified form of this method but latter discontinued because of pollution problem (Homb, 1984).

The original method was replaced by a modification of the dip-treatment method (Sundstol, 1981). The organic matter digestibility of dip treated straw was found to increase 45-50% to 70-75% (Sundstol, 1988). This method further solved the pollution problem and dry matter loss.

#### *Dry Treatment*

The industrial procedures for dry treatment of straw with NaOH has been developed in many countries (Rexen and BachKundsen, 1984). Here, the fodder material is chopped and milled before a solution of NaOH is sprayed on it. After thorough mixing, the material is passed through a press for pellet formation. Ultimately, by the rise in temperature in the press microenvironment, immediate reaction takes place between material and the alkali. This treatment can be better characterized as a combination of physical and chemical methods.

## AMMONIA TREATMENT

Among the other chemicals, $NH_3$ is the most suitable because, in addition to hydrolyzing the lignocellulosic linkages, it also contributes to nitrogen fraction. Ammonia treatment for low quality fodder has been extensively reviewed by Sundstol and Coxworth (1984).

The stack method for ammonia treatment of straw was developed in Norway in 1970-1975. Stack of straw are wrapped with polyethylene and injected with 3% of anhydrous ammonia (Sundstol *et al.*, 1978). The liquid ammonia evaporates gradually and the ammonia gas diffuses the stack. One of the greatest advantages of this method is that ammonia being fungicidal so the drying step can be skipped before ammonia is added. Ammonia treatment of straw can also be carried out within 24 hours in an oven at temperature of about 90°C, but the cost increases.

## UREA TREATMENT

Urea is widely used in fodder treatment because of its safe application and handling. Its supplementation usually results in higher dry matter intake and digestibility, however some workers have found no effect of urea treatment in animal performance (Ibrahim *et al.,* 1987). Supplementation with concentrates or with legume hay has been reported to improve straw utilization, although results may vary depending on the level of supplementation (Nianogo *et al.,* 1999). Urea works better at high temperature or otherwise demands natural ureases for the reaction to start and form ammonia as decomposed product. This limits its application to tropical countries, which is generally overcome by the supplementation of raw soybean powder. Moreover, the undecomposed or partially decomposed urea leads to an increase in ammonia production, which is generally toxic to cattle.

The chemical treatment of fodders poses various environmental pollution problem and are not amenable to industrial production.

## BIOLOGICAL METHODS

Concept of microbial treatment for delignification of lignocellulosic feed got attraction due to the traditional feeding practice of decomposed wood to the animals in South Chile (Zadrazil, 1982). The fungi decomposes the forest wood and the end product known as *Palo-podrido* is used as feed for the ruminants. The healthy wood of the forest has an *in vitro* digestibility from 0-3%, while the treated wood can be degraded upto 60-77% and has lignin content as low as 1% of total dry matter of partially degraded wood. In similar way, the lignocellulosic feeds can be delignified under laboratory conditions if proper environment could be provided.

White-rot fungi, primarily basidiomycetes, are the major degraders of the lignified wood material. Zadrazil (1985) tested more than two hundred strains of white-rot fungi out of which some (Table 3) were able to improve the digestibility from 18 to 52% cellulose digestibility, accompanied with only 12% loss of dry matter. Easily utilizable energy source is essential for the mycelial growth of the fungus and production of hydrogen peroxide, which is an essential prerequisite for lignin degradation (Eriksson *et al.*, 1986). As soon as the easily utilizable nutrients are exhausted from the medium, the fungus shifts towards secondary metabolism as a survival strategy to degrade and utilize lignin. Nitrogen, carbon or sulphur limitation in the medium is reported to trigger the secondary metabolism of fungus (Jefferies *et al.*, 1981).

Considerable amount of work has been done in the past for developing new chemical and biological methods which can be utilized in developing protienaceous foods and feeds (Solomons, 1983; Kuhad *et al.*, 1997). Of the biological methods, bacterial fermentation has for some time been the one most developed method for the production of this kind of feed and food. The special interest shown in obtaining protein by means of unicellular bacteria, also called "single cell protein (SCP)", is particularly due to the fact that algae, yeasts and bacteria grow fast and have a high doubling rate. Genetic manipulation of these microorganisms is also easier, due to rapid sequence of generations, than with the higher plants, and it is possible to influence growth rate, adaptability and sometime chemical composition. Alternative fermentation products of this kind can be economically used as source of protein for farm animals (Cooney *et al.*, 1980).

**Table 3. Influence of fungal species on biodelignification of wheat straw at 25°C in 60 days of treatment (Zadrazil, 1985)**

| *Fungus* | *Dry matter loss (%)* | *Lignin loss (%)* | *Change in IVDMD (%)* |
|---|---|---|---|
| *Abortiparus biennis* | 24.2 | 7.1 | +23.4 |
| *Agrocybe aegerita* | 9.6 | 0.0 | –29.0 |
| *Collybia radicata* | 12.3 | 2.0 | –20.0 |
| *Dichomitus squalens* | 43.2 | 13.5 | +28.3 |
| *Flammulina velutipus* | 11.3 | 1.7 | –18.3 |
| *Fomes fomentarius* | 18.5 | 12.0 | –14.0 |
| *Ganoderma applanatum* | 12.7 | 2.9 | +5.4 |
| *Ganoderma lucidum* | 25.2 | 10.8 | +25.7 |
| *Lentinus edodes* | 15.4 | — | +29.7 |
| *Lenzites betulina* | 22.5 | –11.2 | +8.6 |
| *Meripilus giganteus* | 19.3 | 9.9 | +18.6 |
| *Phlebia livida* | 8.3 | 2.6 | –2.1 |
| *Phellinus robustus* | 13.9 | 15.8 | +11.8 |
| *Pleurotus cornucopiae* | 32.2 | 11.3 | +8.9 |
| *P. corticatus* | 21.0 | 7.1 | +14.5 |
| *P. eryngii* | 14.3 | — | +23.2 |
| *P. flabellatus* | 30.8 | 14.2 | +21.2 |
| *P. ostreatus* | 15.4 | 7.0 | +22.4 |
| *P. sajorcaju* | 17.6 | 14.4 | +18.6 |
| *P. serotinus* | 25.0 | 10.7 | +33.3 |
| *Polyporus arcularius* | 21.6 | 12.1 | +20.6 |
| *P. brumalis* | 21.6 | 9.0 | +20.0 |
| *P. galactinus* | 11.3 | 5.8 | 25.7 |
| *P. versicolor* | 24.0 | 7.3 | –6.7 |
| *Poria expensa* | 25.2 | 6.1 | +21.1 |
| *Sporotrichum* | 55.8 | 22.0 | +15.9 |
| *Stropharia rugosoannulata* | 30.6 | 11.7 | +25.7 |
| *Trametes hirsute* | 25.6 | 9.0 | +20.0 |
| *T. versicolor* | 27.5 | 9.6 | +11.2 |
| *Volvariella volvacea* | 17.9 | — | –25.4 |
| *Xylaria polymorpha* | 20.2 | — | –29.8 |

## EXOGENOUS ENZYMES

During the 1960s, evidences became available that show the direct application of enzymes to forage before consumption can improve cattle performance (Burroughs *et al.,* 1960). In comparison to controls, cattle gained 6.4-24.0% more weight and converted feed 6.0-21.2% more efficiently when ground maize, oat silage, maize silage, or lucerne hay was treated with an enzyme cocktail containing amylolytic, proteolytic, and cellulolytic enzymes. Unfortunately, some other reports with enzyme cocktails have resulted in quite a significant decrease in animal performance when applied to forage feed (Perry *et al.,* 1960; Clark *et al.,* 1961; Clark *et al.,* 1966). Various studies thereafter, were designed to address issues related to the inconsistency of results by investigating the effects of feed type (Beauchemin *et al.,* 1995), application levels (Beauchemin and Rode, 1996), enzyme products (Feng *et al.,* 1996; Lewis *et al.,* 1999), and enzyme applications (McAllister *et al.,* 1999). Other factors, such as dry forage or silage (Beauchemin *et al.,* 1995), and the enzyme is infused directly into the rumen (McAllister *et al.,* 1999), have also been investigated. Multifactor involvement makes it obvious that mechanistic understandings of synergism between exogenous and autochthonous microbiotas (*in vitro* or *in vivo*), as well as in rumen, is still to be clearly understood.

## Ligninolytic enzymes

Lignin degradation by white-rot fungi is an oxidative process and phenol oxidases are the key enzymes (Kuhad *et al.,* 1997; Leonowicz *et al.,* 1999). Of these, lignin peroxidase (LiP), manganese peroxidase (MnP) and laccase especially from white-rot fungi *(Phanerochaete chrysosporium*, *Pleurotus ostreatus* and *Trametes versicolor*) have been studied in more detail. LiP and MnP oxidizes the substrate by two consecutive one-electron oxidation steps with intermediate cation radical formation. Laccase has broad substrate specificity and oxidizes phenols and lignin substructures with the formation of oxygen radicals. Other enzymes that participate in the lignin degradation processes are $H_2O_2$-producing enzymes and oxidoreductases, which can be located either intra- or extracellularly. Bacterial and fungal feruloyl, and p-coumaroyl esterases are relatively novel enzymes capable of releasing feruloyl and p-coumaroyl, and play an important role in biodegradation of cell walls in grasses (Kuhad *et al.,* 1997). These enzymes act synergistically with xylanases to disrupt the hemicellulose-lignin association, without mineralization of the lignin *per se* (Borneman *et al.,* 1990). The accessibility of cell wall polysaccharides from the plant to microbial enzymes is dictated by the degree to which they are associated with phenolic polymers (Chesson 1981; Kuhad *et al.,* 1997). White-rot fungi are well adapted to degrade wood by a simultaneous attack on lignin and cellulose/hemicellulose and some are reported to selectively degrade far more lignin than polysaccharides (Eriksson *et al.,* 1990; Kuhad *et al.,* 1997). The model fungus for lignin degradation is *P. chrysosporium* (Kirk, 1990) but recently certain other lignin degrading fungi have also been studied, *viz.*, *C. subvermispora*, *Phlebia radiata*, *Pleurotus eryngii*, *Cyathus bulleri*, (Lundell, 1993; Martinez *et al.*, 1994; Hatakka, 2001; Vasdev and Kuhad, 1995; Vasdev *et al.*, 2005). The lignolytic enzymes are directly involved not only in the degradation of lignin in the natural lignocellulosic substrates but also in the degradation of various xenobiotic compounds (Barr and Aust, 1994; Hatakka, 1994; Pointing, 2001) including dyes (Glenn *et al.,* 1983; Pasti-Grigsby *et al.,* 1992; Vasdev and Kuhad 1994a; Vasdev *et al.,* 1995). Some white-rot fungi produce all five enzymes reported in Table 1, while others produce either one or two of them (Hatakka, 1994; Kuhad *et al.*, 1997).

## MODE OF ACTION

An early criticism of enzyme inoculation into the rumen was that the enzymes were likely to undergo inactivation by the proteolytic activity of rumen fluid. However, it was subsequently been demonstrated that CMCase and xylanase activities contributed from endogenous sources can remain active in the rumen (Hristov *et al.,* 1998) and all exogenous enzyme activity is not necessarily inactivated once introduced into the rumen (Hristov *et al.,* 1998). The stability of exogenous enzyme activity in the rumen may increase the possibility of improvement in feed digestion *in situ* (Hristov *et al.,* 1996).

Release of reducing sugars by exogenous enzymes is probably an important mechanism by which exogenous enzymes operate (Beauchemin *et al.,* 1996). The degree of sugar release is dependent on the feed type as well as the type of enzyme. The release of sugars from feed is partially the result of solubilization of neutral detergent fibre (NDF) and acid detergent fibre (ADF) (Hristov *et al.,* 2000; McAllister *et al.,* 2000) and is consistent with the observed increase in the soluble fraction and rate of *in situ* digestion (Hristov *et al.,* 2000). Additionally, although exogenous enzymes affect the release of soluble carbohydrates, the amount liberated represents only a minute portion of the total carbohydrate present in the diet.

A more likely explanation for the mode of action of exogenous enzymes is that they act synergistically with the rumen microbiota. Fiberolytic enzymes normally act as member of a cellulosome complex, presumably because there are a range of glycosidic bonds that need to be hydrolyzed in a defined fashion. A logical extension of this reasoning would therefore be to add enzymes that are not produced by rumen microorganisms and attack structures in forage that result in an increased extent of digestion. Ligninolytic enzymes justify the noted prerequisite for increasing forage digestibility.

The Messerschmidt mechanism is based upon extensive studies of ascorbate oxidase, and it suggests that the resting enzyme has all the 4 Cu atoms oxidized as Cu (II) species with the T3 pair bridged by a hydroxide ligand. The first step is the reduction of a single molecule of the reducing substrate at the T1 centre. The electron is transferred to the trinuclear site, along a Cys–His pathway. Transfer of two electrons gives rise to a 'hydro-peroxide' intermediate, which is further reduced in two reductive events. This ultimately leads to two molecules of water, via an intermediate radical or transition state. An alternative study done on *Rhus* laccase by Reinhammar (Reinhammar, 1984) says that the T1 and T2 Cu atoms are reduced, either in an independent manner or via an intramolecular electron transfer. Various arguments led Reinhammar to conclude that two independent substrate molecules reduce the two sites. When both T1 and T2 sites are reduced, they simultaneously transfer one electron each to the T3 pair. Thereafter, the newly deoxidized T1 and T2 atoms are reduced again and the enzymes become totally reduced. The details of the reoxidation of this fully reduced species are unclear but are again likely to be involved in the intermediate steps with the first of these being the binding of $O_2$ as a peroxide to the T3 pair. Transfer of an electron from T1 generates paramagnetic oxygen intermediate (Fig. 2).

Copper atom in laccase plays an important role in the stabilization of the structure besides their role in the enzymatic catalysis. It has very little effect on secondary structure and plays an important role in protein stability but not on the ability of protein to exchange charge. About 35% of the polypeptide chains are buried and constitute a particular compact structure, probably composed of β-sheets. Laccase shows a high content of β-structures and a small content of α-helices, a typical secondary structure composition of these enzymes. On the other hand, another small part of the structure is particularly sensitive to high temperature and it is probably the reason for the loss of enzyme activity (Ragusa *et. al.,* 2002).

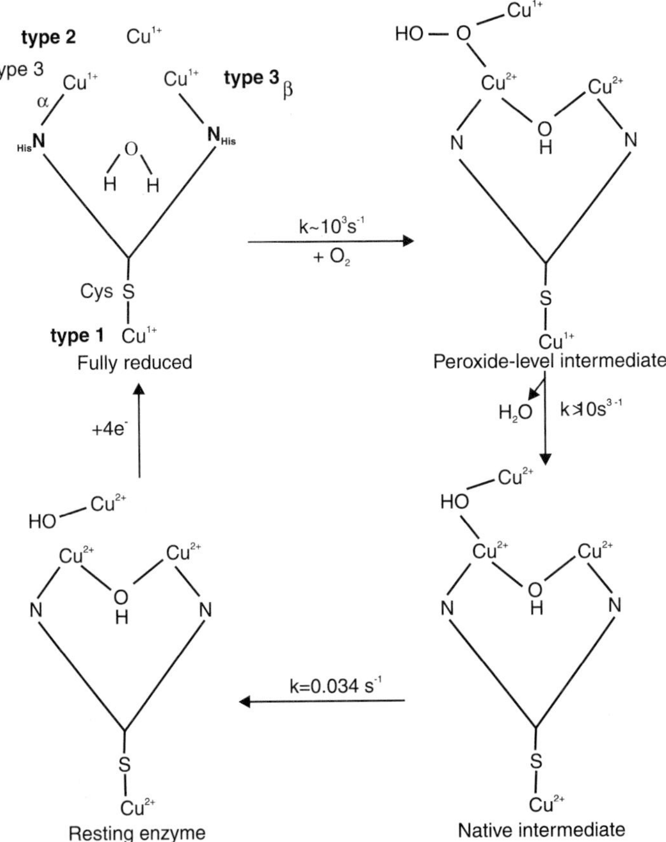


**Fig. 14.2.** Proposed reaction mechanism of dioxygen reduction in the multicopper oxidases.

## SOLID-STATE BIOCONVERSION OF LIGNOCELLULOSICS

White-rot fungi degrade wood by simultaneous attack on the lignin, cellulose and hemicelluloses (Kuhad *et al.,* 1997). A number of white-rot fungi have been studied for their ligninolytic activity, some of which are *Trametes versicolor, Cyathus bulleri, C. stercoreus, Phlebia radiata, Pycnoporus sanguineus, P. cinnabarinus, Pleurotus ostreatus, P. eryngii, P. sajorcaju, P. florida, Phanerochaete chrysosporium, Cereporiopsis subvermispora, Sporotrichum thermophile, Coriolus pruinosum, Dichomitus squalens, Lentinus edodes, Panus tigrinus, Rigidoporus lignosus, Oudemansiella radicata, Polyporus pletenis, P. brumalis, Phlebia tremellosus* and *P. ochraceofulva.*

Agro-byproduct treatment with ligninolytic fungus under SSF condition has been considered a way to produce materials with improved properties for animal fodder, such as higher digestibility and partial defibriation, respectively (Kamra *et al.*, 1988). It mimics part of natural system with many interactions. The efficiency in lignin removal depends upon the fungal species and varies from one substrate to another. Conditions that favour the growth of white-rot fungi include adequate substrate, moisture, aeration and temperature (Breen and Singleton, 1999). The increase in digestibility depends on the fungal species, cultivation time, temperature, water/air ratio in the substrate and on the preparation, bulk density, and finally composition of substrate. Also, the effect of seed culture on SSF of wheat straw by *P. chrysosporium* for animal feed development has been reported (Basu *et al.*, 2002).

For SSF cultures, geometrical and physical restrictions are quite different to those observed in Submerged fermentation (SmF) cultures. Oxygen diffusion is perpendicular to large cell aggregates, which are dispersed in thin layer of broth. In this sense, it may be that SSF cultures are in better physiological condition and for such a reason, produce less proteolytic enzymes than in SmF cultures. Few limitations of SSF are hindrance of microbial growth due to complex substrate, problems of mass-transfer and regulation of some physical factor.

There have been few reports on wheat-straw treatment with *Pleurotus* species and *P. chrysosporium* to convert the locked renewable material to more digestive feed for ruminants (Basu *et al.*, 2002; Kamra *et al.*, 1988). While studying on the growth of white-rot fungi on wheat straw by SSF, Moyson and Verachtert (1991) found that digestibility improves with increasing lignin degradation. For large-scale production of animal feed, it is mandatory to attain a high level of lignin degradation with minimum cellulose utilization, which is the obvious prerequisite for fodder development. A reduction in lignin content in *Pleurotus sajor-caju* from 12.6% to 5.6% improved the digestibility of wheat-straw from 29.8% to 59.2%, leaving behind most of the desired cellulose for ruminant.

The specificity of biological reactions and ecofriendly approach of *in-vivo* enzymatic digestion, (IVED) which helps in overcoming substrate accessibility constraints for lignocellulosics by laccases and thus making it an interesting alternative to chemical, physical and mediator based treatments, has been reported by Sharma *et al.* (Sharma *et al.*, 2005). Sharma *et al.* envisaged a novel system for laccase production by *Ganoderma* sp. under SmF and SSF conditions. The experimental strategy consisted of two separate phases, (i) xylanase production from *Bacillus pumilus* strain MK001 under SSF using wheat bran as the solid substrate and (ii) laccase production by *Ganoderma* sp. kk-02 using bacterial fermented wheat bran (BFWB) as substrate under SmF and SSF conditions. Moreover effect of amino acids and their analogues, vitamins, surfactants and polyethylene glycol has also been studied on laccase production by the fungus when cultivated on BFWB (Sharma *et al.*, 2005). The above-mentioned method can be intelligently extrapolated and applied in the modification of the lignocellulose materials for cattle feed production.

## MOLECULAR METHODS

Since the introduction of the term lignin (from the Latin *lignum*, meaning wood) by the French scientist De Candolle (De Candolle *et al.,* 1832) to describe the non-cellulosic encrusting substances present in the wood, an impressive amount of data has been accumulated on the chemistry, biosynthesis, distribution, and orientated modification of lignin in the wood. Interest in lignin has, for many years, lagged behind the interest in other plant polymers such as cellulose and hemicelluloses, which are closely associated with lignins in the cell wall. This limited attention can be attributed to the chemical complexity of the polymer and lack of convenient analytical techniques. In addition, in contrast to other plant polymers which has important nutritional (e.g. starch) or agro-industrial (e.g. cellulose) applications, the practical uses of lignins and of their byproducts are increasing, however, have not yet been fully achieved. Removal of lignin for cellulose production requires large amounts of chemical and energy input, resulting in financial and ecological costs. Efforts are being made worldwide towards the development of tree varieties with less or modified lignin which would enhance wood-pulp production efficiency. Transgenic trees with enhanced or reduced expression of different gene encoding enzymes of the phenylpropanoid metabolism show that it is possible to develop varieties with more easily extractable lignin (Pilate *et al.*, 2002; Baucher *et al.*, 2003).

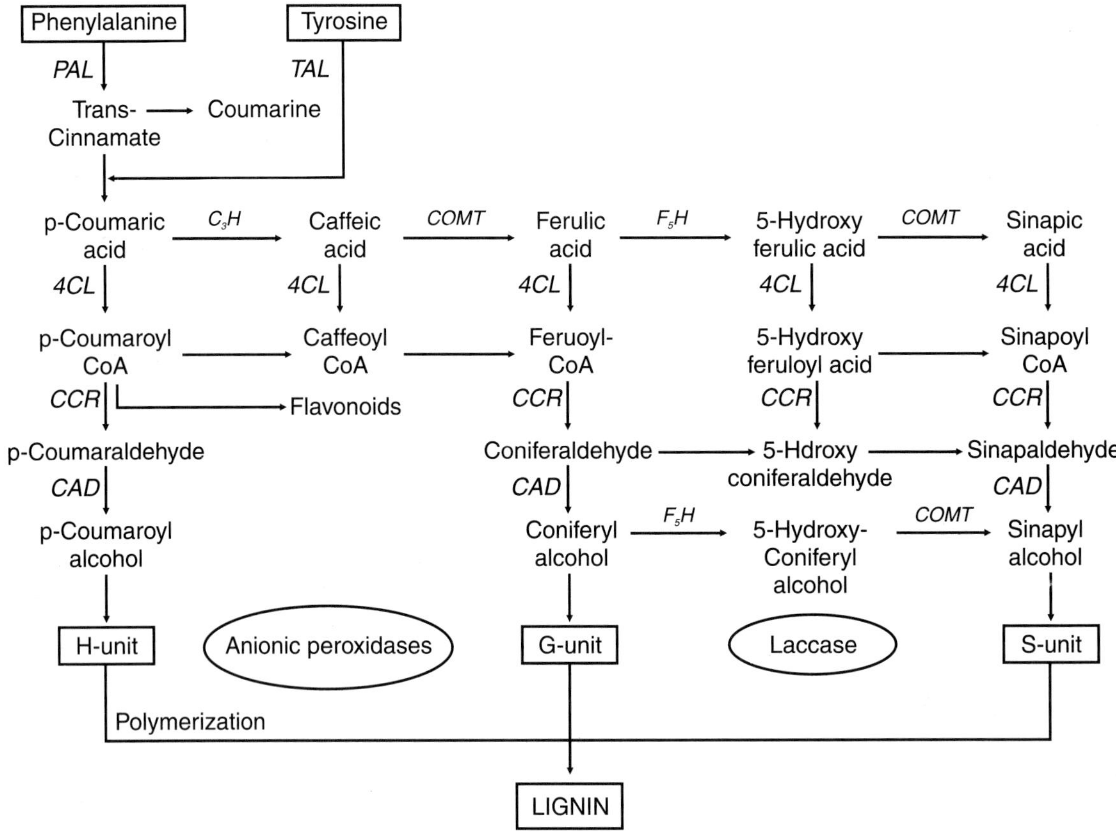


**Fig. 3.** The lignin biosynthetic pathway. The lignin biosynthetic enzymes are, PAL= phenylalanine ammonia lyase, TAL=tyrosine ammonialyase, C4H =cinnamate 4-hydroxylase, C3H= 4-hydroxycinnamate 3-hydroxylase, COMT= caffeic acid 3-O-methyltransferase, F5H =ferulate 5-hydroxylase, 4CL =4-coumarate, CoA ligase, CCoA-3H = coumaroylcoenzyme A 3-hydroxylase, CCR=cinnamoyl-CoA reductase, and CAD=cinnamyl alcohol dehydrogenase.

## DOWN REGULATION OF LIGNIN BIOSYNTHETIC PATHWAY

Lignin is a complex, aromatic polymer that limits plant cell wall degradation by ruminants and reduces the nutritional value of forages. Genetic engineering, using an antisense strategy, offers the potential to modulate enzymes in the lignin biosynthetic pathway (Fig. 3) to reduce lignin, thereby improving forage quality and animal performance. Antisense sorgam *O*-methyltransferase gene (*omt*) to downregulate maize OMT by 60% to modulate enzymes in lignin biosynthetic pathway has been reported (He *et al.,* 2003; Atanassova *et al.,* 1995). Transgenics with downregulated OMT averaged 20% less lignin in stems and 12% less lignin in leaves which increases mean whole-plant digestibility from 72 to 76%.

Down regulation in expression of L-Phenylalanine Ammonia-Lyase (PAL) or Cinnamate 4-Hydroxylase (C4L) has been reported to reduce the lignin content and altered lignin composition in transgenic tobacco (Sewalt *et al.,* 1997). Anomalous behaviour was seen in the ratio of syringyl-guaiacyl, which explains the alternate and multiple phenylpropanoid pathways, prior to the formation of specific monolignol precursors. These experiments have produced somewhat conflicting results showing change in cell wall digestibility due to alteration in both lignin content and composition. Transgenic tobacco plants with reduced expression of cinnamaldehyde dehydrogenase (CAD) produced lignin with a decrease in S/G ratio and increased degradability even though there was no change to the total amount of lignin (Bernard *et al.,* 1998).

Laccase antisense poplars were also analysed for total soluble phenolic content. In one line of antisense laccase transformants (poplars underexpressing the *lac3* gene), a two-to three-fold increase in soluble phenolic compounds was observed. Interestingly, Dean *et al.* (1998) showed that in yellow poplars transgenic lines over-expressing the *Acer pseudoplatanus* laccase gene, there was a nominal increase in the deposition of phenolic material or change in wall phenolic components. The over-expressed laccase isoform could conceivably play a role in the polymerization and cross-linking of simple phenolic units.

The presence of laccases, which in contrast to peroxidases are specifically located in the cell wall and do not require a toxic electron acceptor, is likely to be related to the processing of wall associated phenolic substrates. Future investigations will help us to identify their specific functions on modification of cell wall properties through the polymerization of monolignols or oxidative cross-linking of hydroxycinnamic acids.

## COMBINATORIAL MODIFICATION OF MULTIPLE LIGNIN TRAITS

Many reports now describe the manipulation of plant metabolism by suppressing the expression of single genes. The potential of such work could be greatly expanded if multiple genes could be coordinately suppressed. Abbott and coworkers (2002) had tested a novel method for achieving this by using single chimeric constructs incorporating sense sequences for multiple genes to target suppression of three lignin biosynthetic enzymes. The cDNA for CAD, COMT, and CCR were fused and transformed to tobacco seedlings using *Agrobacterium tumefaciens* (Abbot *et al.*, 2002). They compared this method with a more conventional approach by crossing the plants harbouring different antisense transgenes. Simultaneous suppression of multiple genes on single constructs can offer a great potential for rapid and coordinated suppression of multigene on diverse biochemical pathways, therefore this can be recommended for adoption.

Appreciable work has been done to develop genetically engineered trees, with the goal of reducing lignin content to enhance wood-pulp production efficiency (Li *et al.*, 2003). But, no report has been noticed for the reduced lignin content in the fodder crop to facilitate the digestibility of cattle feed.

Considerable evidence is now available which shows that in angiosperm tree, the syringyl monolignol pathway branches out from the guaiacyl pathway through coniferaldehyde and is regulated in sequence by three genes. Through multigene cotransformation [i.e. 4-coumarate-CoA ligase (*4CL*) and coniferaldehyde 5-hydroxylase (*Cald5H* )], upto 52% less lignin, a 64% higher S/G-ratio, and 30% more cellulose has been reported (Li *et al.*, 2003). The result of multigene cotransfer system motivates us to think a very broad application in fodder-crop genetic engineering and functional genomics. After the report of highly efficient and convenient agrobacterial trans-formation system in white-rot fungi (Sharma *et al.*, 2006), a multigene cotransfer system can be extrapolated to maintain the ligninolytic enzyme cocktail in organism used in cattle feed development.

A recent study confirmed that different alterations to lignin biosynthetic enzymes could cause different changes to spatial organization of lignin and cell wall polysaccharides at the cellular and subcellular levels (Chabannes *et al.,* 2001). The interactions between lignin and cell wall polysaccharides are likely to be a very important determinant of digestibility. An improved understanding of the types and quantity of lignin required by the plant cells for normal development will be important in the genetic manipulation of the plant cell wall for improved digestibility. Large decreases in lignin content appear to cause defect in plant growth and metabolism (Jung *et al.,* 1999; Jones *et al.,* 2001). More subtle changes to lignin concentration and cell wall ultrastructure may be as effective as reducing the over all lignin content. Molecular tools for manipulating plant cell wall architecture are becoming available, and offer new approaches for understanding and improving forage digestibility. However, the advantages of reduced lignin are offset by the disadvantage of plants with reduced lignin, which are more easily attacked by predators such as insects, fungi and bacteria. Indeed, increasing lignin content has been promoted as a defense against pests.

Finally, one of the main concerns about genetically modified trees is their acceptance by public opinion. The fact that these transformed plants are non-food products designed for industrial purposes should facilitate public acceptance. Risk assessment for transgenic trees presents special challenges because of the time scale involved, but also because of the complex interplay of ecological, genetic, agronomic and social factors. At the ecological level, altering the lignin profiles of plants should not provide a competitive advantage to the transformed lines.

The process of genetically modifying trees is very much in its infancy. The public perception of the new products is a critical aspect of its wide development but the potential long term benefits are obvious.

## GENETIC MANIPULATION IN POTENTIAL LIGNIN DEGRADING MICROBES

### Rumen Bacteria

Fibre degrading rumen microbes have been conceptualized and developed as a genetically modified super bug. The idea of genetically modified rumen bacteria is based on the assumption that the rumen microbiota does not produce the correct biocatalytic cocktail to enhance fibre degradation. For example, *Ruminococcus* and *Fibrobacter* do not produce exocellulases and are active against crystalline cellulose to make them more potent. Published studies (Cotta *et al.,* 1997; Krause *et al.,* 2001; Gobius *et al.,* 2002) have demonstrated that modification of *B. fibrisolvens* with xylanase has been successful and *in vitro* digestibility of fibre can be improved, but unfortunately this still does not overpower the native *Fibrobacter* and *Ruminococcus* species.

Initially *B. fibrisolvens* was modified with a xylanase (family 10 GH) from *N. patriciarum*, a GH family different from the family 11 typically present in *B. fibrisolvens* (Gobius *et al.,* 2002). GH10 and GH11 xylanases differ in their catalytic properties (Baldwin and Palmquist, 1965) and was speculated that introduction of a GH11 xylanase would increase its ability to digest fibre (Gobius *et al.,* 2002). Fibre digestibility could be improved *in vitro* by more than 28% in comparison with the native untransformed strain (Gobius *et al.,* 2002). Subsequent studies demonstrated that the recombinants were unable to compete with native cellulolytic *Ruminococcus* strains *in vitro* and inhabit rumen for a very short time (Krause *et al.,* 2001).

An alternative approach is to create recombinant bacteria that can degrade fibre at low pH. It is well known that fibre digestion goes down in animals when pH declines to 6.5 due to grain diet feeding (Goad *et al.,* 1998). Russell and Wilson (1996) have proposed that the addition of fibrolytic to an acid-resistant species such as *Prevotella* would create far more competitive 'acidic niche' that the native bacteria is unable to fill.

## Fungus

Laccases, like other ligninolytic enzymes, are generally difficult to express in non-fungal as well as fungal system. Different codon usage, missing chaperons, and post-translational modifications such as disulfide bridges or glycosylation can all cause low expression levels and misfolded proteins that are degraded or driven into inclusion bodies (Bulter *et al.,* 2003). Still there is very few successful reports on laccase over expression and increase in enzyme efficiency ($k_{cat}$), which is a prerequisite for its wider application in recalcitrant substrate like lignin rich cattle feed. Laccase from *Myceliopthora thermophila* (MtL) was expressed in the functional form in *Saccharomyces cerevisiae.* Directed evolution improved the expression 8-fold, highest till now reported for laccase in yeast *i.e.* 18 mg/l (Bulter *et al.,* 2003). Together with a 22-fold increase in $k_{cat}$, the total activity was enhanced 170-fold. Expression in *S. cerevisiae* has also been reported for other laccase genes (Kojima *et al.,* 1990; Cassland *et al.,* 1999). Kojima *et al.* (1990) demonstrated expression of laccase from *Coriolus hirsutus* qualitatively. Yasuchi reported a 5 mg/l expression of *C. hirsutus* laccase (Yasuchi *et al.,* 1990). Even, *M. thermophila* laccase and *Rhizoctonia solani* laccase were mutated on a pentapeptide segment believed to be near type-1 Cu site (Xu *et al.,* 1998). The single mutation did not yield significant change in the enzyme, were as triple mutation resulted in several key alterations. The observed effects are interpreted as possible mutation-induced structural perturbations on the molecular recognition between the reducing substrate and laccase and on the electron transfer from the substrate to the type-1 Cu centre (Xu *et al.,* 1998). A cDNA encoding LccIV, a previously uncharacterized laccase isozyme of the white-rot basidiomycete *Trametes versicolor*, was expressed in the methylotrophic yeast *Pichia pastoris.* The LccIV isozyme is not expressed by *T. versicolor* under normal culture conditions and the enzyme was, therefore, investigated to determine whether it had any unusual properties. The native signal peptide of LccIV directed efficient secretion and correct proteolytic processing of LccIV to the mature form, whereas, substitution with the *Saccharomyces cerevisiae* mating factor signal peptide led to retention of an additional tetrapeptide at the amino-terminus of the secreted enzyme and 25%

lower specific activity in fermentor medium. Active LccIV was purified to homogeneity by sequential steps of ion-exchange, size-exclusion and hydrophobic interaction chromatography (Brown *et al.,* 2001). Collins and Dobson has investigated the regulation of laccase gene (*lcc*) expression in *T. versicolor* by using reverse transcription (RT)-PCR as a tool to determine the effects on *lcc* mRNA levels when the fungus was cultured under a variety of physiological conditions (Collins and Dobson, 1997). They demonstrated that *lcc* transcription was activated by copper and nutrient nitrogen. They also demonstrated that laccase induction by two aromatic compounds, 2,5-xylidine (XYL) and 1-hydroxybenzotriazole (HBT), occurred at the level of gene transcription. The addition of either copper or XYL to 10-day-old cultures grown in the absence of copper resulted in the rapid appearance of *lcc* mRNA, but a corresponding increase in laccase activity occurred only when copper was present.

The presence of multiple gene families for the secreted laccases requires systematic genetic analysis to elucidate their functions. Gene families probably produce closely related proteins that are subtly different in their activities, allowing the transformation of a wider range of substrates or showing differential regulation (Mansur *et al.,* 1997).

Transcription of *P. chrysosporium* strain BKMF-1767, LiP genes are dramatically modulated by culture conditions. Holzbaur and Tien (1988) examined transcript levels of the genes encoding isozymes H8 (*lip*A) and H2 (*lip*D) by northern blot hybridization. Under carbon limitation, *lip*D transcripts dominated and *lip*A transcripts were not detected. Under nitrogen limitation, *lip*A is the most abundant transcript and *lip*D expression was relatively low. Quantitative RT-PCR and nuclease protection assays have been developed to detect all known LiP transcripts (Stewart *et al.*, 1992; Reiser *et al.*, 1993). The genes *lip*A, *lip*B and *lip*I were expressed at similar levels in both C-and N-limited cultures. In contrast, *lip*C and *lip*J transcript levels were substantially increased under N-limitation. Consistent with Holzbaur and Tien (1988), *lip*D transcripts were the most abundant under C-limitation. Immunogold labelling experiments have identified LiPs and MnPs in *P. chrysosporium* colonized wood (Blanchette *et al.*, 1989; Daniel *et al.*, 1989, 1990) and Datta *et al.* (1991) partially purified MnP, LiP, and GLOX from colonized Aspen pulp. However, enzyme yields from solid wood samples are too low for precise isozyme identification, which become a bottleneck for the basic lignin degradation studies under controlled condition.

Aryl-alcohol oxidase (AAO), an extracellular flavoenzyme involved in the lignin biodegradation by some white-rot fungi. Varela *et al.* for the first time reported the expression of AAO from *P. eryngii* and in the ascomycete *Aspergillus nidulans.* The activity of the recombinant enzyme in *A. nudulans* cultures is much higher than found in the extracellular fluid of *P. eryngii* (Varela, 2001). Two roles have been proposed for AAO in degradation. One of them is in $H_2O_2$ production, which is generally used by peroxidases. In the case of *P. pulmonaris,* where the high rate of lignin mineralization of wheat straw correlates with the high level expression of both AAO and Mn-oxidizing peroxidase (Varela, 2001). Alternatively, $H_2O_2$ has also been suggested to be the precursor to hydroxyl radical, also proposed to be involved in the initial attack on lignin. The cooperation of AAO and laccase for the production of hydroxyl radical is also reported. The reduction of phenoxy radicals via AAO in turn prevents the polymerization of lignin during its biodegradation by ligninolytic enzymes as reported in *P. ostreatus* (Varela, 2001).

## GENOMICS AND METAGENOMIC APPROACHES

High-throughput DNA sequencing, modern instrumentation era and the advent of the '-omics' disciplines now offer the potential to obtain a complete blueprint for the microbial niche and to assess its genetic potential in a comparative and functional manner. The genome sequence of white-rot fungi (e.g. *P. chrysosporium*) and rumen microbes (e.g. *F. succinogens*, *R. albus*, and *P. ruminicola* strain 23) with reference to fibre degrading gene sequence is available (Krause *et al.*, 2003).

The use of PCR in conjunction with phylogenetically stable molecular markers like 16S rDNA and Internal Transcribing Spacers (ITS) as species-specific identifiers has provided us with the ability to detect single cell in microbial community, irrespective of their culturability. The recent surge of research in molecular microbial ecology provides compelling evidence for the existence of many novel types of organism and provides an entirely different approach for tapping into the potentially limitless resource of uncultured bacteria. The termite nest and the rumen is no different, and clone libraries based on 16S rDNA and ITS clearly suggest that a vast reservoir of physiologies are present in the rumen and termite nest (Froster *et al.*, 1998; Tajima 1999; Whitford *et al.*, 2001).

Handelsman *et al.* (1998) first coined the term 'metagenome' to describe genomic and associated functional analysis at the community level. The technology is still in its infancy and issues such as representative cloning, quantitative lysis and expression still need to be addressed. We have developed a totally benign technology in our laboratory that allows us to isolate community DNA from the diverse soil sample rich in humic acid and other inhibitors (Sharma *et al.*, unpublished). We have also developed a highly efficient, convenient and expeditious genetic transformation system for lignin degrading fungus (Sharma *et al.,* 2006). Our basic understanding on the topic overcomes the hurdles in lignin modification by collecting the methods that have been successfully applied to other species and determine how best to apply them to the target experiments. The transformation techniques described by Sharma *et al.* (2006) provides a reliable, quick, and simple method for using transgenic technology in the genetic improvement of industrially important white-rot fungi. The high transformation frequency, together with the precision and simplicity of T-DNA integration, makes T-DNA a suitable DNA element for genome mutagenesis approaches in white-rot fungi such as gene tagging, promoter entrapment, and gene activation, which can be exploited and implemented in cattle feed development. Consequently, *A. tumefaciens* mediated transformation of whole genome sequenced *P. chrysosporium,* may prove to be a valuable tool for the functional analysis of white-rot fungi.

The total number of prokaryotic cells on the earth has been estimated to be 4-6 $\times$ $10^{30}$ (Cowan *et al.*, 2005), thought to comprise between $10^6$ and $10^8$ separate genospecies (distinct taxonomic groups based on gene sequence analysis) (Cowan *et al.*, 2005). This diversity presents an enormous, largely untapped genetic and biological pool that can be exploited for the recovery of novel genes, entire metabolic pathways and their products (Cowan, 2000).

Recent advances in direct DNA cloning methodology have enabled the comprehensive expression of genes from the majority of genomes collected from complex consortia. Direct DNA extraction and generation of environmental DNA libraries provide a format for high-throughput expression screening for targeted activities and capture of unique enzymes. This unique method has resulted in the discovery of thousands of new enzyme with novel sequences and often with novel activities, stabilities, and operational optima. There have been reports of direct DNA extraction from insect gut to target the discovery of new xylanases with a view to understand their evolutionary relationship to other known xylanases and to explore its potential in industrial hemicellulose degradation, which can be applied in fodder improvement (Brennan *et al.*, 2004).

Libraries prepared from the intestinal tracts of termites and moths were screened to obtain unusual microbial xylanases. Four unique clones that generated clearing zone on azo linked xylan agar plate were isolated. Biochemical analysis revealed that they were true xylanases, though their sequences were remarkably different from the known xylanases (Brennan *et al.,* 2004).

Ferrer *et al.,* (2005) has reported novel hydrolase diversity retrieved from a metagenome library of bovine rumen microflora. The phylogenetic novelty of the hydrolases suggests that some of them may have potential for new applications in biocatalysis (Ferrer *et al.,* 2005). A very recent report on laccase in rumen may play an important role in the ryegrass lignin digestion, which has biotechnological potential for exploitation in pasture-fed animals (Beloqui *et al.,* 2006). Future of metagenomics is still due and needed to be explored for novel biocatalyst, which can overcome the lignin load from cattle feed.

## RUMEN ECOLOGY AND DIGESTIVE PHYSIOLOGY

*In vivo* trials have reported to find more microbial protein contents in rumen of cattle relative to buffalo, which was found to be stimulated by inorganic sulphur in diet. The ruminants did not differ for dry matter digestion, though buffalo digested more crude protein than cattle feed supplemented and concentrated with oat hay. It may presumably reflect more ruminal proteolytic activity in buffalo which was reported to have more of protozoa ($9.5 \times 10^5$/ml) and total viable bacterial count ($37.5 \times 10^8$/ml), however, ruminant species did not differ for concentration of cellulolytic bacteria and anaerobic fungi (Table 4). Improved rumen ecology in terms of protozoa and viable bacteria (cellulolytic), by the addition of sulphur do not confirm the increase in the digestibility of fibrous component. Previous reports have confirmed that physiological factor in addition to microbial profile may exert influence on digestive functions of ruminants (Sangwan *et al.,* 2001).

## CONCLUSION

This review has discussed the major problems we face during delignification of lignin rich conventional cattle feed with an holistic attempt to improve plant cell wall digestion *in vitro.* A semi-historical survey conserved the various attempts and strategies that have been employed to improve fibre digestion in the rumen. It seems clear during the literature survey that several major challenges the plant improvement of fibre is possible, provided the approach is based on a rational scientific basis and combinatorial effort, overlooked till now. Genomics technologies, including metagenomes, now provide microbiologists with their best opportunity, to date, to see the iceberg in its entirety, rather than just its tip, both a comparative and functional system. Clearly, genomics will provide a massive increase in the rate of information acquisition, and both novel and conventional methodologies and techniques must flourish for the potential of the genomics to be completely realised. The development of a feasible biological delignification process should be possible if ecophysiological requirements and optimal bioreactor design are effectively coordinated for efficient lignin-degrading microorganisms. Applying sufficient knowledge of both genetic tools as well as conventional biology of higher fungi and solid state fermentation, a low cost technology for the conversion of lignocellulosic into value added cattle feed could be developed. Finally our submission and proposal is a strategic, idealistic rationale to overcome the lignin load from cattle feed. Deciphering the complete lignin synthesis pathway along with other noted and benign approach in the review can offer a long awaited solution to lignocellulose problem.

## REFERENCES


Abbott, J.C., Barakate, A., Pincon, G., Legrand, *et. al.* (2002). Simultaneous suppression of multiple genes by single transgenes. Down regulation of three unrelated lignin biosynthetic genes in tobacco. *Plant Physiology*, 128, 844-853.

Atanassova, R., Favet, N., Matz, F., Chabbert, B., Tollier, M.T., Monties, B., Fritig, B., and Legrand, M. (1995). Altered lignin composition in transgenic tobacco expressing *O*-methyltransferase sequences in sense and antisense orientation. *Plant Journal*, 8, 465-477

Barr, D.P. and Aust, S.D. (1994). *Environmental. Science and Technology,* 28, 78-87.

Basu, S., Gaur, R., Gomes, J., Sreekrishnan, T.R., and Bisaria, V. (2002). Effect of seed culture on solid-state bioconversion of wheat straw by *Phanerochaete chrysosporium. J. Bioscience and Bioengneering,* 93(1), 25-30.

Baucher, M., Halpin, C., Petit-Conil, M., and Boerjan, W. (2003). Lignin: Genetic engineering and impact on pulping. *Critical Review in Biochemistry and Molecular Biology,* 38, 305-350.

Baldwin, R.L. and Palmquist, D.L. (1965). Effect of diet on the activity of several enzymes in extracts of rumen microorganisms. *Applied Microbiology,* 13, 194-200.

Beauchemin, K.A. and Rode, L.M. (1996). Use of feed enzymes in ruminant nutrition. In, *Animal Science Research and Development-Meeting Future Challenges,* (Rode, L.M., Ed.)., pp. 103-131. Minister of Supply and Services Canada, Ottawa, ON.

Beauchemin, K.A., Rode, L.M. and Sewalt, V.J.H. (1995). Fibrolytic enzymes increase fiber digestibility and growth rate of steers fed dry forages. *Canadian Journal of Animal Science,* 75, 641-644.

Bernard, L., Chaise, J.P., Delval, E. and Poncet, C. (1998). Validation of the main modeling methods for the estimation of marker mean retention times in the different compartments of the gastrointestinal tract in sheep. *Journal of Animal Science,* 76, 2485-2495.

Beloqui, A., Pita, M., Polaina, J., Martinez, A., Golyshina, O.V., Zumarraga, M., Yakimov, M.M., Garcia-Arellano, H., Alcalde, M., Fernandez, V.M., Elborough, K., Andreu, J.M., Ballesteros, A., Plou, F.J., Timmis, K.N., Ferrer, M. and Golyshin, P.N. (2006). Novel polyphenol oxidase mined from a metagenome expresion library of bovine rumen: Biochemical properties, structural analysis and phylogenetic relationships. *Journal of Biological Chemistry,* (in press).

Burroughs, W., Woods, W., Ewing, S.A., Greig, T. and Theurer, B. (1960). Enzyme additions to fattening cattle rations. *Journal of Animal Science,* 19, 458-464.

Borneman, W.S., Hartley, R.D., Morrison, W.H., Akin, D.E. and Ljungdahl, L.G. (1990). *Applied Microbiology and Biotechnology,* 33, 345–351.

Boudet, A.M., Chabannes, M., Barakate, A., Lapierre, C., Marita, J.M., Ralph, J., Pean, M., Danoun, S., Malphin, C. and Grima-Pettenai, J. (2001). Strong decrease in lignin content without significant alteration of plant development is induced by simultaneous down-regulation of CCR and cinnamyl alcohol dehydrogenase in tobacco plants. *Plant J.* 28. 257-270.

Breen, A., Singleton, F.L. (1999). Fungi in lignocellulose breakdown and biopulping. *Current Opinion Biotechnology,* 10, 252-258

Brennan, Y., Callen, W.N., Christoffersen, L., *et al.*, (2004). Unusual microbial xylanase from insect guts. *Applied and Environmental Microbiology,* 70(6)., 3609-3617.

Brown, M.A., Zhao, Z., Mauk, A.G. (2002). Expression and characterization of a recombinant multi-copper oxidase, laccase IV from *Trametes versicolor Inorganica Chimica Acta,* 331 232–238

Boudet, A.M. (2000). Lignins and lignification, Selected issues. *Plant Physiology Biochemistry,* 38, 81-96.

Chabannes, M., Ruel, K., Yoshinaga, A., Chabbert, B., Jauneau, A., Joseleau, J.P. and Boudet, A.M. (2001). In situ analysis of lignins in transgenic tobacco reveals a different impact of individual transformation on the spatial patterns of lignin deposition at the cellular and subcellular levels. *Plant Journal,* 28, 271-282.

Clark, J.D., Dyer, I.A. and Templeton, J.A. (1961). Some nutritional and physiological effects of enzymes for fattening cattle. *Journal of Animal Science,* 20, 928-938.

Cotta, M.A., Whitehead, T.R. and Rasmussen, M.A. (1997). Survival of the recombinant *Bacteroides thetaiotaomicron* strain BTX in *in vitro* rumen incubations. *Journal of Applied Microbiology,* 82, 743-750.

Collins, P.J. and Dobson, A.W. (1997). Regulation of Laccase Gene Transcription in *Trametes versicolor Applied and Environmental Microbiology,* 63 (9). 3444-3450.

Chesson, A. (1981). *J. Sci. Food Agric.* 32, 745–758.

Cowan, D.A. (2000). Microbial genomes – the untapped resource. *Trends in Biotechnology,* 18, 14-16.

Cowan, D., Meyer, Q., Stafford, W., Muyanga, S., Cameron, R., and Wittwer, P. (2005). Metagenomic gene discovery, past, present and future. *Trends in Biotechnology,* 23 (6), 321-329.

Cullen, D. (1997). Recent advances on the molecular genetics of lignolytic fungi. *Journal of Biotechnology,* 53, 273-289.

Cooney, C.L., Rha, C., and Tannenbaum, S.R. (1980). Single–cell protein, engineering, economics and utilization in foods. *Advances in food research,* 26, 1-52.

Dean, J.F.D., La Fayette, P.R., Rugh, C., Tristam, A.H., Hoopes, J.T., Eriksson, K.E.L., and Merckle, S.A. (1998). Laccases associated with lignifying vascular tissues, in, Lewis, N.G., Sarkanen, S. (Eds.)., Lignin and lignin biosynthesis, *American Chemical Society,* pp. 96-108.

De Candolle, A.P., De la lignine, in, *Physiologie vegetale, Bechet Deterville,* Paris, (1832). pp. 194-201.

Dixon, R.A., Guo, D., Chen, F., Inou, K. and Blound, J.W. (2001). Down-regulation of caffeic acid 3-O-Methyltransferase and Caffeoyl CoA 3-O-Methyltransferase in transgenic alfalfa: Impacts on lignin structure and implications for the biosynthesis of G and S lignin. *Plant Cell,* 13. 73-88.

Eriksson, K.-E.L., R.A. Blanchette, and P. Ander. (1990). Microbial and enzymatic degradation of wood and wood components**.** *Springer Verlag, Berlin, Germany.*

Ferrer, M., Golyshina, O.V., Chernikova, T.V., Khachane, A.N. *et al.*, (2005). Novel hydrolase diversity retrieved from a metagenome library of bovine rumen microflora. *Environmental Microbiology,* 7(12)., 1996-2010.

Forster, R.J., Whitford, M.F., Teather, R.M. and Krause, D.O. (1998). Investigations into rumen microbial diversity using molecular cloning and probing techniques. *In, Genetics, Biochemistry, and Ecology of Cellulose Degradation* (Onodera, R., Itabashi, H., Ushida, K., Yano, H. and Sasaki, Y., Eds.)., pp. 16-24. Sukuka, Japan.

Feng, P., Hunt, C.W., Pritchard, G.T. and Julien, W.E. (1996). Effect of enzyme preparations on in situ and in vitro degradation and in vivo digestive characteristics of mature cool-season grass forage in beef steers. *Journal of Animal Science,* 74, 1349-1357.

Goad, D.W., Goad, C.L. and Nagaraja, T.G. (1998). Ruminal microbial and fermentative changes associated with experimentally induced subacute acidosis in steers. *Journal of Animal Science,* 76, 234-241.

Gobius, K.S., Xue, G.P., Aylward, J.H., Dalrymple, B.P., Swadling, Y.J., McSweeney, C.S. and Krause, D.O. (2002). Transfromation and expression of an anaerobic fungal xylanase in several strains of the rumen bacterium *Butyrivibrio fibrisolvens. Journal of Applied Microbiology,* 93, 122-133.

Glenn, J.K. and Gold, M.H. (1985). *Archives of Biochemistry and Biophysics,* 242, 329-341.

Glenn, J.K., Morgan, M.A., Mayeld, M.B., Kuwahara, M. and Gold, M.H. (1983) *Biochem. Biophys. Res. Comm.,* 114, 1077-1083.

Hristov, A.N., McAllister, T.A. and Cheng, K.J. (1998). Stability of exogenous polysaccharide-degrading enzymes in the rumen. *Animal Feed Science Technology,* 76, 161-168.

Hristov, A.N., McAllister, T.A. and Cheng, K.J. (2000). Intraruminal supplementation with increasing levels of exogenous polysaccharide-degrading enzymes, effects on nutrient digestion in cattle fed a barley grain diet. *Journal of Animal Science,* 78, 477-487.

Hristov, A.N., Rode, L.M., Beauchemin, K.A. and Wuerfel, R.L. (1996). *Effect of a commercial enzyme preparation on barley silage in vitro and in sacco dry matter degradability.* Proc. West. Sec. Am. Soc. Anim. Sci. Rapid City, South Dakota 47, 282-284.

Handelsman, J., Rondon, M.R., Brady, S.F., Clardy, J. and Goodman, R. M. (1998). Molecular biological access to the chemistry of unknown soil microbes, a new frontier for natural products. *Chem. Biol.* 5, R245-R249.

Hatakka, A. (1994). *FEMS Microbiology Reviews.* 13,125-135.

Hatakka, A. (2001). In *Biodegradation of lignin.* (eds M. Hofrichter and A. Steinbüchel). Lignin, *Humic Substances and Coal,* Vol. 1, Wiley-VCH, Weinheim, Germany. pp.129-180.

Hatakka, A.I. and Uusi-Rauva, A.K. (1983). *European Journal of Applied Microbiology and Biotechnology.* 17, 235-242.

He, X., Hall, M.B., Gallo-Meagher, M., and Smith, R.L (2003). Improvement of forage quality by downregulation of maize *O*-methyltransferase. *Crop Science.* 43, 2240-2251

Homb, T. (1984). In: F. Sundstol and E. Owen (Ed.) *Straw and other fibrous by-products as feed.* Elsevier Science Publishers, Amesterdam.

Hofrichter, M., Scheibner, K., Schneegass, I., and Fritsche, W. (1998). *Applied and. Environmental Microbiology,* 64, 399-404.

Ibrahim, M.N.M., Ketelaar, R.S., Tamminga, S. and Zemmelink, G. (1987). Degradation characteristics of untreated and urea-treated rice straw in the rumen. *In* RM. Dixon, ed. Ruminant feeding systems utilizing fibrous agricultural residues. *Proceedings of the Seventh Annual Workshop of the Australian-Asian Fibrous Agricultural Residues Research Network.* Thailand, p. 123–126.

Jones, L., Ennos, A.R. and Turner, S.R. (2001). Cloning and characterization in irregular xylem 4 (irs4). a severely lignin deficient mutant of Arabidopsis. *Plant Journal,* 26, 205-216.

Jung, H.J., Ni, W., Chapple, C.C.S. and Meyer, K. (1999). Impact of lignin composition on cell wall degradability in an Arabidopsis mutant. *J. Sci. Food Agric.* 79, 922-928.

Krause, D.O., Bunch, R.J., Dalrymple, B.D., Gobius, K.S., Smith, W.J.M., Xue, G.P. and McSweeney, C.S. (2001). Expression of a modifed Neocallimastix patriciarum xylanase in *Butyrivibrio fibrisolvens* digests more fiber but cannot effectively compete with highly fibrolytic bacteria in the rumen. *Journal of Applied Microbiology,* 90, 388-396.

Kamra, D.N., and Zadrazil. (1988). Microbiological improvement of Lignocellulosics in animal feed production, p. 56-63. In F. Zadrazil and P. Reiniger (ed.)., *Treatment of Lignocellulosics with white-rot fungi.* Elsevier Applied Science, London, United Kingdom.

Kapoor, R.K., Sharma, K.K., Kuhar, S., and Kuhad, R.C. (2005). Diversity of lignin degrading microorganisms, ligninolytic enzymes and their biotechnological applications. In T. Satyanarayana and B.N. Johri (ed). I.K. International Pvt. Ltd.., New Delhi. 815-846.

Kirk, T.K. (1990). *Biochemistry Journal 268,* 475–480.

Krause, D.O., Denman, S.E., Mackie, R.I., Morrison, M., *et al.*, (2003). Opportunities to improve fiber degradation in the rumen, microbiology, ecology, and genomics. *FEMS Microbiology Reviews.* 27, 663-693.

Kuhad, R.C., Singh, A., Tripathi, K.K., Saxena, R.K. and Eriksson, K.E. (1997). Microorganisms as an alternative source of protein. In, Topics in food science and nutrition, *Nutrition Reviews,* 55(3)., 65-75.

Kuhad, R.C., Singh, A., and Eriksson, K.E. (1997). In K.E. Eriksson (ed). *Advances in Biochemical Engineering Biotechnology,* Springer-Verlag, Germany. 46-125.

Leonowicz, A., Sklarz, G. and Wojtas-Wasilewska, M. (1985). *Phytochem.* 24, 393–396.

Li, L., Zhou, Y., Cheng, X., Sun, J., Marita, J.M., Ralph, J., and Chiang, V. L. (2003). Combinatorial modification of multiple lignin traits in trees through multigene cotransformation. *PNAS.* 100(8)., 4939-4944.

Lundell, T. (1993). Ligninolytic System of the White-rot Fungus *Phlebia radiata, Lignin Model Compound Studies.* University of Helsinki, Helsinki, Finland.

Lewis, G.E., Sanchez, W.K., Hunt, C.W., Guy, M.A., Pritchard, G.T., Swanson, B.I. and Treacher, R.J. (1999). Effect of direct-fed fibrolytic enzymes on the lactational performance of dairy cows. *Journal of Dairy Science.* 82, 611-617.

McAllister, T.A., Oosting, S.J., Popp, J.D., Mir, Z., Yanke, L.J., Hristov, A.N., Treacher, R.J. and Cheng, K.J. (1999). Effect of exogenous enzymes on digestibility of barley silage and growth performance of feedlot cattle. *Canadian Journal of Animal Science,* 79, 353-360.

McAllister, T.A., Stanford, K., Bae, H.D., Treacher, R.J., Hristov, A.N., Baah, J., Shelford, J.A. and Cheng, K.J. (2000). Effect of a surfactant and exogenous enzymes on digestibility of feed and on growth performance and carcass traits of lambs. *Canadian Journal of Animal Science,* 80, 35-44.

Mansur, M., Suarez, T., Fernandez-Larrea, J.B., Brizuela, M.A. & Gonzalez, A. E. (1997). Identification of laccase gene family in the new lignin degrading basidiomycete CECT 20197. *Applied and Environmental Microbiology,* 63, (7). 2637-2646.

Martinez, A.T., Camarero, S., Guillen, F., Gutierrez, A., Munoz, C., Varela, E., Martinez, M.J., Barrasa, J.M., Ruel, K. and Pelayo, J.M. (1994). *FEMS Microbiology Reviews.* 13, 265-274.

Messerschmidt, A., Ladenstein, R., Huber, R., Bolognesi, M. Avigliano, L., Petruzelli, R., Rossi, A. & Finazzi-Agro, A. (1992). Refined crystal structure of ascorbate oxidase at 1.9 A resolution. *Journal of Molecular Biology.* 224, 179-205.

Nianogo, A.J., Louis, S.L., Solaiman, S., Ouédraogo, C.L., and Siaway, A. (1999) Effect of urea treatment on digestibility and utilization of sorghum straw. *Biotechnol. Agro. Soc. Environ.* 3 (2), 78-85.

Pilate, G., Guiney, E., Holt, K., Petit-Conil, M., *et al.,* (2002). Field and pulping performances of transgenic trees with altered lignification. *Nature Biotechnology.* 20, 607-612.

Pointing, S.B. (2001). *Applied Microbiology and Biotechnology,* 57, 20-33

Perry, T.W., Cope, D.D. and Beeson, W.M. (1960). Low vs. high moisture shelled corn with and without enzymes and stilbestrol for fattening steers. *Journal of Animal Science,* 19, 1284-1294.

Perry, T.W., Purkhiser, E.D. and Beeson, W.M. (1966). Effects of supplemental enzymes on nitrogen balance, digestibility, or energy and nutrients, and on growth and feed efficiency of cattle. *Journal of Animal Science,* 25, 760-764.

Rexen, F.P. and Bach-Knudsen, K.E. (1984). In: F. Sundstol and E. Owen (Ed.) Straw and other fibrous by-products as feed. *Elsevier Science Publishers,* Amesterdam.

Russell, J.B. and Wilson, D.B. (1996). Why are ruminal cellulolytic bacteria unable to digest cellulose at low pH? *Journal of Dairy Science,* 79, 1503-1509.

Ragusa, S., Cambria, M.T., Pierfederici, F., Scire,A., Bertoli, E., Tanfani, F. & Cambria, A. (2002).. Structure-activity relationship on fungal laccase from *Rigidoporus Lignosus*, a Fourier-transform infrared spectroscopic study. *Biochimica et Biophysica Acta,* 1601, 155-162.

Reinhammar, B. (1984). *In copper proteins and copper enzymes* Lontie, R., Ed.; CRC Press Vol. III.

Sangwan, D.C., Bhatia, S.K., Kumar, S., and Singh, S. (2001). Degradative and metabolic activity of rumen fungi and cellulolytic bacteria in cattle and buffalo. *Technical Bulletin,* 55-58.

Sharma, K.K., Gupta, S and Kuhad, R.C. (2006). *Agrobacterium*-mediated delivery of marker genes to *Phanerochaete chrysosporium* mycelial pellets, A model transformation system for white-rot fungi. *Biotechnology and Applied Biochemistry,* 43, 181-186.

Sharma, K.K., Kapoor, M., and Kuhad, R.C. (2005). *In vivo* enzymatic digestion, *in vitro* xylanase digestion, metabolic analogues, surfactants and polyethylene glycol ameliorate laccase production from *Ganoderma* sp. kk-02. *Letters in Applied Microbiology,* 41, 24-31.

Solomons, G.L. (1983). Single cell protein. *Critical Reviews in Biotechnology,* 1, 21-58.

Sundstol, F. (1988). Livestock production science, 19: 137-158.

Theander, O. and Aman, P. (1984). In: F. Sundstol and E. Owen. *Straw and other by-product as feed.* Elsevier, Amsterdam pp. 45.

Tajima, K., Aminov, R.I., Nagamine, T., Ogata, K., Nakamura, M., Matsui, H. and Benno, Y. (1999). Rumen bacterial diversity as determined by sequence analysis of 16S rDNA libraries. *FEMS Microbiology and Ecology,* 29, 159-169.

Vasdev, K. and Kuhad, R.C. (1994). *Folia Microbiologia,* 39(1)., 61-64.

Vasdev, K., Kuhad, R.C. and Saxena, R.K. (1995). *Current Microbiology,* 30,1-4.

Viniegra-González, G., Favela-Torres, E., Aguilar, C.N., Rómero-Gomez, S.D.J., Díaz-Godínez, G. and Augur, C. (2002). Advantages of fungal enzyme production in solid state over liquid fermentation conditions. *Biochemical Engineering Journal,* 3643, 1-11.

Varela, E., Guillen, F., Martinez, A.T., and Martinez, M.J. (2001). Expression of Pleurotus eryngii aryl-alcohol oxidase in *Aspergillus nidulans*, purification and characterization of the recombinant enzyme. *Biochimica et Biophysica Acta,* 1546, 107-113.

Whitford, M.F., Teather, R.M. and Forster, R.J. (2001). Phylogenetic analysis of methanogens from the bovine rumen. *BMC Microbiology,* 1, 5.

Zadrazil, F. (1984). In: In: F. Sundstol and E. Owen. Straw and other by-product as feed. *Elsevier,* Amsterdam. pp. 276.

Zadrazil, F., Grinbergs, J. and Gonzalez, A. (1982). Palo-podrido decomposed wood used as feed. *European Journal of Applied Microbiology and Biotechnology,* 15, 167-171.

# PART IV

## BIOTECHNOLOGICAL APPLICATIONS

# 15

# Bioethanol from Crop Residues, Production Forecasting and Economics: An Indian Perspective

**Rajeev Kumar Kapoor[1], Anuj Kumar Chandel[1], Sarika Kuhar[1], Rishi Gupta[1], and Ramesh Chander Kuhad[1]**

*[1]Lignocellulose Biotechnology Laboratory, Department of Microbiology, University of Delhi, South Campus, New Delhi-110021, India.*

Sugarcane, wheat, rice, maize, oat, sorghum and barley are the major crops grown in India. About 432 million tons (Mt) agroresidue was generated in the year 2004 and the annual ethanol production from these crop residues is estimated to be 123.4 Gigalitre (GL). However, only approximately 313.0 Mt of crop residue is left after their use as fodder, which can be converted to 89.42 GL of bioethanol. In terms of availability, sugarcane bagasse topped the figure and has been found the most promising feedstock, which can produce 50.1 GL of bioethanol followed by rice straw (35.0 GL) and wheat straw (27.3 GL). In addition to this, a total of 11.02 GL of bioethanol can be produced from the other crop residues like Sorghum (4.57 GL), Maize (3.56 GL), Oat (2.36 GL) and Barley (0.53 GL). Since India produces only 27% of its total crude oil requirement and oil reserves are depleting globally, the exploitation of these crop residues for producing fuel ethanol will be favourable for improving not only the national economy but environment safety also. Indian government has taken initiatives to explore the bioethanol production from agricultural feedstock and make it mandatory to blend 5% in fossil fuel for the supply in nine states and four union territories.

## INTRODUCTION

Developments in biotechnology have made bioethanol production feasible not only from conventional substrate like sugarcane molasses but also from lignocellulosics. Bioethanol from cellulose generates 8 to 10 time net energy as is required for its production (Griffiths and Atlas, 2005). The burning of fossil fuels at the current rate is likely to create an environmental crisis globally. According to CSE (Center for Science and Environment, Government of India) report, vehicular pollution

**Keywords:** Bioethanol, crop residues, forecasting, India

contributes to a shocking 64% of the total pollution in Delhi, 52% in Mumbai and 30% in Kolkatta. In just last seven years, the number of vehicles in India has touched to 17 million and surprisingly 48% of these vehicles are in three metros alone. Use of fossil fuel generates carbon dioxide ($CO_2$), methane ($CH_4$), and a significant quantity of nitrous oxide ($N_2O$). Moreover, it produces nitrogen oxides ($NO_2$), hydrocarbons, and carbon monoxide (CO), which, though not greenhouse gases themselves, influence chemical cycles in the atmosphere that produce or destroy greenhouse gases, such as tropospheric ozone. Most of these harmful gases are formed due to incomplete combustion of fossil fuel, while ethanol contains 35% oxygen that helps the complete combustion of fuel and thus reduces harmful tailpipe emissions. It also reduces particulate emissions that pose a health hazard. Usage of bioethanol can play a role in reducing the harmful gases in the environment reducing ozone precursor by 20-30%. The increased usage of the fossil fuel cannot be contained although this is a well-known fact that the decline in the worldwide crude oil production would start before 2010 and can be exhausted completely by year 2050 (Campbell and Laherrere, 1998). Hence, there is a great need to explore new renewable energy sources.

Advances in technology have paved the way for conversion of biomass into fuel ethanol as a potential source of energy. Lignocellulosics being the most abundant biomass available on earth, offers a potential source of carbon substrates for the production of ethanol by fermentation (Lynd, 1990; Wyman, 1994), and have attracted considerable attention as an alternate feedstock and energy resource (Kuhad and Singh, 1993). Researchers in the last decades have demonstrated that lignocellulosics require extensive processing / hydrolysis of the raw material into fermentable sugars and its subsequent fermentation into a myriad of fuels and chemicals (Knauf and Moniruzzaman, 2004). Development of a bioconversion process, which utilizes low value lignocellulosics, offers a great potential for reducing the production cost and increasing the use of ethanol as a fuel additive. The world ethanol production in 2001 was 31 GL (Berg, 2001). The main producers of ethanol are Brazil and USA, which accounts for 62% of the world production. The major feedstock for ethanol production in Brazil is sugarcane, while in USA it is corn stover. In India, lignocellulosic biomass as crop residues is available in plenty. In most of the developed countries, infrastructure enabling a fast market penetration of low blended ethanol fuels exists (Sanchez *et al.,* 2004). Renewable fuels particularly ethanol should get more attention in India. This paper describes the availability of residues from major agriculural crops in India and a preliminary data of their conversion into ethanol as biofuel and its prospects.

## EFFECT OF ETHANOL ON INDIAN ECONOMY

India is not self-sufficient in petroleum and imports about two-thirds of its requirement. In the year 2003-04, India produced 33.4 million tons of oil and imported 90.4 million tons. The import cost in 2003-2004 was around Rs. 81,000 crores (Table 1). If the oil prices goes up the import cost also goes up immediately, apart from the demand pull price increase due to the continuing increase in crude oil purchase, is in the range of Rs. 800 billion/year influencing in a big way the countries foreign exchange reserves. US light crude oil touched a recent peak of $70.80 on 31st August 2005. A blend of ethanol in fossil fuel will, therefore, make a big difference both in our import as well as in environmental management.

**Table 1. Production and import of crude oil in India**

| Year | Production (Mt) | Import (Mt) | Import as % of total | Total (Mt) | Import value Crore (Billion) | U.S. Dollar ($) |
|---|---|---|---|---|---|---|
| 1971 | 6.8 | 11.7 | 18.5 | 63 | 107 | 0.024 |
| 1981 | 10.5 | 16.2 | 26.7 | 61 | 3349 | 0.744 |
| 1991 | 33 | 20.7 | 53.7 | 39 | 6118 | 1.360 |
| 2000 | 32 | 57.9 | 89.9 | 64 | 30,695 | 6.821 |
| 2003-04 | 33.4 | 90.4 | 123.8 | 73 | 81,000 | 18 |
| 2004-05 | 33.4 | 100 | 133.4 | 75 | 1,21,500 | 27 |

Source: Ministry of Petroleum and Natural gas

The benefits of converting biomass to ethanol as stated previously include increased national energy security, reduction in greenhouse gas emissions, and use of renewable resources, foundation of a carbohydrate-based chemical process industry, macroeconomic benefits for rural communities and society at large (Wyman, 2003; Subramanian *et al.,* 2005). The energy content of ethanol is higher than the energy required for producing ethanol (Griffiths and Atlas, 2005; Shapouri *et al.,* 2002) however, one liter of E-85 a mixture of 85% ethanol and 15% gasoline by volume, when used in a mid size passenger car could replace 0.72 L of gasoline (Kim and Del, 2004).

In India, lignocellulosics are abundantly available, which may be successfully converted to bioethanol for partially supplementing gasoline or other petroleum products. Theoretically, one dry ton of bagasse can generate 112 gallons of ethanol (Knauf and Moniruzzaman, 2004). Apart from the conventional crop residues, new energy crops (sunflower stalks, alfalfa fibers) and weeds (*Eichhornia sp.*) have been identified as potential bioethanol production sources (Kim and Del, 2004; Sharma *et al.,* 2002; Nigam, 2002). More recently, wild plants like *Lantana camara*, *Prosopis juliflora* and *Saccharum spontaneum* have been found as potential source of fermentable sugars, which can be subsequently converted into fuel ethanol (unpublished data).

## DATA ACQUISITION

The facts and figures for the production of ethanol from potential crops were from the Department of Agriculture and Cooperation, Ministry of Agriculture, Government of India. The estimation of crop residues from total grains production was worked out using the ratio of grain / crop residue (Kim and Del, 2004). The ethanol production from different crop residues was calculated / ascertained from US Department of Energy website, which provides "Theoretical ethanol yield calculator" at http://www.eere.energy.gov/biomass/ethanol_yield_calculator.html.

## COMPOSITION OF LIGNOCELLULOSIC BIOMASS

Basically, the lignocellulosic biomass comprises of cellulose, hemicellulose and lignin (Hayn *et al.,* 1993; Kuhad *et al.,* 1997), Cellulose is a polymer of glucose with a repeating unit of $C_6H_{10}O_5$ strung together by β-glycosidic linkages. Hemicellulose consists of short, highly branched chains of sugars. In contrast to cellulose, which is a polymer of only glucose, a hemicellulose is a polymer of five different sugars. It contains five-carbon sugars (usually D-xylose and L-arabinose) and six-carbon sugars D-galactose, D-glucose, and D-mannose. Table 2, shows the composition of the selected crop residues. Carbohydrates (hexosans and pentosans) account for approximately 50-70% with lignin 9-19%.

**Table 2. Composition of lignocellulosic biomass of selected crop.**

| *Crop residue* | *Residue/crop ratio* | *Dry matter %* | *Lignin %* | *Carbohydrates %* | *Ash %* | *Ethanol yield l/kg of dry biomass* |
|---|---|---|---|---|---|---|
| Barley straw | 1.2 | 81.0 | 9.00 | 70.00 | 11.00 | 0.31 |
| Corn stover | 1 | 78.5 | 18.69 | 58.29 | 6.00 | 0.29 |
| Oat straw | 1.3 | 90.1 | 13.75 | 59.10 | 12.00 | 0.26 |
| Rice straw | 1.4 | 88.0 | 7.13 | 49.33 | 12.00 | 0.28 |
| Sorghum straw | 1.3 | 88.0 | 15.00 | 61.00 | 10.00 | 0.27 |
| Wheat straw | 1.3 | 90.1 | 16.00 | 54.00 | 10.00 | 0.29 |
| Sugarcane bagasse | 0.6 | 71 | 14.5 | 67.15 | 4.00 | 0.28 |

## TECHNOLOGICAL OPTIONS FOR EFFICIENT CONVERSION OF LIGNOCELLULOSICS INTO ETHANOL

The bioconversion of lignocellulosics to ethanol includes three processes: (a) depolymerization of structural polysaccharide into fermentable reducing sugars, (b) fermentation of these sugars subsequently into ethanol and (c) ethanol recovery.

### Pretreatment

The lignin hemicellulosic network retards cellulose biodegradation by cellulolytic enzymes. In order to remove the protecting lignin and to increase the accessibility of the cellulolytic enzymes to cellulose, the lignocellulosic biomass must be pretreated. Pretreatment affects the structure of biomass by solubilizing or otherwise altering hemicellulosic, lignin structure, reducing cellulose crystallinity and increasing the available surface area and pore volume of the substrate. Pretreatment can be carried out in different ways such as mechanical comminution, steam explosion, ammonia fibre explosion, acid or alkaline pretreament and biological treatments (Table 3).

**Table 3. Pretreatment technologies available for the hydrolysis of lignocellulosics.**

| *Class* | *Method known as* | *References* |
|---|---|---|
| Physical method | Mechanical comminution | 18 |
| | Pyrolysis | 19 |
| Physico-chemical pretreatment | Steam explosion (auto hydrolysis) | 20 |
| | Ammonia fibre explosion (AFEX) | 21 |
| | Ammonia recycled percolation (ARP) | 22 |
| | $CO_2$ Explosion | 23 |
| Chemical pretreatment | Oxidative delignification | 24 |
| | Ozonolysis | 25 |
| | Organosolve process | 26 |
| | Alkaline hydrolysis | 27 |
| | Acid hydrolysis | 28, 29 |
| Biological pretreatment | Pretreatment with white-rot fungi | 30 |
| | Fungal pretreatment | 31 |

Acid hydrolysis of lignocellulosics releases xylose as the main sugar fraction in hydrolysates. Unfortunately, these hydrolysates also contain several inhibitors like; furan derivatives resulted from degradation of sugars, aliphatic acids released from hemicellulosic acetyl groups, and phenolic's from lignin. In order to remove the inhibitors and increase the hydrolysate fermentability, several chemicals, biological, physical methods have been used. These methods include overliming (Martinez *et al.,* 2000), charcoal adsorption (Rodrigues *et al.,* 2001), ion exchange (Nilvebrant *et al.,* 2001), enzymatic detoxification with laccase (Martin *et al.,* 2002), and biological detoxification (Lopez *et al.,* 2004).

### Enzymatic Hydrolysis

The acid, alkaline or fungal pretreated lignocellulosics can be saccharified enzymatically to get fermentable sugars (Kuhad *et al.,* 1997; Itoh *et al.,* 2003; Ghose and Bisaria, 1979; Tucker *et al.,* 2003). Bacteria and fungi are the good sources of cellulases, hemicellulases that could be used for the hydrolysis of pretreated lignocellulosics. The enzymic cocktails are usually mixtures of several hydrolytic enzymes comprising cellulases, xylanases and mannanases (Duff and Murray, 1996). The most important process improvement made for the enzymatic hydrolysis of biomass was the introduction of simultaneous saccharification and fermentation (SSF), which has been improved to include the co-fermentation of multiple sugar substrates. This new variant of SSF is now known as simultaneous saccharification and co-fermentation (SSCF).

## FERMENTATION OF LIGNOCELLULOSIC HYDROLYSATE

Bioconversion of various agroresidues into ethanol by different yeasts has been summarized in Table 4. The sugar syrup obtained after thermochemical or enzymatic hydrolysis is used for ethanol fermentation. The ability to ferment pentoses along with hexoses is not widespread among microorganisms (Toivola *et al.,* 1984). The ideal organism for the production of ethanol from lignocellulosic hydrolysate would be the one, which can utilize various forms of sugars generated by lignocellulose hydrolysis. *Saccharomyces cerevisiae* is capable of converting only hexose sugars to ethanol. Of the various xylose-fermenting yeasts, *Candida shehatae* has shown greater ethanol production than *Pichia stipitis.* This was due to increased uptake of xylose, glucose, mannose, and galactose (Abbi *et al.,* 1996a; Abbi *et al.,* 1996b). However, genetically engineered organisms are now being employed for ethanol fermentation. These organisms can greatly improve ethanol production efficiency and reduce the cost of operation (Toivola *et al.,* 1984; Dien *et al.,* 2000). The recombinant strains of *E. coli* with the genes from *Zymomonas mobilis* for the conversion of pyruvate to ethanol have been constructed. The recombinant plasmids with xylose reductase and xylitol dehydrogenase genes from *P. stipitis* and xylulokinase gene from *S. cerevisiae* have been transformed into *Saccharomyces* sp. for the co-fermentation of glucose and xylose (Sun and Cheng, 2002). Though new technologies have greatly improved bioethanol production, yet there are still a lot of problems that have to be solved. The major problems include maintaining a stable performance of genetically engineered yeast in commercial scale fermentation operations (Ho *et al.,* 1998), developing more efficient pretreatment technologies for lignocellulosic biomass, and integrating optimal components into an economic ethanol production system (Dien *et al.,* 2000).

**Table 4. Various agro residues for which ethanol production has been reported**

| *Agro residue* | *Microorganism used* | *References* |
|---|---|---|
| *Barley husk* | *S. cereviseae* | 40 |
| *Corn stover* | *Z. mobilis* | 41 |
| *Rice straw* | *C. shehatae NCL-3501* | 42, 43 |
| *Sugarcane bagasse* | *S. cereviseae* | 35 |
| *Sorghum* | *S. cereviseae* | 44 |
| *Sun flower stalk* | *S.cereviseae var.ellipsoideus* | 45 |
| *Wheat straw* | *P. stipitis NRRL-Y-7124* | 46 |
| *Sugarcane leaves* | *K. fragilis NCIM-3358* | 47 |

## PRELIMINARY DATA FOR ETHANOL PRODUCTION FROM INDIVIDUAL CROP RESIDUE

Depending upon the ratio of hexosans (glucan, galactan and mannan) and pentosans (xylan, arabinan) ethanol yield from undermentioned crop residue varies. The availability of various crop residues in India is depicted in Table 5, along with the estimated production of ethanol from each of the residue. The following section describes the availability of major crop residues and the estimated production of bioethanol.

**Table 5. Crop residue availability in India.**

Grain production and crop residue: Million tones
Area: Million Hectares
Crop residue and Grain yield: kg / hectare
Ethanol production: GL

| *Crop* | *Grain production* | *Crop residue production* | *Area* | *Grain Yield* | *Crop residue yield* | *Ethanol production* |
|---|---|---|---|---|---|---|
| Rice | 89.3 | 125 | 44.8 | 1991 | 2787 | 35.0 |
| Wheat | 72.6 | 94.38 | 26.4 | 2753 | 3579 | 27.3 |
| Sorghum | 8.0 | 10.4 | 10.0 | 799 | 1039 | 4.57 |
| Oat | 7.0 | 9.1 | 9.4 | 739 | 961 | 2.36 |
| Maize | 12.3 | 12.3 | 6.6 | 1878 | 1878 | 3.56 |
| Sugarcane (bagasse) | 298.5 | 179.0 | 4.3 | 69202 | 41498 | 50.1 |
| Barley | 1.43 | 1.71 | 7.6 | 188 | 225 | 0.53 |

### Sugarcane bagasse

Sugarcane bagasse is the most abundant lignocellulosic feedstock in India, second after Brazil, the largest producer with 27% of total global production. Nearly, 179 million tons of bagasse is annually produced in India, cultivated on 4.3 million hectare area with the yield, 41498.0 kg/hectare. India's leading sugarcane producing states are Uttar Pradesh (UP), Maharashtra and Tamil Nadu, which account for about 70% of the national output (Smouse *et al.,* 1998). Most of the bagasse is burnt for generating power for boilers, which are highly inefficient. In the year 2001 to 2002, UP was found to be the largest producer (69.73 million tons) contributing 38.95% of total bagasse production. The next was Maharashtra, which produced 27.08 million tons followed by Tamil Nadu, producing 21.8 million tons (Table 6). Based on waste availability, it was estimated that annually 50.1 GL of ethanol could be produced if sugarcane bagasse is used judiciously.

## Rice Straw

In India total annual production of rice straw is about 125 million tons, cultivated on 44.8 million hectare agricultural land with the yield of 2787 kg/hectare. This represents second largest feedstock, however it is not preferred as cattle feed and is also used for making cardboard and low quality paper. West Bengal is the major producer (21.36 millions tons) contributing 17.08% to the total Indian production. UP comes next with 17.44 million tons annual production, followed by Andhra Pradesh, where the total production is 15.94 million tons (Table 6). The total ethanol production could be around 35 GL, if complete rice straw is used for ethanol production.

## Wheat Straw

The annual production of wheat straw is about 94.38 million tons, cultivated on 26.4 million hectares of land with the average yield of 3579 kg/hectare. Uttar Pradesh (UP) alone produces a maximum, 32.52 million tons wheat straw, which is 34.45% of the total production in India. Punjab comes second with 20.15 million tons followed by Haryana, 12.29 million tons (Table 6) although this feedstock is mainly used as cattle feed but if used for bioethanol production, it can yield 27.37 GL of ethanol.

## Shorghum Straw

The annual production of sorghum straw in India is 10.4 million tons and cultivated on 10 million hectares land with yield of 1038.7-kg/ hectare. Maharashtra is the largest producer of shorghum (5.08 million tons), contributing 48.84% of the total production followed by Karnataka (1.87 million tons) (Table 6). The fullest utilization of sorghum residues can produce 4.57 GL of ethanol.

## Oat Straw

The annual Indian production of oat straw residues is 9.1 million tons and cultivated on 9.4 million hectares land with the yield 960.7 kg/hectare. Rajasthan is the major state where the total production of oat is 4.94 million tons, which is 54.28% of the total production followed by Gujarat (1.63 million tons) and UP (1.23 million tons) (Table 6). The potential amount of bioethanol from oat residue is expected to be 2.36 GL.

## Maize straw

About 12.3 million tons of maize crop residue are produced annually in India. This crop is cultivated on 6.6 million hectares agricultural land with a yield of 1878.0 kg/hectare. After harvesting the maize grain, the crop residue is totally waste, which is least used as cattle feed and is mostly burnt. Madhya Pradesh is the biggest producer giving 1.71 million tons with 13.90% of the total production followed by Bihar, 1.56 million tons and Karnataka 1.51 million tons (Table 6). If complete residue is subjected to use, 3.56 GL ethanol could be produced.

**Table 6. Data showing the production of major crop residues by three largest producing states of India.**

| *Crop residue* | *States* | *Production (million tons)* | *All India Share of Production (%)* |
|---|---|---|---|
| Rice | West Bengal | 21.36 | 17.08 |
| | Uttar Pradesh | 17.44 | 13.94 |
| | Andhra Pradesh | 15.94 | 12.74 |
| Wheat | Uttar Pradesh | 32.52 | 34.45 |
| | Punjab | 20.15 | 21.34 |
| | Haryana | 12.29 | 13.02 |
| Maize | Madhya Pradesh | 1.71 | 13.90 |
| | Bihar | 1.56 | 12.68 |
| | Karnataka | 1.51 | 12.27 |
| Sugarcane | Uttar Pradesh | 69.73 | 38.95 |
| | Maharashtra | 27.08 | 15.12 |
| | Tamil Nadu | 21.80 | 12.17 |
| Oat | Rajasthan | 4.94 | 54.28 |
| | Gujrat | 1.63 | 17.91 |
| | Uttar Pradesh | 1.23 | 13.51 |
| Sorghum | Maharashtra | 5.08 | 48.84 |
| | Karnataka | 1.87 | 17.98 |
| | Andhra Pradesh | 0.81 | 7.78 |

## Barley Straw

In India, annual production of barley residue is 1.71 million tons. Barley is mainly grown in Rajasthan, Madhya Pradesh and Andhra Pradesh, on an area of 7.6 million-hectare giving an of 225 kg/ hectare. After recovering the grains, whole of the crop residue is a waste, which can be converted to ethanol. The potential amount of ethanol production from this crop residue is expected to be 0.53 GL.

# FACTUAL AVAILABILITY OF CROP RESIDUE FOR BIOETHANOL PRODUCTION IN INDIA

Despite enormous benefits of the crop residues, most of crop residues are removed and / or burnt in India. In Punjab and Haryana state, for instance, about 70% of the total rice straw produced from combine harvest is burnt in open fields. Similarly, after combine harvesting of wheat, 70% of wheat straw is converted to *bhusa* (chopped straw) for feeding animals and the remaining is burnt in the field. Unlike wheat straw, rice straw is not used as a cattle-feed mainly because of high lignin, silica and oxalate contents. Besides the loss of organic matter and plant nutrients, burning of crop residues also causes atmospheric pollution that poses a threat to human health and ecosystem.

India has a large cattle population of 294 million. Even though India has 67 Mha of grazing land, but grass productivity is low due to land degradation, leading to near total dependence of cattle on crop residues of cereals and pulses. The estimated total residues utilized as fodder was 301 Mt in 1996–97 and is projected to be 385 Mt for 2010, accounting for about 47% of total residues generation (Ravindranath *et al.,* 2005). Amongst the major crop residues (Table 5), sugarcane baggasse can't be used as fodder. If we deduct the share of cattle feed (fodder) from the available crop residue, still 89.42 GL of ethanol can be produced.

## ADDITIONAL POTENTIAL SUBSTRATES FOR ETHANOL PRODUCTION THEIR AVAILABILITY AND ECONOMICS

The economics of ethanol production using different available raw materials is compared in Table, 7 (http://www.renewingindia.org/newsletter). The cost of ethanol today is \$ 1.16 / gal. According to U.S. Department of energy analysis, if the enzymes necessary to convert biomass to ethanol can be bought for less than 10¢ per gallon of ethanol, the cost of making ethanol could drop as low as 75¢ a gallon (Griffiths and Atlas, 2005). In India ethanol is mainly produced from molasses, a byproduct of sugarcane industry, but the other crop, which requires less inputs than sugarcane is sweet sorghum, which is a multipurpose crop. This crop produces grain, sweet juice and baggasse, which is an excellent feed for cattle. It uses almost 50% less water than sugarcane and is a 4-month crop. Ethanol from damaged foodgrains has the lowest cost but the availability is insufficient to meet the demand for 5% ethanol-petrol blends. Another low cost option is ethanol from surplus molasses and sugarcane juice, but again its availability is low to meet the blending demand (http://www.renewingindia.org/newsletters/ethanol). Corn in India is mostly used as corn flour/ flakes for human consumption thus none of it is available for ethanol production. Therefore the most promising alternative is agro-residues which are available plentiful in India, while unfortunately the production of ethanol from agro residues is nil. Recently the research under Department of Biotechnology (DBT), Government of India, has been directed to use wild plant material as substrate for ethanol production. Two such plants material *Lantana* sp. and *Proposis* sp. have been identified and are being worked out at Department of Microbiology, University of Delhi South Campus, New Delhi. Similarly, other plant species, which are largely spread in uncultivated wastelands, could be a good source of raw material for ethanol production. Fuel ethanol production will remain a significant industry and become a potentially self-sustainable agriculture based system for the 22$^{nd}$ century, if the utilization of lignocellulosic becomes a commercial reality. Research and development work for converting lignocellulosic biomass into ethanol is continuing in various laboratories and more research in this area needs to be encouraged for reducing the cost of ethanol production.

**Table 7. Comparative cost of production from different raw materials available for ethanol production.**

| *Raw Material* | *Ethanol yield (Litre / ton)* | *Landed price (Rs./ ton)* | *Net ethanol price Rs.* |
|---|---|---|---|
| Sugarcane molasses | 205 | 2,100 | 15.36 |
| Sugarcane juice | 70 | 800 | 14.89 |
| Corn | 400 | 4,850 | 16.36 |
| Damaged food grain | 360 | 745 | 6.27 |
| Agroresidue | 250 | 1,000 | 23.64 |

## CURRENT STATUS OF BIOETHANOL IN INDIA

Ethanol availability needs to be enhanced to meet the auto-biofuel requirements of India. America including Brazil produce 65% of the total ethanol produced worldwide (Vijayaraghavan, 2002). Ethanol has proved its potential as a biofuel for transport and industrial sectors in several countries especially in Brazil. In India, the future requirements and availability of ethanol to be blended with gasoline (10%) up to year 2007 is calculated in Table 8. The total demand of ethanol is about 4.64 GL against present/planned availability of about 0.744 GL, resulting in scarcity of 3.89 GL ethanol for the year 2005–2006. Present availability of ethanol derived from sugarcane molasses would only be sufficient for 5% gasoline-ethanol programme.

**Table 8. Ethanol availability based on production/ planned from molasses and uses in India**

| *Year* | *Molasses production (mL)* | *Production of alcohol (GL)* | *Industrial use (mL)* | *Potable use (mL)* | *Other uses (mL)* | *Surplus availability of alcohol (mL)* |
|---|---|---|---|---|---|---|
| 1998-99 | 7.00 | 1.41 | 534.4 | 584.0 | 55.2 | 238.2 |
| 1999-00 | 8.02 | 1.65 | 518.9 | 622.7 | 57.6 | 454.8 |
| 2000-01 | 8.33 | 4.69 | 529.3 | 635.1 | 58.8 | 462.7 |
| 2001-02 | 8.77 | 1.77 | 539.8 | 647.8 | 59.9 | 527.7 |
| 2002-03 | 9.23 | 1.87 | 550.5 | 660.7 | 61.0 | 597.5 |
| 2003-04 | 9.73 | 1.97 | 578.0 | 693.7 | 70.0 | 627.5 |
| 2004-05 | 10.24 | 2.07 | 606.9 | 728.3 | 73.5 | 665.8 |
| 2005-06 | 10.79 | 2.19 | 619.0 | 746.5 | 77.2 | 744.3 |
| 2006-07 | 11.36 | 2.30 | 631.4 | 765.2 | 81.0 | 822.8 |

Source: Winrock international India, A sugar industry perspective, New Delhi August: 2002, page 5.

As per United States Department of Energy (USDE) report. In US 75 ethanol plants are operating and 13 more are under construction, these plants produce more than 10.6 billion liters of ethanol per year (http://www.eere.energy.gov/biomass/.) Considering the above facts, it is necessary to switch over from conventional feedstock sugarcane molasses / juice to new lignocellulosic feedstock for fulfilling the increased ethanol demand in India. This strategy not only will fetch the ethanol demand but will also be helpful in lignocellulosic feedstock management and eventually will aid to environmental improvement. The area under sugarcane is presently less than 2% of total cultivable area in the country. Therefore, area under sugarcane cultivation should be increased. This would meet all the ethanol requirements of automotive industry as well as other uses. Therefore, it is important to exploit the potential of presently available sugar and starch-based energy crops for increased ethanol production.

The total bioethanol production from crop residues is estimated to be 491 GL/yr globally (Shapouri *et al.,* 2002). India alone has the capacity to produce 25% i.e. 123 GL/yr of the total production, if the entire crop residue as shown in Table 5 is available for ethanol production. Despite realizing these benefits at bench scale, commercialization and widespread application of lignocellulosic biomass utilization has not yet occurred. Feedstock availability, its location and transport to the site of treatment, pretreatment strategies, efficient hydrolytic agents, availability of robust fermentative microorganisms and process options, all affect the production cost and in turn commercialization of ethanol. Recent technology developments like using cellulolytic enzymes in (SHF) and (SSF) have the potential to remove these economic performance obstacles and make commercialization possible.

## INITIATIVES TAKEN BY GOVERNMENT OF INDIA

Government of India resolved on 1st January 2003 that 5% ethanol-doped-petrol may be supplied in the following nine states (1. Andhra Pradesh 2. Goa 3. Gujarat 4. Haryana 5. Karnataka 6. Maharashtra 7. Punjab 8. Tamilnadu 9. Uttar Pradesh) and four contiguous union territories (1.Damman and Diu 2. Dadra and Nagar Haveli 3. Chandigarh 4. Pondicherry). For this Indian Government has launched three pilot projects, two in Maharashtra (Miraj, Manmad) and one in UP (Bareilly) during April and June 2001, since then, these pilot projects have been supplying 5% ethanol-doped-petrol only to the retail outlets under their respective supply areas.

Many government/university/industry partnerships have been formed to develop compatible processes and enable an integrated approach towards biomass conversion to achieve the required cost reduction. According to Bureau of Indian Standards, only 5% mixing of ethanol with gasoline is allowed which needs to be increased to 10% to make E-10, which is a high octane clean burning fuel.

## CONCLUSION

In short, renewable energy resources pose less environmental risk than conventional sources. With careful management, we can use renewable energy without perturbing natural ecological activity in a harmful way, although no energy technology has zero environmental cost. By contrast, reliance on conventional energy sources seems by definition to alter the balance of material and energy in the ecosystem in a dangerous manner. For these reasons, renewable energy resources represent a sustainable alternative for future strategy to produce low cost ethanol as biofuel.

## ACKNOWLEDGEMENTS

The authors thank the Department of Biotechnology, Government of India, Ministry of Science and Technology, for funding a nationwide project on fuel ethanol production from lignocellulosic waste and Dr. Sanjiv Chauhan from Indian Agricultural Research Institute (IARI) for helping in collecting the data. Authors are also indebted to Dr. Renu Swarup, Director, DBT (Govt. of India) and Prof. K.K. Kathuria, Dept. of English K.U.K for their help during the prepration of the article.

## REFERENCES

Abbi, M., Kuhad, R.C., and Singh, A. (1996). Bioconversion of pentose sugars to ethanol by free and immobilized cells of *Candida shehatae* (NCL-3501): fermentation behaviour. *Process Biochemistry,* 31, 555-560.

Abbi, M., Kuhad, R.C., and Singh, A. (1996). Fermentation of xylose and rice straw hydrolysate to ethanol by *Candida shehatae* NCL-3501. *Journal of Industrial Microbiology,* 17, 20-23.

Agricultural statistics at a glance August (2004), Department of Agriculture & Co-operation, Ministry of Agriculture, Government of India, www:agricoop.nic.in/ statatglance2004/AtGlance.pdf.

Ali, M. (2004). Performance of a newly developed integrant of *Zymomonas mobilis* for ethanol production on corn stover hydrolysate. *Biotechnology Letters,* 26, 321-325.

Balu, V. (2003). Energy security and sustainable alternatives: National policy on biofuel from non-edible oils, Conference on scientific strategies for production of non-edible oil for use as biofuels. *Sustainable Transformation of Rural Areas (SuTRA),* Bangalore.

Berg, C. (2001). World Ethanol Production. The Distillery and Bioethanol Network. Available at http://www.distill.com/ world ethanol production.htm.

Bustos, G., Ramirez, J.A., Garrote, G., and Vazquez, M. (2003). Modeling of the hydrolysis of sugarcane bagasse with hydrochloric acid. *Applied Biochemistry and Biotechnology* 104, 51-67.

Cadoche, L. and Lopez, G.D. (1989). Assessment of size reduction as a preliminary step in the production of ethanol from lignocellulosic wastes. *Biology of Wastes,* 30, 153–157.

Campbell, C.J. and Laherrere, J.H. (1998). The end of cheap oil. *Scientific American,* 3, 78–83.

Damaso, M.C.T., Castro, M. de., Castro, R.M., Andrade, M.C., and Pereira, N. (2004). Application of xylanase from *Thermomyces lanuginosus* IOC-4145 for enzymatic hydrolysis of corncob and sugarcane bagasse. *Applied Biochemistry and Biotechnology,* 115, 1003-1012.

Dien, B.S., Nichols, N.N., O'Bryan, P.J., and Bothast, R.J. (2000). Development of new ethanologenic *Escherichia coli* strains for fermentation of lignocellulosic biomass. *Applied Biochemistry and Biotechnology,* 84/86, 181-196.

Dipardo, J. (2000). Outlook for biomass ethanol production. Energy information administration. Available from http:// www.eia.doe.-gov/oiaf/analysispaper/biomass.html. Chemical Society, Washington, DC. p. 55–90.

Draude, K.M., Kurniawan, C.B. and Duff, S.J.B. (2001). Effect of oxygen delignification on the rate and extent of enzymatic hydrolysis of lignocellulosic material. *Bioresource Technology,* 79(2), 113-120.

Duff, S.J.B. and Murray, W.D. (1996). Bioconversion of forest products industry waste cellulosics to fuel ethanol: a review. *Bioesource technology,* 55, 1-33.

Fan, L.T., Gharpuray, M.M., and Lee, Y.H. (1987). In: *Cellulose Hydrolysis Biotechnology Monographs,* Springer, Berlin, p. 57.

Ghose, T.K., and Bisaria, V.S. (1979). Studies on mechanism of enzymatic hydrolysis of cellulosic substances. *Biotechnology Bioengineering,* 21, 131–146.

Goel, A.K. (2002). A sugar industry perspective. *Ethanol: Winrock International India,* New Delhi. p. 5.

Gregg, D.J. and Saddler, J.N. (1996). Factors affecting cellulose hydrolysis and the potential of enzyme recycle to enhance the efficiency of an integrated wood to ethanol process. *Biotechnology and Bioengineering,* 51, 375–383

Griffiths, M. and Atlas, R.M. (2005). Preemptive bioremediation: applying biotechnology for clean industrial products and processes. In R.M. Atlas and J. Philp (eds.). *Bioremediation: Applied microbial solutions for real-world environmental cleanup,* ASM press, Washington, DC. p. 318-356.

Hayn, M., Steiner, W., Klinger, R., Steinmuller, H., Sinnar, M., and Esterbauer, H. (1993). Basic research and pilot studies on the enzymatic conversion of lignocellulosics In: Saddler, J.N. (eds.), *Bioconversion of forest and agricultural plant residues,* CAB International, Wallingford, U.K, p. 33-72.

Ho, N.W.Y., Chen, Z., and Brainard, A.P. (1998). Genetically engineered *Saccharomyces* yeast capable of effective cofermentation of glucose and xylose. *Applied and Environmental Microbiology,* 64, 1852–1859.

Holtzapple, M.T., Lundeen, J.E. and Sturgis, R. (1992). Pretreatment of lignocellulosic municipal solid waste by ammonia by explosion (AFEX). *Applied Biochemistry and Biotechnology,* 34/35, 5–21.

Itoh, H., Wada, M., Honda, Y., Kuwahara, M. and Watanabe, T. (2003). Bioorganosolve Pretreatments for simultaneous saccharification and fermentation of beech wood by ethanolysis and white-rot fungi. *Journal of Biotechnology,* 103 (3), 273-280.

Keller, F.A., Hamilton, J.E., Nguyen, Q.A. (2003). Microbial pretreatment of biomass: potential for reducing severity of thermochemical biomass pretreatment. *Applied Biochemistry Biotechnology,* 27-41, 105-108.

Kim, H.T., Kim, J.S., Sunwoo, C. and Lee, Y.Y. 2003. Pretreatment of corn stover by aqueous ammonia. *Bioresource Technology,* 90(1), 39-47.

Kim, S. and Dale, E.B. (2004). Global potential bioethanol production from wasted crops and crop residues. *Biomass and Bioenergy* 26, 361–375.

Knauf, M. and Moniruzzaman, M. (2004). Lignocellulosic biomass processing: A perspective. *International Sugar Journal,* 106, 147-150.

Krishna, S.H., Reddy, T.J and Chowdary, G.V. (2001). Simultaneous saccharification and fermentation of lignocellulosic wastes to ethanol using thermotolerant yeast. *Bioresource Technology,* 77(2), 193-196.

Kuhad, R.C. and Singh, A. (1993). Lignocellulose biotechnology: current and future prospects. *Critical Reviews in Biotechnology,* 13, 151–172.

Kuhad, R.C., Singh, A., and Eriksson, K-EL (1997). Microorganisms and enzymes involved in the degradation of plant fiber cell walls. *Advances in Biochemical Engineering and Biotechnology,* 57, 47-125.

Lopez, M.J., Nichols, N.N., Dien, B.S., Moreno, J., Bothast, R.J. (2004). Isolation of microorganisms for biological detoxification of lignocellulosic hydrolysates. *Applied Microbiology and Biotechnology,* 64(1), 125-131.

Lynd, L.R. (1990). Large scale fuel ethanol from lignocellulose. *Applied Biochemistry and Biotechnology,* 24/25, 695-719.

Mamma, D., Christakopoulos, P., Koullas, D., Kekos, D., Macris, B.J. and Koukios, E. (1995). An alternative approach to the bioconversion of sweet sorghum carbohydrates to ethanol. *Biomass and Bioenergy,* 8(2), 99-103.

Martín, C., Galbe, M., Wahlbom, C.F., Hägerdal, B.H. and Jönsson, L.J. (2002). Ethanol production from enzymatic hydrolysates of sugarcane bagasse using recombinant xylose-utilising *Saccharomyces cerevisiae. Enzyme and Microbial Technology,* 31(3), 274-282.

Martinez, A., Rodriguez, M.E., York, S.W., Preston, J.F. and Ingram, L.O. (2000). Effects of $Ca(OH)_2$ treatments ("overliming") on the composition and toxicity of bagasse hemicellulose hydrolysates. *Biotechnology and Bioengineering,* 69, 526-536.

Nigam, J.N. (2001). Ethanol production from wheat straw hemicellulose hydrolysate by *Pichia stipitis. Journal of Biotechnology,* 87, 17-27.

Nigam, J.N. (2002). Bioconversion of water-hyacinth (*Eichhornia crassipes*) hemicellulose acid hydrolysate to motor fuel ethanol by xylose-fermenting yeast. *Journal of Biotechnology,* 97, 107-116.

Nilvebrant, N., Reimann, A., Larsson, S., and Jonsson, L.J. (2001). Detoxification of lignocellulose hydrolysates with ion exchange resins. *Applied Biochemistry and Biotechnology,* 91-93, 35-49.

Palmarola-Adrados, B., Galbe, M., Zacchi, G. (2005). Pretreatment of barley husk for bioethanol production. *Journal of Chemical Technology and Biotechnology,* 80(1), 85-91.

Ravindranath, N.H., Somashekar, H.I., Nagaraja, M.S., Sudha, P. Sangeetha, G., Bhattacharya, S.C., Abdul Salam, P. (2005). Assessment of sustainable non-plantation biomass resources potential for energy in India. *Biomass and Bioenergy,* 29, 178-190.

Rodrigues, R.C.L.B., Felipe, M.G.A., Almeida e silva, J.B., Vitola, M., Gomez, P.V. (2001). The influence of pH, temperature and hydrolysate concentration on the removal of volatile and nonvolatile compounds from sugarcane bagasse hemicellulosic hydrolysate treated with activated charcoal before or after vacumn evaporation. *Brazilian journal of chemical engineering,* 18, 299-311.

Sanchez, G., Pilcher, L., Roslander, C., Modig, T., Galbe, M., and Liden, G. (2004). Dilute-acid hydrolysis for fermentation of the Bolivian straw material *Paja Brava. Bioresource Technology,* 93, 249-256.

Sawada, T., Nakamura, Y., Kobayashi, F., Kuwahara, M., and Watanabe, T. (1995). Effect of fungal pretreatment and steam explosion pretreatment on enzymatic saccharification of plant biomass. *Biotechnology and Bioengineering,* 48, 719-724.

Shapouri, H., Duffield, J. A. and Wang, M. (2002). *The energy balance of corn ethanol: an update,* AER-813. Washington, DC. USDA.

Sharma, S.K., Kalra, K.L. and Grewal, H.S. (2002). Fermentation of enzymatically sacchharified sunflower stalks for ethanol production and its scale up. *Bioresource Tecnology,* 85, 31-33.

Sharma, S.K., Kalra, K.L. and Kocher, S. (2004). Fermentation of enzymatic hydrolysate of sunflower hulls for ethanol production and its scale up. *Biomass and Bioenergy,* 27, 399-402.

Smouse, S.M., Staats, G.E., Rao, S.N., Goldman, R., and Hess, D. (1998). Promotion of biomass cogeneration with power export in the Indian sugar industry. *Fuel Process Technology,* 32, 227-247.

Sreenath, S.K., Koegel, R.G., Moldes, A.B., Jeffries, T.W and Straub, R.J. (2001). Ethanol production from alfalfa fiber fractions by saccharification and fermentation. *Process Biochemistry,* 36(12), 1199-1204.

Subramanian, K.A., Singal S.K., Saxena, M., and Singhal, S. (2005). Utilization of liquid biofuels in automotive diesel engines: An Indian perspective. *Biomass and Bioenergy,* 29, 65-72.

Sun, Y. and Cheng, J. (2002). Hydrolysis of lignocellulosic materials for ethanol production. *Bioresource Technology,* 83, 1–11.

Toivola, A., Yarrow, D., Van-den-bosch, E., Van-dijken, J.P., Sheffers, W.A. (1984). Alcoholic fermentation of D-xylose by yeasts. *Applied Microbiology and Biotechnology,* 47, 1221-1223.

Tucker, M.P., Kim, K.H., Newman, M.M. and Nguyen, Q.A. (2003). Effects of temperature and moisture on dilute-acid steam explosion pretreatment of corn stover and cellulase enzyme digestibility. *Applied Biochemistry and Biotechnology,* 105–108, 165-177.

Vidal, P.F. and Molinier, J. (1988). Ozonolysis of lignin – improvement of in vitro digestibility of poplar sawdust. *Biomass,* 16, 1–17.

Vijayaraghavan, R. (2002). Ministry of Petroleum and Natural Gas Resolution—The Gazette of India. Part-II, Section-3. *Ministry of Petroleum & Natural Gas,* Government of India 19, 1–15.

Wyman, C.E. (1994). Alternative fuels from biomass and their impact on carbon dioxide accumulation. *Applied Biochemistry and Biotechnology,* 45/46, 897-915.

Wyman, C.E. (2003). Potential synergies and challenges in refining cellulosic biomass to fuels, chemicals and power. *Biotechnology Progress,* 19, 254-262.

Zheng, Y.Z., Lin, H.M., and Tsao, G.T. (1998). Pretreatment for cellulose hydrolysis by carbon dioxide explosion. *Biotechnology Progress,* 14, 890–896.

# 16

# Enzyme Prebleaching of Pulp : Perspectives in Indian Paper Industry

**R.K. Jain, R.M. Mathur, Vasanta V. Thakur and A.G. Kulkarni**

*Central Pulp & Paper Research Institute, Post Box No. 174, Himmat Nagar, Saharanpur-247 001(U.P.), India*

With the current trends to employ environmentally benign technologies, xylanase prebleaching could prove to be one of the options before paper industry in reducing the generation of chloro-organic compounds while improving the final brightness of the bleached pulps. The technology has been successfully adopted commercially in number of paper mills in the developed countries but still it is in the developing stage as far as Indian paper Industry is concerned.

Central Pulp & Paper Research Institute (CPPRI) is an apex organization dedicated to applied research in the areas of pulp and paper in India and South-East Asian region. The Institute has carried out extensive studies in the recent past in the area of enzymatic prebleaching of pulps and evaluated several Xylanase enzymes (indigenously developed as well as commercially available) on wood and non-wood based pulps at the laboratory of the institute and at various pulp and paper mills in different parts of the country.

The five samples of xylanase enzymes have been evaluated for their bleach response on hard-wood and non-wood based pulps, the response of two xylanase enzymes was found to be most encouraging in terms of reduction in demand of elemental chlorine, which was reduced by 15-17%. The other widespread benefits were improved effluent characteristics specifically the AOX levels which were reduced by 25 – 30%. Other benefits of the process reported included gain in bleached pulp brightness by 2 – 2.5% ISO and improved tear strength by 5% – 10% at similar level of bleach chemical demand.

The present study highlights Indian experience on application of enzyme pre-bleaching process for wood and non-wood based paper mills covering some of the important considerations during commercialization of the process. This includes the prevalent mill conditions (temperature, pH and retention time) and the equipment required for effective dispersion of the enzymes to achieve desired efficiency level to make the technology a success in Indian paper industry.

## INTRODUCTION

Due to concerns regarding the release of chlorinated organic material in pulp mill effluents into receiving rivers and streams, the pulp and paper industry in India is active in adopting the environmentally benign technologies that reduce or eliminate the use of elemental chlorine in pulp bleaching. There are several technologies used by the pulp and paper mills in the area of pulping and bleaching to meet the environmental challenges including modified cooking methods and oxygen delignification. Both of these methods focus on removing more of the lignin prior to the bleaching stages, which reduces the chemical demand in the bleaching stages. However, these technologies are very effective but sometimes they do require substantial capital investments.

Bleaching of kraft pulp using a combination of chlorination and alkaline extraction has been practised in India for many years. The effluent from chlorination and alkaline extraction stages cannot be recycled back to the chemical recovery furnace due to its high level of corrosive chloride. In addition, the effluent contains large amounts of chlorinated organic compounds some of which are known to have toxic, mutagenic, and carcinogenic effects. The growing public concerns regarding the environmental effects of these chlorinated organic compounds have driven pulp mills to seek new bleaching technologies that reduce or eliminate the use of elemental chlorine in pulp bleaching.

The pulp and paper industry, the largest processor of renewable raw materials by single category, has so far been a scarce user of enzymes. There has been a traditional application of amylases in starch modification for paper coating and a growing use of lipases for pitch removal. The major application target of enzymes as authentic tools, however, the separation of lignin from cellulose in pulp processing like pulping and bleaching, is still in its infancy.

Cellulose and lignin are rather rigid organic polymers, which have been invented by nature for constructive and long-term preservation purposes. Harsh physiochemical conditions have to be applied to attack or modify these two compounds. Given the mild reaction of enzymatic catalysis it is not surprising that it took biochemists and enzymologists a long time to find an application for their alternative approach. The task of enzyme application in pulping is lignin removal. Previous attempts often suffered from contaminating amounts of cellulases in test preparations, so the compatibility of enzyme application and fibre quality has been questioned. When talking about enzymatic lignin degradation, lignin removal and enzymatic bleaching are faced with severe technological constraints, but also with some pretty nice discoveries.

Enzymatic prebleaching of pulp employing xylanase enzymes has been an area of major interest for the paper industry not only in the developed countries but also in developing countries. The process employing the use of the identified xylanase enzymes could help in phasing out the chlorine and chlorine based bleaching chemicals during bleaching of pulps, ultimately resulting in lowering the toxicity (AOX) levels in the bleach plant effluents (Wong *et al.,* 1997).

Xylanase enzymes used in bleaching of chemical pulps are being developed by companies, which claim their products suitable as prebleaching agents for the pulps produced in the paper industry employing variety of fibrous raw materials. However, these enzymes, which are marketed by several suppliers, are found to be highly sensitive to the conditions like temperature, pH & doses of the enzymes besides the nature of pulp. Therefore, it becomes difficult for the pulp and paper mills to decide which particular enzyme should suit to the specific requirement of their industry in order to achieve desired effect.

In view of the above, Central Pulp & Paper Research Institute (CPPRI) has been engaged for the last many years in conducting studies on evaluation of various xylanase enzyme preparations which are available globally and also being developed indigenously by several reputed biotechnological laboratories to study their response on the type of pulps produced in Indian paper mills. Having successfully carried out exhaustible enzymatic evaluation studies on wood and non-wood based pulps, the Institute conducted studies at the mill sites under the condition prevalent at the mills, employing various identified enzymes in order to explore the possibility to demonstrate the technology on commercial scale in Indian paper industry (Jain *et al.,* 2001).

The present paper highlights studies carried out at the Institute on enzymatic prebleaching of pulp on wood and non-wood pulps covering some of the important considerations which are important during commercialization of the process to achieve desired efficiency level, so that technology could be explored successfully in Indian paper industry.

## WHAT ARE XYLANASE?

Xylanase is an enzyme that catalyzes the hydrolysis of 1,4-β–D-xylosidic linkages in xylan, a hemicellulose in pulp. Several industrial xylanase preparations are commercially available, with many other natural xylanases having been identified. Variations in xylanases can occur between size, structure, amino acids, position attacked on the xylan, attachment to xylan, stability in the pulp slurry and stability while acting on the pulp. While all xylanases release xylose from natural xylan, the differences between commercial xylanase preparations can significantly impact the bleach boosting effect.

Xylanases can be classified structurally into two major groups: Family F or 10, and Family G or 11. Family F xylanases are of relatively higher molecular weight and Family G are of relatively lower molecular weight. The Family G enzymes can be further divided into those with high and low isoelectric points. Liberation of reducing sugars from purified xylan by the PI. 5.5 xylanase from *T. reesei* correlates well with its bleaching effect on fibers, but the behaviour of the PI 9.0 xylanase is more complex. It appears to be affected not only by its catalytic activity, but also by electrostatic interactions with its substrate. The effect of the PI 9.0 xylanase on bleachability of pulp increased more at pH 7.0 than would be expected from its activity on purified xylan (Kirk and Jeffries, 1996).

Enzymes are natural, highly specific protein catalysts. Xylanase enzymes are produced commercially from microorganisms (bacteria, yeast or fungi), and specifically hydrolyse xylan without hydrolyzing cellulose. Enzymes of a similar type have been commercially produced for at least 20 years and are used in application areas as varied as food processing, animal feed, textiles etc.

## DESIRABLE CHARACTERISTICS OF XYLANASE AS BLEACH BOOSTER

Effective xylanases should have several properties:

- First, they should be stable on kraft pulps. Some xylanase preparations non-specifically absorb on to pulp fibres and are inactivated by degradation products from kraft pulping;
- Second, they should have a neutral to alkaline pH optimum. Residual alkali leaks out of the pulp during enzyme treatment, and the pH of even well washed pulp stocks can shift upwards dramatically;
- Third, they should have good thermal stability.

## MECHANISM OF XYLANASE ACTION IN KRAFT PULP BLEACHING

Xylanases such as endoxylanases are xylan-specific enzymes. They catalyze the hydrolysis of xylose-xylose bonds within the xylan chain and only solubilize a fraction of the total xylan present. However, the actual enzymatic mechanism in bleaching is not yet well understood.

Several hypotheses have been proposed to explain the mechanism of bleach boosting by xylanase. Similarities exist between the possible explanations and it may be that a combination of factors accounts for the bleach boosting effect (Kantelinen *et al.,* 1993; Wong *et al.,* 1997; Viikari *et al.,* 1996a; Viikari *et al.,* 1996b; Jean *et al.,* 1994).

One possible mechanism is that xylan reabsorbed on the fibre surface during kraft pulping and creates a barrier against lignin removal. The xylanase treatment may remove the xylan barrier allowing better access by bleaching chemicals to residual lignin covered up by the xylan barrier. Contrary to this suggested mechanism, Filice *et al.* demonstrated that only a partial de-polymerization of the xylan and not complete removal was needed to facilitate bleach boosting.

It has also been suggested that readsorbed xylan may be chemically bonded to residual lignin in pulp. These lignin-carbohydrate complexes (LCC's) are thought to be difficult to be removed due to diffusion limitations. According to this hypothesis, by cleaving xylan chains involved in these complexes, smaller residual structures of increased mobility are formed. The increased mobility allows for easier removal of these structures from the fibre. Kantelinen and co-workers (1993) determined that instead of a decrease in the size of the material removed from the fibre, the average molecular mass increased as a result of the enzyme treatment. This result is in contrast with the proposed mechanism.

It has also been suggested that the xylanase treatment results in the removal of Hexenuronic acid, a degradation product of 4-0-methylglucuronic acid. It is one such component that is thought to contribute to the kappa number and brightness of pulps. The lack of methoxy groups from material solubilized by the enzyme treatment indicates that the nature of the material removed is derived from carbohydrates and not lignin.

Skjold-Jorgensen found that xylanase treatment decreased the demand for active chlorine ($aCl_2$) for a batch: kraft pulp by 15%, but decreased $aCl_2$ of pulp from a continuous process by only 6 to 7%. They also showed that DMSO extraction of residual xylan does not lead to an increase in bleachability, but that xylanase treatment does. This indicates that it is the DMSO-insoluble xylan fraction that is chemically bound to the chromophores. Paice *et al.,* (1992) have shown that the prebleaching effect on black spruce pulp is associated with a drop in the degree of polymerization, even though the xylan content decreases only slightly. Prebleaching thus appears to be associated with xylan depolymerization, even though not necessarily with solubilization of the xylan-derived hemicellulose component.

## MATERIALS AND METHODS

### Sources of Xylanase Enzymes

Several xylanase enzyme preparations used in the present investigations were procured from various sources including indigenously developed in laboratory and commercial scale as well as the commercial imported enzymes which are denoted as A, B, C, D and E. The details about the xylanase preparation are given in Table 1.

**Table 1. Sources of xylanase enzymes**

| *Xylanase enzyme* | *Source* |
|---|---|
| A | Indigenously developed enzyme from an indigenous fungal strain. |
| B | Indigenously manufactured commercial enzyme from bacterial strains |
| C | Imported commercial enzyme |
| D | Imported commercial enzyme |
| E | Indigenously manufactured commercial enzyme from bacterial strain |

## PULP SAMPLES

Wood kraft pulp samples were procured from a large integrated pulp and paper mill in U.P. employing eucalyptus as major raw material where as kraft bagasse pulp was obtained from an agro-based mill in Maharasthra state equipped with chemical recovery system.

## ENZYME ASSAY TECHNIQUES

All the enzymes were tested for filter paper activity for cellulase contamination by the method of Mandals & Weber and xylanase activity was measured by the method of Bailey *et al.* (1992).

## XYLANASE PRETREATMENT OF PULPS

Enzymatic pretreatment of the pulps was carried out under the optimized conditions as mentioned in Table 2.

**Table 2. Enzyme pretreatment conditions**

| | *Wood Pulp treated with Enzymes* | | | | |
|---|---|---|---|---|---|
| *Parameters* | *A* | *B* | *C* | *D* | *E* |
| Dose of enzyme, IU/g | 7.0 | 5.0 | 5.0 | 5.0 | 10 |
| Consistency of pulp, % | 10 | 10 | 10 | 10 | 10 |
| Retention time, min | 120 | 120 | 120 | 120 | 120 |
| pH of pulp at 25 deg. C | 6.2 | 8.5-9.0 | 8.5 | 8.5-9.0 | 8.5 |
| Temperature of pulp, deg. C | ≅ 50 | ≅ 50 | ≅ 50 | ≅ 50 | ≅ 50 |

## RESULTS AND DISCUSSION

While optimizing the dosages of enzymes during enzyme treatment of wood kraft and bagasse kraft pulps, it was observed that excess dosages of xylanase enzymes for a longer time has resulted in decreased pulp yield and lower strength properties of the pulps. This could probably be due to loss of hemicelluloses by the drastic action of enzyme. However under optimized dosages of enzymes ≅ 5 IU/gm of pulps and optimized treatment conditions (Table 2), no significant loss in pulp yield could be observed except with enzyme preparation E wherein, the treatment of the pulp at optimized conditions of time and temperature could result in 1.5% pulp yield loss along with drop in viscosity which was reduced from 560 to 510 $cm^3/g$ of the pulp leading to loss in strength properties (Table 3). Reasons for the same could be attributed to contamination of xylanases with cellulase activity, which could be ascertained by determining the cellulase activity in the xylanase enzyme preparation E by method of Mandals & Weber (Wong *et al.,* 1997). Therefore this enzyme named as E was found to be unsuitable for its application in bleaching of pulps.

**Table 3: Characterization of unbleached pulp for yield and other parameters of wood kraft pulp before and after enzyme treatment**

| | *Pulp treated with enzymes* | | | | | |
|---|---|---|---|---|---|---|
| *Parameters* | *Control* | *A* | *B* | *C* | *D* | *E* |
| Pulp Yield, % | 99.89 | 99.82 | 99.62 | 99.80 | 99.74 | 98.10 |
| Kappa number of pulp | 18.0 | 16.5 | 16.7 | 16.6 | 17.0 | 16.0 |
| Brightness of pulp, %ISO | 27.5 | 28.5 | 29.0 | 28.4 | 28.0 | 29.0 |
| CED Viscosity of pulp, $cm^3/g$ | 560 | 620 | 590 | 557 | 604 | 510 |

## EFFECT OF XYLANASE TREATMENT ON BLEACH CHEMICALS REQUIREMENT

Bleaching of the xylanase enzyme treated kraft pulps; both from wood and bagasse have been found to respond in different manner than untreated pulps, while bleaching with conventional CEH bleach sequence being adopted in majority of Indian paper mills. Based on optimization studies of the chlorine dosages during chlorination stage, it was observed that significantly less chlorine is required to bleach kraft wood and kraft bagasse pulps while bleaching to similar targeted brightness level of 80.0% ISO.

Table 4 shows the effect of various xylanase preparations on the bleach chemical requirement during conventional CEH bleach sequence of wood kraft pulps with a targeted brightness level of 80% ISO. From the results shown it is clearly seen that there is a reduction of nearly 14 - 15% of chlorine demand with all the enzymes preparations in case of wood kraft pulps except enzyme preparation D, where chlorine saving of only 9.3% could be noticed. Further the enzyme treated pulps could be bleached to higher brightness with a gain in brightness level of 2-3% ISO in all the cases while using similar chlorine dosages as compared to the control pulp sample.

Reduction in chlorine requirement or improved pulp brightness in case of enzyme treated pulps was reflected from the reduction in kappa number of the unbleached pulps after enzyme treatment (X stage) and also chlorine treatment and also the pulp after enzyme treatment followed by chlorine and alkali extraction stage (XCE). The reduction in kappa number by 7-8% was noticed after enzyme treatment of all pulp samples treated with various enzymes sample and also the reduction of kappa number of more than 25% was noticed after chlorine treatment and alkali extraction stage in enzyme treated pulps (Table 4).

**Table 4. Effect of enzyme pretreatment on bleach chemical requirement using various xylanase preparations on kraft wood pulp**

| | | *A* | | *B* | | *C* | | *D* | | *E* | |
|---|---|---|---|---|---|---|---|---|---|---|---|
| *Parameters* | *Control* | *ET1* | *ET2* | *ET1* | *ET2* | *ET1* | *ET2* | *ET1* | *ET2* | *ET1* | *ET2* |
| % Savings in elemental chlorine | - | - | 15 | - | 15 | - | 14 | - | 9.3 | - | - |
| % Reduction in Kappa No. after X stage | - | 8.0 | | 7.2 | | 7.3 | | 5.5 | | 11.1 | |
| % reduction in Kappa No. after XCE stage | - | 28.0 | ND | 25.9 | 14.5 | 26.0 | 16.0 | 27.0 | 23.0 | ND | ND |
| Final brightness of the pulp, % ISO | 80.0 | 82.5 | 80.0 | 82.5 | 81.3 | 83.0 | 80.2 | 82.0 | 80.0 | 83.0 | 80.0 |

X : Enzyme treatment
XCE : Enzyme treated pulp after chlorine and alkali extraction treatment
ETI : Enzyme treated pulp treated with similar chlorine dose
ET2 : Enzyme treated pulp treated with optimized less chlorine dose

Similar observations were made in case of kraft bagasse pulps where nearly 18% reduction in chlorine demand could be observed in enzyme treated pulps with targeted brightness level of 83.0% whereas brightness gain of 2.0% ISO could be achieved i.e. from brightness level of 83% ISO to 85% ISO when the enzyme treated pulp was bleached with similar chlorine dose of 5.1% as in case of the control pulp. Results are shown in Table 7. This reduction in chlorine demand and gain in pulp brightness also reflects from reduction in kappa number of the enzyme treated pulp where 20% reduction in kappa number in enzyme treated pulp followed by chlorine treatment and alkali extraction stage could be noticed. (Table 6).

## EFFECT OF ENZYME TREATMENT ON UNBLEACHED PULP

Keeping in view the availability of enzymes for commercial applications, cost and the results of efficiency levels, it is evident from the evaluation studies that enzymes B and enzymes D could screened to be better ones. Therefore, studies were continued with these enzymes, one indigenous and the other imported one. Table 5 shows the bleach chemical demand, strength and optical properties of kraft wood pulp treated with these two identified xylanase enzyme preparations viz. B & D against control. From the results shown in Table 5, it is clearly evident that with a savings of nearly 15% chlorine demand employing enzyme preparation B, no loss in strength properties of the pulps like burst, tensile and tear could be noticed, however the tear index was improved significantly from 5.3 Mn $m^2/g$ to 6.0 Mn $m^2/g$. In case of pulps treated with enzyme D, the strength properties of the pulp, though were at par with control pulp but the chlorine savings were comparatively lower i.e nearly 10.0% only. With regards to the optical properties of the enzyme treated pulps, there was an improvement in the post colour number (P.C.No.) of the pulps both with enzyme preparations B & D, as the P. C. No. of the enzyme treated pulps were decreased from 14.94% to 9.79% and 10.65% respectively.

**Table 5. Enzymatic prebleaching of kraft wood pulp using xylanase**

| *Bleaching of Pulp using Conventional CEH Sequence Before & After Enzyme Treatment* | | | | | |
|---|---|---|---|---|---|
| *Particulars* | *Control* | *Enzyme B* | *Enzyme D* | | |
| **Chlorination** | | | | | |
| % Chlorine Applied | 3.9 | 3.9 | 3.3 | 3.9 | 3.54 |
| % Chlorine savings | - | - | 15 | - | 10.0 |
| Final brightness of pulps, % | 80.0 | 82.5 | 80.5 | 82.0 | 80.5 |
| Brightness gains, % | - | 2.5 | 0.5 | 2.0 | 0.5 |

**Strength and optical properties of the bleached wood Kraft pulps, before & after enzyme treatment using two identified xylanase enzymes B and D**

| *Particulars* | *Control* | *Enzyme B* | *Enzyme D* |
|---|---|---|---|
| Revolution (PFI) | 4000 | 4000 | 4000 |
| Freeness (CSF) | 220 | 250 | 250 |
| Apparent density, .g/m$^3$ | 0.79 | 0.76 | 0.75 |
| Burst index Kpa.m2/g | 4.43 | 4.49 | 4.51 |
| Tensile Index , Kpa.m$^2$/g | 64.4 | 64.3 | 64.4 |
| Tear Index, Mn m$^2$/g | 5.3 | 6.0 | 5.3 |
| **Optical properties** | | | |
| Opacity, % | 92.2 | 91.2 | 91.0 |
| P.C.No., % | 14.94 | 9.79 | 10.65 |

Similar observations could be made in case of kraft bagasse pulps. Treatment of the kraft bagasse pulp with xylanase enzyme B preparation showed no loss in pulp yield or pulp viscosity (Table 6). With regards to the strength properties of the enzyme treated pulps properties like burst, tear and tensile could be maintained at par with control pulps. However, there was improvement in optical properties particularly in respect of P.C.No. of the pulps since P.C.No. of enzyme treated pulps were dropped, from 13.5 to 9.2 and 9.6% respectively when treated with xylanase enzyme preparation B and D respectively (Table 7).

**Table 6. Yield and other characteristics of unbleached kraft bagasse pulp before and after enzyme treatment using mentified xylanase (b) enzyme**

| *Parameters* | *Pulp treated with enzymes* | |
|---|---|---|
| | *Control* | *Enzyme treated* |
| Pulp yield, % | 99.20 | 99.02 |
| Kappa number of unbleached Pulp | 26.2 | 25.5 |
| Kappa No. of pulp after XCE Stage* | 5.07 | 4.04 |
| % Reduction in kappa after XCE stage* | - | 20% |
| Brightness, % ISO | 30.3 | 3.08 |
| CED Viscosity of pulps, Cm$^3$/gm | 535 | 520 |

* XCE enzyme treated pulp followed by chlorine treatment and alkali extraction stage

**Table 7. Xylanase pretreatment of kraft bagasse pulps**

| *Bleaching of Pulp using Conventional CEH Sequence Before and After Enzyme Treatment* | | | |
|---|---|---|---|
| *Particulars* | *Control* | *ET-1* | *ET-2* |
| Chlorination | | | |
| % Chlorine Applied | 5.1 | 5.1 | 4.2 |
| Final Brightness of pulp % ISO | 83.0 | 85.0 | 84.0 |
| Gain in pulp brightness, % | - | 2.0 | 1.0 |
| **Strength and optical properties** | | | |
| Parameters | Control | ET-1 | ET-2 |
| Freeness CSF | 335 | 350 | 355 |
| Burst index Kpa.m2/g | 2.60 | 3.05 | 2.55 |
| Tensile Index , Kpa.$m^2$/g | 45.5 | 56.0 | 45.0 |
| Tear Index, Mn $m^2$/g | 4.05 | 4.40 | 4.20 |
| | **Optical properties** | | |
| Brightness of pulp, % ISO | 83.0 | 85.0 | 84.0 |
| P.C.No., % | 13.5 | 9.2 | 9.6 |

. ET1 — Pulp treated with similar chlorine dose
. ET2 — Pulp treated with less optimized Chlorine dose

## ENVIRONMENTAL EFFECT OF ENZYME TREATMENT

Reduction in total chlorine demand of around 15% during enzyme treatment of pulps, followed by CEH bleach sequence resulted in remarkable reduction in the toxicity (AOX) of the bleach plant effluents. From the results shown in Table 8, it is evident that the AOX level in the bleach effluent generated from enzyme treated pulps was reduced to nearly 25% to 30%. Further reduction in ratio of COD to BOD in bleach effluent from kraft wood and kraft bagasse (pulp) after enzyme treatment is an indication in improved biological treatability of the bleach effluents.

**Table 8. Characteristics of bleach effluents of kraft wood and kraft bagasse pulps before and after enzyme treatment using xylanase enzyme B**

| | *Wood Kraft Pulp* | | *Bagasse Kraft Pulp* | |
|---|---|---|---|---|
| *Parameters* | *Control* | *Enzyme treated* | *Control* | *Enzyme Treated* |
| % reduction in AOX | - | 30.0 | - | 25.0 |
| AOX, Kg/tp | 2.29 | 1.60 | 4.13 | 3.13 |
| COD, Kg/tp | 23.52 | 28.5 | 34.6 | 38.2 |
| BOD, Kg/tp | 11.2 | 14.2 | 14.4 | 19.0 |
| COD: BOD ratio | 1:2.1 | 1:1.97 | 1:2.4 | 1:2.0 |

## IMPORTANT CONSIDERATIONS DURING COMMERCIALIZATION OF ENZYMATIC PREBLEACHING PROCESS

Process flow sheet of mill scale trial on the process of enzyme prebleaching of pulp is shown in Figure 1. Based on the present studies, it has been observed that the key factor influencing the success of mill scale trial on enzymatic prebleaching are:

## Selection of Appropriate Enzyme

It has been observed that efficiency of the xylanase enzymes available commercially and/or produced indigenously differ greatly and depend upon molecular size, overall structure, the number and identity of amino acids and the structure of the active site. These differences influence the ability of the enzyme to penetrate into the pulp and the nature of the action against the pulp. Thus it would be desirable to evaluate and identify an appropriate xylanase enzyme having optimum efficiency levels under the condition prevalent in the mill in respect:

- Temperature
- pH
- Retention time

## EFFICIENT DISPERSION OF ENZYME WITH PULP

The studies revealed that reasons for poor efficiency during enzyme prebleaching has been the inadequate mixing of the enzyme with pulp. Enzyme mixing devices being tried in most of the mills did not proved to be adequate. However, installation of an MC pump and proper dilution of the enzyme as well as the pulp to desired consistency with paper machine back water could facilitate proper mixing and dispersion of enzyme with pulp and in achieving desired pH levels. This facilitated in achieving the desired efficiency levels.

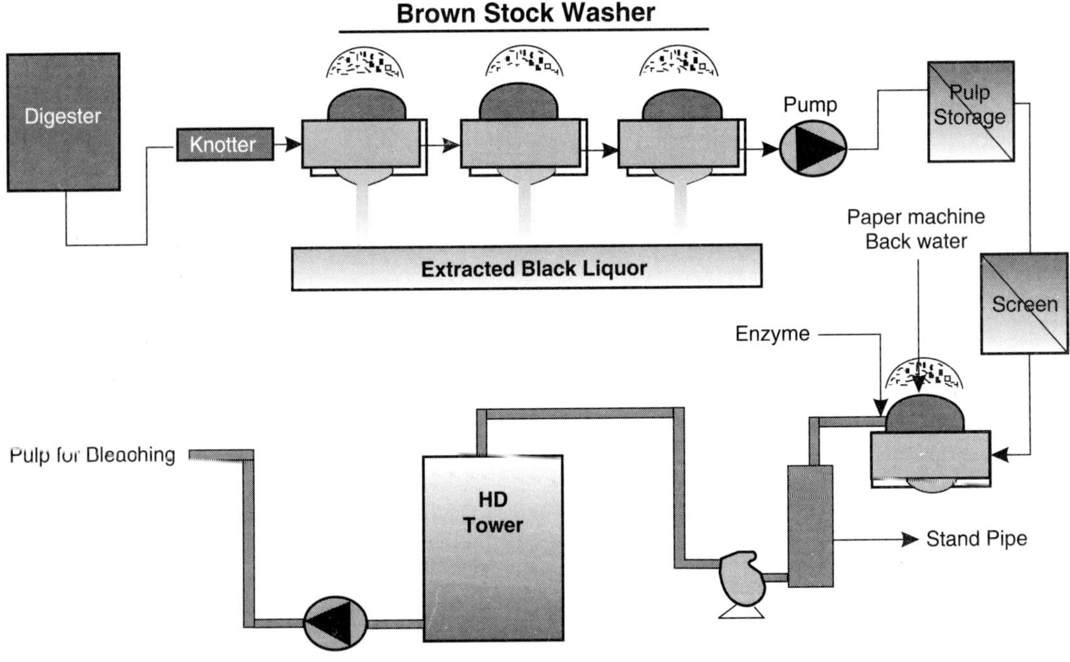


**Fig. 1:** Process flow sheet of commercial mill trial on enzymatic prebleaching of kraft pulp in paper mill in India.

## DECISION OF THE MILL ABOUT OBJECTIVE OF THE ENZYME PREBLEACHING

Mill should decide the clear-cut objective of the xylanase prebleaching of pulps, if the objective is the improvement in brightness of the bleached pulp, or it is the reduction in the usage of elemental chlorine. In this case the chlorine charge is required to be reduced based on the requirement assessed during the evaluation studies. This has to be carried out in the initial startup of the xylanase treatment trial. It has been experienced that there is possibility of reduction in chlorine charge at hypo stage also.

## RETENTION TIME OF ENZYME TREATED PULP IN H.D. TOWER

It has been experienced that due to channeling of the pulp in HD tower, the contact time between enzyme and pulp sometimes is not adequate because of the reduced retention time of the pulp. In this case, the desired efficiency levels are not obtained in respect of brightness gain or reduction of chlorine charge. The problem could be overcome by raising the HD tower level.

## CONCLUSIONS

(1) Enzymatic prebleaching of chemical pulps using xylanase enzyme could prove to be an effective technology in reducing the chlorine demand by 15% - 18% with gain in bleached pulp brightness of 2 to 3% ISO.

(2) Before introducing enzyme prebleaching technology in any mill, it is important to evaluate a particular enzyme preparation for its response towards pulp for various parameters like temperature, pH, enzyme activity and cellulose contamination.

(3) AOX level in bleach effluent could be reduced to a level of at least 25% in case of kraft pulps from wood and non-wood based raw materials employed in Indian paper industry.

(4) The technology needs to be boosted in Indian paper industry in order to explore possibilities of desired standard norms, in respect of the AOX levels.

## REFERENCES

Bailey, M.J., Bailey, P. and Poutanen, K. (1992). Interlaboratory testing methods for assay of xylanase activity. *Journal of Biotechnology,* 23, 257-278.

Bissoon, S., Singh, S. and Christov, L. (2002). Evaluation of the bleach enhancing effect on xylanases on bagasse pulp", In: *Biotechnology in the Pulp and paper Industry,* Eds Viikari, L. and Lantto, R. Pp 247-254.

Jain, R.K, Mathur, R.M. and Kulkmi, A.G. (2001). Enzymatic prebleaching of pulps: challenges & opportunities in Indian pulp & paper industry. *8th International Conference on Biotechnology in the Pulp & Paper Industry 8th (ICBPPI),* Helsinki, Finland, June 4 -8.

Jain, R.K., Mathur, R.M., Vasanta, V., Thakur, Kulkarni, A.G. (2005). Enzymatic Prebleaching of Pulp – A Case Study on Mill Experience", *Proceedings of the International Conference on Pulp and Paper,* Paperex Dec. 3-6, New Delhi India.

Jean, P., Hamilton, J. and Senior, D.J. (1994). Mill Trial Experiences with Xylanase: AOX and Chemical Reductions. *Pulp & Paper,* Canada 95, 12.

Kantelinen, A. *et. al.,* (1993) Proposed mechanism of the enzymatic bleaching of Kraft pulp with xylanases. *Holzforschung,* 47, 318-324.

Kent, K.T. and Thomas, J.W. (1996). Roles for Microbial Enzymes in Pulp and Paper Processing, *ACS Symposium Series,* 655, American Chemical Society Washington DC, Chapter-1, Pp 1-12.

Madlala, A.M., Bissoon, S., Singh, S. and Christov, L. (2001). Xylanase-induced reduction of chlorine dioxide consumption during elemental chlorine-free bleaching of different pulp types *Biotechnology Letters,* 23, 345-351.

Viikari, L., Suurnakki, A., and Buchert, J. (1996b). In: *Enzymes for Pulp and Paper Processing,* Eds Jeffries, T. W. and Viikari, L. ACS Sym. Ser. 655, American Chemical Society: Washington, Pp 15.

Viikari, Liisa, Suurnakki, A. and Buchert, J. (1996a). Enzyme Aided bleaching of Kraft Pulp: Fundamental Mechanism and Practical Applications P.18-19, *ACS Symposium Series,* 655, American Chemical Society Washington DC.

Wong, K.K.Y, De Jong, E., Saddler, J. N. and Allison, R. W. (1997). Mechanism of xylanase aided bleaching of Kraft pulp. Part-II Target substrates *APPITA J,* 50, 509-518.

Wong, K.K.Y, Martin, L.A, Garna, F, M, Saddler, J.N. and De Jong, E. (1997). Bleaching boosting and direct brightness by multiple xylanase treatments during peroxide leaching of Kraft pulps. *Biotechnology Bioengineering,* 54, 312-318.

# 17

# Biotechnological Applications of Thermostable and Alkalistable Microbial Xylanolytic Enzymes

**S. Adhikari and T. Satyanarayana**

*Department of Microbiology, University of Delhi, South Campus, New Delhi-110 021, India*

Cellulose, hemicellulose and lignin are the major polymeric constituents of plant biomass. Xylans are the most common and complex hemicellulosic polysaccharides, and are composed of β-(1, 4)-linked D-xylopyranosyl units, which may be substituted by the side chains of acetic acid, arabinose and 4-O-methyl-D-glucuronic acid. Xylanases [1, 4-β-D xylan xylanohydrolase (EC 3.2.1.8)] are hemicellulases that hydrolyze xylan into xylose, xylobiose and xylooligosaccharides. Xylanases are secreted by bacteria, fungi and protozoa. Biotechnological applications of xylanases include bioconversion of lignocellulosic materials to fermentative products, clarification of juices, improvement of consistency of beer and digestibility of animal feedstock. Nowadays, a considerable attention is being focused on the thermostable and alkalistable xylanases because of their potential applications in pulp and paper industries. The use of xylanases in bleaching step has been shown to reduce the amount of chlorine required to achieve comparable levels of paper brightness, and further, make the process eco-friendly.

## INTRODUCTION

Lignocellulosics are one of the natural resources that have the potential to support increasing energy demands worldwide. Lignocellulosic biomass includes materials such as agricultural and forestry residues, municipal solid waste and other industrial wastes. The chemical properties of the components of lignocellulosics make them a substrate of enormous biotechnological value (Malherbe and Cloete, 2003). These materials are composed of three main components: cellulose, hemicellulose and lignin. Xylan is one of the major components of hemicellulose in agricultural and forests residues. The conversion of xylan to useful products serves to strengthen the overall economics of the processing of lignocellulosic biomass and also to develop new ways of energy production from renewable resources. Biotechnological application of lignocellulosics is a difficult task, and the commercial utilization of this technology has not progressed as fast as one would have desired. The reasons could be the lack of knowledge of enzymes, and their mechanisms involved in the decomposition and conversion of lignocellulosic materials, and their components (Dekker and Richards, 1976; Ljungdahl and Eriksson, 1985; Godden *et al.,* 1989; Mishra *et al.,* 1990).

The wood cells are composed of different layers, which differ from one another with respect to their structure and chemical composition. Basically, cellulose forms a skeleton, which is surrounded by other substances functioning as matrix (hemicelluloses) and encrusting (lignin) materials. Cellulose, hemicellulose and lignin are closely associated, and covalent cross linkages have been suggested to occur between lignin and polysaccharides (lignin-carbohydrate complexes, LCC). The side-groups arabinose, galactose and 4-O-methylglucuronic acids are most frequently perceived as connecting links to lignin (Fengel and Wegener, 1983). It is generally agreed that the hemicellulose molecules are oriented parallel to the cellulosic fibrils.

Biodegradation of xylan requires endo-1, 4-β–xylanases (1, 4-β–D-xylan xylanohydrolase EC 3.2.1.8), β–xylosidase (1, 4 β–D-xylan xylohydrolase EC 3.2.1.37) and several other accessory enzymes to completely hydrolyze xylan (Shah, 2000). These enzymes are considered important tools in several industrial sectors. They are used in saccharification of agricultural, industrial and municipal wastes, pig and poultry cereal based diets, clarification of juices, improvement of bakery products (Gilbert and Hazlewood, 1993), processing of pulp and fibres (Yang *et al.,* 1995; Hespel *et al.,* 1997; Bhat, 2000), thus improving the bleachability of cellulose pulps and decreasing the consumption of chlorine chemicals in pulp bleaching process (Purttula *et al.,* 1993; Viikari *et al.,* 1994; Roncero *et al.,* 2000; Clarke *et al.,* 2000). Since 1986, after Viikari *et al.* (1986) reported the use of xylanases in pre-bleaching of kraft pulp, the xylanases have become even more a good option for an environment-friendly biotechnological application. Most of the xylanases known to date are optimally active below 50°C and active in acidic or neutral pH. However, in the process of enzyme-assisted kraft pulp bleaching, the incoming pulp has a high temperature and alkaline pH, and therefore, the search for thermostable and alkaline xylanase is necessary. Few xylanases are reported to be active and stable at alkaline pH and elevated temperatures (Gessesse and Mamo, 1998; Rani and Nand, 2000; Sharma, 2000). This chapter is aimed at reviewing the state of art on microbial thermostable and alkalistable xylanases and their applications.

## HEMICELLULOSE: OCCURRENCE AND STRUCTURE

Hemicelluloses are non-cellulosic polysaccharides found in plant tissues (Woodward, 1984). The cell wall of land plants has xylan as the most common hemicellulosic polysaccharide representing more than 30% of dry weight (Joseleau *et al.,* 1992). Hemicelluloses are made up of a relatively limited number of sugar residues like D-xylose, D-mannose, D-glucose, D-galactose, 4-O-methyl-D-glucuronic acid, D-galacturonic acid, D-glucuronic acid, and to a lesser extent, L-fucose. Hard-wood and softwood hemicelluloses are characterized on the basis of composition and structure of the individual sugars present in them.

Hemicellulose of hardwood is composed of O-acetyl-4-O-methylglucurono-β-D-xylan often called glucuronoxylan. The xylan content in different hardwood species varies from 15-30%. The backbone consists of β-D-xylopyranose units linked by β-1, 4-bonds having average degrees of polymerization (DP) 150-200, while the softwood hemicellulose is an O-acetylgalactoglucomannan with various structural variations. Arabino-(4-O-methylglucurono) xylan forms a smaller but still substantial part of this wood. The relative proportion of O-acetyl (4-O-methylglucurono) xylan varies significantly, even among species belonging to the same genus (Timell, 1964).

## STRUCTURE AND COMPOSITION OF XYLAN

Xylan, the major component of the hemicellulose, is composed of 1, 4-β–D-xylopyranosyl residues (Whistler and Richards, 1970). Most of the xylans occur as heteropolysaccharides, containing different substituent groups in the backbone and side chains (Biely, 1985; Puls and Poutanen, 1989). Common constituents found on the backbone of xylan are acetyl, arabinosyl and glucuronosyl residues (Whistler and Richards, 1970). Homoxylans, on the other hand, exclusively consist of xylosyl residues. This type of xylan has been isolated from esparto grass (Chanda *et al.,* 1950), tobacco stalks (Eda *et al.,* 1976) and guar seed husk (Montgomery *et al.,* 1956). Xylan from different sources, such as grasses, cereals, softwood and hardwood differ in composition. Birchwood xylan contains 89.3% xylose, 1% arabinose, 1.4% glucose, and 8.3% anhydrouronic acid (Kormelink and Voragen, 1993). Neutral xylan of rice bran contains 46% xylose, 44.9% arabinose, 6.1% galactose, 1.9% glucose and 1.1% anhydrouronic acid (Shibuya and Iwasaki, 1985). Wheat arabinoxylan contains 65.8% xylose, 33.5% arabinose, 0.1% mannose, 0.1% galactose and 0.3% glucose (Gruppen *et al.,* 1992). Corn fiber xylan is one of the complex heteroxylans containing β–(1,4)-linked xylose residues (Shah *et al.,* 1999). It contains 48-54% xylose, 33-35% arabinose, 5-11% galactose, and 3-6% glucuronic acid.

The xylans from hardwood are O-acetyl-4-methylglucouronoxylans. This polysaccharide consists of at least 70 β-xylopyranose residues with an average degree of polymerization (DP) between 150 and 200, and linked by β-1, 4 glycosidic bonds (Timell, 1964). Every tenth xylose residue carries a 4-O-methylglucuronic acid attached to the second position of xylose (Woodward, 1984). Hardwood xylans are highly acetylated. Birchwood xylan contains more than 1 mole acetic acid per 2 moles of xylose (Dekker, 1989). Acetylation generally occurs more in C-2 than C-3 position, but acetylation at both of these positions has been reported (Bouveng, 1961; Lindberg *et al.,* 1973). Acetyl groups are responsible for partial solubility of xylan in water (Whistler and Richards, 1970) and these groups can be easily removed by alkali extraction (Dekker, 1989).

Softwoods are composed of arabino-4-O-methylglucuronoxylans. They have a higher 4-O-methylglucuronic acid content than hardwood xylans. The 4-O-methylglucuronic residues are attached to the C-2 position. Softwood xylans are non-acetylated, and instead have α–1, 3-glycosidic bonds in the C-3 position of xylose (Puls and Schuseil, 1993). Almost 12% of xylose residues have arabinosyl substituents (Wong *et al.,* 1988). The ratio of β–D-xylopyranose, 4-O-methyl- α–D-glucuronic acid and L-arabinofuranose is 100:20:13 (Puls and Schuseil, 1993). Softwood xylans are less branched and shorter than hardwood xylans with DP of 70-130.

Xylan from grasses is highly branched, and contains large proportions of L-arabinofuranosyl units but have smaller proportions of uronic acids. Uronic acids are probably linked to the C-2 position of xylose. An L-arabinose side chain substitutes every seventh xylose unit of xylan from the internodes of wheat. Higher arabinose content in xylan with greater ratio of xylose: arabinose in the nodes of both wheat and barely has been reported (Puls, 1992). Xylan from graminaceous plants contains O-acetyl groups, and 1-2% phenolic acid substituents (Bacon *et al.,* 1975). One in every 31-arabinose residues of barley straw xylan has been shown to be esterified with p-coumaric acid, and one in every 15 with ferulic acid (Mueller *et al.,* 1986).

## CRYSTAL STRUCTURE XYLAN

The X-ray polarization and infrared techniques have been utilized to study the conformation of the backbone in (4-methylglucurono) xylan. True xylan from esparto grass and a (4-O-methylglucurono) xylan from birch have been found to exist in two crystallographic forms, one a "hydrate" and the other "dry", one form being converted into the other by drying or wetting. Fibre structure of a (4-O-methylglucurono) xylan from birchwood revealed that the xylan backbone possesses a three-fold screw axis, with a rotation of 120 degree for each xylose residue and a repeating length of 15Å. This conformation has led to the conclusion that the intermolecular hydrogen bond O-3H...O'-5, previously shown in cellulose, was probably also present in xylan, although with a somewhat greater oxygen-oxygen distance. Energy calculations suggest that the D-xylose ring exists in the common '$C_1$ conformation', which indicates that in both the 1-4 and 1-3 glycosidic linked xylans, the bonding is diequatiorial (1e-4e). The intra-chain hydrogen bond O(2') H...O(6) reinforces the O(3)...O(5') hydrogen bond to support a two-fold extended helical structure like a ribbon and O(6) group hydrogen bonds to O(3) atoms in adjacent chains to form sheets (Atkins, 1992).

Xylan from white birchwood was one of the first polymers to be obtained in the form of single crystals. The crude xylan was purified by boiling 0.2% oxalic acid solution, to give a highly swollen, water-insoluble fraction in a yield of 7.4% of wood. Further purification was effected by autoclaving in water at 120°C and crystallization. The final crystallization was carried out with a 0.2% aqueous solution, that when cooled to 60-70°C, deposited crystals in the form of hexagonal pellets. The average thickness of the layers is 50Å, and the average length of fully extended xylan is 105Å (Timell, 1964).

## XYLANOLYTIC ENZYMES

Due to the heterogeneity of xylan, its hydrolysis requires the action of a complex xylanolytic enzyme system comprising β–1,4-endoxylanase, β–xylosidase, α–L-arabinofuranosidase, α–glucuronisidase, acetyl xylan esterase and phenolic acid esterases. These entire enzymes act together cooperatively to convert xylan to its constituent sugars. Hence, xylanolytic enzymes refer to a multienzyme complex.

Heteroxylan consists of a variety of glycosidic bonds due to different substituents in side chain and backbone, and hence, it requires a synergistic action of a number of main-chain and side-chain cleaving enzyme activities for complete hydrolysis (Biely, 1985). Thus the degradation of such a complex polysaccharide may require the synergistic action of different components of the xylanolytic enzyme system (Bailey *et al.,* 1992; Sunna and Antranikian, 1997) (Figure 1). The presence of such

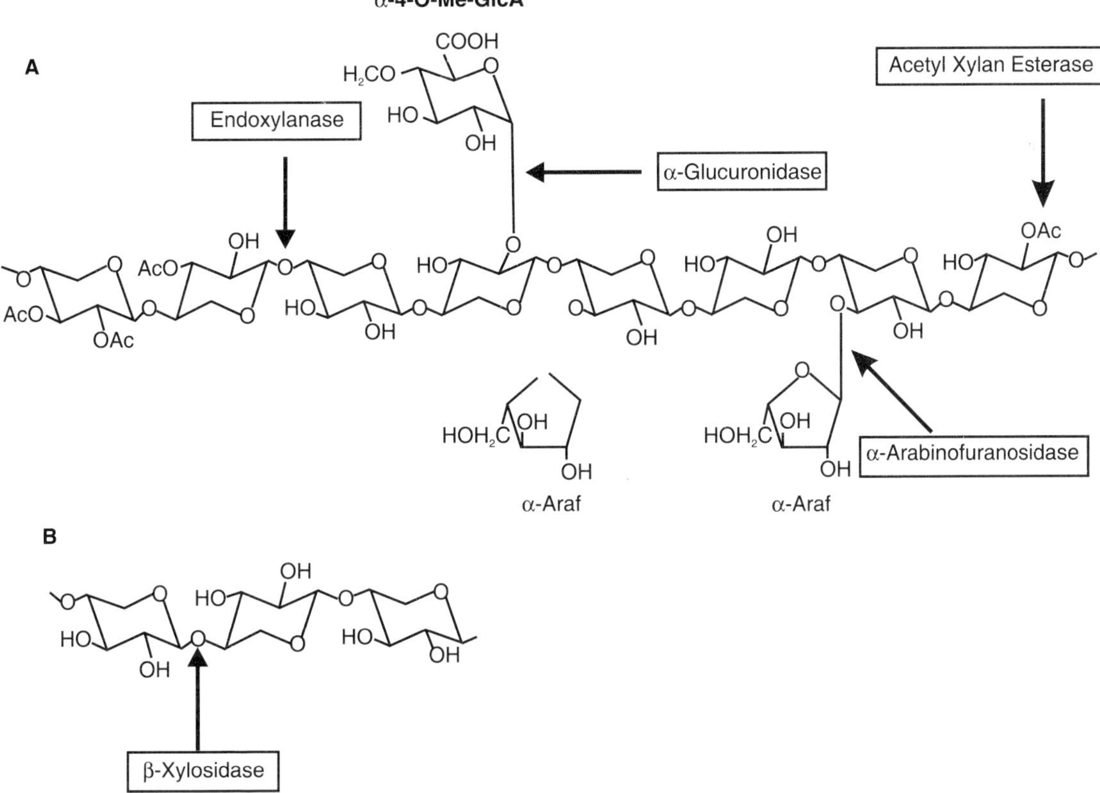


**Fig. 1(a & b).** The xylanolytic enzymes involved in the degradation of xylan (Sunna and Antranikian, 1997)

a multicomponent xylanolytic enzyme system is quite widespread amongst bacteria and fungi (Woodward, 1984; Wong *et al.,* 1988; Coughlan and Hazlewood, 1993). Following are the components of xylanolytic enzyme systems.

## ENDOXYLANASE [β–1,4-ENDOXYLANASE(1,4-β–D-XYLAN XYLOHYDROLASE) [EC 3.2.1.8)]

This enzyme cleaves the internal glycosidic linkages of the heteroxylan backbone resulting in a decreased DP of the substrate. Substrate attack is not random, and the bonds to be hydrolyzed depend on the nature of the substrate i.e. its length, degree of branching or the presence of substituents (Reilly, 1981). During initial hydrolysis of xylan, the main products formed are xylooligosaccharides. As hydrolysis proceeds, these xylooligosaccharides are further hydrolyzed to xylotriose, xylobiose, and xylose (Dekker and Richards, 1976; Wong *et al.,* 1988; Debeire-Gosselin *et al.,* 1992). The end products released from the hydrolysis of xylan (e.g., xylose, xylobiose and xylotriose, and /or arabinose) have been used to differentiate endo-acting xylanases. Xylanases, thus, may be classified as non-debranching (arabinose-nonliberating) or debranching (arabinose-liberating) enzymes (Dekker and Richards, 1976; Reilly, 1981). Arabinose-cleaving endoxylanases have been purified from *Aspergillus niger* (Takenishi and Tsujisaka, 1975), *Neurospora crassa* (Mishra *et al.,* 1984), *Streptomyces roseislerotiacus* (Grabski and Jeffries, 1991), *Talaromyces byssochlamydoides* (Yoshioka *et al.,* 1981), and *Trichoderma koningii* (Wood and McCrae, 1986). Many organisms are able to produce both debranching and non-debranching xylanases, leading to a maximum efficiency of xylan hydrolysis (Takenishi and Tsujisaka, 1975; Wong *et al.,* 1988).

## XYLOSIDASES [β–D-XYLOSIDASE (β–D-XYLOSIDE XYLOHYDROLASE) [EC 3.2.1.37]

They are exoglycosidases that hydrolyze short xylooligosaccharides and xylobiose from the nonreducing end to liberate xylose (Wong *et al.,* 1988). True β–xylosidases have been reported to be able to cleave artificial substrates like p-nitrophenyl β–D-xyloside (Coughlan and Hazlewood, 1993). β–xylosidases are larger enzymes with molecular weights between 60 and 360 kD, and they may be mono or dimeric proteins and have been reported both in bacteria and fungi. In bacteria and yeasts, β–xylosidase appears to be mainly cell associated (Biely, 1985), although extracellular β–xylosidase activity has also been reported (Deleyn *et al.,* 1978; Zimmermann *et al.,* 1988; Dobberstein and Emeis, 1991). In the yeast *Cryptococcus albidus,* xylobiose and xylotriose enter the cells through a β–xyloside permease transport system and are hydrolyzed by β–xylosidase to xylose (Biely, 1985).

In general, purified β–xylosidases are unable to hydrolyze xylan. However, there are some reports of β–xylosidases that are able to attack xylan slowly to liberate xylose (Dekker and Richard, 1976). Among the xylooligomers, xylobiose is usually the best substrate. The affinity of the enzyme towards xylooligosaccharides decreases with increasing DP (Van Doorslaer *et al.,* 1985). A number of β–xylosidases has been reported to possess α–arabinosidase activity, e.g. the enzymes from *A. niger* (Rodionova *et al.,* 1983), *Trichoderma reesei* (Poutanen and Puls, 1988), *T. ethanolicus* (Shao and Wiegel, 1992), and *Penicillium wortmanni* (Deleyn *et al.,* 1978). Most of the β–xylosidases studied are reported to be completely inhibited by their hydrolysis product xylose (Poutanen and Puls, 1988; Dobberstein and Emeis, 1991). Important role of β–xylosidase appears to be relieving the end product inhibition of endoxylanase. *Thermomonospora fusca* produces a cell-associated β–xylosidase that was found to enhance the saccharification of xylan (Bachmann and McCarthy, 1991). Many β–xylosidases have transxylosidation (transferase) activity, especially at high substrate concentrations resulting in products of higher molecular weight than the substrate (Conard and Noethen, 1984).

## ARABINOFURANOSIDASE

Arabinofuranosidases have great role in the degradation of xylan and very few have been isolated from the organisms and characterized. Arabinofuranosidases are of two types: exo-acting, α–L-arabinofuranosidase (EC 3.2.1.55), which is active against p-nitrophenyl-α–L-arabinofuranosides and branched arabinans, and the endo-1,5-α–L-arabinase (EC 3.2.1.99) is active only towards linear arabinans (Van der Veen *et al.,* 1992). Endoarabinases have been reported in *Bacillus subtilis* (Kaji and Saheki, 1975), *Clostridium felsineum* (Kaji, 1984), and various fungi (Kaji and Tagaawa, 1970; Poutanen, 1988; Bachmann and McCarthy, 1991; Kaneko *et al.,* 1993). These enzymes hydrolyze 1, 5-α–L-arabinans, but not the chromogenic substrate phenyl-α–L-arabainofuranoside or gum arabic. Most of the arabinan-degrading enzymes studied are of the exo-acting type (Dekker and Richards, 1976). The size of native arabinofuranosidases may reach up to 495 kD and are found in mono-, di-, tetra-, hexa-, and octameric forms. The production of arabinofuranosidase in several actinomycetes seems to be induced by arabinan, xylan, and wheat bran (Mackenzie *et al.,* 1984; Zimmermann *et al.,* 1988). α–L-arabinofuranosidases capable of hydrolyzing 1, 3-and 1, 5-α–L-arabinofuranosyl linkages in arabinoxylan have been reported in *A. niger* (Kaji and Tagawa, 1970) and *B. substilis* (Kaji, 1984). The *A. niger* enzyme first attacks the α–L-1,3 linked arabinofuranosyl residues in arabinan to an extent of 30% and then proceeds with the slow attack of the α–L-1,5-arabinan, which will then be completely converted to arabinose (Kaji and Tagawa, 1970). A novel purified arabinosidase, 1, 4-β–D-arabinoxylan arabinofuranohydrolase from *A. awamori* is highly specific for arabinoxylans, and it is able to release only arabinose from arabinoxylan, and xylan backbone is not degraded, and thus, xylooligosaccharides are not produced (Kormelink *et al.,* 1991a). Synergism between α–L-arabinofuranosidases and xylanases has been reported. An increase in xylose, xylobiose and arabinose production was observed when both enzymes were used simultaneously (Greve *et al.,* 1984; Poutanen *et al.,* 1988).

## α–GLUCURONIDASE : α–D-GLUCURONIDASES (EC 3.2.1. 139)

This enzyme hydrolyzes α–1, 2 linkages between glucuronic acid and xylose residues in glucuronoxylan. The substrate specificity of α–glucuronidase depends on the enzyme source. The enzymes from *Agaricus bisporus* and *Streptomyuces olivochromogens* require a low-molecular weight glucuronoxylan substrate (Puls *et al.,* 1987). They release 4-O-methylglucuronic acid from 4-O-methyl-glucuronose-substitued xylooligomers, but not from the polymer (Puls and Schuseil, 1993). The ruminal bacterium *Fibrobacter succniogenes* produces a α–glucuronidase that is unable to release 4-O-methylglucuronic acid from intact xylan.

## ACETYLXYLAN ESTERASES (EC 3.1.1.6)

Acetylxylan esterases remove the O-acetyl substituents at the C-2 and C-3 positions of xylose residues in acetylxylan. Their production in fungi and bacteria was first reported in 1985 (Biely, 1985). Very few acetylxylan esterases have been purified and little is known about their physiochemical properties. *Trichoderma reesei* produces two monomeric isoenzymes that are glycosylated and have molecular mass of 34 KDa (Sundbey and Poutanen, 1991). The acetyl esterase from *T. reesei* liberates acetic acid from acetylated xylooligomers. The newly described acetylxylan esterase I and II from *Thermoanaerobacterium* sp JW/SL-YS 485 deacetylates acetylated xylan. The enzyme I acts on short acetylated xylooligosaccharide intracellularly (Shao *et al.,* 1995), while the II removes the acetyl group from xylan extracellularly.

## Production of Xylanases

Development of suitable medium and culture conditions is obligatory in order to obtain maximum yield of an enzyme (Narang and Satyanarayana, 2001). Thermostable and alkalistable xylanase production has been studied in both submerged as well as solid state fermentations by bacteria and fungi.

### SUBMERGED FERMENTATION (SmF)

Submerged fermentation involves the growth and fermentation by microorganisms in the presence of sufficient water so as to dissolve the whole medium components in it and is well characterized with homogenous conditions maintained throughout the processing. The physical parameters such as pH, temperature and dissolved oxygen can be effectively controlled and scaling up of the process is feasible due to defined conditions. The choice of carbon and nitrogen sources varies from organism to organism. Xylanase production by microbes may be growth associated or growth-independent. Dey *et al.* (1992) reported non-growth associated xylanase production by *Bacillus* sp., while Esteban *et al.* (1982) and Nakamura *et al.* (1993a) observed xylanase production coupled with bacterial exponential phase of growth. Bocchini *et al.* (2002) described the production of thermostable and cellulase-free xylanase from *Bacillus circulans* through SmF process. There are very few reports on the production of xylanases that are cellulase-free, thermostable and alkalistable. Sharma (2000) produced thermoalkalistable xylanase by *Geobacillus thermoleovorans* AP7 (syn. *Bacillus thermoleovorans*) in submerged fermentation. Techapun *et al.* (2002) have also reported thermostable, alkalitolerant and cellulase-free xylanase by *Streptomyces* sp. 106 in submerged fermentation. Most of the bacterial xylanases are produced by this process, and the production is optimized by monitoring various physical and nutritional parameters.

### PHYSICAL PARAMETERS

The most important physical parameters suitable for the growth and production of industrially important xylanases are temperature, pH, agitation, and dissolved oxygen.

#### Temperature

Xylanases have been reported from both mesophiles and thermophiles. More reports pertain to thermophilic bacteria because thermostable enzymes are needed in paper bleaching processes. Intensive investigations have been carried out on xylanolytic enzymes of mesophilic bacteria and fungi (Wong *et al.,* 1988; Coughlan and Hazelwood, 1993). Most of the xylanases from fungi and bacteria show their growth and optimum activity between 40-70°C, while extremophiles like *Dictyoglomus*, *Geobacillus* and *Thermotoga* show their optima beyond this temperature.

### pH

pH has a profound effect on the production as well as stability and activity of the extracellular xylanases. In general, fungal xylanases are less stable under alkaline conditions. The pH optima for the production of enzyme and its activity depend on its source. Alkaliphiles have been reported to frequently produce xylanases with pH optima near neutral range (Horikoshi and Atsukawa, 1973; Okazaki *et al.,* 1984; Tsujibo *et al.,* 1990).

### Agitation and Aeration

Mechanical agitation and aeration are crucial factors for the production of xylanases, since this aid in uniform distribution of the nutrients and supply oxygen for the growth of the organisms. Agitation rate of 250 rpm supported higher xylanase production by *Bacillus* sp. (Okazaki *et al.,* 1984). Xylanase production as well as growth rate were favoured at 150 rpm and 1 vvm aeration in *Thermomyces lanuginosus* (Purkarthofer *et al.,* 1993). Agitation at 200 rpm was optimum for xylanase production in *Bacillus* Sam-3 strain (Shah *et al.,* 1999) and *B. licheniformis* (Archana and Satyanarayana, 1998). Techapun *et al.* (2003) recorded 1 vvm of aeration and agitation of 150 rpm for the production of thermostable and cellulase-free xylanase by *Streptomyces* sp. Ab 106.

## NUTRITIONAL PARAMETERS

### Carbon Sources

The different sources of carbon as well as their concentrations are important factors not only for the optimal production of enzymes but also their localization (extracellular or cell bound). Xylan has been widely reported to support xylanase production (Esteban *et al.,* 1982; Okazaki *et al.,* 1984 Nakanishi *et al.,* 1992). Among lignocellulosic substrates, wheat bran was reported to be a good carbon source for xylanase production (Wong *et al.,* 1988; Dey *et al.,* 1992; He *et al.,* 1993). Rice straw was found to be the best carbon source for xylanase production in *Bacillus* AB 16 (Dhillon *et al.,* 2000).

### Nitrogen sources

The source of nitrogen in the culture medium is another important parameter influencing the growth and production of metabolites. Organic form of nitrogen such as yeast extract, tryptone, peptone, and soybean meal and corn steep solids has been extensively used for the production of xylanases. Yeast extract in combination with peptone enhanced xylanase production in *Bacillus* sp.

SSP-34 significantly (Subramaniyan and Prema, 1998). Similarly, tryptone supported maximum xylanase secretion in *Bacillus* sp. B16 (Dhillon *et al.,* 2000). Soybean meal increased the production by 25% in *Bacillus* Sam-3 (Shah *et al.,* 1999).

## Trace elements and vitamins

Trace elements and vitamins have been reported to be important for the production of enzymes, especially by thermoanaerobes (Antranikian *et al.,* 1987) and some Bacilli (Ara *et al.,* 1996). Besides carbon and nitrogen sources, hyperthermophiles usually require the addition of different trace elements and vitamins to the medium. *Thermotoga* sp. FjSS3-B.1 required casamino acid in the culture medium. A requirement for amino acids was reported for the production of xylanase by *Streptomyces* sp QG-11-3 (Beg *et al.,* 2000a). Vitamin solution and trace elements were required for xylanase production in *Bacillus* sp. AB16 (Dhillon *et al.,* 2000). Dey *et al.* (1992) reported a high xylanase production in a medium containing wheat bran and yeast extract, which provided required trace elements and vitamins.

## SOLID STATE FERMENTATION (SSF)

In contrast to submerged fermentation, the solid state fermentation involves the growth and fermentation by microorganisms, especially fungi, on moist, water insoluble, and solid substrate in the absence or near absence of free water. It offers distinct advantages as it requires no fermentation controls, and lesser water volumes lead to greater product titre and is considered the self-control process. The SSF is useful in the area of solid waste management, biomass energy conservation and production of various primary and secondary metabolites. In SSF, the solid substrate particles serve as source of carbon, nitrogen, salts and growth factors and also provide mechanical support and anchorage to microbial cells.

There are several factors (physical, chemical and biochemical), which one needs to consider for any SSF (Pandey *et al.,* 2001). The major factors that affect microbial growth and activity in SSF include selection of suitable microorganism and substrate, pretreatment of the substrate, particle size (inter particle space and surface area) of the substrate, water content and water activity ($a_w$) of the substrate, relative humidity, type and size of inoculum and temperature of the fermenting matter/removal of metabolic heat, period of cultivation, maintenance of uniformity in the environment of SSF, and the gaseous atmosphere (oxygen consumption rate and carbon dioxide evolution rate).

Alam *et al.* (1994) reported production of a thermostable and cellulase-free xylanase by *T. lanuginosa* in SSF stable at 70°C. A thermostable xylanase preparation from *Humicola* sp. exhibited temperature optima at 75°C (Da-Silva *et al.,* 1994). Similarly, xylanase from *Thermomonospora* sp. produced through SSF was stable at 80°C (Srivastava, 1993). Bacteria in general are considered to be less amenable to SSF (Lonsane and Ramesh, 1990). Few bacteria belonging to genera *Bacillus, Streptomyces, Clostridium* and *Cellulomonas* have, however, been shown to produce thermostable hemicellulases (Uchino and Nakane, 1981; Kumar and Singh, 1992; Kanko *et al.,* 1993; Shao *et al.,*

1995; Babu and Satyanarayana, 1996). Dunlop (1995) described a bacterium, isolated from wood compost, to produce xylanase that was active at 80°C.

Agro-industrial residues are generally considered to be the best substrates for the SSF processes (Pandey *et al.,* 2001). A number of such substrates have been employed for the cultivation of microorganisms to produce enzymes. Some of the substrates that have been used included sugar cane bagasse, wheat bran, rice bran, maize bran, gram bran, wheat straw, rice straw, rice husk, corn cobs and others. Wheat bran was the best among the various substrates used in SSF for the production of xylanases. This could be attributed to its different and more suitable physiochemical structure than other lignocellulosics (Krik *et al.,*1983; Jaleel *et al.,* 1992). Some *Bacillus* spp. have been shown to grow and produce enzymes in SSF (Archana and Satyanarayana, 1997; Lonsane and Ramesh, 1990). Xylanase production by SSF led to 22-fold increase in enzyme levels in *B. licheniformis* (Archana and Satyanarayana, 1997). High-level xylanase production was achieved by an alkaliphilic *Bacillus* sp. using wheat bran as the substrate (Gessesse and Mamo, 1999). Narang *et al.* (2001) and Kamra and Satyanarayana (2004) reported a high enzyme titres using wheat straw and wheat bran as the substrates by thermophilic fungi, *Melanocarpus albomyces* IIS68 and *Humicola lanuginosa,* respectively. The production of a thermostable xylanase in SSF using wheat bran and eucalyptus kraft pulp as primary substrate was reported in *Streptomyces* sp. QH-11-3 (Beg *et al.,* 2000a). Extremely thermophilic bacteria were not able to grow in solid state fermentation due to requirement of higher levels of available water, than is possible in SSF, and also due to high evaporation rate at elevated temperatures.

## NOVEL XYLANASES

Xylanases have been reported to be produced by a vast variety of prokaryotes and eukaryotes including bacteria, fungi, protozoa, insects, crustaceans and germinating seeds (Dekker and Richards, 1976). To develop novel process in microbial biotechnology, organisms with outstanding properties are required, and much emphasis has been laid on extremophiles for specialized industrial applications (Satyanarayana *et al.,* 1988; Satyanarayana, 1994; Adams and Kelley, 1998; Edwards, 1999). Xylanase is one of the industrial enzymes with a high demand particularly in pulp and paper industry. The demand for microbial xylanases used in pulp and paper processing has increased tremendously, making it a sensible economic choice in bleaching. In keeping with the requirements of the pulping operations, the search for xylanases with novel properties was felt necessary. Generally, the xylanase for pulping should have low molecular weight to ensure the enzyme access to the pulp matrix, thermal stability to ensure activity at or beyond 70°C, activity at alkaline pH to ensure the action on kraft pulps, specificity for xylan and cellulase-free nature to avoid cellulose viscosity losses, affinity for acidic side chains to maximize HexA (hexenuronic acid) removal and alkaline isoelectric point to bind enzyme more readily to the pulp fibres. The alkaliphilic and thermophilic microorganisms have been found to produce alkalistable and thermostable xylanases that do not require adjustment of temperature and pH, which are better suited to make the process cost effective. A list of microorganisms producing alkalistable and thermostable xylanases is given in Table 1.

**Table 1. Properties of alkaliphilic and thermophilic microbial xylanases**

| *Microorganisms* | *Optimum conditions* | | | | *pI* | *MW (kD)* | *Reference* |
|---|---|---|---|---|---|---|---|
| | *(Growth)* | | *(Enzyme activity)* | | | | |
| | *pH* | *Temp(°C)* | *pH* | *Temp(°C)* | | | |
| *Bacillus amyloliquefaciens* | 7 | 45 | 6.8-7 | 80 | 10.1 | 19.6 | Breccia *et al.*, 1998 |
| *Bacillus circulans* AB16 | 8 | 55 | 6 | 80 | | | Dhillon *et al.*, 2000; Dhillon and Khanna, 2000 |
| *Bacillus pumilus* | 10 | 45 | 8-9 | 55-60 | | | Duarte *et al.*, 1999; Duarte *et al.*, 2000; Santos *et al.*, 2002. |
| | | | | | | *45* | |
| *Bacillus licheniformis* A99 | 5-10 | 50 | 6-7.5 | 60 | | | Archana and Satyanarayana 2003 |
| *B. stearothermophilus* | 9.0 | 60 | 6.0-7.0 | 60-70 | | | Rajaram and Varma 1990 |
| *B. stearothermophilus* T-6 | 7.0-7.3 | 60 | 6.5 | 75 | 9 | 43 | Kashin *et al.*, 1993 |
| *Geobacillus thermoleovorans* AP7 | 7.5 | 70 | 8.5 | 80 | | | Sharma 2000 |
| *B. thermoleovorans* K-3d | 7.0 | 70 | 6.0-9.0 | 30-90 | | | Sunna *et al.*, 1997 |
| *Bacillus* sp. | 8 | 37 | 10 | 60 | | | Zheng and Zhang 2000 |
| *Bacillus* sp. | 9 | 37 | 9 | 50 | | | Yang *et al.*, 1995 |
| *Bacillus* C-59-2 | 8.0 | 37 | 5.5-9.0 | 60 | | | Horikoshi and Atsukawa 1973 |
| *Bacillus* C-125 | 10.5 | - | 6.0-10 | 70 | | | Honda *et al.*, 1985 |
| *Bacillus* 41 M-1 | 10.3 | 37 | 9.0 | 50 | | | Nakamura *et al.*, 1993a |
| *Bacillus* sp. NCIM 2128 | 9 | 28 | 8.0 | 60 | | | Balkrishnan *et al.*, 1992 |
| *Bacillus* sp. NCIM 59 | 10 | 50 | 7 | 60 | 15.8, 35 | 4,8 | Kulkarni and Rao 1996; Nath and Rao 2000 |
| *Bacillus* sp. NG-27 | 9-10 | 27 | 7-8.4 | 70 | | | Gupta *et al.*, 1992 |
| *Bacillus* sp. SPS-0 | 8.2 | 60 | 6 | 75 | | 99 | Bataillon *et al.*, 2000 |
| *Bacillus* TAR-1 | 10.5 | 50 | 9.0 | 70 | >9.3 | 23 | Takahashi *et al.*, 2000 |
| *Bacillus* sp. | 9-10 | 45-50 | | | | | Okazaki *et al.*, 1984 |
| W1 | | | 6.0 | 65 | 8.5 | 21.5 | |
| W2 | | | 6.0 | 65 | 8.3 | 22.5 | |
| W3 | | | 7.0-9.5 | 70 | 3.7 | 50 | |
| W4 | | | 6.0-7.0 | 70 | 3.6 | 49.5 | |
| *Clostridium absonum* CFR-702 | 8.5 | 75 | 8.5 | 75 | | | Rani and Nand, 2000 |
| *Clostridium* | | | | | | | Berenger *et al.*, 1985 |
| *stercorarium* A | 7.0 | 65 | 5.5-7.0 | 75 | 4.5 | 4.4 | |
| B | | | | | 44 | 72 | |
| C | | | | | 4.3 | 62 | |
| *Dictyoglomus thermophilum* B1 | 7.0 | 68 | 7.0 | 80 | | | Mathrani and Ahring 1992 |
| *Dictyoglomus* sp. B1 | 8.0 | 70 | 8 | 73 | | | Adamsen *et al.*, 1995 |
| *Dictyoglomus thermophilum* | | | 6.5 | 85 | | 39.8 | Morris *et al.*, 1998 and Gibbs *et al.*, 1995 |
| *Streptomyces thermoviolaceus* OPC520 I | 7 | 50 | 7 | 70 | 4.2 | 54 | Tsujibo *et al.*, 1992 |
| Streptomyces *viridosporus* T7A | 7-8.0 | 70 | 5-9.0 | 30-80 | | | Timothy 1997 |
| *Streptomyces* sp. 106 | 7.0 | 50 | 7-9.0 | 50-80 | | | Techapun *et al.*, 2003 |

| | | | | | | |
|---|---|---|---|---|---|---|
| *Thermoactinomyces thalophilus* sub gr.C | 8.5-9 | 50 | 8.5-9 | 65 | | Kohli *et al.*, 2001 |
| *Thermomonospora fusca* YX | 8.0 | 50 | 6.0-8.0 | 70 | | Irwin *et al.*, 1994 |
| *Thermomonospora sp.* | 9 | 50 | 7 | 70 | | George *et al.*, 2001 |
| *Thermotoga chromogena* MT 814 | 8.0 | 50 | 5.0-8.0 | 75 | | McCarthy *et al.*, 1985 |
| *T. fusca* BD 21 | 8.0 | 50 | 6.0-8.0 | 65 | | Bachmann and Mc Carthy 1991 |
| *Thermotoga* spFjSS3-B1 | | | 5.3 | 100 | 31 | Leuschner and Antranikian 1995 |
| *Thermotoga maritima* MSBS | | | 6.2, 5.4 | 92,105 | 120,40 | Winterhalter and Liebl 1995 |
| *Thermotoga thermamum* | | | 6.0 | 90 | 40 | Leuschner and Antranikian 1995 |

## THERMOPHILIC XYLANASES

Majority of the xylanases studied are of fungal or bacterial origin and most of them were found to be optimally active at or near mesophilic temperatures (approx. 40-60°C) (Sunna and Antranikian, 1997; Subranmaniyan and Prema, 2002). However, the mesophilic enzymes have limitations due to high temperature used in bleaching (Viikari *et al.,* 1994; Gibbs *et al.,* 1995; Saul *et al.,* 1995; Morris *et al.,* 1998; Sunna *et al.,* 2000). Thermophilic bacteria received considerable attention as sources of highly active thermostable xylanolytic enzymes. Many advantages such as reduced contamination risk and faster reaction rates have been reported for thermostable enzymes of thermophiles. Thermally stable enzymes can be used at high temperature for efficient hydrolysis of substrates, and further may be effectively recycled (George *et al.,* 2003). Xylanase treatment of the pulp at elevated temperatures opens up cell wall structure, thereby facilitating lignin removal in the subsequent bleaching stages in pulp and paper processing. Several investigations have been carried out on the production of thermostable xylanases by thermophiles. A number of thermophilic (optimal growth at 50-80°C) and hyperthermophilic (optimal growth > 80°C) xylanase producing microorganisms have been isolated from a variety of sources including thermal springs, hot pools and self-heating decaying organic debris (Harris *et al.,* 1997; Sunna *et al.,* 1997; Vieille and Zekus, 2001; Singh *et al.,* 2003) and their enzymatic properties were studied. The xylanase from thermophilic bacteria such as *Thermomonospora fusca* (McCarthy *et al.,* 1985), thermophilic *Bacillus* sp. (Gruninger and Fiechtter, 1986) and *B. stearothermophilus* (Khasin *et al.,* 1993) show an optimum temperature in the range of 65-80°C. Xylanase A from the thermophilic anaerobe *Clostridium stercorarium* has a temperature optimum of 70°C and has a half-life of 90 min at 80°C. Xylanase from *Thermotoga* sp. is one of the most thermostable xylanases reported to date with an optimum temperature of 105°C at pH 5.5 with a half-life of 90 min at 95°C (Simpson *et al.,* 1991). Xylanase from *Dictyoglomus* sp. exhibited a half-life of 90 min at 80°C (Mathrani and Ahring, 1992). A thermophilic mould, *Thermoascus aurantiacus* has been reported to be stable at 70°C for 24 hours with a half-life of 54 min at 80°C (Yu *et al.,* 1987). *Bacillus stearothermophilus* is one of the aerobic, thermophilic bacteria shown to actively degrade plant polymers like xylan (Gilead & Shoham, 1995). Anaerobic cells that grow and thrive at higher temperature and degrade xylan also include *Thermoanaerobacterium ethanolicus,*

*Thermoanaerobacterium acetigenum* (Nielsen *et al.,* 1993), *Thermoanaerobium brockii, Thermoanaerobacterium* sp. strain JW/SL-YS485 (Shao *et al.,* 1995) and *Thermobacteriodes* sp. (Biely, 1985). Endo-xylanases from the thermophilic actinomycete *Microtetraspora flexuosa* SIIX were reported to have an optimum temperature of 80°C at pH 6.0 (Berens *et al.,* 1996). Hyperthermophilic archaea including *Thermococcus ziilligi* (Uhil and Damiel, 1999), *Sulfolobus solfafaricus* (Uhil and Damiel, 1999), *Sulfolobus solfafaricus* (Cannio *et al.,* 2004) *Pyrodictium abyssi* (Andrade *et al.,* 1999; Niehans and Antranikian, 1999) and a number of *Thermophilum* strains (Bragger *et al.,* 1989) have been reported to produce xylanases.

Analysis of crystal structure, sequence alignments and site directed mutagenesis studies have shown similarities between mesophilic and thermophilic xylanases, but their enhanced thermal stability could be due to certain variations, such as increase in the number of salt bridges and H-bonds (Gruber *et al.,* 1998; Hakulinen *et al.,* 2003), an improved internal packing (Hakulinen *et al.,* 2003), an increased number of charged surface residues (Turenen *et al.,* 2002) often as tandem repeats, thermostabilizing domains (Winterhailter *et al.,* 1995; Fontes *et al.,* 1995; Zverlov *et al.,* 1996), the introduction of disulphide bridges in particular at the N- or C-termini or in the α–helix region (Wakarchuk *et al.,* 1994; Kumar *et al.,* 2000; Turunen *et al.,* 2001). The thermostability of a thermostable xylanase from *Thermoanaerobacterium saccharolyticum* has been correlated with the N-terminal domains of the enzyme (Lee *et al.,* 1993). *Clostridium thermocellum* also contained a homologue of this domain that appeared to confer stability. There is a 28% sequence homology between the thermostabilizing domains of xylanase Y from *C. thermocellum* and xylanase A from *T. saccharolyticum* (Fontes *et al.,* 1995). Recently, the thermostabilizing role of calcium on modular family 10 xylanase has been demonstrated (Abou-Hachen *et al.,* 2003). *Bacillus* D3 xylanase was also shown to have unique adaptation strategy, where a series of surface aromatic residues from a cluster of "sticky patches" between pairs of molecules, and these intermolecular hydrophobic interactions are believed to contribute to the thermostability of xylanase (Harris *et al.,* 1997; Connerton *et al.,* 1999).

Detailed structural analysis of some thermostable xylanases of family 10 and 11 has shown some structural adaptation strategies for their thermostability. The crystal structure of the thermostable xylanase (family 11/G) from *Thermomyces lanuginosus,* as determined by single-crystal X-ray diffraction (Figure 2a) showed that it is a compact, globular protein with two heavily twisted β-sheets. Its structure is similar to other family 11 xylanases. One sheet forms the "outer" surface of the enzyme and consists of five antiparallel strands while the other β-sheet consists of nine antiparallel strands. The most prominent feature of the structure is a long cleft spanning the whole molecule and containing the active site. The reasons for thermostablity of *Thermomyces lanuginosus* xylanase were analyzed by comparing its crystal structure with known structures of mesophilic family 11 xylanases. The thermostability of this xylanase is due to the existence of a disulphide bridge, which connects the C-terminus of the β-strand B9 (residue 110) with the N-terminus of the α-helix (residue 154). The presence of electrostatic interactions between ion pairs has also been implicated in the stabilization of the protein structure (Gruber *et al.,* 1998). *Thermoascus aurantiacus* xylanase, belonging to family 10/F of glycoside hydrolases is a single polypeptide chain that assumes $(\alpha/\beta)_8$ TIM barrel fold (Figure 2b). The presence of a disulphide bridge between C-terminus of helix α 7 and the N-terminus of the β 8 and the presence of more than ten salt bridges might be contributing to the thermostability and activity of this protein at elevated temperatures. Salt bridges between oppositely charged groups are an important aspect to protein structure stability and, when

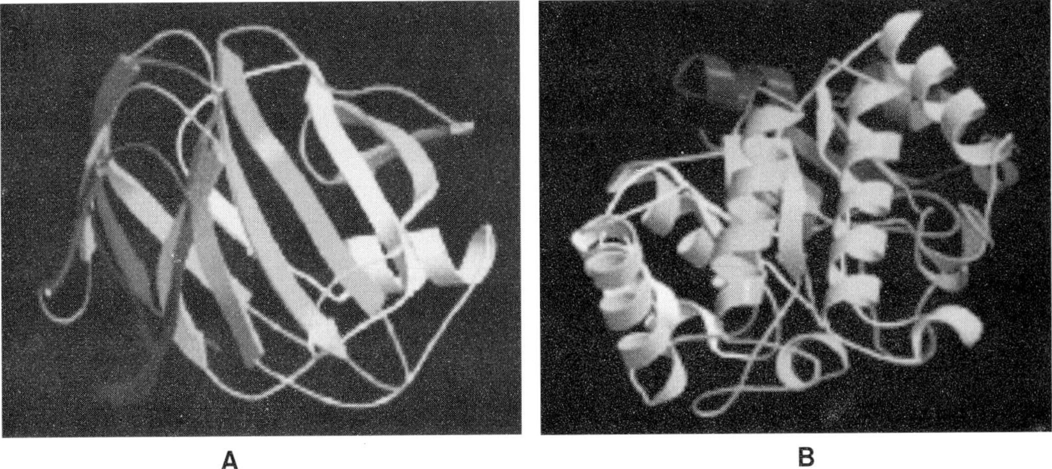


**Fig. 2a.** Crystal structure of thermostable xylanase from *Thermomyces lanuginosus* (Resolution: 1.55 Å ) from George, 2001

**Fig. 2b.** Crystal structure of thermostable xylanase from *Thermoascus aurantiacus* (Resolution 1.8 Å ) from George, 2001

buried, contribute 3-5 kcal/mol in its stability. The presence of six short helices termed as "thermo helix" also contributes to the stability of the protein (Auerbach *et al.,* 1998). The presence of a proline residue in the middle of an α-helix may also contribute to better packing in the protein and the formation of a more stable structure (Natesh *et al.,* 1999). In contrast, a comparative study of the 5 thermophlic and 7 mesophilic 'family 11' enzymes suggested that a general thermostabilizing adaptation in this family is a higher threonine to serine ratio with high β-strands at N-terminus (Hakulinen *et al.,* 2003). Again the thermostability adaptation between family 10 and 11 has also been described with structural differentiation. Family 10 enzymes have a high α–helix content (approx. 40%) while family 11 enzymes have a high β-sheet content (greater than 50%) (Hakulinen *et al.,* 2000).

## ALKALINE XYLANASES

Alkaline xylanases have gained importance due to their application in the development of ecofriendly technologies used in paper and pulp industries. The enzyme is able to hydrolyze xylan which is soluble in alkaline solution (Horikoshi, 1996). The first report on xylanase from alkaliphilic bacteria was published in 1973 by Horikoshi and Atsukawa (1973).

Many of the alkaliphilic xylanases have been produced by alkaliphilic microorganisms such as *Bacillus* sp. (Okazaki *et al.,* 1984) and *Aeromonas* sp. 212 (Ohkoshi *et al.,* 1985) with optimum growth at pH 10. The xylanase showed stability at pH 9-10. The enzymes from *Bacillus* sp. TAR-1(Nakamura *et al.,* 1994), C-125 (Honda *et al.,* 1985) and alkaliphilic *Bacillus* sp. (NCL-86-6-10) (Balakrishnan *et al.,* 1992) were optimally active at pH 9-10. Few fungal species produced alkaline xylanases. *Aspergillus fischeri* (Chandra and Chandra, 1996) was reported to secrete a xylanase with remarkable stability (pH 9.0). The xylanase of *Cephalosporium* was the only one reported from an alkaliphilic fungus that exhibited activity in the pH range between 6.5 and 9.0 (Bansod *et al.,* 1993). Xylanases from four alkaliphilic and thermophilic *Bacillus* strains (W1, W2, W3 and W4) with optimum pH 6.0-7.0 and temperature 65-70°C have been reported (Okazaki *et al.,* 1984). Another thermophilic *Bacillus* sp. (NCIM 59) producing cellulase-free xylanase at pH 10 and active at 50°C has also been isolated (Ishaque and Kluepfel, 1981). XylB of *Bacillus* sp. AR-009 was optimally active at pH 9-10 (Gessease, 1998). Xylanase J from *Bacillus* sp. strain 41M-1 (Campbell *et al.,* 1993) and a xylanase from *Bacillus pumilus* 13a (Rao *et al.,* 1996) were active at pH 9.0. Khasin *et al.* (1993) reported that alkaliphilic *B. stearothermophilus* T-6 produced an extracellular xylanase that optimally bleached pulp at pH 9 and 65°C. Subsequently, many thermostable alkaline xylanase producing alkaliphiles have been isolated from geothermal areas (Dimitrov *et al.,* 1997; Sunna *et al.,* 1997). An alkaline xylanase secreted by *Bacillus* sp. strain TAR-1 showed optimum activity at 70°C at pH 9 and 75°C at pH 7 (Nakamura, 1995).

The pH activity profiles of enzymes are highly dependent on the $pK_a$s of the catalytic residues of the enzymes. A recent study of a family 11 xylanase showed that, residues that contribute positive charges and hydrogen bonds serve to lower the $pK_a$ values. It is noted that the asparagine residue in the xylanases makes the enzyme active in alkaline condition (Torronen and Rouvinen, 1995; Fushinobu *et al.,* 1998; Sapag *et al.,* 2002) rather than aspartic acid residue. Stability at the extreme pH values appears to be characterized by a spatially biased distribution of charged residues. It has also been noted that enzymes stable in alkaline conditions are typically characterized by a decreased number of acidic residues and an increased number of arginines. Furthermore, a recent comparative structural study of family 11 enzymes suggested a correlation between pH activity/ stability and the number of salt bridges, with acidophilic xylanases having much less of these interactions than their alkaliphilic homologs (Hakulinen *et al.,* 2003). Indeed it was even suggested that adaptation to high pH may occur via a mechanism similar to that of adaptation to high temperatures (Hakulinen *et al.,* 2003).

## CLASSIFICATION OF XYLANASES

The heterogeneity and complexity of xylan has resulted in diversification of xylanases with varying specificities, primary sequences and folds necessitating their classification into different forms. Wong *et al.* (1988) classified xylanases on the basis of their physiochemical properties and proposed two groups: those with molecular weight <30 kDa and basic pI, and those with a high molecular weight (>30kDa) and acidic pI (Collins *et al.,* 2005). High molecular weight endoxylanase with low pI values belong to glucanase family 10 formerly known as family F, while low molecular mass endoxylanase with high pI value are classified as glucanase family 11, formerly known as family G. The endoxylanase of family 10 are capable of attacking the glycosidic linkage next to the branch point and towards the non-reducing end and also require two unsubstituted xylopyranosyl units. Endoxylanase of family 10 liberate terminal xylopyranosyl residues attached to a substituted xylanopyranosyl residue and also exhibit aryl-D-xylosidase activity. Family 11 xylanases are true xylanases. They do not have cellulase activity and are produced by bacteria as well as fungi. The catalytic domains consist principally of β–pleated sheets formed into a two layered trough that surrounds the catalytic site (Miao *et al.,* 1994; Withers and Aebersold, 1995). However, several exceptions to this pattern have been recorded (Matte and Forsberg, 1992; Sunna and Antranikian, 1997), and approximately 30% of presently identified xylanases in particular, fungal xylanases cannot be classified by this system.

Later, Henrissat *et al.* (1989) introduced a more complete classification system based on amino acid sequence similarities and hydrophobic cluster analysis (Liu *et al.,* 2004). According to this system, cellulases and xylanases are classified into six families (A-F); subsequently, the families were upgraded to 11 (Gilkes *et al.,* 1991; Henrissat, 1991); xylanases were then subdivided into F and G families, which were analogous to glycohydrolase families 10 and 11 (Henrissat and Bairoch, 1993), comprising high and low molecular weight xylanases, respectively. Subsequently, this method was widely used as a classification system for glycosyl hydrolases to complement the I.U.B. (Anonymous, 1984, International Union of Biochemistry on Nomenclature and Classification of Enzyme).

## CLONING AND EXPRESSION OF XYLANASE GENES

Enzymes of industrial significance are cloned and expressed in suitable microbial hosts for increasing the yields in order to develop a cost-effective and market-friendly process. Recombinant DNA technology offers means to enhance the enzyme production. There are several reports regarding the genetic manipulation on xylanase producing microorganisms. Bacterial and fungal xylanase genes have been cloned and overexpressed in *E. coli*, *Bacillus* and yeast systems (Araki *et al.,* 2000). Isolation and cloning of xylanase gene represents an essential step in engineering of the most efficient microorganisms. Enzyme synthesis can be influenced directly by increasing the level of transcription of the gene encoding the enzymes, and the transcription level can be manipulated directly by promoter choice, promoter manipulation, gene dose and inducer dosage by specific mutagenesis and by affecting the levels of the transcriptional regulator. Various molecular and biotechnological aspects of xylanases produced from microorganisms are described below.

## HETEROLOGOUS CLONING

Thermostable and alkalistable xylanase genes have been isolated from different genera and expressed in *E. coli*. Extracellular activity has been reported in recombinant *E. coli* for the xylanase from alkaliphilic *Aeromonas* sp. (Kudo *et al.,* 1985), alkaliphilic *Bacillus* (Honda *et al.,* 1985), alkaliphilic and thermophilic *Bacillus* sp. (Shendye and Rao, 1993). Cloning and expression of a xylanase gene from the extreme thermophile *Dictyoglomus thermophilum* Rt 46 B1 in *E. coli* has been reported (Stutzenberger and Bodine, 1992), and the recombinant enzyme was found to have an optimum temperature of 85°C. The xylanase cloned from *Thermotoga maritima* showed a maximum temperature of 90°C for its activity (Chen *et al.,* 1997). The recombinant xylanase from *T. neapolitana* was found to be stable at 90°C for 4 hours with a $T_{1/2}$ of 2 hours at 100°C (Veoikodvorskaya, 1997). Thermostable xylanolytic enzymes from thermophilic bacteria (*Dictyoglomus thermophilum* and *Thermotoga maritima*) were expressed in fungal hosts (*Kluyveromyces lactis* and *Trichoderma reesei*) (Bergquist *et al.,* 2002). The efficacy of heterologous protein production and thermostability were comparable with other expression systems developed for thermophilic enzymes.

## HOMOLOGOUS CLONING

Many homologous proteins cannot be efficiently expressed in *E. coli* due to the presence of rare codons in the cloned genes, the need for the post-translational modifications, toxicity of the coded protein to the host cells, and susceptibility of the foreign protein to protease of *E. coli*. The level of expression of the xylanses from *Bacillus* sp. is relatively high in the homologous host system than that in *E. coli*. In *Bacillus pumilus* IPO, extracellular secretion of the xylanase was achieved using the homologous host *B. subtilis* (Panbangred *et al.,* 1995). The homologous expression of the xylanase gene from an alkaliphilic thermophilic *Bacillus* sp. has been demonstrated in $Xyn^-$ mutant *B. subtilis* A8 (Shendye and Rao, 1993) and *B. subtilis* M5111 (Shendye *et al.,* 1994). Enhanced production of xylanases has also been reported by chromosomal gene integration in an alkaliphilic and thermophilic *Bacillus* sp. (Shendye and Rao, 1994).

## OVEREXPRESSION in *E. coli*

For the commercial success of the cloned gene it is imperative to overexpress the protein. Copious amounts of pure protein is required from the overexpressed cloned gene for the crystallographic study of the enzyme. Koo *et al.* (1996) have reported over expression of xylanase from *Clostridium thermocellum* in *E. coli*. The recombinant enzyme had an optimum temperature 60°C and pH 5.4. A thermostable xylanase from *Caldocellum saccharolyticum* has also been cloned and overexpressed in *E. coli*. The enzyme expressed by the recombinant had a $T_{1/2}$ of 2-3 min at 80°C (Luthi *et al.,* 1990). Moreau *et al.* (1994) have cloned and overexpressed a thermostable xylanase from *C. saccharolyticum* in *E. coli*. High-level expression and single step purification of a thermostable xylanase from *Bacillus stearothermophilus* T-6 has been reported (Lapidot *et al.,* 1996). The xylanase gene was cloned into T-7 polymerase expression vectors. The enzyme was found to constitute over 70% of the cell protein. Xylanase gene from *Bacillus* sp. strain NG-27 was cloned in *E. coli* and the xylanase expression resulted into a 47 kDa protein active at 70°C and pH 8 (Gupta *et al.,* 2000).

Besides molecular cloning, protein engineering also plays a significant role in the preparation of enzymes with desired properties by examining the active site of an enzyme and redesigning the protein by identifying the roles of specific residues in catalysis by X-ray crystallographic data and site-directed mutagenesis, providing the basis of protein engineering for the manipulation of the gene for desired enzymatic properties.

Certain characteristic features of the enzyme are responsible for the thermostabilization of the enzyme. Thermophiles contain proteins, which are thermostable, resist denaturation and proteolysis (Tahir *et al.,* 2000; Kumar, 2001) and are called as chaperons that help in restoring the functional native form after denaturation (Everly and Alberto, 2001). The cell membrane of thermophiles contains fatty acids, which provide the thermostability and rigidity to enable the organism to live and remain active at high temperature. Cell wall of archaebacteria and some thermophilic microorganisms contain lipids that are linked with ether providing more resistance than fatty acids (Haki and Rakshit, 2003). Besides these the thermophiles contain a reverse DNA gyrase that induces positive supercoils in the DNA which elevates the melting temperature (Tm) of DNA and hence, DNA remains unaffected at high temperature required for the growth of organism. Cloning and comparison of the gene sequence of the thermostable proteins indicated that a number of salt bridges, increase in the hydrophobicity index and rigidity in the helix and loop regions affect thermostability to a greater extent (Hongbattarakere, 2002). A comparative analysis of xylanase gene from *Thermotoga maritima* showed that decrease in asparagines and glutamine residues contribute towards increase in enzyme thermostability (Viikari *et al.,* 1994).

The presence of repeated domain is the characteristic feature of the bacterial xylanase but the thermostability of the enzyme is characterized by the presence of the N-terminal domains as in *Thermoanaerobacterium saccharolyticum* (Leathers *et al.,* 1984). A similar thermostabilizing domain has been reported from thermostable xylanases produced by a mesophilic *Bacillus* sp. strain NG-27n (Gupta *et al.,* 2000). Thermostabilizing domains are found mainly in the thermophilic xylanases of family 10. There are some factors, which are responsible for the thermostability of the enzymes when structurally analysed. An increase in hydrophobicity in a region of a protein should reduce the number of water molecules having access to this locus or create a more compact protein as a result of secondary structure interaction. This phenomenon could eventually increase protein stability as shown in Ni73D xylanase. In *Bacillus circulans* mutant xylanase proteins, disulphide bridges conferred thermoprotection as observed by 15°C increase in thermostability (Wakurchuk *et al.,* 1994). The crystal structure analysis of the thermophilic xylanase gives more information on its thermostability properties. *Thermoascus aurantiacus* xylanase belonging to family 10 of glycoside hydrolysis is a single polypeptide chain that assumes $(\alpha/\beta)_8$ TIM barrel fold. The presence of a disulphide bridge between C-terminus of helix $\alpha$ 7 and N-terminus of $\beta$ 8 and the presence of more than ten salt bridges might be contributing to the thermostabililty and activity of the protein at elevated temperatures. The presence of six short helices termed as "thermohelix" may also contribute to the stability of the protein (Auerbach *et al.,* 1998). Xyn B from *Dictyoglomus thermophilum* Rt 46B.1 is the most thermostable xylanase, which is folded into a single domain based on two $\beta$-sheets that pack against each other. An increased proportion of polar surface correlates strongly with increased thermostability (Vogt *et al.,* 1997). Xyn B contains an extended B3-A5 connecting loop, which results in enhanced H-bonding and hence thermostability.

## APPLICATIONS OF THE HEMICELLULOLYTIC ENZYMES

The main aim of hemicellulolytic enymes would be to convert polymeric and oligomeric substances from the hemicelluloses into monomeric sugars. Utilization of hexoses for conversion into alcohol, and pentoses as substrates for microorganisms for the conversion into rich protein fodder (Eriksson, 1985). They are also used as food and beverage additives to improve the properties of dough and the quality of bakery products as well as clarification of juices and wines (Rouau *et al.,* 1994; Hang and Woodams, 1997). In addition they are also being used in the poultry industry to increase feed efficiency and improvement of the nutritional value of silage and green feed (Zheng *et al.,* 2000; Classen, 1996) and in textile industry along with pectinase for accumulated retting of hemp, flux, jute, sisal and bast fibre.

Considerable interest has now been focused in the potential use for xylanolytic enzymes for the treatment of cellulosic pulps for the removal or partial hydrolysis of residual xylans. The need for the removal of hemicellulose and lignin component is essential because these are the major chromophoric substances present in the plant cell wall and removal of these is prerequisite for processing of the cellulosic constituent of plant cell wall for pulp and paper industries (Singh *et al.,* 2005). Several attempts have been made to remove xylan using xylanolytic enzymes for upgrading the dissolving pulps (Viikari *et al.,* 1986, 1987; Chauvet *et al.,* 1987; Paice *et al.,* 1988). Xylanase application in the pulp and paper industry was commenced in 1986, when Viikari *et al.* (1986) first demonstrated its utility in pulp and paper industry. Enzymatic applications are generally considered to be environmentally benign. The limitations of the use of enzymes in pulping and paper making processes are related to the size and properties of the enzymes. The application of xylanases may facilitate the removal of residual lignin by removing the residual and reprecipitated xylan (Iversen and Wännström, 1986). The incorporation of xylanases into pulp and paper mill bleaching sequence may be more effective and will prove to be an ecofriendly process. The conditions used in many stages in pulping and paper manufacture have led to the development of enzyme preparations having pH and temperature ranges suitable for this target process. Xylanase that are optimally active and stable at alkaline pH and elevated temperature have potential for application in enzyme-assisted bleaching of kraft pulps and other biotechnological process (Wong and Saddler 1993; Viikari *et al.,* 1994; Deneault *et al.,* 1994; Hespell *et al.,* 1997) because of the cellulose extraction in the kraft process, and the pulp before enzyme treatment has an alkaline pH and high temperature (Zamost *et al.,* 1991). The use of thermostable alkaline xylanase for enzyme assisted pulp bleaching could greatly reduce the need for pH and temperature readjustment, thus offering enormous technical and economic advantages (Zamost *et al.,* 1991; Harris, 1997).

Many of the xylanases used in industry today appear to be of mesophilic and neutrophilic origin. The enzymes from extremophilic sources may be of tremendous utility in many biotechnological processes (Collins *et al.,* 2005). Thermophilic enzymes could be used in applications where a cooling step would be uneconomical or where high temperatures are required to increase the bioavailability or solubility of substrates to reduce viscosity or to reduce the risk of contamination. Alkalophilic enzymes would obviously be beneficial in processes where extreme pH condition is required. The major current application of xylanases is in the pulp and paper industries where the high temperature (55-70°C) and alkaline pH of the pulp substrate requires thermo-alkaliphilic enzymes for efficient biobleaching (Derewenda *et al.,* 1994; Viikari *et al.,* 1994; Beg *et al.,* 2001).

## PULP AND PAPER BLEACHING

The pulp and paper industry is emerging as one of the largest areas for enzyme application in the world. Traditional bleaching methods in the pulp and paper industry rely on chlorine and its derivative and thus place heavy burden on the environment. The effluent generated from the bleaching section contains oxygenated halogens and other associated compounds. The process based on chlorine or peroxide bleaching increases the adsorbable organic halides (AOX) discharge, biological/chemical oxygen demand (BOD/COD) and also adds colour and toxicity to the effluent generated. Thus, the alternative strategies based on the use of biological agents may prove to be a better and more suitable process. Xylanases are the key enzyme applied for the enzyme aided bleaching as it removes xylan and consequently enhances the removal of lignin in the subsequent bleaching process. Enzyme application also improves pulp fibrillation and water retention, reduction of beating times in virgin pulps, restoration of bonding and increased freeness in recycled fibres, and selective removal of xylan from dissolving pulps (Srinivasan and Rele, 1999). It also helps in the significant reduction in the use of chlorine compounds by the conventional bleaching process. Since chlorine compounds are toxic, mutagenic, persistent, bioaccumulating and potentially carcinogenic, the reduction in the level of chlorine consumption by the use of biocatalyst would be the significant step towards alleviating environmental degeneration and degradation.

Kraft process is the most common and initial stage of pulp processing. The kraft process involves the cooking of the wood chips or raw materials in a solution of sodium sulphide and sodium hydroxide at 170°C for 2 hours resulting in the degradation and solubilization of lignin. The resulting pulps has characteristic brown colour primarily due to the residual lignin and lignin derivatives. Here in this stage most of the lignin is removed but 3-5% of the lignin remains in the residual form. The residual lignin gives brown or dark colour, which is due to extensive oxidation and modification to form lignin-carbohydrate complex during cooking process. It is very difficult to remove this complex easily due to its covalent bonding to hemicellulose and cellulose. The removal of this residual lignin after kraft process is called bleaching where the pulp is purified by destruction, alteration and solubilization of the lignin coloured organic matters and the undesirable residues on the fibres of the pulp.

Conventionally, the bleaching process is carried out by the use of chlorine and its derivatives and converts the residual lignin in the pulp to water of alkaline soluble products by any one of the bleaching sequences as oxygen based totally free (TCF) and chlorine dioxide based elemental chlorine (ECF) sequences. The effluents produced during the bleaching process contribute to waste water pollution from the pulp and paper industries, which are toxic, mutagenic, persistent and bioacculmulating that cause numerous harmful disturbances in biological system (Kulkarni and Rao, 1996). This classical process based on chlorine, chlorinated compounds $H_2O_2$, Oxygen and Ozone is being replaced by the new bleaching method called the biobleaching process (Viikari *et al.,* 1994; Jimenez *et al.,* 1999; Beg *et al.,* 2001). It involves use of microorganism or their enzymes to bleach pulp. Ligninases and xylanases are the two enzymes that can be used in biobleaching but the ligninases having adverse effect on pulp properties, xylanase have more preferably used during enzymatic bleaching. Two types of phenomenon have been shown to involve in enzymatic pretreatment of the pulp. One is the hydrolysis of reprecipitated and readsorbable xylan and the other is the separation of xylan-lignin complex produced during kraft process (Yang and Eriksson, 1992). In addition, the hemicellulase treatment removes chromophoric group from the pulp (Wong *et al.,* 1996). Xylanase pretreatment has a lot of advantages over bleaching of the pulp. Since xylan does not form tightly packed structure, it is more accessible to hydrolytic enzymes than the other enzymes. In pulping process, the resultant pulp has a characteristic brown colour owing to the presence of residual lignin and its derivative. In order to obtain white and bright pulp suitable for manufacturing good quality paper, it is necessary to bleach the pulp to remove the constituents such as lignin and its degradative products (Lundgren *et al.,* 1994). Biobleaching of pulp is reported to be more effective with xylanase than with lignin-degrading enzymes. This is because the lignin is cross-linked mostly to the hemicellulose and the hemicellulose is more rapidly depolymerized than lignin (Sabramaniyan and Prema, 2000). The degradation of the cellulose is the major problem associated with the conventional pulping process, which invariably affects the cellulose fibres and affects the quality of paper (Shoham *et al.,* 1992).

The removal of xylan by the xylanases from the pulp leads to decrease in energy demand during bleaching (Bajpai and Bajpai, 2001) and also lead to reduction in AOX and dioxin concentration due to reduced chlorine requirement to achieve a given brightness (Murata *et al.,* 1992; Iimori *et al.,* 1994; Viikari *et al.,* 1994; Iimori *et al.,* 1996).

The results from laboratory studies and mill trials have shown about 35-41% reduction in active chlorine in the chlorination stage for hardwood and 10-20% for softwood pulp, whereas the saving in total active chlorine were found to be 20-25% for hardwood and 10-15% for softwood pulping.

A number of paper mills in North America, Scandinavia and Finland have used xylanase enzyme after several successful mill trials and in other countries it is still in practice. The potential economic benefits of enzyme bleaching are significant to the pulp and paper industry. A simple calculation of relative economic benefit in an ECF sequence reveals that the reduction of approximately 5 kg chlorine dioxide per ton of pulp, assuming a chlorine dioxide cost of US$ 70 per kg leads to saving of about US$ 2 per ton of pulp in chlorine dioxide costs alone. Similarly in TCF, the cost of oxygen based chemicals (ozone, peroxide) is even higher and the respective saving even more pronounced. Additional saving in alkali can also be expected by the alkaline xylanase (Farrel *et al.,* 1996).

## PULP BLEACHING PROCESS

### Pulping

Pulping is the process in which wood or other cellulosic raw materials are digested with chemical under high temperature and pressure. The process of cellulose pulp production involves an initial step of debarking and chipping of the raw materials (logs). The purpose of pulping is to remove lignin in order to facilitate fibre separation and to improve the paper making properties of the pulp. There are three major types of pulping process employed according to the raw materials and finished products. These constitute mechanical, chemi-mechanical and chemical. Chemical processing is generally in use for the pulping process. Most chemical pulping is carried out according to kraft (sulphate) process or sulphite pulping and semichemical pulping. The kraft process is predominant for pulping of the wood during paper manufacturing process. It entails treating wood chips (160-180°C) with liquid containing in sodium hydroxide and sodium sulphide, which promote cleavage of the various ether bonds in the lignin. The lignin degradation products so formed are dissolved in the alkaline pulping liquid and 90-95% of the lignin is removed from the wood at this stage.

The fibres after the chemical treatment are subjected to washing. The washed pulp is then partially dewatered by decker, a cylindrical screen revolving across the path of the pulp. The pulp at this condition called decker pulp, which contains 5-10% lignin. The residual lignin is difficult to remove due to its covalent bonding to the hemicellulose and perhaps to cellulose fibres. The residual lignin in the kraft pulp is responsible for the dark colour, which should be removed further by multistage bleaching process.

## Bleaching

Bleaching is the treatment of cellulosic fibres with chemicals to increase brightness by removing the low amount of residual lignin present after cooking without decreasing the molecular weight of the cellulose. The bleaching of the pulp involves five successive stages of treatment as: i) Chlorine (C) ii) Alkali extraction (E) iii) Chlorine dioxide (D) iv) Alkali extraction (E) and v) Chlorine dioxide (D). Sometime alkali extraction stages can be used in the presence of oxygen (O) or peroxide (P) or hypochlorite (H).

Conventionally, the bleaching of digested pulp by chlorine has implied the generation of adsorbable organic halogen (AOX) which is problematic for biological wastewater treatment due to its high toxicity and low biodegradability (Sierra *et al.,* 1994). Consequently, due to market pressure and government norms, the pulp and paper industry has made substantial changes to conventional pulping and bleaching technologies. The cleaner processes were focusing on bleaching within chlorine-free sequence to eliminate emission of organic halogen compounds by other alternative bleaching processes.

The other bleaching process introduced in bleaching plant is the replacement of elemental chlorine-by-chlorine dioxide, the so-called 'Elemental Chlorine Free (ECF)' bleaching process. This alternative leads to high brightness pulp with acceptable strength. Again, due to the response of market demands 'Total Chlorine Free (TCF)' bleaching has been used for non-chlorine bleached pulp, which was made possible after the action of a pre-delignification step with pressurized oxygen which leads to a pulp with considerably lower kappa number. This process is based on the action of the oxygen, ozone, hydrogen peroxide or enzyme that replaces chlorine and chlorine dioxide bleaching (Byrd *et al.,* 1992) and devoid of organohalogens in the effluent. To obtain an efficient operation, both ECF and TCF have additional requirements such as improved quality of raw materials, extended cooking post-digested oxygen delignification, optimum washing to remove high lignin fractions etc.

## Biobleaching

Though, new environmentally benign elemental chlorine free (ECF) and totally chlorine free (TCF) bleaching technologies are necessary for minimizing the hemicellulose content in dissolving pulps, the brightness at a high level and the quality of the effluents in term of toxicity and adsorbable organic halogens (AOX) is not improving. Enzymes (xylanases) can be used as biotechnological alternative in the bleaching of pulp (Viikari *et al.,* 1994; Suurnakki *et al.,* 1996) in order to improve delignification by degradation of hemicelluloses (Senior and Hemilton, 1993). The process of xylanase treatment for bleaching is simple with pH and temperature adjustment. When the xylanases are mixed into the pulp kept in a holding tank for 1-3 hours, the pulp is bleached in the normal way. It has been postulated that during enzyme treatment the xylanase hydrolyzes xylan slightly and that caustic extraction allows for removal of short xylan fragments, which are covalently attached to lignin fragments. Again, the xylanases can also remove reprecipitated xylan on the surface of the kraft pulp fibre allowing for better chemical penetration and lignin accessibility, in turn improving lignin extractability. However, due to the removal of some of the residual lignin and opening of the structure of the pulp, the chlorine consumption is much lower. Xylanase treatment will be more advantageous if the adjustment of temperature and pH are eliminated completely. The use of enzyme in the prebleaching stage leads to a reduction in chemical consumption and improved brightness, maintaining fibre strength and saving chemical cost. The amount of enzyme needed for the bleaching depends on the type of pulp and bleaching sequence. Availability of the enzyme that is cellulase-free, working at temperature above 80°C and pH >8.0 would be very beneficial. In this respect, the thermostable and alkalistable xylanases have high potential in pulp bleaching process in paper industry. The sequential process of the pulping and bleaching is illustrated in the flow chart (Figure 3).

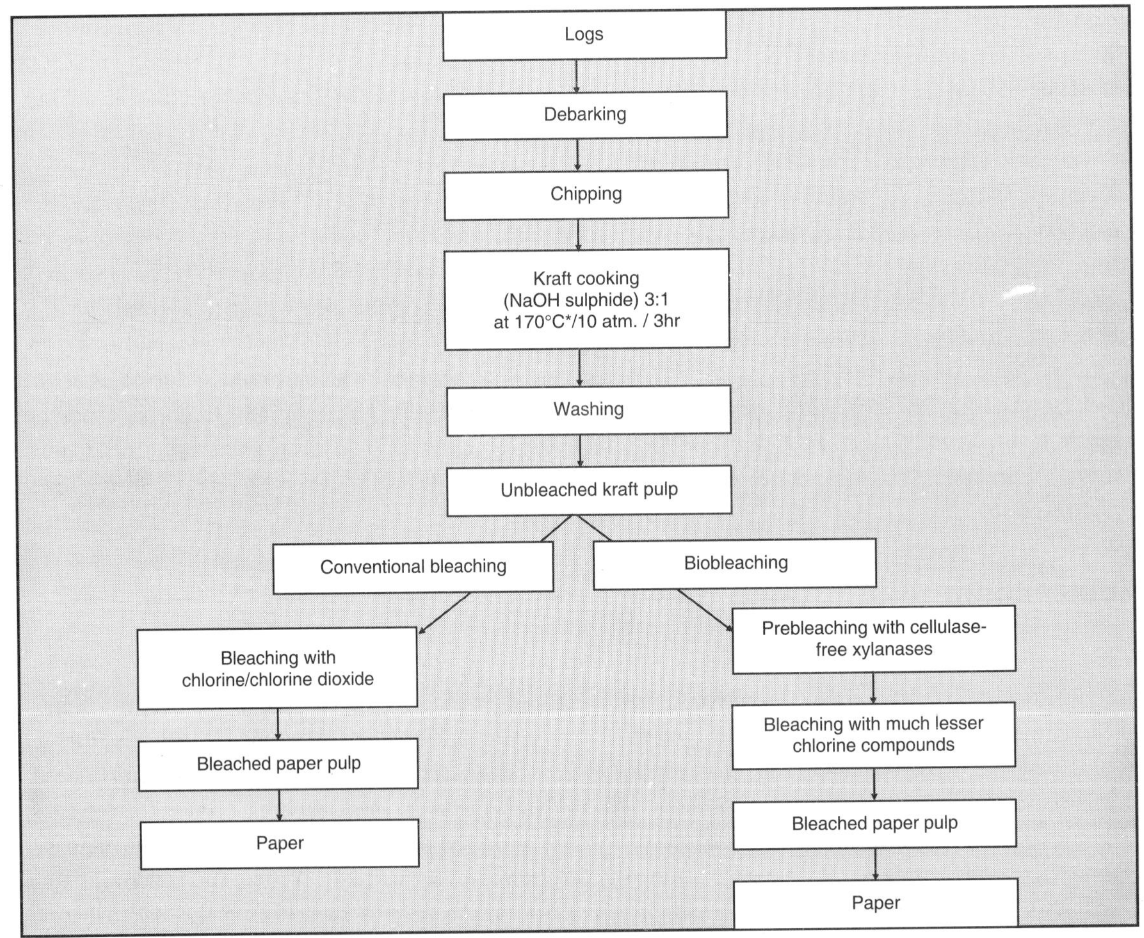


**Fig. 3.** Flow chart showing kraft pulping and bleaching process.

## CONCLUSIONS

Xylan, a hemicellulose, is the second most abundant polysaccharide on the earth and it is a major component in the secondary cell wall. Microbial xylanases have their potential uses in various biotechnological processes. Environmental pressure has led the pulp and paper industry to develop new technologies in order to reduce the presence of various pollutants in effluents of bleaching plants. One of the choices for this purpose is enzyme-based biotechnology using a xylanase. Xylanases with thermostability and alkalistability are suitable candidates for biobleaching as such conditions are typical in kraft pulps. Thus, the xylanases with these properties are needed for application in paper and pulp industry. The molecular cloning and protein engineering are expected to aid in the development of enzymes with the desired properties. In the context of global market for industrial applications, xylanases would help in developing an eco-friendly process for bleaching of paper pulps.

## REFERENCES

Abou-Hachem, M., Olsson, F. and Nordberg-Karlsson, E. (2003). Probing the stability of the modular family 10 xylanase from *Rhodothermus marinus. Extremophiles,* 7, 483–491.

Adams, M.W.W. and Kelly, R.M. (1998). Hyperthermophilic enzymes finding them and using them. *Trends in Biotechnology,* 16,329-332.

Alam, M., Gomes, I., Mohiuddin, G. and Hoq, M.M. (1994). Production and characterization of thermostable xylanase by *Thermomyces lanuginosus* and *Thermoascus aurantiacus* grown on lignocellulose. *Enzyme and Microbial Technology,* 16, 298-302.

Andrade, C.M.M.C., Pereira, N. and Antranikian, G. (1999). Extremely thermophilic microorganisms and their polymer hydrolytic enzymes. *Reviews in Microbiology,* 30, 287–298.

Antranikian, G., Zablowski, P. and Gottschalk, G. (1987). Conditions for the overproduction and excretion of thermostable α-amylase and pullulanase from *Clostridium themohydrosulfuricum* DSM 567. *Applied Microbiology and Biotechnology,* 27, 75-81.

Ara, K., Igarashi, K., Hagihara, H., Sawada, K., Kobayashi, T. and Ito, S. (1996). Separation of functional domains for the α-1-4 and α-1, 6 hydrolytic activities of a *Bacillus* amylopullulanase by limited proteolysis with papain. *Bioscience, Biotechnology and Biochemistry,* 60(4), 634-639.

Araki, T., Hashikawa, S. and Morishita, T. (2000). Cloning, sequencing, and expression in *Escherichia coli* of the new gene encoding β–1, 3-xylanase from a marine bacterium, *Vibrio* sp. strain XY-214. *Applied and Environmental Microbiology,* 66, 1741–1743.

Archana, A. and Satyanarayana, T. (1997). Xylanase production by thermophilic *Bacillus licheniformis* A99 in solid-state fermentation. *Enzyme and Microbial Technology,* 21 (1), 12-17.

Archana, A. and Satyanarayana, T. (1998). Cellulase-free xylanase production by thermophilic *Bacillus licheniformis* A99. *Indian Journal of Microbiology,* 38, 13-19.

Archana, A. and Satyanarayana,T. (2003). Purification and characterization of a cellulase-free xylanase of a moderate thermophile *Bacillus licheniformis* A99. *World Journal of Microbiology and Biotechnology,* 19, 53-57.

Atkin, E.D.T. (1992). Three dimensional structure, interactions and properties of xylans. In: *Xylan and xylanases.* Visser, J., Beldman, G., Someren, MAK, and Voragen, A.G.J., Eds), pp. 21-39, Elseiver, Amsterdam.

Auerbach, G. Ostendrop. R. Praade. L., Korndorfer, I. Dams,T., Huber, R. And Jaenicke, R. (1998). Lactate dehydrogenase from the hyperthermophilic bacterium *Thermotoga maritima*: the crystal structure at 2.1 Å resolutions reveals strategies for intrinsic protein stabilization. *Structure,* 6, 769-781.

Babu, K.R. and Satyanarayana, T. (1996). Production of bacterial enzymes by solid-state fermentation. *Journal of Scientific and Industrial Research,* 55, 564-567.

Bachmann, S.L. and McCarthy, A.J. (1991). Purification and cooperative activity of enzyme constituting the xylan-degrading system of *Thermomonospora fusca. Applied and Environmental Microbiology,* 57, 2121-2130.

Bacon, J.S.D., Gordon, A.H., Morris, E.J. and Farmer, V.C. (1975). Acetyl groups in cell wall preparations from higher plants. *Biochemistry Journal,* 149, 485-487.

Bailey, M.J., P. Biely, and K. Poutanen. (1992). Interlaboratory testing of methods for assay of xylanase activity. *Journal of Biotechnology,* 23, 257-270.

Bajpai, P. and Bajpai, P.K. (2001). Development of a process for the production of dissolving pulp using xylanase enzyme. *Appita Journal,* 54, 381.

Balakrishnan, H., Dutty-Choudhury, M., Srinivasan, M.C., Rele, M.V. (1992). Cellulase-free xylanse production from an alkalophilic *Bacillus* species. *World Journal of Microbiology and Biotechnology,* 8, 627 -31.

Bansod S.M., Chaudhary, M.D., Srinivasan, M.C., and Rele, M.V. (1993). Xylanase active at high pH from alkalotolerant *Cephalosporium* species. *Biotechnology Letters,* 15, 965-970.

Beg, Q.K., Bhushan, B., Kapoor, M., Hondal, G.S. (2000a). Production and characterization of thermostable xylanase and pectinase from a *Streptomyces* sp. QG-11-3. *Journal of Industrial Microbiology and Biotechnology,* 24, 396-402.

Beg, Q.K. Kapoor, M., Bhushan, B. and Hondal, G.S. (2000b). Enhanced production of a thermostable xylanase from *Streptomyces* sp. QG-11-3 and its application in biobleaching of eucalyptus kraft pulp. *Enzyme and Microbial Technology,* 27(7), 557-466.

Beg, Q.K. Kapoor, M., Mahajan, L. and Hondal, G.S. (2001). Microbial xylanases and their industrial applications. *Applied Microbiology and Biotechnology,* 56, 326-338.

Berenger, J.F., Frixon, C., Bigliardi, J. and Creuzet, N. (1985). Production, purification and properties of thermostable xylanase of *Clostridium sterocorarium. Canadian Journal of Microbiology,* 31, 635-643.

Berens, S., Kaspari, H. and Klemne, J. (1996). Purification and characterization of two different xylanase from the thermophilic Actinomycete *Microteraspora flecuosa* SIIX. *Antonie van Leeuwenhock,* 69, 235-241.

Bergquist, P., Te'o, V., Gibbs M., Cziferszky, A., Paula de Faria, F., Azevedo, M. And Nevalainen, H. (2002). Expression of xylanase enzymes from thermophilic microorganisms in fungal hosts. *Extremophiles,* 177-184.

Bhat, M.K. (2000). Research review paper: Cellulases and related enzymes in biotechnology. *Biotechnology Advances,* 18,355-383.

Biely, P. (1985). Microbial xylanolytic systems. *Trends in Biotechnology,* 3, 286.

Bocchini, D.A., Alves-Prado, H.F., Baida, L.C., Roberts, I.C., Gomes, E. and Da-Silva, R. (2002). Optimization of xylanase production by *Bacillus circulans* D 1 in submerged fermentation using RSM. *Process Biochemistry,* 38, 727-731.

Bouveng, H.O. (1961). Phenyliscocynate derivatives of carbohydrates. II. Location of Q- acetyl groups in birch xylan. *Acta Chem. Scand,* 15, 96-100.

Bragger, J.M., Daniel R.M., Corlbear, T. and Morgan, H.W. (1989). Very stable enzymes from extremely thermophilic archaebacteria and eubacteria. *Applied Microbiology and Biotechnology,* 31, 556-61.

Breccia, J.D., Sineriz, F., Baigori, M.D., Castro, G.R. and Hatti-kaul, R.(1998). Purification and characterization of a thermostable xylanase from *Bacillus amyloliquefaciens. Enzyme and Microbial Technology,* 22(1), 42-49.

Byrd, M.V., Gratzi, J. S. and Singh, R.P.(1992). Delignification and bleaching of chemical pulps with ozone: a literature reviews. *TAPPI Journal.,* 75, 207-213.

Campbell, R.L., Rose, D.R. Wakarchuk, W.W., To, R., Sung, W. and Yaguchi, M. (1993). A comparision of the structure of the 20 kD xylanase from *Trichoderma harzianum* and *Bacillus circulans. Proceedings of second TRICEL, symposium Trichoderma reesei cellulases of other hydrolases,* Espoo 8, 63-72.

Cannio, R., Di Prizito, N., Rossi, M. and Morana, A. (2004). A xylan-degrading strain of *Sulfolobus solfataricus*: isolation and characterization of the xylanase activity. *Extremophiles,* 8, 117–124.

Chanda, S.K. Hirst, E.L., Jones, J.K.N. and Percival, E.G.V. (1950). The constitution of xylan from esparto grass (*Stipa tenacissima* L.). *Journal of Chemical Society,* 1950, 1289-1297.

Chandra Raj, K. and Chandra, T. S. (1996). Purification and characterization of xylanse from alkali tolerant *Aspergillus fischeri* FXnl. FEMS *Microbiology Letters,* 145, 457-461.

Chen, C., Adolphson, R., Dean, J.F.D., Eriksson K.L., Adams M. W.W. and Westpheling. J. (1997). Release of lignin from kraft pulp by a hyperthermophilic xylanase from *Thermotoga maritima. Enzyme and Microbial Technology,* 20, 39-45.

Clarke J.H., Davidson, K., Rixon, J.E. Halstead J.R. Fransen, M.P., Gilbert, H.J. and Hazlewood, G.P. (2000). A comparison of enzyme aided bleaching of softwood pulp using a combination of xylanase, mannanase and α-galactosidase. *Applied Microbiology and Biotechnology,* 53, 661-667.

Classen, H.L. (1996). Cereal grain starch and exogenous enzymes in poultry diets. *Animal Feed Science and Technology,* 62, 21–27.

Collins, T., Gerday, C. and Feller, G. (2005). Xylanases, xylanase families and extremohilic xylanases. *FEMS Microbiology Reviews,* 29, 3-23.

Connerton, I., Cummings, N., Harris, G.W., Debeire, P. and Breton, C. (1999). A single domain thermophilic xylanase can bind insoluble xylan: evidence for surface aromatic clusters. *Biochimica et Biophysica Acta,* 1433, 110–121.

Coughlan, M.P. and Hazlewood, G.P. (1993). β-1-4-D-Xylan degrading enzyme systems: Biochemistry, molecular biology and application. *Biotechnology and Applied Biochemistry,* 17, 259-289.

Da-Silva, R., Yim, D.K and Park, Y.K.J. (1994). Application of thermostable xylanase from *Humicola* sp. for pulp improvement. *Journal of Fermentation and Bioengineering,* 77,109-111.

Debeire-Gosselin, M., Loonis, M., Samain, E. and Degbeire, P. (1992). Purification and properties of a 22 kD endoxylanase excreated by a new strain of themophilic Bacillus. In: *Xylan and Xylanases.* Eds. Visser, J., Beldmen, G., Kusers-van Someren, M.A.,Voragen, A.G.J. Elsevier Science Publishers, Amsterdam pp. 463-466.

Dekker, R.F.H. and Richard, G.N. (1976). Hemicellulases, their occurrence, purification, properties and mode of action. *Advances in Carbohydrate Chemistry,* 22, 227-252.

Dekker, R.F.H. (1989). Biodegradation of the hetero-1,4- linked xylans. *ACS Symposium Ser.,* 399, 619-629.

Deleyn, F., Claeyssens, M.m Van Beelmen, J. and De Bruyne, C.K. (1978). Purification and properties of a β-xylosidasse from *Penicillium wortmannii Canadian Journal of Microbiology,* 56, 43-50.

Derewenda, U., Swenson, L., Green, R., Wei, Y., Morosoli, R., Shareck, F., Kluepfel, D. and Derewenda, Z.S. (1994). Crystal structure, at 2.6Å resolution, of the *Streptomyces lividans* xylanase A, a member of the F family of β–1,4-D-glycanases. *Journal of Biological Chemistry,* 269, 20811–20814.

Dey, D., Hinge, J., Shendye, A. and Rao, M. (1992). Purification and properties of extracellular endoxylanases from alkalophilic thermophilic *Bacillus* sp. *Canadian Journal of Microbiology,* 38(5),436-442.

Dhillon A., Gupta, J.K., Khanna, S. (2000). Enhanced production, purification and characterization of a novel cellulase – poor thermostable, alkalitolerant xylanase from *Bacillus circulans* AB16. *Process Biochemistry,* 35, 849-856.

Dhillon, A, Gupta J.K, Jauhari, B.M, Khanna S. (2000). A cellulase-poor, thermostable, alkalitolerant xylanase produced by *Bacillus circulans* AB16 grown on rice straw and its application in biobleaching of eucalyptus pulp. *Bioresource Technology,* 73, 273-277.

Dimitrov, P.L., Kambourova, M.S., Mandeva, R.D. and Emanuilova, E.I. (1997). Isolation and characterization of xylan-degrading alkali-tolerant thermophiles. *FEMS Microbiology Letters,* 157, 27-30.

Dobberstein, J. and Emeis, C.C. (1991). Purification and characterization of beta-xylosidase from *Aureobasidium pullulans. Applied Microbiology and Biotechnology,* 35, 210-215.

Duarte M.C.T., Portugal, E.P., Ponezi. A.N., Bim, M.A., Tagliari, C.V., Franco, T.T. (1999). Production and purification of alkaline xylanases. *Bioresource Technology,* 68, 49-53.

Duarte, M.C.T., Pellegrino, A.C.A., Portugal, E.P., Ponezi, A.N., and Franco, T.T. (2000). Characterization of alkaline xylanases from *Bacillus pumilus. Bioresource Technology,* 31, 90-/4.

Dunlop, R.W., Wang, B. and Ball, D. (1995). PN-WO 9527779.

Eda, S. Ohnishi, A. and Kato, L. (1976). Xylan isolated from the stalk of *Nicotiana tabacum. Agricultural and Biological chemistry,* 40, 359.

Eriksson, K.E. (1990). Biotechnology in the pulp and paper industry. *Wood Science Technology,* 24, 79–101.

Esteban, R.Villanueva, J.R. and Villa, T.G. (1982). β-D-Xylanases of *Bacillus circulans* WL-12. *Canadian Journal of Microbiology,* 28, 733-739.

Everly C. Alberto J. (2000). Stressors, stress and survival: overview. *Frontier Bioscience,* 5, 780 -786.

Farrell, R.L., Viikari, L., and Senior, D.J. (1996). Enzyme treatment of pulp. In: *Pulp bleaching: Principle and practice.* Eds. Jeffries, T.W. and Viikari L. ACS Symp. Ser. 655, Atlanta, G.A. TAPPI Press pp 363-377.

Fengel, D. and Wegener, G. (1983). Wood: chemistry ultrastructure and reactions. *Walter de Gruyter and Co. Berlin,* pp 613.

Fontes, C.M., Hall, J., Hirst, B.H., Hazlewood, G.P. and Gilbert, H.J. (1995). The resistance of cellulases and xylanases to proteolytic inactivation. *Applied Microbiology and Biotechnology,* 43, 52–57.

Fontes,C.M.G.A., Hazlewood, G.P., Morang, E., 2Hall, J., Hirst, B.A. and Gilbert, H.J. (1995). Evidence for a general role for non-catalytic thermostabilizing domains in xylanases from thermophilic bacteria. *Biochemical Journal,* 307, 151-158.

Fushinobu, S., Ito, K., Konno, M., Wakagi, T. and Matsuzawa, H. (1998)._Crystallographic and mutational analysis of an extremely acidophilic and acid-stable xylanase: biased distribution of acidic residues and importance of Asp 37 for catalysis at low pH. *Protein Engineering,* 11, 1121-1128.

George, S.P. (2001). *Molecular and Biochemical aspects of extremophilic Actinomycete.* Ph. D. Thesis. Division of Chemical science, National Chemical Laboratory, India.

George, S.P., Ahmad, A., Rao M.B. (2001). A novel thermostable xylanase from *Thermomonospora* sp: influence of additives on thermostability. *Bioresource Technology,* 78, 221-224.

George, S.P., Nath, D.S. and Rao M.B. (2003). Thermophilic xylanases- Molecular and biotechnological perspectives. In: *Advances in Microbiology.* Ed Trivedi, P.C. Scientific Publishers, India.

Gessesse, A. (1998). Purification and properties of two thermostale alkaline xylanase from an alkaliphilic *Bacillus* sp. *Applied and Environmental Microbiology,* 64, 3533-3535.

Gessesse, A. and Mamo, G. (1998). Purification and characterization of an alkaline xylanase from alkaliphilic *Micrococcus* sp. AR-135. *Journal of Industrial Microbiology and Biotechnology,* 20**,** 210-214.

Gibbs, M.D., Reeves, R.A. and Bergquist, P. K.(1995). Cloning, sequencing and expression of a xylanase gene from the extreme thermophilic *Dictyoglomus thermophilum* Rt 46B.1 and activity of the enzyme on fiber-bound substrate. *Applied and Environmental Microbiology,* 61, 4403-4408.

Gilbert H.J.,and Hazlewood, G.P.((1993). Bacterial cellulases and xylanases. *Journal of General Microbiology,* 139, 187-194.

Gilead, S., and Shohanm, Y. (1995). Purification and characterization of L-arabinofuranosidase from *Bacillus stearothermophilus* T-6. *Applied and Environmental Microbiology,* 61, 170-174.

Gilkes, N.R., Claeyssens, M., Aebersold, R., Henrissat, B., Meinke, A., Morrison, H.D., Kilburn, D.G., Warren, R.A. and Miller Jr., R.C. (1991). Structural and functional relationships in two families of beta-1,4-glycanases. *European Journal of Biochemistry,* 202, 367–377.

Godden, B., Legon, T., Helvenstenin, P. and Pennincks, M. (1989). Regulation of the production of hemicellulolytic and cellulolytic enzymes by a *Streptomyces* sp. growing on lignocellulose. *Journal of General Microbiology,* 135, 285-292.

Grabski, A.C. and Jeffries, T.W. (1991). Production, purification and characterization of beta-(1-4)-endoxylanase of *Streptomyces roseisclerotiaus. Applied and Environmental Microbiology,* 57,987-992.

Greve, J.C. Labavitch, J. M. And Hungate, R.E. (1984). α-L- Arabinofuranosidase from *Ruminococcus albus* 8: purification and possible role in hydrolysis of alfalfa cell wall. *Applied and Environmental Microbiology,* 47, 1185-1140.

Gruber, K., Klintschar, G., Hayn, M., Schlacher, A., Steiner, W. and Kratky, C. (1998). Thermophilic xylanase from *Thermomyces lanuginosus*: high-resolution X-ray structure and modeling studies. *Biochemistry,* 37, 13475–13485.

Gruninger, H. and Fiechter, A. (1986). A novel, highly thermostable D-xylanase. *Enzyme and Microbial Technology,* 8, 309-314.

Gruppen, H., Hammer, R.J., Voragen, A.G.I. (1992). Water-unextractable cell wall material from wheat flour. 2. Fractionation of alkali-extracted polymers and comparison with water-extractable arabinoxylans. *Journal of Cereal Science,* 16, 53-67.

Gupta, N., Vohra, R.M. and Hoondal, G.S. (1992). A thermostable extracellular xylanase from alkaliphilic *Bacillus* sp NG-27**.** *Biotechnology Letters,* 14**,** 1045-1046.

Gupta, N. Reddy, V.S. Maiti, S. and Ghosh, A. (2000). Cloning expression, and sequence analysis of the gene encoding the alkali-stable thermostable endoxylanassse from alkalophilic mesophilic *Bacillus* sp. Strain NG-27. *Applied and Environmental Microbiology 66,* 2631-2635.

Haki, G.D., and Rakshit, S.K. (2003). Developments in industrially important thermostable enzymes: a review. *Bioresource Technology,* 89(1), 17-34.

Hakulinen, N., Turunen, O., Janis, J., Leisola, M., and Rouvilnen, J. (2003). Three-dimensional structures of thermophilicβ–1, 4-xylanase from *Chaetomium thermophilum* and *Nonomuraea flexuosa.* Compararison of twelve xylanases in relation to their thermal stability. *European Journal of Biochemistry,* 270, 1399-1412.

Hang, Y.D., and Woodams, E.E. (1997). Xylanolytic activity of commercial juice-processing enzyme preparations. *Letters in Applied Microbiology,* 24, 389-392.

Harris, G.W., Pickersgill, R.W., Commerton, I., Debeire, P., Touzel. J.P., Breton, C. and Perez, S. (1997). Structural basis of the properties of an industrially relevant thermophilic xylanase. *Proteins,* 29, 77-86.

He, L., Bicherstaff, G.F., Pateerson, A. and Buswell, A. (1993). Purification and paritial characterization of two xylanases that differ in hydrolysis of soluble and insoluble xylan fractions. *Enzyme and Microbial Technology,* 15, 13-18.

Henrissat, B. Cleeyssens, M., Tomma, D., Lemesle, L. and Mornon J.P. (1989). Cellulase families revealed by hydrophobic cluster analysis. *Gene,* 81, 83-95.

Henrissat, B. (1991). A classification of glycosyl hydrolases based on amino acid sequence similarities, *Biochemical Journal,* 280, 309-316.

Henrissat, B., Bairoch, A. (1993). New families in the classification of glycosyl hydrolases based on amino acid sequence similarities *Biochemical Journal,* 293, 781-788.

Hespell, R.B., O'Bryan, P.I, Moniruzzaman, M., and Bothast, R. J. (1997). Hydrolysis by commercial enzyme mixture of AFTEX-treated com fiber and isolated xylans. *Applied Biochemistry and Biotechnology,* 62, 87-97.

Honda, H, Kudo, T. and Horikoshi, K. (1985). Molecular cloning and expression of the xylanase gene from alkalophilic *Bacillus* strain C-125 in *Escherichia coli. Journal of Bacteriology,* 161,784-785.

Hongpattarakere, T. (2002). Hyperthermostable cellulolytic and hemicelluloltic enzymes and their biotechnological applications. *Songklanakarin Journal of Science and Technology,* 24(3), 481-491.

Horikoshi, K. and Atsukawa, Y. (1973). Xylanase produced by alkaliphilic *Bacillus* No.C. 59-2. *Agricultural and Biological Chemistry,* 37, 2097.

Horikoshi, K. (1996). Alkaliphiles-from an industrial point of view. *FEMS Microbiology Reviews,* 18, 259-270.

Iimori, T., Kaneko, R., Yoshikawa, H.,Machida, M., Yoshioka H., Murakami, K.(1994). Screening of pulp bleaching fungi and bleaching activity of newly isolated fungus SKB–1152. *Mokuzai Gakkaishi,* 40(7), 733-737.

Iimori, T., Yoshikawa, H., Kaneko R., Miyawaki S., Machida M., Murakami K. (1996). Effects of treatment conditions on treatment times for biobleaching by SKB–1152. *Mokuzai Gakkaishi,* 42,313-317.

Irwin, D., Jung, E.D. and Wilson. D.B. (1994). Characterization and sequence of *Thermomonospora fusca* xylanase. *Applied and Environmental Microbiology,* 60, 763-770.

Ishaque M, Kluepfel D (1981). Production of xylanolytic enzymes by *Streptomyces flavogriseus. Biotechnology Letters,* 3, 481-486.

Iversen, T., and Wannstrom, S. (1986). Lignin-carbohydrate bonds in a residual lignin isolated from pine kraft pulp. *Holzforschung,* 40, 19-22.

Jaleel, S.A., Srikanta, S. and Karanth, N.G. (1992). Production of fungal amyloglucosidase by solid-state fermentation: Influence of some parameters. *Journal of Microbiology and Biotechnology,* 7, 1-8.

Jimenez L. Martinez C. Perez I. Lopez F. (1999). Biobleaching of cellulose pulp from wheat straw with enzymes and hydrogen peroxide. *Process Biochemistry,* 35, 149-57.

Joseleau, J. P., Comtat. J. and Ruel, K. (1992). Chemical structure of xylans and their interaction in the plant cell walls. In: *Xylans and Xylanases,* Eds..Visser, J. Beldman, G., Kusters-van Someren, M.A. and Voragen, A. G. J. Elsevier Science Publishers, Amsterdam, Pp1-15.

Kaji, A. and Tagawa, K. (1970). Purificatin, crystallization, and amino acid composition of alfa-L- arabinofuranosidase from *Aspergillus niger. Biochemica et Biophysica Acta,* 207,456-464.

Kaji, A. and Saheki, T. (1975). Endo-arabinase from *Bacillus subtilis* F-11. *Biochemistry and Biophysics Acta,* 410, 354-360.

Kaji, A. (1984). L- Arabinosidases. Advances in Carbohydrayte Chemistry. *Annual Review of Biochemistry,* 42, 383-394.

Kamra, P. and Satyanarayana, T. (2004). Xylanase production by *Humicola lanuginosa* in solid-state fermentation. *Applied Biochemistry and Biotechnology,* 119,145-157.

Kaneko, S., Shimasaki, T. and Kusakabe, I. (1993). Purification and some properties of intracellular α–L-arabinofuranosidase from *Aspergillus niger* 5-16 *Bioscience, Biotechnology and Biochemistry,* 57, 1161-1165.

Khasin, A., Alchanati, I. and Shoham, Y. (1993). Purification and characterization of a thermostable xylanase from *Bacillus stearothermophilus* T-6. *Applied and Environmental Microbiology,* 59(6), 1725-1730.

Kohli, N., Singh, P., and Chaudhary, D. (2001). Thermostable, alkaliphilic and cellulase free xylanase production by *Thermoactinomyces thalophilus* sub group C. *Enzyme and Microbial Technology,* 28, 606-610.

Koo, B.J., Oh, H.G. Chu, K.H. Yung, C.K. Jung, C.K., Jung and Kyu, D. Y. (1996). Purification and characterization of the *Clostridium thermocellum* xylanase from recombinant *Escherichia coli. Journal of Microbiology and Biotechnology,* 6, 414-419.

Kormelink, F.J.M., Searle-van Leeuwen, M.J.F., Wood, T.M. and Vorgane, A.G.J. (1991a). Purification and characterization of a (1,4)-beta-D-arabinofuranohydrolase from *Aspergillus awamori. Applied Microbiology and Biotechnology,* 35,231-232.

Kormelink F.J.M, and Voragen AG (1993). Degradation of different (glucurono) arabino) xylans by a combination of purified xylan-degrading enzymes. *Applied Microbiology and Biotechnology,* 38, 688-695.

Krik, T.K., Jeffries, T.W. and Leathem, G.F. (1983). Biotechnology, application and implications for the pulp and paper industry. *TAAPI,* 66, 45-51.

Kudo, J., Ohkoshi, A. and Horikoshi, K. (1985). Molecular cloning and expression of a xylanase gene of alkalophilic *Aeromonas* sp. No. 212 in *Escherichia coli. Journal of General Microbiology,* 13, 2825-2830.

Kulkarni, N. and Rao, M. (1996). Application of xylanases from alkaliphilic thermophilic *Bacillus* sp NCIM 59 in biobleaching of bagasse pulp. *Journal of Biotechnology,* 51,167-173.

Kulkarni, N., Shendye, A. and Rao, M. (1999). Molecular and Biological aspects of xylanases. *FEMS Microbiology Reviews,* 23,411-456.

Kumar, N. and Singh, K. (1992). Chemical and microbial changes during solid state fermentation of wheat straw with *Coprinus fimetarius. Biological Wastes,* 33, 231-242.

Kumar, P.R., Eswaramoorthy, S., Vithayathil, P.J. and Viswamitra, M.A. (2000). The tertiary structure at 1.59 Å resolution and the proposed amino acid sequence of a family-11 xylanase from the thermophilic fungus *Paecilomyces variotii* Bainier. *Journal of Molecular Biology,* 295, 581–593.

Kumar, S., Nussinov, R. (2001). How do thermophilic proteins deal with proteins? A review. *Cellular and Molecular Life Science,* 58, 1216-1233.

Lapidot, A., Mechaly, A. and Shoham. Y. (1996). Overexpression and single-step purification of a thermostable xylanase from *Bacillus stearothermophilus* T-6. *Journal of Biotechnology,* 51, 259-264.

Leathers, T.D. Kurtzman, C. P. and Detroy, R.W. (1984). Overproduction and regulation of xylanase in *Aureobasidium pullulans* and *Cryuptococcus albidus. Biotechnology and Bioengineering,* 14, 225-250.

Lee, Y.E. Lowe, S.E., Henrissat. B. ans Zikus. J.B. (1993). Characterization of the active site and thermostability regions of endoxylanase from *Thermoanaerobacterium saccharolyticum* B6A-RI. *Journal of Bacteriology,* 175, 5890-5898.

Leuschner, C. and Antranikian, G. (1995). Heat-stable enzymes from extremely thermophilic and hyperthermophilic microorganisms. *World Journal of Microbiology and Biotechnology,* 11, 95-114.

Lindberg, B. Rosell, K.G. and Sevensson, S. (1973). Position of the O-acetyl groups in birch xylan. *Svensk. Papperstid,* 76, 30-32.

Liu, L., Zhang, J, Chen, B., Shao, W. (2004). Principle component analysis in F/10 and G/11 xylanase. *Biochemical and Biophysical Research Communications,* 322, 277–280.

Ljungdahl, L.G. and Eriksson, K.E. (1985). Ecology of microbial cellulose degradation. In: *Advances in Microbial Ecology.* Ed Marshal, K.C. Plenum Press, New York 8, 237-299.

Lonsane, B.K. and Ramesh, M.V. (1990). Production of bacterial thermostable α-amylase by SSF, a potential tool for achieving economy in enzyme production and starch hydrolysis. *Advances in Applied Microbiology,* 35, 1-56.

Lundgren, K.R., Bergkvist, L., Hogman, S., Joves, H., Eriksson, G., Bartfai, T., Laan, J.V.D., Rosenberg, E. and Shoham, Y. (1994).TCF Mill Trial on soft wood pulp with Korsnas thermostable and alkaline stable xylanase T6. *FEMS Microbiology Reviews,* 13, 365.

Luthi, E., Bhana, J. M. and Bergquist, P.L. (1990). Xylanase from the extremely thermophilic-bacterium *Caldocellum saccharolyticum*: Overexpression of the gene in *Escherichia coli* and characterization of the gene product. *Applied and Environmental Microbiology,* 2672-2683.

Mackenzie, C.R. Bilours, D., Schneider, H. And Johnson, K.G. (1984). Induction of cellulolytic and xylanolytic enzymes systems in *Streptomyces* spp. *Applied and Environmental Microbiology,* 53, 2835-2839.

Malherbe, S. Cloete, T.E. (2003). Lignocellulose biodegradation fundamentals and applications. A review. *Environmental Science and Biotechnology,* 1, 105-114.

Mathrani, I.M. and Ahring, B.K. (1992). Thermophilic and alkaliphilic xylanase from several *Dictyoglomus* isolates. *Applied Microbiology and Biotechnology,* 38, 23-27.

Matte, A. and Forsberg, C.W. (1992). Puri?cation, characterization, and mode of action of endoxylanases 1 and 2 from *Fibrobacter succinogenes* S85. *Applied and Environmental Microbiology,* 58, 157–168.

Mc Carthy, A.J., Peace, E. and Broda, P. (1985). Studies on the extracellular xylanase activity of some thermophilic Actinomycetes. *Applied Microbiology and Biotechnology,* 21, 238-244.

Miao, S.C., Ziser, L., Aebersold, R., and Withers, S.G. (1994). Identification of glutamic acid 78 as the active site nucleophile in *Bacillus subtilis* xylanase using electrospray tandem mass spectrometry. *Biochemistry,* 33, 7027-7032.

Mishra, C. Keshar, S. and Rao, M. (1984). Production and properties of exracellular endoxylanase from *Neurospora crassa. Applied and Environmental Microbiology,* 48,224-228.

Mishra, C., Forrester, I.T., Kelley, B.D., Burgess, R.R. and Leathem, G.F. (1990). Characterization of a major xylanase purified from *Lentinula edodes* culture grown on a commercial solid lignocellulosic substrate. *Applied Microbiology and Biotechnology,* 33, 226-232.

Montgomery, R., Smith, F. and Srivastava, H.C. (1956). Structure of corn hull hemicellulose. I. Partial hydrolysis and identification of 2-O (α-D- glucopyranosyluronicc acid) –D-xylopyranose. *Journal of Chemical Society,* 78, 2837-283.

Moreau, A., Roberge, M., Manin, C. Sharreck, F., Kluepfel, D. and Morosoli, R. (1994). Identification of two acidic residues involved in the catalysis of xylanase A from *Streptomyces lividans. Biochemical Journal,* 302, 291-295.

Morris, D.D., Gibbs, M.D., Chin, C.W.J., Koh, M.H., Wong, K.K.Y., Allison, R.W., Nelson, P.J. and Bergquist. (1998). Cloning of the *xyn*B gene from *Dictyoglomus thermophilum* Rt46B.1 and action of the gene product on kraft pulp. *Applied and Environmental Microbiology,* 64, 1759-1765.

Muller, H.I., Hartley, R.D., Harris, P. J. and Curzon, E.H. (1986). Linkage of p-coumaryl and feruloyl groups to cell wall polysaccharides of barley straw. *Carbohydrate Research,* 148, 71-85.

Murata, S., Kondo, R., Sakai K., Kashino, Y., Nishida, T., Takahara, Y. (1992). Chlorine-free bleaching process of kraft pulp using treatment with the fungus IZU-154 *TAPPI Journal,* 75(132), 91-94.

Nakamura, S., Wakabayashi, K., Nakai. R., Aono, R., and Horikoshi, K. (1993a). Production of alkaline xylanase by a newly isolated alkaliphilic *Bacillus* strain 41 M-1. *World Journal of Microbiology and Biotechnology,* 9, 221-224.

Nakamura, S., Wakabayashi, K., Nakai. R., Aono, R., and Horikoshi, K. (1993b). Purification and some properties of an alkaline xylanase from alkaliphilic *Bacillus* sp. strain 41 M-1. *Applied and Environmental Microbiology,* 59(7), 2311-2316.

Nakamura, S., Nakai, R., Wajabatacgu, K., Ishiguro, Y., Aono, R. and Horikoshi, K. (1994). Thermophilic alkaline xylanase from newly isolated alkaliphilic and thermophilic *Bacillus* sp. strain TAR-I. *Bioscience, Biotechnology and Biochemistry,* 58, 78-81.

Nakamura, S., Nakai. R., Kubo, R., Wakabayashi, K., Aono, R., and Horikoshi, K. (1995). Structure function relationship of the xylanase from alkalophilic *Bacillus* sp. strain 41 M 41 M-1. *Nucleic Acids Symposium Ser.,* 34, 99-100.

Nakanishi, K., Marui, M. and Yasui, T. (1992). Comparison of xylan and methyl β–xyloside induced xylanases from *Streptomyces* sp. *Journal of Fermentation and Bioengineering,* 74(6), 392-394.

Narang, S. and Satyanarayana. T. (2001). Thermostable α-amylase production by an extreme thermophile *Bacillus thermoleovorans. Letters in Applied Microbiology,* 32, 31-35.

Narang, S., Sashi, V. and Bisaria, V.S. (2001). Optimization of xylanase production by *Melanocarpus albomyces* IIS 68 in solid-state fermentation using Response Surface Methodology. *Journal of Bioscience and Biotechnology,* 91, 425-427.

Nath, D.and Rao, M. (2000). pH dependent conformational and structural changes of xylanase from an alkalophilic thermophilic *Bacillus* sp. (NCIM59). *Enzyme and Microbial Technology,* 28, 397-/403.

Niehaus, F., Bertoldo, C., Kahler, M. and Antranikian, G. (1999). Extremophiles as a source of novel enzymes for industrial application. *Applied Microbiology and Biotechnology,* 51, 711–729.

Nielsen, P., Mathrani, I.M. and Ahring, B.K. (1993). *Thermoanaerobium acetigenum* sp. nov., a new anaerobic, extremely thermophilic, xylanolytic non-spore-forming bacterium isolated from an Icelandic hot spring. *Archives of Microbiology,* 159, 460-464.

Ohkoshi, A., Kudo, T. and K. Horikoshi (1985). Purification of three types of xylanases from an alkalophilic *Aeromonas* sp. *Agricultural and Biological Chemistry,* 49, 3037-3038.

Okazaki, W., Akiba. T., Horikoshi, K. and Akahoshi, E. (1984). Production and properties of two types of xylanases from alkaliphilic thermophilic *Bacillus* sp. *Applied Microbiology and Biotechnology,* 19,335-340.

Paice, M.G., Bernier, R., Jr Jurasek, L. (1988). Viscosity enhancing bleaching of hardwood kraft pulp with xylanase from a cloned gene. *Biotechnology and Bioengineering,* 32, 235-239.

Panbangred, W., Fukusaki, E., Epifanio, E.C. Shimmyo A. and Okada, H. (1995). Expression of a xylanases gene of *Bacillus subtilis. Applied Microbiology and Biotechnology,* 22, 259-264.

Pandey, A., Soccol, C.R., Rodriguez-Leon, J.A. and Nigam, P. (2001). In: *Solid state fermentation in biotechnology: fundamentals and applications.* Asiatech Publishers, Inc. New Delhi.

Perttula, M., Ratto, M., Kondradsdothir, M., Kristjansson, J.K. and Viikari, L. (1993). Xylanases of thermophilic bacteria from Icelandic hot springs. *Applied Microbiology and Biotechnology,* 38,592-595.

Poutanen, K. and Puls, J. (1988). Characterictics of *Trichoderma reesei* β-xylosidase and its use in the hydrolysis of solubilized xylans. *Applied Microbiology and Biotechnology,* 28,425-432.

Poutanen, K. (1988). An alfa-L- arabinofurnosidase of *Trichoderma reesei. Journal of Biotechnology,* 7, 271-282.

Puls , J., Schmidt, O. and Ganzow, C.(1987). α–Glucuronidase in two microbial systems. *Enzyme Microbial Technology,* 9, 83-88.

Puls, J., Poutanen, K. (1989). Mechanisms of enzymic hydrolysis of hemicelluloses (xylans) and procedures for determination of the enzyme activity involved. In: *Enzyme systems for lignocellulose degradation.* Ed. Coughlan, M. P. Elsevier Applied Science, London, Pp 151-2165.

Puls, J. (1992). α–Glucuronidases in the hydrolysis of wood xylans. In: *Xylan and Xylanases.* Prog. Biotechnology Vol 7, Eds. Visser, J., Kusers-van Someren, MA Voragen, A. G. J. Elsevier Science Publishers, Amsterdam BV. Pp 213-234.

Puls, J. and Schuseil, J. (1993). Chemistry of hemicelluloses: relationship between hemicellulose structure and enzyme required for hydrolysis. In: *Hemicellulose and Hemicellulases.* Eds. Coughlan, M.P, Hazlewood, G.P. Portland Press, London, Pp 1-28.

Purkarthofer, H., Sinner, M. and Steiner, W. (1993). Cellulase-free xylanase from *Thermomyces lanuginosus*: Optimization of production in submerged and solid-state culture. *Enzyme and Microbial Technology,* 15, 677-682.

Rajaram, S. and Varma, A. (1990). Production and characterization of xylanase from *Bacillus thermoalkalophilus* grown on agricultural wastes. *Applied Microbiology and Biotechnology,* 34, 141-144.

Rani D.S, Nand K. (2000). Production of thermostable cellulase-free xylanase by *Clostridium absonum* CFR-702. *Process Biochemistry,* 36, 355-362.

Rao, M., Khadilkar, S., Bandivadekar, K. and Deshpande, V. (1996). Structural environment of an essential cysteine residue of xylanase from *Chainia* sp. (NCL 82.5.1). *Biochemical Journal,* 316, 771-775.

Reilly, P.J. (1981). Xylanases: structure and function. In: *Trends in the Biology of Fermentations for Fuels and Chemicals.* Ed. Hollaender, A.E. Plenum Press, New York, pp. 111-129.

Rodionova, N.A. Tavobilov, I.M. And Bezborodov, A.M. (1983). β-xylosidase from *Aspergillus niger* 15: purification and properties. *Journal of Applied Biochemistry,* 5, 300-312.

Roncera, M.B., Torres, A.L. Colom, J.F., Vidal, T. (2000). Using xylanase before oxygen delignification on TCF bleaching influence on fibre surface by SEM. *Process Biochemistry,* 36, 45-50.

Rouau X. El-Hayek NL. Moreau, D. (1994). Effect of enzyme preparation containing pentosanases on bread making quality of flours in relation to change in pentosan properties. *Journal of Cereal Science,* 19, 259-72.

Shah, B.C. and Bothast, R.J. (1999). Pretreatment and enzymatic saccharification of corn fiber. *Applied Biochemistry and Biotechnology,* 76, 75-77.

Shah, B.C. (2000)..α-L-Arabinofuranosidases biochemistry, molecular biology and application in biotechnology. *Biotechnology Advances,* 18, 403-423.

Santos, E.S.D., Guirardello, R., Franco, T.T. (2002). Preparative chromatography of xylanase using expanded bed adsorption. *Journal of Chromatography,* 944, 217-24.

Sapag, A., Wouters, J., Lambert, C., de Ioannes, P., Eyzaguirre, J. and Depiereux, E. (2002). The endoxylanases from family 11: computer analysis of protein sequences reveals important structural and phylogenetic relationships. *Journal of Biotechnology,* 95, 109–131.

Satyanarayana, T., Jain, S. and Johri, B.N. (1988). Cellulases and xylanases of thermophilic moulds. In *'Perspectives in Mycology and Plant Pathology'.* Eds. Agnihotri, V.P., Sarbhoy, A.K. and Malhotra, D.K. Publishing House, New Delhi, pp 24-60.

Satyanarayana, T. (1994). Production of bacterial extracellular enzymes by solid-state fementation. In: *Solid State Fermentation.* Ed Pandey, A. Wiley eastern limited New Delhi, India. Pp 122-129.

Saul, D.J., Williams, L.C., Reeves, R.A., Gibbs, M.D. and Bergquist, P. L. (1995). Sequence and expression of a xylanase gene from the hyperthermophile *Thermotoga* sp. Strain GJSS3-B.1 and characterization of the recombinant enzyme and its activity on kraft pulp. *Applied and Environmental Microbiology.* 61, 4110-4113.

Shao, W., DeBlosis, S. and Wiegel, J. (1995). A high–molecular-weight, cell–associated xylanase isolated from exponentially growing *Thermoanaerobacterium* sp. strain JW/SL-YS485. *Applied Microbiology and Biotechnology,* 61, 937-940.

Sharma, A. (2000). Thermo-alkali stable cellulase-free xylanase of an extreme thermophilic *Bacillus thermoleovorans* AP7. Ph.D. Thesis. University of Delhi, South Campus New Delhi, India.

Shendye, A. and Rao, M. (1993). Cloning and extracellular expression in *Escherichia coli* of xylanases from an alkalophilic thermophilic *Bacillus* sp. NCIM I59. *FEMS Microbiology Letters,* 108, 297-302.

Shendye, A. and Rao, M. (1993). Molecular cloning and expression of xylanases from an alkalophilic thermophilic *Bacillus* (NCIM 59) in *Bacillus subtilis* A8. *Enzyme and Microbial Technology,* 15, 343-347.

Shendye, A., Gaikaiwari, R. and Rao, M. (1994). Expression of the cloned xylanases from an alkalophilic thermophilic *Bacillus* in *Bacillus subtilis. World Journal of Microbiology and Biotechnology,* 10, 414-416.

Shibuya, N. and Iwasaki, T. (1985). Structural feature of rice bran hemicellulose. *Phytochemistry,* 24, 285-288.

Shoham, Y., Schwartz, Z. Khasin, A., Gat, O., Zosin , Z., and Rosenberg, E. (1992). Delignification of wood pulp by a thermostable xylanase from *Bacillus stearothermophilus* strain T-6. *Biodegradation,* 3, 207-218.

Simpon, H.D., Haufler, U.R. and Daniel, R.M. (1991). An extrremly thermostable xylanase from the thermophilic eubacterium *Thermotoga. Biochemical Journal,* 277, 413-417.

Singh, R.P., Kapur, N. and Singhal, V. (2005). Engineering of xylanolytic organisms and enzymes for pulp and paper industries. In: *Biotechnological Applications of Microbes.* Eds. Varma, A. and Podila, G. K. I.K. International Pvt. Ltd. Pp 221-237.

Singh, S., Madlala, A.M. and Prior, B.A. (2003). *Thermomyces lanuginosus*: properties of strains and their hemicellulases. *FEMS Microbiology Reviews,* 27, 3-16.

Srinivasan, M.L. and Rele, M.V. (1999). Microbial xylanases for paper industry. *Current Science,* 77(1), 137-142.

Srivastava, K.C. (1993). Properties of thermostable hemicellulolytic enzymes from *Thermomonospora* strain 29 grown in solid-state fermentation on coffee processing solid waste. *Biotechnology Advances,* 11(3), 441-465.

Stutzenberger F.J, Bodine, A.B. (1992). Xylanase production by *Thermomonospora curvata. Journal of Applied Bacteriology,* 72, 504–511.

Subramaniyan, S. and Prema, P. and Ramakrishan, S.V. (1998). Isolation and screening for alkaline thermostable xylanases. *Journal of Basic Microbiology,* 37(6), 431-437.

Subramaniyan, S. and Prema, P. (2000). Cellulase-free xylanases from *Bacillus* and other microorganisms. *FEMS Microbiology Letters,* 183(1), 1-7.

Subramaniyan, S. and Prema, P. (2002). Biotcohnology of microbial xylanases: molecular biology, and application. *Critical Reviews in Biotechnology,* 22, 33–64.

Sunna, A. and Antranikian, G. (1997). Xylanolytic enzymes from fungi and bacteria. *Critical Reviews in Biotechnology,* 17 (1), 39-67.

Sunna, A., Prowe, S.G. Stoffregen, T. and Antranikian, G. (1997). Characterization of the xylanases from the new isolated thermophilic xylan-degrading *Bacillus thermoleovorans* strain K-3d and *Bacillus flavothermus* strain LB3A, *FEMS Microbiology Letter,* 148, 209-216.

Sunna,A., Gibbs,M.D. and Bergquist,P.L. (2000). The thermostabilizing domain, XynA, of *Caldibacillus cellulovorans* xylanase is a xylan binding domain. *Biochemical Journal,* 3,583-586.

Suurnakki, A., Clark, T.A. Allison, R.W., Viikari, L. and Buchert, J. (1996). Xylanase and mannose-aided ECF and TCF bleaching. *Tappi Journal,* 79, 111-117.

Tahir, T.A., Berrin, J.G. Flatman R., Ruussel, A., Roepstorff, P. Williamson, G., Juge, N. (2000). The thermostabilizing domain of the modular xylanase XynA of *Thermotoga maritima* represents a novel type of binding domain with affinity for soluble xylan and mixed linkage beta-1,3/beta-1, 4-glucan. *Journal of Biological Chemistry* 277(46), 44035-44043.

Takenishi, S. and Tsujisaka, Y. (1975). The modes of action of three xylanases produced by a strain of *Aspergillus niger* van Tieghem. *Agricultural and Biological Chemistry,* 39, 2315-2323.

Techapun, C., Sinsuwongwat, S., Poosaran N., Watanabe, M., and Sasaki, K. (2001). Production of a cellulase-free xylanase from agricultural waste materials by a thermotolerant *Streptomyces* sp. *Biotechnology Letters,* 23, 1685-1689.

Techapun, C., Chareonrat, T., Watanabe, M., Poosaran, N., Sasaki, K. (2002). Thermostable and alkaline-tolerant cellulase-free xylanases produced from thermotolerant *Streptomyces* sp. Ab106. *Journal of Bioscience and Bioengineering,* 93(4), 431-433.

Techapun C., Poosaran N., Watanabe M., and Sasaki, K. (2003). Thermostable and alkaline-tolerant microbial cellulose free xylanase produced from agricultural wastes and the properties required for use in pulp bleaching processes: a review. *Process Biochemistry,* 38, 1327-1340.

Timell, T.E. (1964). Wood hemicelluloses: Part I., *Advances in Carbohydrate Chemistry,* 19, 247-302.

Timothy, S.M. Don, L.C. (1997). Purification and characterization of an alkaline xylanase from *Streptomyces viridosporus* T7A. *Enzyme Microbial Technology,* 21, 160-164.

Torronen, A. and Rouvinen, J. (1995). Structural comparison of two major endo-1,4-xylanases from *Trichoderma reesei. Biochemistry,* 34, 847–856.

Tsujibo, H., Sakamoto, T., Nishino, N. and Hasegawa, T. (1990). Purification and properties of three types of xylanases produced by an alkaliphilic Actinomycete. *Journal of Applied Bacteriology,* 69, 398-405.

Turunen, O., Ethuatio, K., Fenel, F., Vehmaanpera,J.,Wu, X.,Rouvinen and Leisola, M. (2001). A combination of weakly stabilizing mutation with a disulphide bridge in the $\alpha$ helix region of *Trichoderma reesei* endo-1, 4-$\beta$-xylanase II increases the thermostability through synergism. *Journal of Biotechnology,* 88, 37-46.

Turunen, O., Vuorio, M., Fenel, F. and Leisola, M. (2002). Engineering of multiple arginines into the Ser/Thr surface of *Trichoderma reesei* endo-1,4-beta-xylanase II increases the thermotolerance and shifts the pH optimum towards alkaline pH. *Protein Engineering,* 15, 141–145.

Uchino, F. and Nakane, T. (1981). A thermostable xylanase from a thermophilic *Bacillus* sp. *Agricultural and Biological Chemistry,* 45, 1121-1227.

Uhil, A.M. and Daniel, R.M. (1999). The ?rst description of an archaeal hemicellulase: the xylanase from *Thermococcus zilligii* strain AN1. *Extremophiles,* 3, 263–267.

Van der Veen, P., Flipphi, M.J.A., Voragen, A.G.J. and Visser J. (1992). Induction of arabinases from *Aspergillus niger* on monomeric substrates. In *Xylan and Xylanases.* Eds. Visser, J., Beldmen, G., Kusers-van Someren, M.A.,Voragen, A. G. J. Elsevier Science Publishers, Amsterdam. Pp. 497-500.

Van Doorslaer, E., Kersters- Hilderson, H., and De Bruyne, C.K. (1985). Hydrolysis of beta-D- xylosidase from *Bacillus pumilus. Carbohydrate Research,* 140, 342-346.

Veoikodvorskaya, T.V. Volkov, I.Y. Vasilevko, V.T. Zverlov, V.V. and Piruzian, E.S. (1997). Purification and some properties of *Thermotoga neapolitana* thermostable xylanase B expressed in *E. coli* cells. *Biochemistry,* (Moscow) 62, 66-70.

Vieille, C. and Zeikus, G.J. (2001). Hyperthermophilic enzymes: sources, uses, and molecular mechanisms for thermostability. *Microbiology and Molecular Biology Reviews,* 65, 1–43.

Viikari, L., Ranua, M., Kantelinen, A., Sunduist, J., and Linko, M. (1986). Bleaching with enzymes. In: *Proceedings of 3rd International Conference on Biotechnology in the Pulp and Paper Industry,* STFI, Stockholm, Sweden. 67-69.

Viikari, L., Ranua, M., Kantelinen A., Linko, M. and Sundquist, J. (1987). Application of enzymes in bleaching. In: *Proceedings of 4th International Symposium on Wood and Pulping Chemistry,* Vol.1 Pp 69-90.

Viikari, L. Kantelinen, A., Sundquist J., Linko, M.(1994). Xylanases in bleaching from an idea to the industry. *FEMS Microbiology Reviews,* 13, 335-350.

Vogt, G., Woell, S. and Argos, P. (1997). Protein thermal stability, hydrogen bonds, and ion pairs. *Journal of Molecular Biology,* 269, 631-643.

Wakarchuk, W.W., Sung, W.L., Campbell, R.L., Cunningham, A., Watson, D.C. and Yaguchi, M. (1994). Thermostabilization of the *Bacillus circulans* xylanase by the introduction of disulphide bonds. *Protein Engineering,* 7, 1379-1386.

Whistler. R.L, and Richards, E.L. (1970). Hemicelluloses. In: *The carbohydrates.* Eds. Pigman, W. and Horton, D. Academic Press, New York, Pp 447-469.

Winterhalter, C., Heinrich, P., Candussio, A., Wich, G. and Liebl, W. (1995). Identi?cation of a novel cellulose-binding domain within the multidomain 120 kDa xylanase XynA of the hyperthermophilic bacterium *Thermotoga maritima. Molecular Microbiology,* 15, 431–444.

Withers, S.G. and Aebersold, R. (1995). Approaches to labeling and identification of active-site residues in glycosidases. *Protein Science,* 4, 361-372.

Wong, K.K., Tan, L.U. Saddler, J. (1988). Multiplicity of β–1-4 xylanase in microorganism: functions and applications. *Microbiology and Molecular Biology Reviews,* 52, 305-317.

Wong, K.K.Y. and J.N. Saddler (1992). *Trichoderma* xylanases, their properties and application. *Journal of Critical Reviews in Biochemistry,* 12, 413-435.

Wong, K.K.Y., Clarke, P. and Nelson, S.L. (1996). Possible roles of xylan derived chromophores in xylanase prebleaching of soft wood kraft pulp. *ACS Symposium Ser.,* 618, 352.

Woodward, J. (1984). β-xylanases: functions, properties and applications. *Enzyme and Fermentation Biotechnology,* 12, 413-435.

Yang J.L, Lou G., Eriksson KE. (1992). The impact of xylanase on bleaching of kraft pulps. *TAPPI Journal,* 75(12), 95-101.

Yang, V.W., Zhuang, Z., Elegir, G., Jeffries, T.W. (1995). Alkaline active xylanase produced by an alkalophilic *Bacillus* sp. isolated from kraft pulp. *Journal of Microbiology,* 15,434-441.

Yoshioka, H., Nagato, N., Chavanich, S., Nilubol, N. and Hayashida (1981). Purification and properties of thermostable xyalanase from *Talaromyces byssochlamydoides* YH 50. *Agricultural and Biological Chemistry,* 45, 2425-2432.

Yu, E.K.C., Tan, L.U.L., Chan M.H.K., Deschatelets, L. and Saddler, J.N. (1987). Production of thermostable xylanase by a thermophilic fungus *Thermoascus aurnatiacus. Enzyme and Microbial Technology,* 9, 16-25.

Zamost, B.L., Nielsen, H.K., and Starnes, R.L (1991). Thermostable enzymes for industrial applications. *Journal of Industrial Microbiology,* 8, 71-78.

Zheng L, Du Y, and Zhang J. (2000). Biobleaching effect of xylanase preparation from an alkalophilic *Bacillus* sp. on ramie fibers. *Biotechnology Letters,* 22(17), 1363-1367.

Zheng W, Schingoethe DJ, Stegeman GA. Hippen AR, Treachert R. (2000). Determination of when during the lactation cycle to start feeding a cellulose and xylanase enzyme mixture to dairy cows. *Journal of Dairy Science,* 83(10), 2319-2325.

Zimmermann, W., Winter, B. and Broda, P. (1988). Xylanolytic enzyme activities produced by mesophilic and thermophilic Actinomycetes grown on graminaceous xylan and lignocellulose. *FEMS Microbiology Letters* 55, 181-186.

Zverlov, V., Piotukh, K., Dakhova, O., Velikodvorskaya, G. and Borriss, R. (1996). The multidomain xylanase A of the hyperthermophilic bacterium *Thermotoga neapolitana* is extremely thermo-resistant. *Applied Microbiology and Biotechnology,* 45, 245–247.

# 18

# Application of Xylanases in the Pulp and Paper Industry: An Appraisal

**Mukesh Kapoor[1], Ajay Singh[2] and Ramesh Chander Kuhad[1]**
[1]*Department of Microbiology, University of Delhi, South Campus, Benito Juarez Marg, New Delhi-110021, India*
[2]*Department of Biology, University of Waterloo, Waterloo, Ontario, Canada*

The pulp and paper industries process huge quantities of lignocellulosic biomass every year to keep pace with ever rising demand for paper which is used in education, civilization and spread of knowledge on a global scale. The organochlorine compounds produced during pulping process have been a matter of utter concern in the paper industry for the past two decades. Application of xylanases has proven to be a cost effective way for mills to realize a variety of bleaching benefits, including: reducing or eliminating $Cl_2$, decreasing AOX discharges, freeing up chlorine dioxide generating capacity without expensive capital investments. Programs involving microbiologists, biochemists, process engineers have to be coordinated and in collaboration with the paper industry the problem be looked at in total perspective to obtain xylanases with novel properties and competitive cost that can be exploited commercially. An overview presentation is made on the potential application and current global status on the use of xylanases in paper and pulp biotechnology along with industrial benefits of enzyme treatment and the issues that must be considered to get the most benefit from using xylanase treatment in a mill.

## INTRODUCTION

Historically, the pulp and paper industry has been considered to be a major consumer of natural resources (wood, water) and energy (fossil fuels, electricity) and a significant contributor of pollutant discharges to the environment (air, water and land) (Thompson *et al.,* 2001). Most of the integrated pulp and paper mills in India and abroad adopt the most cost-effective kraft process for pulping where wood chips are cooked in a solution of sodium sulfide and sodium hydroxide and is bleached in a multistep sequence. The first steps (delignification) remove the bulk of the residual lignin and the final steps (brightening) develop high pulp brightness. Delignification is usually accomplished by chlorine water followed by alkaline extraction. Chlorine reacts with the residual lignin by oxidation and substitution; NaOH subsequently replaces most of the chlorine substituents with hydroxyl group, ionizes the phenolic groups, and dissolves the degraded lignin. Brightening is most often done with chlorine-dioxide, which oxidatively degrades any remaining lignin and destroys any other chromophores in the pulp (Bajpai, 2004). During chlorination, there is generation of chlorinated phenols, resins and fatty acids, chlorinated hydrocarbons, hypochlorite, carbohydrates and their degradation products. These compounds exhibit low molecular weight, high toxicity, mutagenicity and bacteriostatic properties. During alkali extraction, there is generation of chlorinated and oxidized kraft lignin, hemicelluloses and their degradation products. The presence of these compounds increases the BOD and COD values of the effluents (Eriksson and Kirk, 1985). The discharge of pulp and paper mill wastewater, without any treatment, results in toxicity to aquatic life, and increase in the cost of water treatment.

The technology of chemical pulp bleaching has entered a period of extremely rapid change driven by growing environmental consciousness not only in the market place but also by increasingly stringent government regulations over pulp and paper industry's waste streams, processes and products. During the last decade, the paper industry has upgraded and modernized its pulping operations based on the use of new alternative bleaching techniques such as prolonged kraft cooking, oxygen, ozone and hydrogen peroxide bleaching. However, these commercial alternatives are eclipsed by poor selectivity, economics and the products generated in chemical delignification process. As a result of these limitations, pulp and paper industry still finds itself in the focus of a debate over possible negative environmental impacts from pulp bleaching processes (Gregory *et al.,* 1998). Moreover, elemental chlorine free (ECF) paper and totally chlorine free (TCF) papers have been projected as targets to be achieved all over the developed world and it is possible that the restrictions may extend even to paper products imported from developing countries. It shall become imperative for developing nations including India, who are signatories to GATT and WTO to fall in line and develop technologies alternative to traditional bleaching processes in pulp and paper industries (Srinivasan, 1998).

Xylanase prebleaching technology has become one of the solutions considered by the pulp and paper industry to give an innovative, environmentally and economically acceptable answer to the pressures exerted on chlorine bleaching by regulatory authorities in western countries and by more demanding, environmentally minded consumers. The xylanase aided bleaching method thus represents a new type of generic technology and interest in this type of technology is already shown by the increasing numbers of papers and reviews (Biely, 1985; Bajpai and Bajpai, 1992; Viikari *et al.,* 1986; Viikari *et al.,* 1994; Suurnakki *et al.,* 1997; Bajpai, 1999; Beg *et al.,* 2001; Bajpai, 2005) published during recent years describing numerous xylanases from new sources, as well as bleaching results obtained using various xylanases, pulps and bleaching sequences. The present review summarizes the major achievement in this field.

## XYLANASES

Endo-xylanases (1,4-β–D-xylan xylanohydrolases, EC 3.2.1.8) catalyzes the random hydrolysis of 1,4-β–D-xylosidic linkages in xylans (Wong and Saddler 1992). Most xylanase are rather small proteins (molecular mass, around 20 KDa) with a basic isoelectric point (pI 8-10). Another group of xylanases have been identified. These are somewhat larger proteins (molecular mass > 40 KDa), with acidic isolelectric point (pI 3-5) (Wong *et al.,* 1988). The grouping based on physiochemical properties correlates well with the classification of xylanases into glycosyl hydrosyl families 10 and 11 (previously

F and G), based on the amino acid similarities (Henrissat and Bairoch, 1993). The family 10 xylanases, with high $M_r$ and low pI, seem to exhibit greater catalytic versatility than those of family 11, with low $M_r$ and high pI and thus are able to hydrolyze highly substituted xylans more efficiently.

## SUBSTRATE HYDROLYSIS

Most of the xylanases are able to hydrolyze different type of xylans, showing differences only in the spectrum of end products. The main products formed from the hydrolysis of xylans are xylobiose, xylotriose and substituted oligomers of two to five xylosyl residues. The chain length and the structure of the substituted products depend on the mode of action of the individual xylanases (Patel *et al.,* 1993; Stutzenberger and Bodine, 1992).

## PROPERTIES

Most xylanases studied are active in slightly acidic conditions between pH 4.0 and 6.0 and temperatures below 70°C. More thermophilic and alkalophilic xylanases are of great importance because of the prevailing conditions in pulp processing. Some of the alkalophilic strains of *Bacillus* sp. have been reported to produce xylanases with relatively high activity at alkaline pH values (Honda *et al.,* 1985; Nakamura *et al.,* 1993). Xylanases which are stable and function efficiently at high temperatures are produced by several thermophilic bacteria (Viikari *et al.,* 1993). The most thermophilic xylanases so far described are produced by an extremely thermophilic bacteria *Thermotoga* sp. *T. maritima* produces at least two hyperthermophilc xylanases, of which one has its temperature optimum at 90°C and other at 105°C (Winterhalter and Libel, 1995). However, microorganisms living in extreme conditions are often difficult to grow; however, there is a wide scope of using novel genetic techniques for screening for xylanases better suited for industrial applications. Thus, several xylanase genes encoding proteins active at temperatures from 75°C to 95°C (pH 6-8) have been isolated from the extremely thermophilic bacteria *Thermotoga* and *Dictyoglomus* without the laborious production and purification of respective enzymes (Berqvist, 1995).

## ACCESSIBILITY OF PULP HEMICELLULOSES

The action of enzymes is affected by the accessibility of substrates in the fibre matrix. The main factors limiting the access of enzymes to woody materials are the specific surface area and the porosity, i.e. the median pore size of fibres (Wong *et al.,* 1988). Other factors such as the molecular organization of the fibre components (Puls *et al.,* 1990) and the linkages between lignin and carbohydrates (Gellerstedt and Lindfors, 1991; Iversen and Wannstrom, 1986) may also have a significant role in the accessibility of fibre-bound substrates, especially in the case of xylans and glucomannans. The molecular size and structure of enzyme can be expected to be important factors when considering the limited accessibility of the substrates in fibre matrices. The molecular size of the xylanase from *T. ressei* (20 KDa, diameter 3-4 nm) (Torronen *et al.,* 1994) should allow this protein to penetrate most of the pores in chemical pulps. Thus, while the native wood is highly inaccessible to enzymes, chemical pulps can be structurally modified by enzymes.

## FACTORS AFFECTING XYLANASE ACTION

In practical process conditions, the various properties of the enzymes, such as substrate specificity, pH and temperature optima, and mode of action are of the utmost importance. Due to high temperature and alkalinity of the pulp, it would be desirable to be able to use enzymes at pH 9-10 at 80-90°C. Because most commercial xylanases of the first generation cannot meet these requirements, adjustment of pH to about 5-7 and cooling of the pulp to 50-60°C is necessary (Suurnakki *et al.,* 1997). The action of enzymes is also influenced by the electrochemical interactions between fibres and enzymes (Buchert *et al.,* 1993; Buchert *et al.,* 1993). The carboxyl groups within the fibre cell wall are mainly responsible for the swelling properties of pulp in water (Scallan, 1977). The surface charge and swelling of fibres have been reported to affect the action of xylanases. The more negative the surface charge is, the less pulp is hydrolyzed. The surface charge was adjusted by choice of the metal substituents of the carboxyl groups in pulp. However, the surface charge or swelling may not be the primary factors affecting the hydrolysis. The type of counter-ion and the degree of substitution of the carboxyl groups in pulp also has a profound effect on the action of xylanases in the pulp matrix (Scallan, 1977). The elucidation of these factors has increased basic knowledge and improved practical pulp treatments.

## PULP BLEACHING PROCESS

Pulping process, primarily aimed at removal of lignin, and to separate the wood fibres from each other in order to render them suitable for further industrial processing (Sjostrom, 1993). This objective is achieved by chemical, thermal, mechanical or by various combinations of these treatments (Smook, 1989). Wood pulp is prepared by three main processes (Thompson *et al.,* 2001):

- Mechanical pulping process involves passing a block of wood, usually debarked, through a rotating grind-stone where the fibres are stripped off and suspended in water.
- Chemical pulping which utilizes significantly large amounts of chemicals to break down the wood in the presence of heat and pressure. The spent liquid is either recycled or disposed of by burning for heat recovery.
- Chemical thermo-mechanical pulping (a combination of the two) where in the wood is first partially softened by chemicals and the remainder of the pulping proceeds with mechanical force.

Apart from wood, other fibrous raw materials being used by the pulp and paper industry, in India, are bamboo, agricultural residues like cereals, straws, sugarcane baggasse and a variety of grasses. The overall worldwide production of cereal straw is estimated around 2,900 million tons each year. In 1970, pulp from non-woody fibres accounted for only 6.7% of the overall worldwide production figure. By 1993, the proportion had risen to 10.6% and, some time in the future, it will predictably double that of pulp made from woody materials. Agricultural fibres thus constitute an alternative to wood for making pulp on account of their high growth rate and adaptability to various soil types. One of the major problems posed by this type of raw material is the low quality of the resulting paper in relation to wood pulp. This is largely due to its high content in non-fibrous material. One other problem arises from the fact that the spent cooking liquid contains much silica and is thus difficult to recycle (Seisto *et al.,* 1997). Wheat straw is seemingly the most suitable among cereal straw materials for manufacturing paper (Hamilton and Leopold, 1969; Jimenez and Lopez, 1993; Watson *et al.,* 1998). The advantages of these raw materials are that they provide excellent fibres for special types of paper and that is the sole source of fibrous raw materials in some geographic areas. This trend is expected to consolidate in forthcoming years owing to the demands of highly populated countries that possess vast agricultural and fast-growing plant resources (Wilson, 1996).

## PULPING

Chemical pulping process can be divided into three categories –sulphite pulping (cooking the pulp under acidic condition), soda pulping and kraft pulping (cooking the pulp under alkaline conditions) (Lopez *et al.,* 2000) as detailed below:

### Soda Pulping Process

An amount of wood is placed in the digester and supplied with the amounts of soda and water required to obtain a soda concentration of 8% (as $Na_2O$) and a liquid/solid ratio of 6:1 along with 0.5% of anthraquinone. The contents subjected to heat treatment at 174°C for 45 min in the digester washing with water and passage through a wet shredder. This is followed by subjecting the pulp to mechanical treatment on a Sprout-Waldron refiner separation and subsequent discard of uncooked material from the soda pulp.

### Kraft Pulping Process

The procedure was similar to the previous one except that it uses sodium sulphide, soda and water in the proportion required to obtain an active alkali concentration of 10% (as $Na_2O$), a sulphidity of 16% and a liquid/solid ratio of 6:1. The mixture is cooked at 165°C for 60 min.

### Sulphite Pulping Process

The sulphite procedure is similar to the previous two but used sodium carbonate, sodium sulphite and water in appropriate amounts to ensure a sulphite concentration of 12.5% (as $Na_2O$) and a liquid/solid ratio of 6:1. The mixture is supplied with 0.5% anthraquinone and cooked at 165°C for 120 min.

The kraft pulping has occupied the dominant position because of advantages in chemical recovery and pulp strength (Smook, 1989; Bajpai and Bajpai, 1992; Kuhad and Singh, 1993; Suurnakki *et al.,* 1997). Recently, modified and totally new kraft pulping processes using altered cooking conditions such as relatively even alkali profile and decreased cooking temperatures have been introduced. The main aim of the pulping methods such as modified continuous cooking (MCC), superbatch cooking and isothermal cooking (ITC) is to produce more easily bleachable pulps having both relatively low lignin content and acceptable strength properties (Suurnakki *et al.,* 1997).

## BLEACHING OF PULP

The bleaching of pulp *i.e.* the purification and whitening of pulp is achieved in a series of stepwise operations. Each step in a bleaching process is referred to as a stage. According to Singh (1979), each bleaching stage essentially consists of three major steps:

1. Mixing of pulp with chemicals and heat.
2. Retention of the mixture in suitable vessels.
3. Washing of the pulp with water, after reaction, to remove the impurities released by chemical reaction together with excess of bleaching chemicals.

Several variables, such as the amount of chemicals added, pulp consistency, temperature, time and pH play a decisive role in control of a bleaching stage (Bajpai, 1999). Bleach plants vary from a single stage to as many as nine or ten stages. Such a combination of stages is called bleaching sequence. Different bleaching stages are commonly represented by symbols as shown in the Table 1.

**Table 1. Commonly used bleaching chemicals and their salient features**

| *Bleaching Agent/Stage* | *Symbol* | *Salient Features* |
|---|---|---|
| Chlorination (chlorine gas or chlorine water) | C | Converts lignin to compounds that are soluble in water and alkali, therefore, used in conjunction with alkali extraction (Anonymous, 1991; Viikari *et al.,* 1994; Bajpai, 1997) |
| Caustic/Alkali Extraction (NaOH solution) | E | Causes darkening of pulp (unlike bleaching), facilitates neutralization of acidic groups, removal of chlorinated resins, lignin and hemicellulose (Bajpai *et al.,* 1994; Christov and Prior, 1996) |
| Hypochlorite (sodium or calcium hypochlorite) | H | Selectively oxidize, to colourless state, any traces of remaining coloured matter (Gellerstedt and Lindfors, 1991) |
| Chlorine dioxide and chlorite (solution form) | D | More specific in its reaction on pulp and attacks lignin and extractive compounds; preferred over hypochlorite treatment. Chlorite treatment is costly and, therefore, used in paper mills in limited quantities (Buchert *et al.,* 1994; Jimenez *et al.,* 1997) |
| Peroxide (50% solution) | P | Preferred for sulphite pulps with low lignin content but is costly and non-specific (Gellerstedt and Lindfors, 1991; Lundgren *et al.,* 1994; Wong *et al.,* 1997) |
| Oxygen/Ozone (Gaseous ozone, $O_3$, *i.e.* 2% in oxygen) | O/Z | In the presence of protectors like Mg compounds (for carbobhydrates), favourable effects are observed. The technology is rapid but costly and non-specific (Bajpai and Bajapi, 1992; Yang *et al.,* 1993; Buchert *et al.,* 1994; Gubitz *et al.,* 1996; Ruiz *et al.,* 1999) |
| Acid souring ($SO_2$ gas)/Acid treatment ($H_2SO_4$ based) | S/A | Reduces any remaining oxidant to acidity the pulp, Increases the brightness and removes metal ions (Christov *et al.,* 1998) |
| Sodium Borohydride | - | Final treatment of pulp to reduce carboxyl groups and to stabilize the pulp against colour reversion (Parson, 1964). |
| Performic Acid/Peracetic Acid/Peroxymonosulphate | - | Non-chlorine bleaching of pulp (Ragauskas *et al.,* 1994; Viikari *et al.,* 1994) |
| Dithionate | - | Increases brightness of the pulp (Rydholm, 1965) |
| Bromine, Fluorine | - | Strong oxidizing properties that facilitates lignin oxidation (Loras, 1980) |

The selection of the proper number of stages or the bleaching sequence is determined by:

1. The nature of fibre species to be bleached (softwood or hardwood) and the desired range of brightness to be obtained.
2. The pulping process applied during the conversion of wood to pulp (chemical, mechanical or combination of both).
3. The end use for which the bleached pulp is designated.

Traditionally, bleaching of chemical pulp has been carried out with elemental chlorine and chlorine dioxide. However, the chlorinated organic compounds formed during chlorine bleaching have attracted negative attention during recent years. Public concern, together with tightened environmental regulations, have driven the pulp and paper industry to seek out and utilize alternative bleaching processes such as oxidizing agent based totally chlorine-free (TCF) and especially the chlorine dioxide-based elemental chlorine-free (ECF) sequences. These sequences are increasingly used in industrial pulp bleaching. Oxygen delignification is commonly used as a prebleaching stage prior to the ECF and TCF bleaching sequences. A reduction of lignin content by about 50% can be achieved using oxygen, with relatively low loss of carbohydrate yield and without impairing the strength properties of pulp (Sjostrom, 1993).

## ALTERATIONS IN HEMICELLULOSES DURING PULPING PROCESSES

Xylan and glucomannans form the basic backbone polymers of wood hemicelluloses. Native harwood and softwood xylans differ in their degree of polymerization, which in hardwood xylans is about 150-200 and in softwood about 70-130 (Zinbo and Timell, 1965, 1967). The most abundant hemicellulose in hardwood is *O*-acetyl-(4-*O*-methylglucurono) xylan. The backbone of this hard wood xylan consists of β–(1-4)-D-xylopyranose residues, with on average, one α–(1-2)-linked 4-*O*-methyl glucuronic acid substituent per 10 to 20 such residues. Approximately 60 to 70% of the xylose units are esterified with acetic acid at the hydroxyl group of carbon 2 and/or 3 and on average every tenth xylose unit carries an α–1,2-linked uronic acid side groups. In the case of softwood plants, xylan is mainly arabino-4-*O*-methyl glucuronoxylan, which in addition to 4-*O*-methyl glucuronic acid is also substituted by α-arabinofuranoside units linked by β-1, 3-linkage to the xylan backbone and the ratio of arabinose side groups to xylose residue is 1:8. Rarely, acetyl groups are attached to the softwood xylan. The reducing ends of the xylan chains are reported to be linked to rhamnose and galacturonic acid in order to make alkali-resistant end groups of xylan chain (Lindberg *et al.,* 1973; Bouveng, 1961; Zinbo and Timell, 1965, 1967).

The hemicellulose components are heavily modified during the pulping processes. At the beginning of the conventional kraft cooking, xylan in wood is partly solubilized by the alkaline cooking liquid, and many of the side groups and acetic acid residues are cleaved off (Hamilton *et al.,* 1958). During kraft pulping, the amount of arabinose and glucuronic acid side groups in xylan is decreased and complete deacetylation takes place (Hamilton *et al.,* 1958). The ester linkages between glucuronic acid and lignin in wood are not believed to survive the alkaline cooking, unlike the other linkages between arabinose or xylose and lignin that have been reported to exist in kraft pulps (Minor, 1986).

As the cook proceeds, the alkali concentration decreases, and degraded, short chain xylan precipitates in a more or less crystalline form on the surface of cellulose microfibrils (Yllner and Enstrom, 1956; Yllner *et al.,* 1957). Apparently the arabinose and glucuronic acid units, protruding from the linear 1, 4-xylopyranose chains, prevents crystallization and their removal during the kraft cook increases the tendency of xylan to crystallize (Yllner and Enstrom, 1956; Yundt, 1951; Immergut and Ranby, 1957). A considerable part of the xylan is readsorbed or reprecipitated onto the cellulose although a part remains undissolved at its original location in the fibre (Croon and Enstrom, 1962; Hartler and Lund, 1962). The reactions of xylans have been verified using the continuous liquor flow digestion method and precipitating xylan on xylan-free materials under alkaline conditions (Dillen and Noreus, 1968; Yllner *et al.,* 1957).

As a result of the solubilization of hemicelluloses during cooking, the distribution and content of xylan and glucomannans in kraft pulp fibres differs from that in the native wood fibres. In softwood kraft fibres, the xylan concentration is generally higher in outer layers, and glucomannans is more concentrated in the middle layers of the fibre. However, due to different analysis methods, variations in the distribution of polysaccharides in softwood kraft fibres have been reported (Luce, 1964). There is however a general agreement that the outer surface layer of hardwood kraft fibres is rich in xylan.

The reprecipitation of xylan is followed by the reprecipitation of dissolved lignin during kraft pulping. These redeposited polymers have also been suggested to be chemically linked to each other (Jansson and Palenius, 1972; Jansson *et al.,* 1975). Hemicelluloses seem to physically restrict the passage of high-molecular mass lignin out of the pulp fiber (Scallan, 1977; Ahlgren *et al.,* 1971).

In sulphite cooking, hemicellulose is extensively solubilized to monomeric and oligomeric compounds, and no reprecipitation occurs. Thus, the distribution of hemicellulose is relatively constant across the pulp fibres. The arabinose side-groups of softwood xylan are completely removed in acidic sulphite cooking, although some acetyl groups are present in sulphite pulps (Suurnakki, 1997).

## ROLE OF XYLANASES IN PULP AND PAPER INDUSTRY

### The Impact of Xylanases on Pulp Bleachability

Crude culture filtrates of xylanases produced by different microorganisms have been used in the first delignification experiments (Viikari *et al.,* 1986, 1987), and even today most of the reports published are based on results with unpurified enzymes. The effect of xylanases has in most cases been independent of the origin of the enzyme, and both fungal and bacterial xylanases have been shown to increase bleachability (Patel *et al.,* 1993). The role of xylanase activity in delignification of kraft pine pulp has been extensively studied with purified xylanases of *T. longibrachiatum* (Buchert *et al.,* 1992; Tenkanen *et al.,* 1992a, b). The purified xylanases of *T. reesei* were shown to have different pI values, pH optima and substrate specificities. The xylanase with pI 9 has been shown to be relatively more specific for non-substituted xylans. However, in limited hydrolysis of xylans of pine kraft pulp, both purified xylanases resulted in approximately the same kappa number reduction and brightness increase in subsequent chemical delignification (Buchert *et al.,* 1992). Interestingly, the pH optimum of the pI 9 xylanase was shifted in pulp treatments to pH 6–7 from the optimum pH of 5.0 determined on isolated xylan (Buchert *et al.,* 1992).

In xylanase prebleaching, the side group cleaving enzymes may enhance the action of xylanases in a synergistic way by removing the side groups. If xylan is attached to lignin molecules through the side groups, hydrolysis of the linkages between the side group and xylan may enhance delignification. Kantelinen *et al.,* (1988) studied the role of side group splitting enzymes with partially purified enzymes—acetyl esterase of *T. reesei*, commercial $\alpha$–arabinosidase of *Aspergillus niger*, $\alpha$–glucuronidase of *Agaricus bisporus*, either alone or in combination with xylanases from *S. olivochromogenes.* In both cases, the $\alpha$–arabinosidase and acetyl esterase were relatively ineffective when used alone. Concurrently, with this work, both enzymes were shown to react mainly with substituted xylooligomers (Poutanen and Sundberg, 1988; Puls *et al.,* 1987). The role of $\alpha$–arabinosidase in pulp treatments was also studied. This enzyme was unable to release arabinose from pine kaft pulp when acting alone but simultaneous hydrolysis with xylanase resulted in the liberation of arabinose with a Xyl:ara ratio of 10:1 (Kantelinen, 1992). Apparently, arabinose was released from the solubilized oligomers but not from the fibres.

Analogously to this, other side group cleaving enzymes when used together with xylanases may also affect already solubilized oligomers. The effect of the $\alpha$– arabinosidase of *A. niger* and the purified $\alpha$–glucuronidase of *T. reesei* used together with xylanase was further studied in the pretreatment of pine pulp in a DEDED bleaching (Kantelinen, 1992). The side group splitting enzymes were first allowed to react alone for two hours to permit the possible reaction with insoluble xylan. After the addition of xylanase, the pulp samples were further incubated for two hours. In all enzymatically pretreated pulps, the final brightness was higher than in the reference pulp bleached with the same chlorine multiple. $\alpha$–arabinosidase had no additional effect on pulp bleachability. The $\alpha$–glucuronidase appeared to enhance the delignification slightly, as shown by the extraction stage kappa number, D1 brightness and final brightness (Table 2). The possible additional effect of $\alpha$–glucuronidase on pulp bleachability was probably due to enhancement of the action of xylanase.

**Table 2. Impact of side group splitting enzymes in conventional bleaching of kraft pulp**

| *Treatment* | *Kappa factor* | *E-Stage kappa number* | *$D_1$ Brightness* | *Final Brightness (% ISO)* | *Viscosity ($dm^3$ $Kg^{-1}$)* |
|---|---|---|---|---|---|
| Reference | 0.18 | 4.5 | 83.1 | 90.5 | 1050 |
| Reference | 0.15 | 6.5 | 73.8 | 88.1 | 1050 |
| Glucuronidase/Xylanase | 0.15 | 4.7 | 81.4 | 90.8 | 1070 |
| Arabinosidase/Xylanase | 0.15 | 4.9 | 80.5 | 90.5 | 1090 |
| Xylanase | 0.15 | 4.9 | 80.5 | 90.5 | 1090 |

Tenkanen (1998) purified two endoxylanases, one $\beta$–xylosidase, one $\alpha$–glucuronidase and two acetyl esterases from *T. reesei* and studied the synergy, for the first time, between all essential xylanolytic enzymes produced by one single organism. Table 3 shows the list of some of the commercial xylanases being used for prebleaching.

**Table 3. Commercial xylanases and their industrial suppliers**

| Name | Supplier |
|---|---|
| Cartazyme SR 10 | Clariant, U.K. |
| Cartazyme HT | Clariant, U.K. |
| Cartazyme HS 10 | Clariant, U.K. |
| Cartazyme PS10 | Clariant, U.K. |
| Cartazyme MP | Clariant, U.K. |
| Cartazyme NS10 | Clariant, U.K. |
| Cartazyme 9407 E | Clariant, U.K. |
| Irgazyme 40x 4 | Genencor, Finland/ |
| Irgazyme 40 | Genencor, Finland/ |
| Irgazyme 10Ax4 | Genencor, Finland/ |
| Ecopulp X-200 | Rohm Enzyme OY, Finland |
| Ecopulp X-200/4 | Rohm Enzyme OY, Finland |
| Ecopulp TX-100 | Rohm Enzyme OY, Finland |
| Ecopulp TX-200 | Rohm Enzyme OY, Finland |
| Ecopulp X-100 | Rohm Enzyme OY, Finland |
| Ecopulp XM | Rohm Enzyme OY, Finland |
| Ecozyme | Thomas Swan Co., U.K. |
| GS 35 | logen Corporattion, Canada |
| HS 70 | logen Corporattion, Canada |
| Optipulp-L 10000 | Solvay Interox |
| Pulpzyme HA | Novo Nordisk, Denmark |
| Pulpzyme HB | Novo Nordisk, Denmark |
| Pulpzyme HC | Novo Nordisk, Denmark |
| Albazyme 40-4X | Ciba Geigy, Switzerland |
| Albazyme 10A | Ciba Geigy, Switzerland |
| VAI Xylanase | Voest Alpine, Austria |
| Alko Rajamaki, Finland | Ecopulp |
| Meito Sankyo, Nogaya Japan | Xylanase |
| Rohm, Germany | Rholase 7118 |
| Thomas Swan, UK | Ecozyme |
| Sankyo, Japan | Sanzyme PX, Alpelase F<br>Sanzyme X |
| Enzyme Development, USA | Enzeko xylanase |

## MECHANISM OF XYLANASE-AIDED BLEACHING

Elucidation of the exact mechanism of the enzyme aided bleaching method is a key question being explored by several researchers worldwide. Xylanases such as endoxylanases are xylan-specific enzymes. They catalyze the hydrolysis of xylose-xylose bonds within the xylan chain and solubilize only a fraction of the total xylan present (Beg *et al.,* 2001). The effect of xylanases is not based on direct attack on residual lignin but on the modification of pulp hemicellulosees, enhancing the removal of lignin in chemical bleaching (Suurnakki *et al.,* 1997). One hypothesis suggests that precipitated xylan blocks or occludes extraction and that xylanase increases the accessibility (Kantelinen *et al.,* 1993). This model is based on reports that xylan reprecipitates on the fibre surfaces (Yllner *et al.,* 1957). Another possible explanation for xylanase action in bleaching is that the disruption of xylan chain by xylanase interrupts lignin-carbohydrate bonds, improves the accessibility of the bleaching chemicals to the pulps and facilitates easier removal of solubilized lignin in bleaching (Paice *et al.,* 1992). Both of these hypotheses would allow the enhanced diffusion of entrapped lignin from the fibre wall. In addition it has been suggested that xylanase treatment removes the chromophoric groups from the pulp (Paice *et al.,* 1992).

The action of xylanase on both reprecipitated and lignin-carbohydrate complexes xylan suggest that it is probably not only the type but also the location of the xylan that is important in the mechanism of xylanase aided-bleaching. The xylanase of *T. reesei* has been observed to act rather uniformly in all accessible surfaces of kraft pulps, indicating that the effect of xylanase on bleachability is not only an outer surface phenomenon. The type and location of the enzymatically-attacked xylan, hindering the leaching of pulp lignin, are thus still questions to be answered (Suurnakki *et al.,* 1997).

DMSO extraction of reprecipitated xylan does not lead to an increase in bleachability but that xylanase treatment does. This shows that DMSO-extractable xylan is not involved in bleach boosting. The removal of pulp xylan by DMSO is dependent on the degree of polymerization (DP) of xylan. As the DP of reprecipitated xylan is not known, the role of reprecipitated xylan in xylanase-aided bleaching cannot be conclusively determined by DMSO extractions (Suurnakki *et al.,* 1990).

## ANALYSIS OF THE XYLANASE ACTION IN PULP

The effect of xylanases on pulp bleachability has been studied by different methods. In pulp applications, the action of xylanases must be compared using the actual substrate, i.e. pulp. In these tests, the liberation of sugars from the pulp and into solution has been measured either by reducing sugar analysis or by HPLC. In addition, increase in the liberation of lignin derived compounds after the enzymatic treatment or after alkaline extraction has been used for the evaluation of different xylanases.

The most reliable methods to compare the effects of different enzymes is, however, to bleach the enzymaticaly treated pulps. In practice, the brightness increases after one or two stage peroxide delignification has been used to predict the effects of enzymatic treatment on pulp bleachability. This method can generally be used for the prediction of the effects obtainable using enzymes in conjunction with any bleaching sequences (Suurnakki *et al.,* 1997).

## PERFORMANCE ASSESSMENT OF XYLANASE IN DIFFERENT BLEACHING SEQUENCES

The use of xylanases in different bleaching sequences of kraft pulp consistently leads to a reduction in chemical consumption. However, the benefits obtained by enzymes are dependent not only on the type of pulp but also on the chemical bleaching sequence used, as well as the final target brightness and environmental goals of the mill (Table 4).

**Table 4. The aims of enzyme-aided bleaching in different bleaching sequences**

| *Bleaching* | *Sequence* | *Aim* |
|---|---|---|
| Conventional | X(C/D)EDED | Reduction of elemental chlorine consumption , AOX reduction |
| ECF | XDEDED | Reduction of chlorine consumption , |
| | XD(EP)DED | AOX reduction |
| | OXDEDED | Increase in productivity, when $ClO_2$ is limited |
| TCF | OXQPP | Brightness increase |
| | OXQPZP | Reduction in chemical consumption |

Originally, xylanases were applied in order to reduce the consumption of chlorine chemicals, especially elemental chlorine. Later, enzymes have been combined with various ECF and TCF bleaching sequences to improve the otherwise lower final brightness value of pulp or to decrease the bleaching cost. Results from laboratory studies and mill trials show about 35–41% reduction in active chlorine at the chlorination stage for hardwoods and 10– 20% for softwoods, whereas savings in total active chlorine were found to be 20– 25% for hardwoods and 10–15% for softwoods (Tolan and Canovas, 1992; Skerker *et al.,* 1992; Yang *et al.,* 1992; Werthemann, 1993; Allison *et al.,* 1993; Bajpai *et al.,* 1993, 1994, 1999; Valchev *et al.,* 1998, 1999; Awakoumora *et al.,* 1999; Senior and Hamilton, 1991, 1992a,b, 1993, 1999; Thibault *et al.,* 1999; Bim and Franco, 2000; Zhan *et al.,* 2000). The effect of xylanase on conventional and ECF bleaching is shown in Tables 5 to 8. The ability of the xylanase enzyme to reduce bleaching chemical consumption makes it possible to consider significant modifications in the bleaching sequence. It is possible to completely exclude the first chlorination stage in a C/D EOPDEPD and replace it with an enzyme stage (X) to become XEOPDEPD. This has been verified in a mill trial in Spain (Turner *et al.,* 1992). The advantage of this modified sequence is that filtrates from the EOP stage can be recirculated to the recovery system without the risk of chloride-initiated corrosion. This approach helps contribute to closing the water circulation system of the pulp mill and minimizes effluent discharge.

**Table 5. Effect of xylanase treatment on chlorine requirement in conventional bleaching.**

| *Total active chlorine* | *Final brightness* | |
|---|---|---|
| | **Control** | **Treated** |
| 7.0 | - | 84.5 |
| 7.5 | - | 90.7 |
| 7.8 | - | 91.6 |
| 8.0 | 84.0 | 92.0 |
| 8.6 | 89.0 | 0 |
| 9.2 | 91.5 | - |

# Bleaching sequence: D/CEoD$_E$D (50% Chlorine dioxide substitution was used in chlorination stage). Based on data from Tolan and Canovas, 1992.

**Table 6. Effect of xylanase treatment on brightness of eucalyptus kraft pulp in conventional bleaching.**

| *Enzyme* | *Bleaching sequence* | *Brightness (% ISO)* | | *Increase in brightness due to enzyme treatment (ISO Points)* |
|---|---|---|---|---|
| | | *Control* | *Enzyme* | |
| Novozyme 473 | CEHH | 80.6 | 83.6 | 3.0 |
| VAI Xylanase | CEHH | 80.4 | 82.5 | 2.1 |
| Cartazyme HS-10 | CEHH | 80.4 | 84.2 | 3.8 |
| | CEH | 78.1 | 83.0 | 4.9 |

Kappa number of unbleached pulp= 25.5. Based on the data from Bajpai *et al.,* 1994.

**Table 7. Effect of xylanase treatment on conventional bleaching***

| Chlorine dioxide substitution (%) | Control | | Treated | |
|---|---|---|---|---|
| | Kappa Factor | Final Brightness (% ISO) | Kappa Factor | Final Brightness (% ISO) |
| 10 | 0.200 | 90.0 | 0.05 | 90.0 |
| | 0.233 | 90.2 | 0.10 | 90.8 |
| | 0.266 | 90.3 | 0.15 | 91.8 |
| 40 | 0.150 | 89.2 | 0.05 | 90.0 |
| | 0.173 | 90.0 | 0.10 | 90.8 |
| | 0.200 | 91.0 | 0.15 | 91.8 |
| 70 | 0.200 | 90.0 | 0.05 | 88.7 |
| | 0.233 | 90.6 | 0.10 | 90.0 |
| | 0.250 | 90.9 | 0.15 | 90.3 |
| 100 | 0.150 | 83.5 | 0.10 | 87.4 |
| | 0.173 | 85.3 | 0.15 | 87.4 |
| | 0.200 | 87.0 | | |

Kappa number of unbleached pulp= 25.5. Based on the data from Bajpai *et al.,* 1994.

**Table 8. Effect of xylanase treatment on bleach chemical requirement**

| *Pulp* | *Bleaching Sequence* | *Chemical Consumption* | | | | *Final Brightness (% ISO)* |
|---|---|---|---|---|---|---|
| | | *aCl* | *NaOH* | $O_2$ | $H_2O_2$ | |
| Hardwood | $DE_{OP}DE_PD$ | 46.0 | 18.0 | 5.0 | 0.5 | 89.5 |
| | $XDE_{OP}DE_PD$ | 37.0 | 16.0 | 5.0 | 0.5 | 91.0 |
| Softwood | $C_{85}D_{15}E_ODE_PD$ | 94.0 | 46.0 | 5.0 | 1.0 | 89.1 |
| | $XC_{85}D_{15}E_ODE_PD$ | 84.0 | 44.0 | 5.0 | 1.0 | 88.9 |

Based on data from Dunlop and Gronberg, 1994

A non-chlorine bleaching process (EnZone process) has been developed. The process consists of xylanase treatment (X) in combination with oxygen (O), ozone (Z), and alkaline peroxide (P) bleaching stages. Hardwood (eucalyptus) kraft pulp delignified *via* an OXZ sequence had consistently lower kappa number and higher brightness than OZ bleached pulp. Pulp delignified via an OXZP sequence readily yielded a pulp of brightness 85.0-90.0% (ISO), compared with a brightness of 78.0-84.7% for the OZP pulp. The EnZone-bleached pulp had higher brightness stability, similar tensile index, and slightly lower tear index (Yang *et al.,* 1992). $XDE_pD$ was found to be he best alternative to the $DE_pD$ sequence as it surpassed the latter in pulp brightness (by 37%), featured a moderate yield loss (9.3%) and provided paper sheets with acceptably smaller breaking length and burst index (20.2 and 13.1% less, respectively). The FDEpD sequence was that resulting in the greatest increase in pulp brightness (8.6%) and the greatest savings in chlorine consumption (10.9%), however, the yield, the breaking length and burst index of the paper sheets were markedly decreased by 25.2, 24.7 and 41.8%, respectively (Jimenez *et al.,* 1997). *Eucalyptus* kraft pulp treated with MED 2D supernatant at 10 IU per gram dry weight pulp resulted in a 10.5% reduction in kappa number. XECEDED-refined bleached pulp resulted in hand sheets with increased brightness compared to the control. Xylanase treatment of pulp release chromophores absorbing at 235 nm (Haarhoff *et al.,* 1999). The use of a novel xylanase A was found beneficial in the elemental chlorine bleaching of oxygen deliginified eucalyptus kraft pulps. The application of the enzyme made possible production of fully bleached pulps with high brightness (89% ISO) and viscosities (above $800 cm^3/g$), without elemental chlorine ($XD_nEpD$ and $D_nEpD$ sequences) at low chlorine dioxide consumption (Torres *et al.,* 2000). Pretreatment of pulp with xylanase and its subsequent treatment with 8% hypochlorite, reduced the kappa number by 30%, enhanced the brightness and viscosity by 11% and 1.8%, respectively, and improved the paper properties such as tensile strength and burst factor up to 10% and 17%, respectively (Gupta *et al.,* 2000).

Xylanase dose of 3.5 U/g moisture free pulps exhibited the optimum bleach boosting of eucalyptus kraft pulp at pH 8.5 and 50°C after 2 h of treatment. When xylanase treated pulp was subsequently treated with 4.5% chlorine, it resulted in reduction of kappa number by 25%, enhanced the brightness (% ISO) by 20% and improved the pulp properties such as tensile strength and burst factor by up to 63% and 8%, respectively (Beg *et al.,* 2000).

Damiano *et al.,* (2003) employed two-stage bleaching using a $ClO_2$ chlorination and NaOH extraction (DE sequence). With the enzymatic treatment, in order to obtain the same value of kappa number and brightness, respectively 28.5 and 30% less $ClO_2$ was required in comparison to the enzymaticaly untreated samples. Two extracellular xylanases (xyl I and xyl II) produced by the thermo tolerant fungus *Aspergillus caespitosus* reduced the kappa number and released large amounts of 237 and 465 nm absorbing materials from eucalyptus kraft pulp as compared to control. Xyl II was more effective for bleaching, because it reduced kappa number by 12.6%, while xyl I only 1.7%. This result was probably related with the observed biochemical properties of xyl II, such as higher thermostability and lower molecular mass, which may allow this enzyme to diffuse more easily into the pulp matrix (Sandrim *et al.,* 2005).

A combined treatment of chemicals and commercially available enzymes like Cartazyme HS-10, Novozyme 473, VAI Xylanase, Pulpzyme HA, pIAF18 have been reported to be more beneficial than chemical treatment alone as it reduces chlorine consumption and the release of total organic chlorine in effluents with enhancement in brightness and reduction in kappa number (Paice *et al.,* 1992; Bajpai *et al.,* 1994; Ragauskas *et al.,* 1994). The reduction in kappa number, bleach boosting, direct brightening of pulp involving the use of xylanase pre-treatment or post-treatment of peroxide bleached pulp has also been reported (Kulkarni and Rao, 1996; Wong *et al.,* 1997).

Cellulase-free xylan-degrading enzyme preparations from *Acrophialophora nainiana, Humicola grisea var. thermoidea* and two *Trichoderma harzianum* strains were used as bleaching agents for eucalyptus kraft pulp, prior to a chlorine dioxide and alkaline bleaching sequence. In comparison to the control sequence (without xylanase pretreatment), the sequence incorporating enzyme treatment was more effective. Removal of residual lignin was indicated by a reduction in kappa number. Overall, enzyme preparations from *T. harzianum* were marginally more effective in reducing pulp viscosity and chlorine chemical consumption and improving the brightness of the kraft pulp. However, the highest reduction in pulp viscosity was mediated by the xylanase preparation from *A. nainiana* (Medeiros *et al.,* 2002).

Recently, Dr. Eriksson's group has reported the use of an extremely high specific activity xylanase from *Orpinomyces* sp. strain PC-2 in ECF and TCF bleaching (Shah *et al.,* 2000). An overall brightness gain of 4.3% ISO at the end of an ECF (OXDP) or a TCF(OXZP) bleaching sequence was achieved. A 2.7% ISO brightness gain was achieved at pH 8.0 with only the xylanase treatment. This increase in brightness gain is higher than for any other xylanase treatment of pulp at pH 8.0 with less than 1 g of enzyme protein used to treat one ton of pulp. Shah *et al.,* (2000) also carried out xylanase treatment from *Thermotga maritima* of oxygen bleached hardwood kraft pulp at 90° C and pH 10. About 25% savings in bleaching chemicals with 90.5% brightness was obtained. The brightness of untreated pulp was only 86% using the same bleaching sequence.

The effects of xylanase and mannanase treatment on the bleachability of oxygen delignified superbatch pulps from different origins have been studied by Suurnakki *et al.,* (1998) using peroxide or ozone based sequences. Xylanase pretreatment increased the final brightness of all the pulps more than the mannanase treatment. The combination of the more neutral *T. reesei* xylanase treatment and chelation was more suitable than the acidic *A. kawachii* xylanase treatment at pH 2.5 for superbatch pulp TCF bleaching.

Yango mill in Japan has recently introduced enzyme bleaching into the hardwood kraft bleaching process (Fukunaga *et al.,* 2000). On-site production of xylanase from *Bacillus* strain S-2113 was carried out. The principal component of culture medium was pulp, which was obtained from the fibre line at the mill. The complete culture broth containing enzyme and bacterial cell was used for enzyme bleaching. It was operated successfully with no adverse effect on pulp quality.

As the TCF market grows, Canadian mills, which are the world's largest exporters of market pulp, have started to investigate chlorine free bleaching with xylanase enzymes. European paper makers are now requesting TCF pulps or bleaching pulps with extremely low AOX and/or TOX level in their effluents (Worster, 1993). Many Finnish companies like Enso Gutzeit OY, Kimi Kymmene, Metsa-Sellu OY, Sunila OY, Veitsiluoto OY and United Paper Millsare conducting research to develop chlorine-free bleaching that involves the use of xylanase enzymes, oxygen, peroxide and ozone (Anon, 1992a).

Most of the research on bleaching with hemicellulase enzymes has so far focused on softwoods and hardwoods. Research efforts in this area on nonwoody plants are scanty, even though pulp production from nonwoody plants and agricultural residues is significant. Nevertheless, application of xylanases in prebleaching of bamboo kraft pulp, jute, maize, cotton stalks and sarkanda straws has been reported (Bajpai and Bajpai, 1996; Chen *et al.,* 1996; Wang and Shih, 1996; Spiridon *et al.,*

1998; Weidong *et al.,* 1999; Bissoon *et al.,* 2001; Shobit *et al.,* 2002; Zhao *et al.,* 2002; Sarwarjahan and Talukdar, 2002; Feng *et al.,* 2002, Roncero *et al.,* 2003, 2005). The effectiveness of various commercial cellulase-less xylanases for prebleaching of bamboo kraft pulp has been reported by Bajpai and Bajpai (1996). The maximum brightening effect was noted in the case of Bleachzyme F and Irgazyme 40S. The results with both these enzymes showed a possible 20% chlorine savings at the same brightness level, although the quantitative requirements of the two enzymes were different (Bajpai and Bajpai, 1996). Xylanase pretreatment of wheat straw and sarkanda pulp reduced total active chlorine by 16–24% and sodium hydroxide by 19–20% (Shobit *et al.,* 2002). Sarwarjahan and Talukdar (2002) studied bleaching of jute, maize and cotton stalk soda- Q(anthraquinone) pulps with xylanase enzyme. It was found that xylanase treatment reduced the kappa number and increased viscosity and brightness. Bisson *et al.,* (2002) reported enhancement in brightness of DED-bleached bagasse pulp by 4.5 points using xylanase from *Thermomyces lanuginosus* SSBP. The tensile strength, viscosity and percentage stretch decreased slightly following pretreatment with increased enzyme charges (Table 9). Treatment of sugarcane baggasse with culture filtrate of marine fungus resulted in 30% reduction in chlorine consumption and a ten-point reduction in kappa number. The unbleached baggasse pulp showed a kappa number of 48 and the enzyme-treated pulp showed a kappa number of 38. The drop in kappa number after biobleaching indicates reduction in chlorine consumption during further bleaching of the pulp (Raghukumar *et al.,* 2005).

**Table 9. Effect of pre-treatment with xylanase on the physical properties of bagasse pulp following DED bleaching.**

| *Enzyme Charge (U/g)* | *Tensile Strength (N/15mm)* | *Stretch (%)* | *TEA ($J/M^2$)* | *Viscosity (mPa.s)* |
|---|---|---|---|---|
| Control | 13.76 | 2.72 | 13.83 | 48.4 |
| 5 | 13.34 | 2.35 | 13.54 | 47.7 |
| 50 | 13.21 | 2.15 | 13.34 | 46.3 |
| 150 | 13.18 | 2.10 | 13.38 | 45.8 |

Xylanases have also been studied in the bleaching of sulfite pulps produced by acid Mg, Mg-bisulfite, two-stage sulfite and alkaline methods (Suurnakki *et al.,* 1997). Due to the different location and limited accessibility of hemicellulose in sulfite pulps, as compared to kraft pulp, lower hydrolysis was obtained with xylanase. Furthermore, the enzymatic pretreatment had no effect on the brightness or the kappa number after one stage peroxide bleaching.

Crude xylanase preparation of *Aureobasidium pullulans* and *Thermomyces lanuginous* were tested for their abilities to remove hemicellulose and lignin from sulphite pulps in conjunction with hydrogen peroxide in one-stage or multistage bleaching to dissolving pulp. Biobleaching of unbleached sulphite pulp with *Aureobasidium pullulans* enzyme preparation produced dissolving pulp with superior quality over the control of $OD_1E_0D_2P$-bleached pulp. The hemicellulose content of the xylanase-prebleached dissolving pulp was reduced by 6-8%, whereas brightness was increased by 2.2 points. Although the *Thermomyces lanuginous* xylanase produced some biobleaching effect in the one-stage hydrogen peroxide treatments, there was no obvious improvement in brightness after a full five-stage bleaching to dissolving pulp (Christov and Prior, 1997).

## IMPACT OF XYLANASE ON PULP YIELD, QUALITY AND MORPHOLOGY

The enzyme-treated pulp show unchanged or improved strength properties (Viikari *et al.,* 1990; Viikari *et al.,* 1991; Yang *et al.,* 1992; Viikari *et al.,* 1993; Bajpai *et al.,* 1994; Dunlop and Gronberg, 1994; Bajpai and Bajpai, 1996; Tolan *et al.,* 1996; Chen *et al.,* 1996; Wang and Shih, 1996; Tolan and Guenette, 1997; You and Wu, 2000). Also, these pulps are easier to refine than the reference pulps (Dunlop and Gronberg, 1994; Viikari *et al.,*, 1993; Wong *et al.,*, 1999). Improved viscosity of the pulp has been noted as a result of xylanase pretreatment (Senior and Hamilton, 1992a,b; Yang *et al.,* 1992; Paice *et al.,*, 1988). This is probably caused by the selective removal of xylan as determined by the pentosan

The effect of xylanase treatment on fibre morphology has been extensively studied (Viikari *et al.,* 1996; Rancero *et al.,* 1999, 2000; Beg *et al.,* 2001; Ander and Nyholm, 2001; Bisson *et al.,* 2002). Roncero *et al.,* (1999, 2000, 2003, 2005) have studied the effects of xylanase treatment on fibre morphology in totally chlorine-free and elementally chlorine-free bleaching of eucalyptus pulp. Scanning electron microscopy was used to examine changes of the surface morphology for enzyme and non-enzyme bleaching sequences. Xylanase treatment caused the appearance of cracks, flakes, filaments and peeling of cell walls, which in turn enabled increased contact between the bleaching agent and the substrate.

## MILL IMPLEMENTATION OF XYLANASE TREATMENT

### Compatibility with Mill Conditions

The points that a mill needs to address regarding enzyme treatment are whether xylanase treatment can meet the mills objectives, and the comparative performance of the various commercial xylanases. These questions are related and can be best approached as a first step by carrying out a laboratory study under ideal conditions. The range and variability of the treatment conditions during actual mill operations must be taken into account. The effectiveness of xylanase depends upon the pH, temperature, and length of time. The optimum pH of the enzyme treatment varies, depending on the enzyme, from about pH 3 to 9. The optimum pH is important because it influences the amount of corrosion, pitch deposition, and other operating problems for the mill. The desired temperature for enzyme treatment varies from about 30 to 60°C, depending on the enzyme. The compatibility of these temperatures with the C-stage temperature, the cooling of the brown system, etc. must be addressed. The minimum reaction time required for enzyme treatment varies from about 10 min to 2 h again depending upon the enzyme. The presence of cellulase enzymes impurities can cause degradation of pulp strength, therefore, the assurance of pulp quality requires setting a specification for the maximum allowable cellulase activity and modification of the standard cellulase assays to ensure detection of the relevant levels of cellulase activity (Tolan and Gunette, 1997).

## SCALE UP OF THE ENZYMATIC STEP

Detailed laboratory work is generally needed to optimize and adapt the enzymatic treatment to individual existing mill conditions (Tolan and Guenette, 1997). Interestingly, however, xylanase bleaching has been scaled up directly from laboratory scale to the large industrial scale (1000 TP/d) without intermediate pilot stages. It has also been observed that even higher brightness values can be reached on the full scale than those attainable in the laboratory, which is due to the more efficient mixing systems and higher pulp consistencies. No expensive capital investments have generally been necessary for full-scale runs. The most significant requirement is the addition of pH adjustment facilities. Xylanase pretreatment has been shown to be easily applicable with existing industrial equipment, which is a considerable advantage of this technology (Koponen, 1991).

## IMPACT OF MILL OPERATIONS ON XYLANASE PERFORMANCE

Mill operations also affect the performance of the xylanase enzyme. The effect of raw material, pulping process, brown stock washing and bleaching sequence should be assessed by laboratory testing prior to mill usage of enzymes (Bajpai, 2004; Tolan and Guenette, 1997).

Among raw materials, the important distinction is between hardwoods and softwoods. The percentage of the bleaching chemicals saved by xylanase treatment is greater on hardwoods than on softwoods. At good treatment conditions, the decrease in chlorine chemicals is about 20% on hardwoods and 15% on softwoods.

The digester operation affects the xylan content of the pulp significantly. For example, sulfite pulping destroys most of the xylan and thus sulfite pulp is not suitable for enhanced bleaching by enzyme treatment. In conventional kraft pulping, the xylan content depends strongly on the effective alkalinity. The lower the alkalinity, the higher the xylan content and the benefits of using xylanase enzymes. At high alkalinity (19–22%), much of the xylan is solubilized, which decreases the benefit of xylanase treatment. At low alkalinity (less than about 18%), the xylan structure is more stable, and the bleaching enhancement by xylanase is greater by up to 2- to 3-fold over the high alkalinity pulp. This is often the case for pulp cooked to higher kappa number (Bajpai, 2004). Kraft pulping at severe conditions, such as conventional cooking of softwood to kappa number less than 23, also destroys much of the hemicellulose that is accessible to the enzyme. On the other hand, MCC or oxygen-delignified pulps with low unbleached kappa number respond well to enzyme treatment (Tolan, 1992).

The bleaching sequence and brightness target influences the enzymes benefit to the mill. As a basis of comparison, about 15% of the $ClO_2$ is saved by using enzyme treatment on softwood with $ClO_2$ bleaching to 90% brightness, with the chemical usages optimized both with and without enzyme treatment. The enzyme benefit is greater at higher brightness targets, especially near the brightness ceiling, and lower at lower brightness targets (Tolan and Gunette, 1997). The results of some published mill trials are presented in Table 9.

**Table 10. Mill scale xylanase-aided bleaching trials**

| *Mill* | *Sequence after xylanase treatment* | *Reduction in AOX (%)* | *Pulp properties* | *Reduction in Chemical consumption* | *Brightness increase* |
|---|---|---|---|---|---|
| Enso Gutzeit, Finland | (D50C50)E1D1E2D1 | 29 | Unchanged | 22% Active chlorine | - |
| Bukoza paper mill, CzechoSlovakia | (CD)EDEHD | ND | Unchanged | 30% $Cl_2$, 30% Hypochlorite | - |
| Tasman Pulp and Paper Co., New Zealand | (DC)EODED | 20 | Unchanged | 20% Hypochlorite | - |
| Canfor's International Mill, B.C., Canada | DEOPDED | ND | Unchanged | 15.6% $ClO_2$ | - |
| Ballarpur Paper Industries, India | CEPH | ND | Unchanged | - | 2-3 |
| Crest brook For.Ind., Canada | DEOPDED | ND | Unchanged | 7-17% Reduction in *K* factor, increased productivity | - |
| Scott paper, Spain | EOPDEPD | - | No attack on cellulose | | 4-6 |
| Munksjo, Sweden | OQE1PE2P | - | Unchanged | | 2-4 |
| Metsa-Botnia, Finland | OPPP | ND | Unchanged | | - |
| Metsa-Sellu, Finland | - | ND | Unchanged | 12% total active chlorine | - |
| Morrum Pulp Mill, Sweden | - | 22.9 | Unchanged | 22.9% $ClO_2$ | - |
| Donohue St. Felicie Mill, Quebec | - | ND | Good strength | 24% reduction in *K* Factor | - |

## ADDITION OF XYLANASE TO THE PULP

The effectiveness of the xylanase treatment depends on maintaining the appropriate pH, temperature, and reaction time. Several factors must be taken into account to ensure that the enzyme is applied to the pulp properly without causing any long-term problems for the mill. The most conventional method is to add xylanase to the brownstock pulp prior to the high density (HD) tower. The enzyme reaction takes place in the tower and the treated pulp then passes into the bleach plant. Various ways to add enzymes have been used, including: (1) spraying on the decker pulp mat; (2) adding to either the decker repulper or discharge chute; (3) adding into the stock of medium consistency pulp leading to the HD tower; and (4) adding directly into the HD tower. Xylanase has also been added later in the bleaching sequence rather than to the brownstock pulp. The adequate dispersion of enzyme and acid into the pulp is important to xylanase performance. Some of the hardware configurations that have been used include: a shower bar for acid, then one for the enzyme on the back of the decker, separate hoses for acid and enzyme into the repulper, injection of enzyme at the suction side of an MC (medium consistency) pump. The degree of mixing that is achieved by any of these equipment configurations depends on the absorbency of the brownstock as well as the equipment used (Tolan and Gunette, 1997).

## ECONOMY OF THE XYLANASE TREATMENT

Xylanases are sold as concentrated liquids and the amount required per metric ton of pulp is very low, less than a liter. The cost of enzyme per ton of pulp varies and depends on the dosage required and the supplier. In 1995, the approximate cost of enzyme treatment was around $ 2/TP (Suurnakki *et al.,* 1997). Due to the low enzyme price and low capital costs of enzyme stage, the potential economic benefits of enzyme bleaching are significant.

## BENEFITS WITH XYLANASE TREATMENTS

Xylanase enzymes have proven to be a cost-effective way for mills to realize a variety of bleaching benefits, including:

1. Reducing AOX discharges, primarily by decreasing chlorine gas usage
2. Debottlenecking mills limited by chlorine dioxide generator capacity
3. Eliminating chlorine gas usage for mills at high chlorine dioxide substitution levels
4. Increasing the brightness ceiling, particularly for mills contemplating ECF and TCF bleaching sequences
5. Decreasing cost of bleaching chemicals, particularly for mills using large amounts of peroxide or chlorine dioxide.

## FUTURE INSIGHTS IN XYLANASE BLEACHING

The globalization of the pulp and paper industry is a relatively new phenomenon that has been a harbinger of change and opportunities. Today's paper industry employs advanced chemical and mechanical based technologies to provide high-quality consumer products that are in worldwide demand and support the lifestyles of our new global economy. And yet, from these successes arise many of the current and future difficulties of the industry. Various paper industry leaders have stated that the capital requirements of manufacturing paper are too high, are limiting creativity, and the entrepreneurial spirit of the industry. Coupled with these challenges, the emergence of low-cost fibre resources outside the northern hemisphere has contributed to further pressures on the paper industry to significantly reduce its manufacturing cost through a major redesign of its core manufacturing technologies. Within these difficulties, are disguised but unparallel opportunities for researchers to effectively develop new biotechnology-based processes. The use of xylanases in bleaching of pulps has already opened up a new era of large-scale industrial biotechnical applications in the pulp and paper industry. The development in xylanase based bleaching is focusing on improved enzyme properties and improved enzyme performance. Several laboratories are carrying out extensive investigations across the globe to identify microbial strains with promise and potential for producing xylanases useful to paper industry. Improved properties include higher pH and temperature tolerance of the enzymes, to make the enzyme treatment operations more compatible with existing mill operations. Improved enzyme performance is being approached by tailoring the enzyme action more closely to the hemicellulose structure of the pulp, to result in a greater bleaching benefit or higher pulp yield. These new technologies must reduce the capital costs of pulp production, be readily implemented in today's mill, and provide exceptional return for the resources invested if they are to be commercially feasible. The pathway for the future will entail the development of new global alliances between industry, government and research organizations. Theses new partnerships will be linked by the web, coordinated by industry and government, and will utilize the best research expertise and facilities available in the world. Theses collaborative efforts will generate higher fibre value resources, lower total manufacturing costs, and develop new materials from which new products can be designed and manufactured. In summary, the business of pulp and paper industry has provided a challenge for change, and biotechnology is destined to address this opportunity.

## REFERENCES


Ahlgren, P.A., Yean, W.Q. and Goring, D.A.I. (1971). Chlorite delignification of spruce wood. Comparison of the molecular weight of lignin dissolved with the size pores in the cell wall. *TAPPI Journal,* 54, 737-740.

Allison, R.W., Clark, T.A. and Wrathall, S.H. (1993). Pretreatment of radiata pine kraft pulp with a thermophillic enzyme. Part II. Effect on oxygen, ozone and chlorine dioxide bleaching. *Appita,* 46(5), 349–353.

Ander, P. and Nyholm, K. (2001). The effect of cellulases, hemicellulases and laccase/HBT on spruce pulp fibre deformations. In: *8th International Conference on Biotechnology in Pulp and Paper Industry.* Pp 71–73, 4–8 June, Helsinki, Finland.

Anonymous. (1991). Tappi test methods. *Tappi Press.* Atlanta, USA.

Anonymous. (1992). Development of bleaching technology in Finland. *Paper Pulp,* 74(2), 102–106.

Awakaumova, A.V., Nikolaeva, T.V., Vendilo, A.G., Kovaleva, N.E. and Sinitzyn, A.P. (1999). ECF bleaching of hardwood kraft pulp: New aspects. In: *13th International Papermaking Conference—Progress-99,* Pp IV-5-1– IV-5-13, 22–24 September, Cracow, Poland.

Bajpai, P. (1999). Application of enzymes in the pulp and paper industry. *Biotechnology Progress,* 15, 147-157.

Bajpai, P. (2004). Biological bleaching of pulps. *Critical Reviews in Biotechnology,* 24 (1), 1-58.

Bajpai, P. and Bajpai, P.K. (1992). Biobleaching of kraft pulp. *Process Biochemistry,* 27, 319-325.

Bajpai, P. and Bajpai, P.K. (1996). Application of xylanases in prebleaching of bamboo kraft pulp. *Tappi Journal,* 79(4), 225–230.

Bajpai, P. and Bajpai, P.K. (1997). Reduction of organochlorine compounds in bleach plant effluents. Advances in Biochemical Engineering/Biotechnology. 57: 213-259.

Bajpai, P., Bhardwaj, N.K., Bajpai, P.K. and Jauhari, M.B. (1994). The impact of xylanases in bleaching of eucalyptus kraft pulp. *Journal of Biotechnology,* 36(1), 1–6.

Bajpai, P., Bhardwaj, N.K., Maheshwari, S. and Bajpai, P.K. (1993). Use of xylanase in bleaching of eucalypt kraft pulp. *Appita,* 46(4), 274–276.

Beg, Q.K., Bhushan, B., Kapoor, M. and Hoondal, G.S.(2000). Enhanced production of a thermostable xylanase from Streptomyces sp. QG-11-3 and its application in biobleaching of kraft pulp. *Enzyme and Microbial Technology.* 27:459-466.

Beg, Q.K., Kapoor, M., Bhushan, B. and Hoondal, G.S. (2001). Microbial xylanases and their industrial applications: a review. *Applied Microbiology and Biotechnology,* 56, 326-338.

Berqvist, P. (1995). Hemicellulolytic enzymes from extremely thermophilc bacteria-applications of molecular genetics to pulp bleaching. In: *Proceedings of the 8th International symposium on wood and pulping chemistry.* vol I: p-263, Helenski,.

Biely, P. (1985). Microbial xylanolytic systems. *Trends in Biotechnology,* 11, 286-290.

Bim, M.A. and Franco, T.T. (2000). Extraction in aqueous two-phase systems of alkaline xylanase produced by *Bacillus pumilus* and its application in kraft pulp bleaching. *Journal of Chromatography,* 743(1), 349–346.

Bissoon, S, Christov, L, Singh, S. (2002) Bleach boosting effects of purified xylanase from Thermomyces lanuginosus SSBP on bagasse pulp. *Process Biochemistry,* 37, 567-572.

Bissoon, S., Singh, S. and Christov, L. (2001). Evaluation of the bleach enhancing effect of xylanases on bagasse pulp. In: *8th International Conference of Biotechnology in Pulp and Paper Industry.* 4–8 June, Helsinki, Finland. Pp 63–64,

Bouveng, H.O. (1961). Phenylisocynate derivatives of carbohydrates. II. Location of the O-acetyl groups in birch xylan. *Acta Chemica Scandinavica,* 15, 96-100.

Buchert, J., Kantelinen, A., Ratto, M., Siika-aho, M., Ranua, M. and Viikari, L. (1992). Xylanases and mannanases in the treatment of pulp. In: *Proeedings of 5th International Conference of Biotechnology in Pulp and Paper Industry,* 27–30 May, Kyoto, Japan. Pp 139–145

Buchert, J., Pitkänen, M. and Viikari, L. (1993). Role of Surface Charge on the Action of Hemicelluloses onKraft Pulps. In: *7th International Symposium on Wood and Pulping Chemistry.* Vol.2: pp. 584-590, Beijing, May 25-28.

Buchert, J., Ranua, M., Kantelinen, A. and Viikari, L. (1992). The role of two *Trichoderma reesei* xylanases in the bleaching of pine kraft pulp. *Applied Microbiology and Biotechnology,* 37, 825–829.

Buchert, J., Tenkanen, M., Kantelinen, A. and Viikari, L. (1994). Application of xylanases in pulp and paper industry. *Bioresource Technology,* 50, 65-72.

Buchert, J., Tenkanen, M., Pitkanen, M. and Viikari, L. (1993). The role of surface charge and swelling on the action of xylanases on birch kraft pulp. *TAPPI journal,* 76, 131-135.

Chen, J-C., Wang, B-M., Qu, Y-B., Gao, P-J., Yang, G-H., Li, Z-C. and Zhang, Q. (1996). Prebleaching of wheat straw pulps with Xylanase An-76. In: *3rd International Non-wood Fiber Pulping and Papermaking Conference.,* 15–18 October. Beijing, China. Pp 393–401,

Christov, L.P. and Prior, B.A. (1996). Repeated treatments with *Aureobasidium pullulans* hemicellulases and alkali enhanced biobleaching of sulphite pulps. *Enzyme and Microbial Technology,* 18, 244-250.

Christov, L.P. and Prior, B.A. (1997). Bleaching response of sulphite pulps to pretrement with xylanases. *Biotechnology progress,* 13, 695-698.

Christov, L.P., Akhtar, M. and Prior, B.A. (1998). The potential of biosulfite pulping in dissolving pulp production. *Enzyme and Microbial Technology,* 23, 70-74.

Damiano, V.B., Bocchini, D.A., Gomes, E. and Da Silva, R. (2003). Application of crude xylanase from *Bacillus lichiniformis* 77-2 to the bleaching of eucalyptus kraft pulp. *World Journal of Microbiology and Biotechnology,* 19, 139-144.

Dunlop, N. and Gronberg, V. (1994). Recent developments in the application of xylanase enzymes in ECF and TCF bleaching. In: *Preprints 80th Annual Meeting Technical Section.* pp. A191–A196. 1–2 February, Montreal, Canada.

Eriksson, K-E.L. and Kirk, M.C. (1985). Progress towards making trees easier to pulp and bleach. *Pulp Paper Magazine,* 2, 5-6.

Feng,W., Li, X., Chang, Q. and Liang, G. (2002). Preparation of xylanases from *Aspergillus niger* and its application in wheat straw pulp bleaching. *China Pulp and Paper,* 21(2), 8–13.

Fukunaga, N., Igarashi, H., Takahashi, M., Iwasaki, Y., Suglura, J. and Iwasaki, M. (2000). On site production of xylanase and its application on hardwood pulp bleaching. In: *International Pulp Bleaching Conference,* 27–30 June, Halifax, Nova Scotia, Canada. Pp 175–180.

Gellerstedt, G. and Lindfors, E.G. (1991). On the structure and reactivity of residual lignin in kraft pulp fibres. In: *Proceedings of international pulp bleaching conference,* Volume I. SPCI. pp: 73-88.

Gellerstedt, G. and Lindfors, E-L. (1991). On the structure and reactivity of residual lignin in kraft pulp fibers. In: *Proceedings of International pulp bleaching conference.* Vol 1, Stockholm, June 11-14, Stockholm, SPCI. Pp 73-78

Gregory, A.C.E., O'Connell, A.P. and Bolwell, G.P. (1998). Xylans. *Biotechnolcgy and Genetic Engineering Reviews,* 15, 439-455

Gupta, S., Bhushan, B. and Hoondal, G.S. (2000). Isolation, purification and characterization of xylanase from *Staphylococcus* sp. SG-13 and its application in biobleaching of kraft pulp. *Journal of Applied Microbiology,* 88, 325-334.

Haarhoff, J., Moes, C.J., Cerff, C., Van wyk, W.J., Gerischer, G. and Janse, J.J.H. (1999). Characterization and biobleaching effect of hemicellulases produced by thermophilic fungi. *Biotechnology Letters,* 21, 415-420.

Hamilton, F. and Leopold, F. (1969). Pulp and Paper Manufacture. Vol. 3. Secondary Fibers and Non-Wood Pulping, third ed. TAPPI Atlanta. CPPA Montreal. The Joint Textbook Committee of the Paper Industry.

Hamilton, J.K., Partlow, E.V. and Thompson, N.S. (1958). The behaviour of wood hemicelluloses during pulping. I examination of the xylose-contaning hemicelluloses associated with hardwood and softwood pulps. *TAPPI Journal,* 41(12), 803-811.

Hartler, N. and Lund, A. (1962). Sorption of xylans on cotton. *Svensk Papperstid,* 65, 951-955.

Henrissat, B. and Bairoch, A. (1993). New families in the classification of glycosyl hydrolases based on amino acid sequence similarities. *Biochemical Journal,* 293, 781-788.

Honda, H., Kudo, T. and Horikoshi, K. (1985a). Purification and partial characterization of alkaline xylanase from *Escherichia coli* carrying pCX311. *Agricultural and Biological Chemistry.* 49:3165-3169.

Immergut, B. and Ranby, B. (1957). Investigations of wood hemicelluloses. 1. Sepration of pulp hemicelluloses into xylan and glucomannans fraction. *Svensk Papperstid,* 60, 573-581.

Iversen, T. and Wannstrom, S. (1986). Lignin-carbohydrate bonds in a residual lignin isolated from pine kraft pulp. *Holzforschung,* 40, 19-22.

Jansson, J. and Palenius, I. (1972). Aspects of the colour of kraft pulp and kraft lignin. *Paper and Timber,* 54, 343-352.

Jansson, J., Palenius, I., Stenkud, B. and Sagfross, P.E. (1975). Differences in color and strength of kraft pulps from batch and flow cooking. *Paper and Timber,* 57, 387-390.

Jimenez, L. and Lopez, F. (1993). Characterization of paper sheets from agricultural residues. *Wood Science and Technology,* 27 (6), 468–474.

Jimenez, L., Martinez, C., Perez, I. and Lopez, F. (1997). Biobleaching procedures for pulp and agricultural residues using *Phanerochaete chrysosporium* and enzymes. *Process Biochemistry,* 33, 297-304.

Kantelinen, A. (1992). Enzymes in bleaching of kraft pulp. *VTT Publications, 114.* Espoo, Finland: Technical Research Centre of Finland.

Kantelinen, A., Hortling, B., Sundquist, J., Linko, M. and Viikari, L. (1993). Proposed mechanism of the enzymatic bleaching of kraft pulp with xylanases. *Holzforschung,* 47, 318–324.

Kantelinen, A., Ratto, M., Sundquist, J., Ranua, M., Viikari, L. and Linko, M. (1988). Hemicellulases and their potential role in bleaching. In: *Proceedings of Tappi International Pulp Bleaching Conference.,* 5–9 June, Orlando. Pp 1.

Koponen, R. (1991). Enzyme systems prove their potential. *Pulp and Paper Interntional,* 33(1), 20-25.

Kuhad, R.C. and Singh, A. (1993). Lignocellulosic biotechnology : current and future prospects. *Critical Reviews in Biotechnology,* 13, 151-172.

Kulkarni, N. and Rao, M. (1996). Application of xylanase from alkalophilic thermophilic *Bacillus* sp. NCIM 59 in biobleaching of bagasse pulp. *Journal of Biotechnology,* 51, 167-173.

Lindberg, B., Rosell, K-G. and Svensson, S. (1973). Position of the *O*-acetyl groups in birch xylan. *Svensk Papperstid,* 70, 30-32.

Lopez, F., Ariza, J., Perez, I. and Jimenez, L. (2000). Comparative study of paper sheets from olive tree wood pulp obtained by soda, sulphite or kraft pulping. *Bioresource Technology,* 71, 83-86.

Loras, V. (1980). Bleaching of chemical pulps. *In*: Pulp and paper chemistry and chemical technology. Casey, J.P. (editor). Third edition. Volume I. John Wiley and Sons Inc., New York. pp: 663-702.

Luce, J.E. (1964). Radial distribution of properties through the cell wall. *Pulp and paper magazine Canada,* 65, 419-423.

Lundgren, K.R., Bergkvist, L., Hogman, S., Joves, H., Erikkson, G., Bartfai, T., Laan, J.V.D., Rosenberg, E. and Shoham, Y. (1994). TCF mill trial on softwood pulp with thermostable and alkalistable xylanase T6. *FEMS Microbiology Reviews,* 13, 365-368.

Medeiros, R.G., Silva Jr, F.G., Salles, B.C., Estelles, R.S. and Filho, E.X.F. (2002). The performance of fungal xylan degrading enzyme preprations in elememntal chlorine free bleaching for eucalyptus pulp. *Journal of Industrial Microbiology and Biotechnology,* 28, 24-206.

Minor, J.L. (1986). Chemical linkage of polysacchrides to residual lignin in loblolly pine kraft pulps. *Journal of wood Chemistry and Technology,* 6, 185-201.

Nakamura, S., Wakabayashi, K., Nakai, R., Aono, R. and Horikoshi, K. (1993). Production of alkaline xylanase by a newly isolated alkalophilic Bacillus sp. strain 41M-1. *World Journal of Microbiology and Biotechnology.* 9: 221-224.

Paice, M.G., Bernier, Jr. R. and Jurasek, L. (1988). Viscosity enhancing bleaching of hardwood kraft pulp with xylanase from a cloned gene. *Biotechnology and Bioengineering,* 32, 235–239.

Paice, M.G., Gurnagul, N., Page, D.H. and Jurasek, L. (1992). Mechanism of hemicellulose directed prebleaching of kraft pulp. *Enzyme and Microbial Technology,* 14, 272-276.

Parson, J.L. (1964). Pulp bleaching. *In*: Handbook of pulp and paper technology. Britt, A.W. (editor). John Willey, London. pp: 266-285.

Patel, A.N., Grabski, A.C. and Jeffries, T.W. (1993). Chromophore release from kraft pulp by purified *Streptomyces roseiscleroticus* xylanase. *Applied Microbiology and Biotechnolog,* 39, 405-412.

Poutanen, K. and Sundberg, M. (1988). An acetyl esterase of *Trichoderma reesei* and its role in the hydrolysis of acetyl xylans. *Applied Microbiology and Biotechnology,* 28, 425-432.

Puls, J., poutanen, K., and Lin, J.J. (1990). Acessibility of hemicellulose in hardwood pulps to enzymes. In: *Biotechnology in pulp and paper manufacture.* Kirk, K.T. andChang, H-M. (editors). Butterworth-Heinemann, Boston, pp. 183-190.

Puls, J., Schmidt, O. and Granzow, C. (1987). a-Glucuronidase in two microbial xylanolytic systems. *Enzyme and Microbial Technology,* 9, 83-88.

Ragauskas, A.J., Poll, K.N. and Cesternino, A. (1994). Effects of xylanase pretreatment procedures on non cholrine bleaching. *Enzyme and Microbial Technology,* 16, 492-495.

Raghukumar, C., Muraleedharan, U., Gaud, V.R. and Mishra, R. (2004). Xylanases of marine fungi of potential use for biobleaching of paper pulp. *J Ind Microbiol Biotechnol.* 31, 433-441.

Roncero, M. B., Torres, A. L., Colom, J. F. and Vidal, T. (2000). Effects of xylanase treatment on fibre morphology in totally chlorine free bleaching (TCF) of Eucalyptus pulp. *Process Biochemistry,* 36(1–2), 45–50.

Roncero, M.B. Torres, J.F., Colom, J.F. and Vidal, T. (2003). TCF bleaching of wheat starw pulp using ozone and xylanase. Part A: paper quality assessment. *Bioresource Technology,* 87, 305-314.

Roncero, M.B. Torres, J.F., Colom, J.F. and Vidal, T. (2005). The effect of xylanase on lignocellulosics components during the bleaching of wood pulps. *Bioresource Technology,* 96, 305-314.

Roncero, M.B., Torres, A. L., Colom, J. F. and Vidal, T. (1999). Study the influence of xylanase on the fiber surfaces by SEM. In: *Proceedings of Microscopy as a Tool in Pulp and Paper Research and Development.* pp. 27–30. 21–22 June, Stockholm, Sweden.

Rydholm, A.A. (1965). Pulping processes. Inter Science, New York.

Sandrim, V.C., Rizzatti, A.C.S, Terenzi, H.F., Jorge, J.A., Milagres, A.M.F. and Polizeli, M.L.T.M. (1997). Purification and biochemical characterization of two xylanases produced by *Aspergillus caespitosus* and their potential for kraft pulp bleaching. *Process biochemistry,* 40, 1823-1828.

Sandrim, V.C., Rizzatti, A.C.S. Terenzi, H.F. Jorge, J.A. Milagres A.M.F. and Polizeli M.L.T.M. (2005). Purification and biochemical characterization of two xylanases produced by *Aspergillus caespitosus* and their potential for kraft pulp bleaching Process Biochemistry, 1823-1828.

Sarwarjahan, M. and Talukdar, S. (2002). Bleaching of non-wood soda-AQ pulps using enzymes from thermophillic fungus. In: *56*[th] *Appita Annual Conference* pp. 159–163. 18–20 March, Rotorua, New Zealand.

Scallan, A.M. (1977). The accommodation of water within pulp fibers. In: *Proceedings of fiber-water interactions in papermaking,* Oxford, p 9-27. Scheper, T., Ed. Springer-Verlag, Berlin.

Seisto, A., Poppius-Levlin, K. and Jousimaa, T. (1997). Peroxyformic acid pulping of nonwood plants by the MILOX method. Part 2: Reed pulp for woodfree .ne papers. *Tappi Journal,* 80 (10), 235–240.

Senior, D.J. and Hamilton, J. (1991). Use of xylanase to decrease the formation of AOX in kraft pulp bleaching. In: *Proceedings of Environment Conference of the Technical Section, Canadian Pulp and Paper Association.* pp. 63–67, 8–10 October, Quebec, Canada.

Senior, D.J. and Hamilton, J. (1992a). Bleaching with xylanases brings biotechnology to reality. *Pulp and Paper,* 66(9), 111–114.

Senior, D.J. and Hamilton, J. (1992b). Reduction in chlorine use during bleaching of kraft pulp following xylanase treatment. *Tappi Journal* 75(11), 125–130.

Senior, D.J. and Hamilton, J. (1993). Xylanase treatment for the bleaching of softwood kraft pulps: The effect of chlorine dioxide substitution. *Tappi Journal* 76(8), 200–206.

Senior, D.J., Hamilton, J., Taiplus, P. and Torvinen, J. (1999). Enzyme use can lower bleaching costs, aid ECF conversions. *Pulp and Paper,* 73(7), 59–62.

Shah, A.K., Eriksson, K.-El., Cooper, D. and Adolphson, R. (2000). Extraction in aqueous two-phase systems of alkaline xylanase produced by *Bacillus pumilus* and its application in kraft pulp bleaching. *Journal of Pulp and Paper Science,* 26(1), 8.

Shobhit, M., Satish, K. and Rao, N.J. (2002). Xylanase prebleaching of wheat straw and sarkanda soda pulps. *IPPTA,* 14(2), 45–52.

Sjostrom, E. (1993). Wood chemistry, fundamentals and application. Second edition. Academic Press, San Diego.

Skerker, P.S., Labbauf, M.M., Farrell, R.L., Beerwan, N. and McCarthy, P. (1992). Practical bleaching using xylanases: Laboratory and mill experience with Cartazyme HS-10 in reduced and chlorine free bleach sequences. In: *Tappi Pulping Conference.* p. 27. Boston, 1–5 November, Tappi Press, Atlanta, GA.

Smook, G.A. (1989). Overview of the pulp and paper industry from a chemical industry perspective. *Journal of Chemical Technology and Biotechnology,* 39, 15-27.

Spiridon, I.. Stewart, D. and Popa, V.I. (1998). The effects of xylanase treatments on straw pulps bleaching. In: *7th Int. Conf. Biotech. Pulp and Paper Industry,* pp. C53–C56. 16–19 June, Vancouver, BC, Canada.

Srinivasan, M.C. (1998). Extremophiles: A biotechnological resource for paper and pulp industry. pp: 13-16.

Stutzenberger, F.J. and Bodine, A.B. (1992). Xylanase production by *Streptomyces curvata. Journal of Applied Microbiology,* 72, 504-51.

Suurnakki, A., Tenkanen, M., Buchert, J. and Viikari, L. (1997). Hemicellulases in the bleaching of chemical pulps. *Advances in Biochemical Engineering/Biotechnology,* 57, 262-287.

Tenkanen, M., Buchert, J., Puls, J., Poutanen, K. and Viikari, L. (1992a). Two main xylanases of *Trichoderma reesei* and their use in pulp processing. In: *Xylans and Xylanases. Progress in Biotechnology.* Vol. 7, p. 547. J. Visser, G. Beldmann, M.A. Kusters-van Someren, and A.G.J. Voragen, Eds. Elsevier, Amsterdam.

Tenkanen, M., Puls, J. and Poutanen, K. (1992b). Two major xylanases of *Trichoderma reesei. Enzyme and Microbial Technology,* 14, 566–572.

Tenkanen, M. (1998). Action of *Trichoderma reesei* and *Aspergillus oryzae* esterases in the deacetylation of hemicelluloses. *Biotechnology and Applied Biochemistry.* 27: 19-24.

Thibault, L., Tolan, J., White, T., Yee, E., April, R., and Sung, W. (1999). Use of an engineered xylanase enzyme to improve ECF bleaching at Weyenhaeuser Prince Albert. In: *85th Annual Meeting.* p. B263. 26–29 January, Montreal, Canada.

Thompson, G., Swain, J., Kay, M. and Forster, C.F. (2001). The treatment of pulp and paper mill effluent: a review. *Bioresource Technology,* 77, 275-286.

Tolan, J.S. (1992). Mill implementation of enzyme treatment to enhance bleaching. In: *Proeedings of 78th CPPA Annual Meeting.* pp. A163–A168. 28–29 January, Montreal, Canada.

Tolan, J.S. and Canovas, R.V. (1992). The use of enzymes to decrease the chlorine requirements in pulp bleaching. *Pulp and Paper magazine Canada,* 93(5), 39-42.

Tolan, J.S., Olson, D. and Dines, R.E. (1996). Survey of mill usage of xylanase In: *Enzymes for pulp and paper processing.* pp. 25–35. T. W. Jeffries, and L. Viikari, Eds. ACS Symposium Series. 655.

Tolan, J.S. and Guenette, M. (1997). Using enzymes in pulp bleaching: mill applications. *Advances in Biochemical Engineering/Biotechnology,* 57, 291-310.

Torres, A.L., Ronceio, M.B., Colom, J.F., Pastor, F.I.J., Blanco, A. and Vidal, T. (2000). Effect of a novel enzyme on fibre morphology during ECF bleaching of oxygen delignified *Eucalyptus* kraft pulps. *Bioresource Technology,* 74, 135-140.

Torronen, A., Harkki, A. and Rouvinen, J. (1994). Three-dimensional structure of endo-1,4-beta-xylanase II from *Trichoderma reesei*: two conformational states in the active site. The *EMBO Journal,* 13, 2493 - 2501.

Turner, J.C., Skerker, P.S., Burns, B.J., Howard, J.C., Alonso, M.A. and Andres, J. L. (1992). Bleaching with enzymes instead of chlorine: Mill trials. *Tappi Journal,* 75(12), 83–89.

Valchev, V., Valchev, V. and Christova, E. (1998). Introduction of an enzyme stage in bleaching of hardwood Kraft pulp. *Cellulose Chemistry and Technology,* 32(5–6), 457–462.

Viikari, L., Kantelinen, A., Poutanen, K. and Ranua, M. (1990). Characterization of pulps treated with hemicellulolytic enzymes prior to bleaching. In: *Biotechnology in Pulp and Paper Manufacture.* Kirk, T. K. and Chang, H. M. (Editors). Butterworth Heinemann, Boston, MA. pp. 145.

Viikari, L., Kantelinen, A., Ratto, M. and Sundquist, J. (1991). Enzymes in pulp and paper processing. In: *Enzymes in Biomass Conversion*Leatham, G. F. and Hammel, K. E. (Editors). ACS Symposium Series 460, American Chemical Society, Washington, DC. pp. 12.

Viikari, L., Kantelinen, A., Sundquist, J. and Linko, M. (1994). Xylanase in bleaching: from an idea to the industry. FEMS Microbiology Reviews 13, 335-350.

Viikari, L., Ranua, M., Kantelinen, A. Linko, M. and Sundquist, J. (1987). Application of enzymes in bleaching. *Proceedings of International Symposium on. Wood and Pulping Chemistry*, 4th.Vol.1: pp. 151, 27–30 April, Paris.

Viikari, L., Ranva, M., Kantelinen, A., Sundquist, J. and Linko, M. (1986). Bleaching with enzymes. *In*: *Biotechnology in pulp and paper industry. Third International Conference.* Stockholm. pp: 67-69.

Viikari, L., Tenkanen, M., Buchert, J., Ratto, M., Bailey, M., Siika-aho, M. and Linko, M. (1993). Hemicellulases for industrial applications. In: *Bioconversion of Forest and Agricultural Plant Residues.* pp. 131–182. Saddler, J.N., Ed. CAB International, Wallingford.

Wang, H. and Shih, C.H. (1996). The application of xylanase in the chlorine free bleaching of bamboo pulps. In: *3rd International Non-wood Fiber Pulping and Papermaking Conference..* pp 344–350. 15–18 October, Beijing, China.

Watson, P.A., Bicho, P.A. and Stumborg, M.A. (1998). Wheat straw: a viable fiber source for Canada? *Pulp & Paper Canada,* 99 (12), T449–T452.

Weidong, Y., Zhonghao, C. and Huisheng, Y. (1999). Bleaching of wheat straw kraft-AQ pulp with Xylanase and active oxygen. *China Pulp and Paper,* 18(4), 28–30.

Werthemann, D. (1993). Prebleaching of Pinus radiata pulp using enzymes—Technology to reduce AOX. *Japanese Journal of Paper Technology,* 10, 15–17.

Wilson, R.A. (1996). Non wood fibres: precocious and criteria required for acceptance and expansion in pulp and paper markets. In: Uses for non-wood fibres-commercial and practical issues for papermaking, Volume II: flax, hemp and cereal straw. Pira International, Surrey, UK. pp: 1-11.

Winterhalter, C. and Libel, W. (1995). Two extremely thermostable xylanase of the hyperthermophilic bacterium *Thermotoga maritima* MSB8. *Applied and Environmental Microbiology,* 61(5), 1810-1815.

Wong, K.K.Y. and Saddler, J.N. (1992). *Trichoderma* xylanases, their properties and purification. *Critical Reviews in Biotechnology,* 12, 413-435.

Wong, K.K.Y., Kibblewhite, R.P. and Signal, F.A. (1999). Effect of xylanase and dosage on the refining properties of unbleached softwood kraft pulp. *Journal of Wood Chemistry and Technology,* 19(3), 203–212.

Wong, K.K.Y., Martin, A.L., Gama, F.M. and Saddler, J.N. (1997). Bleach boosting and direct brightening by multiple xylanase treatment during peroxide bleaching of kraft pulp. *Biotechnology and Bioengineering,* 54, 312-318.

Wong, K.K.Y., Tan, L.U.L. and Saddler, J.N. (1988). Multiplicity of β–1,4-xylanase in microorganisms: functions and applications. *Microbiological Reviews,* 52, 305-317.

Yang, J.L., Lou, G. and Eriksson, K-E.L. (1992). The impact of xylanases on bleaching of kraft pulps. *Tappi Journal,* 75(12), 95-101.

Yllner, S. and Enstrom, B. (1956). Studies of the adsorption of xylan on cellulose fibers dyring the sulphite cook. Part I. *Svensk Papperstid,* 59, 229-232.

Yllner, S. Ostberg, K. and Stockman, L. (1957). A study of the removal of the constituents of pine wood in the sulphite process using a continous liquor flow method. *Svensk Papperstid,* 60, 795-802.

You, J. and Wu, S. (2000). Effect of enzyme treatment on properties of straw pulps before and after beating. In: *4th Int. Nonwood Fibre Pulping and Papermaking Conf.* pp. 626–632. 18–21 September, Jinan, China.

Yundt, A.P. (1951). Crystalline hemicelluloses. II Crystalline xylan from paper birch. *TAPPI Journal,* 34, 91-92.

Zhan, H., Yue, B., Hu, W. and Huang, W. (2000). Kraft reed pulp TCF bleaching with enzyme treatment. *Cellulose Chemistry and Technology,* 33(1–2), 53–60.

Zhao, J., Li, X., Qu, Y. and Gao, P. (2002). Xylanase pretreatment leads to enhanced soda pulping of wheat straw. *Enzyme and Microbial Technology,* 30(6), 734–740.

Zinbo, M. and Timell, T.E. (1965). The degree of branching of hardwood xylans. *Svensk Papperstid,* 68, 647-662.

Zinbo, M. and Timell, T.E. (1967). Studies on a native xylan from Norway spruce (Picea abies). 1. Isolation and constitution. *Svensk Papperstid,* 70, 695-701.

# 19

# Enzymatic Deinking of Mixed Office Waste Paper : An Overview

**R.K. Jain and A.G. Kulkarni**

*Central Pulp & Paper Research Institute, Post Box No. 174, Paper Mill Road, Saharanpur, U.P.*

Mixed office waste presents technical and economic challenge to the recycler and of the wide variety of fibre and contaminants present in the paper stock. Toner and other non-contact polymer ink from laser printing process, is one of the most difficult to deal with. Toner and laser printing ink are synthetic polymers with embedded carbon blacks, they don't disperse readily during conventional repulping processes. Moreover, these are not readily recovered during floatation or washing. Because of these problems recycled papers contaminated with toner have a relatively low value. Most of the deinking chemicals and high-energy dispersion steps employed in current deinking technology are tedious, cost prohibitive and sometimes leads to loss in pulp yield. The enzymatic deinking process employing suitable enzymes cocktail, effective in deinking of laser and xerographic waste paper show promise in Indian paper industry. The present article discusses potential of the enzymatic deinking process for nonimpact printed toners employing identified cellulase based deinking enzyme covering the mechanism and factors influencing the enzyme deinking of waste fibres.

## INTRODUCTION

Recycling of fibre is a rapidly growing segment of the paper industry and will continue to gain share of the fibre furnish worldwide. The major driving force for increased use of the recycled fibre has been the environmental concern of the pollution control authorities as well as the customer's preference for the environmentally benign products, which demand highest quality of DPI because of the high brightness level of the final product.

The main quality issues of paper products using recycled fibres have been runnability and printability that have been limiting the use of these furnish. These are normally related to the instability and cleanliness of the raw materials and strength properties, filler contents and average fibre length of the furnishes. The recovered furnish is actually a mixture in which the different material contents can vary based on the type and quality of collected and mixed paper and boards. These runnability and printability problems have been solved by developing better deinking processes, machinery, screening and sorting technology, that can help in achieving better quality of DPI and will also allow to use more low quality of recycled paper and even unsorted materials in near future.

Though the recycled source of fibre, non-coated printed papers which includes xerographic and laser printed papers are the fast growing sources due to the increased use of office photocopier and computer printouts, however toners used in xerographic and laser printing papers presented a special challenge, due to the difficulties in deinking by conventional deinking methods. Thus the volume of recycling of this high quality fibre is in less use. In order to effectively utilise larger volume of recovered non-contact printed laser papers there is a need to introduce an effective and efficient technology that will deink non-contact ink to acceptable residual ink count in an economically and environmentally acceptable manner. New deinking mills established in response to these projected needs are already competing for the cleanest and most homogeneous post-consumer paper sources e.g. sorted white ledger and soon will have to dip deeper into the post-consumer stream of unsorted mixed office waste (MOW) to remain competitive.

Current deinking technology is being stretched to accommodate both the hard-to-remove toner ink and the coloured dyes and unbleached fibre present in unsorted MOW. Additional chemicals, multiple floatation steps and dispersion alleviate some of the limitations of the heterogeneous paper stocks. While pulp cleanliness resulting from this sequence is good enough, however the process is capital and energy intensive, and sometimes loss in pulp yield also occurs. Moreover they are not readily removed during floatation or washing. Because of these problems, recycled paper contaminated with toners have a relatively low value.

CPPRI having created its vast infrastructural facilities in the area of biotechnology is actively engaged in promotion of biotechnological application in pulp and paper industry. Among various biotechnological applications, enzymatic prebleaching of pulp, biopulping and slime control and enzymatic deinking of waste paper are the major areas of interest where thrust has been laid upon. Evaluation of various enzymes available and being developed indigenously on pulps so that suitable enzymes are made available to the industry which could be effectively utilized under existing mill conditions.

Recently the use of enzymes to deink recycled fibres has been proposed and is being studied, intensively in developed countries, wherein thrust has been laid upon evaluation of the various deinking enzymes available on different papers and inks (Jeffries *et al.,* 1993). But there continues to be a lack of understanding of how these enzymes function during deinking of papers. The present article highlights an overview of enzymatic deinking process: how to remove toner inks from MOW covering the details of the process; conditions under what conditions enzymes work efficiently while performing its specific tasks.

## DISCUSSIONS

### Enzyme Deinking

Laser and xerographic printed paper are the major components of office paper and their use is increasing day to day. Laser and xerographic printing use thermo plastic toners that fuse on to fibre surfaces during high temperature non-contact printing. These oven dispensable inks require special chemicals, thermal and mechanical action to detach the ink particles from fibres so that the ink can be removed by flotation—a deinking process step that separates hydrophobic ink particles from hydrophilic pulp fibres. However, costly dewatering and dispersion steps, in addition to additional flotation and washing are necessary to improve the deinking of these types of waste paper. Microbial enzymes have also been shown to enhance the release of toners from office waste. Cellulases and xylanases when applied to xerographic printed papers in a medium consistency mixer are reported to release toner particles and facilitate subsequent flotation and washing steps.

## DEINKING PRINCIPLES AND CURRENT DEINKING PRACTICES

The common technique used for ink removal from laser and xerographic printing waste is washing and floatation. This kind of paper uses thermoplastic toners that fuse on to the fibre surfaces during high temperature. These non-dispersing inks require special chemicals, thermal and mechanical actions to detach the inks from the fibre so that the ink can be removed by floatation and washing. Floatation is a step in deinking process that separates hydrophobic particles from the hydrophilic fibres as shown in Figure 1. Floatation removes particles that are too small to be removed by screens and cleaners and yet, are too big to be removed by washing.

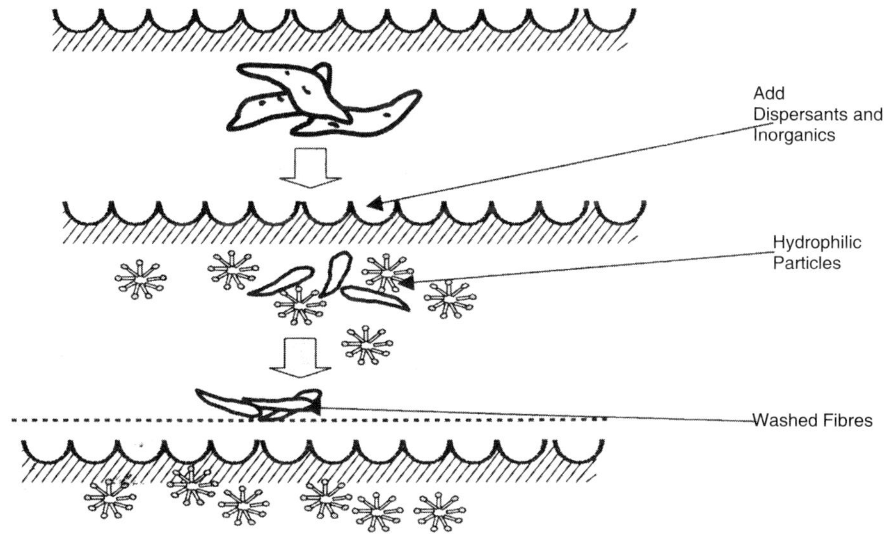


**Fig. 1.** Floatation step

Washing is most efficient for removing the smallest particles of the ink. The objectives are to keep the ink particles finally disperse and agglomerate. Washing requires the ink particles to be rendered hydrophilic so that they remain in the aqueous phase as shown in Figure 2. Both of the processes normally operate at high pH (Pommier *et al.,* 1990; Prasad, 1993) with the use of conventional alkaline deinking agents such as sodium hydroxide, sodium carbonate, sodium silicate and hydrogen peroxide. In this environment, the paper structure collapses rapidly and releases the ink particles into the suspension. A dispersant is added to stabilize the colloidal suspension of ink particles in washing processes.

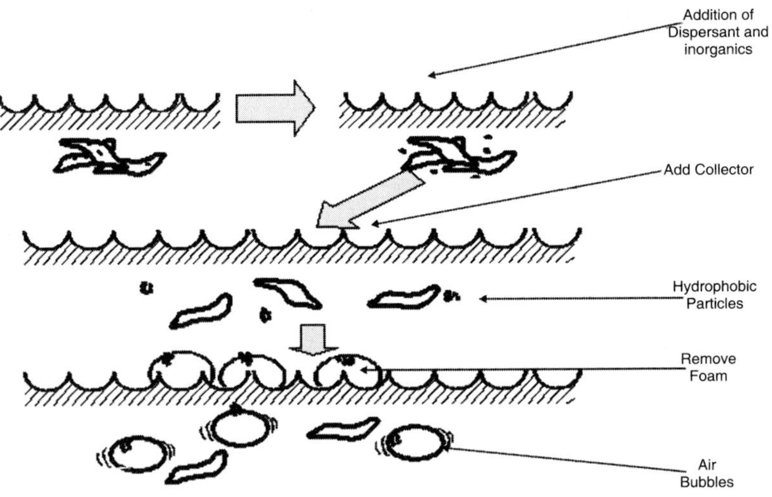


**Fig. 2.** Washing Step

The dispersion is replaced with collector soap in floatation deinking. The optimum size range for the different unit operation is shown in Figure 3.

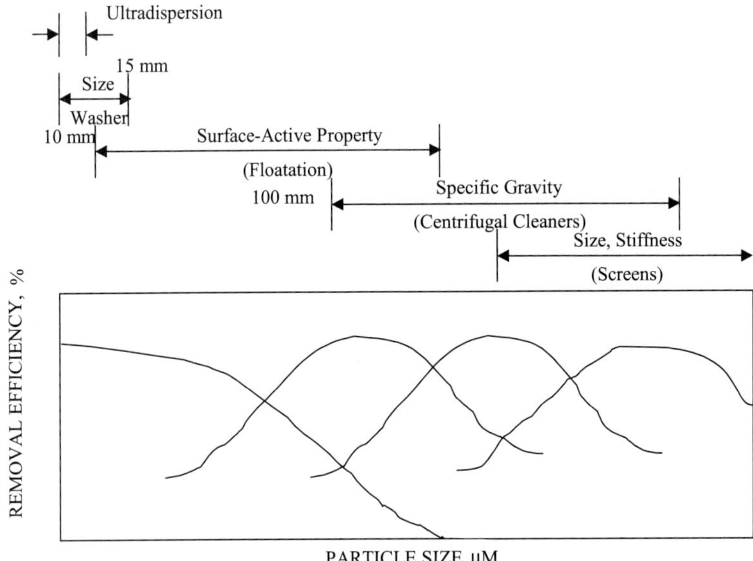


**Fig. 3.** Optimum particle size ranges in various unit operations in deinking

## LIMITATIONS AND PROBLEMS IN CURRENT DEINKING TECHNOLOGY

Laser and xerographic inks consist of thermoplastic polymer of styrene and butadiene acrylate or polyester alongwith carbon black or other pigments. It is designed to be non-tacky at room temperature but it melts at temperature of 70-120°C. During the fusion stage of the copolymer process at high temperature (100°C), these thermoplastic resin binders set to form the printed film by cross-linking.

These hard cross-linked, strongly bonded systems will only be fragmented to minimum size by strong mechanical forces, which in turn lead to fibre degradation. Conventional chemical treatment is not effective in reducing particle size further and particles are visible as dirt in the finished product.

Due to this special treatment of laser and xerographic inks some modified deinking processes have been tried, among which washing / floatation, two-stage floatation and agglomeration and disintegration with subsequent removal by screening and cleaning are prominent to mention. These require some chemicals and the equipment but the process is capital and energy intensive.

## ENZYMATIC DEINKING

Predominant enzymes used for deinking of waste paper are mainly the cellulases and hemicellulases. Cellulases could work in several ways to enhance deinking. These reduce the hydrodynamic drag to increase the filtration and floatation rate. As cellulases are known to enhance drainage rates (Jackson *et al.,* 1993) they can enhance any separation process such as filtration or floatation. Deinking with enzymes involves dislodging of ink particles from fibre surfaces and then separating the disposed ink from the fibre suspension by washing and / or floatation. The first reported use of cellulase for deinking was with newsprint 1991 (Borchardt, 1992; Park *et al.,* 1992) which was reviewed recently (Yong, 1994).

Studies conducted at USDA, fruit products laboratory, Madison indicated that a number of commercially available cellulases alone would facilitate toner removal better than conventional chemical deinking process, used either with or without enzyme (Freiermuth *et al.,* 1994; Pommier *et al.,* 1989). Subsequent studies showed that with alkaline sized papers, it was necessary to employ enzymes active in the neutral to alkaline region, alongwith surfactants to counter the efforts of the sizing. Mechanical action also plays a major role in toner removal.

## MECHANISMS OF ENZYMATIC DEINKING

Deinking can be divided into five different operations occurring partly in consecutive and partly in overlapping stages. These stages are: disintegration, precleaning, chemical or enzyme treatment, floatation and / or washing and finally bleaching.

During enzymatic treatment, the process is preceded by disintegration and followed by a cleaning operation, which makes it very difficult to determine the exact role of the enzyme. Cellulases action may also increase the specific surface area of the fibres and reduce interaction with contaminants. That is to say that there might be microfibrils on the surface of these very frazzled, recycled fibres which could be trapping the ink particles, and by giving the fibre a haircut might then reduce their adhesion. A model of enzymatic deinking is shown in Figure 4.

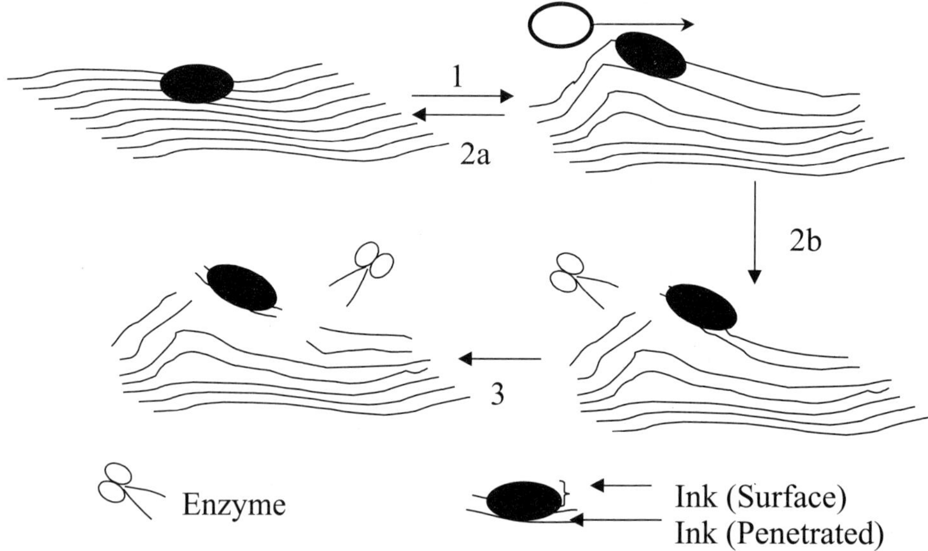


**Fig. 4.** A model of enzymatic deinking

However, the studies have shown that the most important fraction with respect to toner removal is the increased floatation efficiency imparted by cellulase enzyme activity and the increased removal of toner has been observed during floatation stage. During floatation, air bubbles rise to the surface of the floatation tank through a relatively dilute pulp stock, approximately at 1% consistency. The surface of air bubbles being relatively hydrophobic, they carry the toner particle to the top where they are removed by a skimming action. The operating principle of floatation deinking is shown in Figure 5.

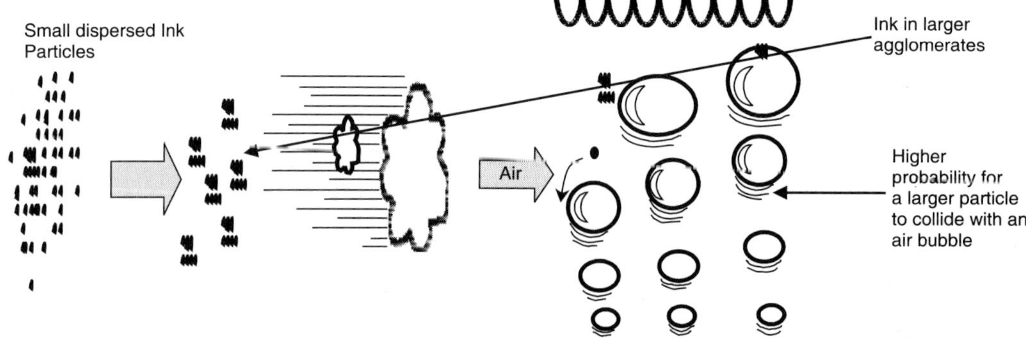


**Fig. 5.** Operating principle of floatation deinking

## Mechanism of Enzymatic Toner Removal

It is interesting to note that most of the ink removal, as reflected by final pulp brightness, occurs during the flotation stage. This no doubt says something about the mode of action of the enzyme on the mixture of ink and pulp. A hypothesis is that cellulase produces larger ink particles, which are more efficiently removed by the floatation process.

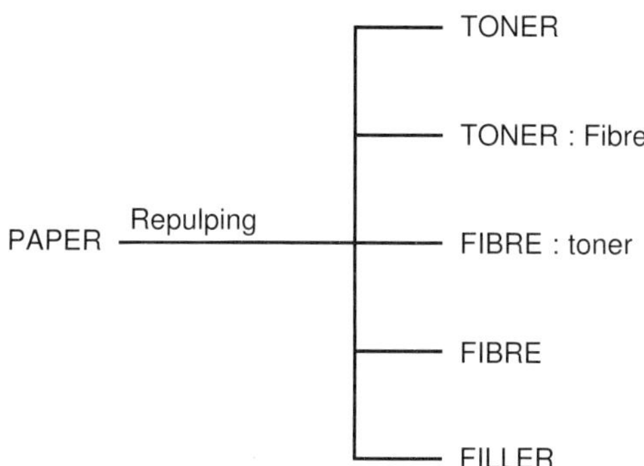


**Fig. 6:** Repulping of non-contact printed paper, potential fractions from repulping

Figure 6 lists the various possible particle types generated during the repulping process. We can speculate on which class of particles the cellulase action has the greatest effect. Without enzyme or chemical action, the toner and toner-fibre particles would be carried over in the flotation process, leaving the latter three classes behind in the flotation cell.

The approach of the enzyme to the surface of the fibre-toner particles (that carry a lower proportion of toner) would probably be where the biggest improvement occurred. This would create more toner and toner-fibre particles, which could be removed by the flotation.

Cellulase action on the toner-fibre particles (that are rich in toner), might also improve the deinking efficiency, and improve yield, since the hydrophobicity might be improved, meaning more efficient flotation and the rejects might contain less cellulosic material.

# PROCESS CONDITIONS DURING ENZYMATIC DEINKING

## Deinking Enzymes and the Order of its Addition

The preferable enzyme used for deinking wastepaper is commonly a mixture of cellulase and hemicellulase. The order for addition of material is to first add the proper amount of waste paper and water in order to achieve desired consistency followed by the diluted solution of enzyme.

Access of cellulase enzyme to cellulase fibre is essential for achieving maximum activity of enzyme. Proper mixing of the enzyme at consistency 11-16% helps in dislodging toner particles from fibre and presence of non-ionic surfactants helps the enzyme to penetrate through the paper additives with the increased availablity of cellulase for cellulose attachment. Smoothing of fibre increases pulp freeness, which prevents toner particles from becoming trapped in the repulped fibre network.

Enzyme dosing point has been of great significance for achieving maximum enzyme efficiency. Enzyme activity being site specific, distribution of enzyme throughout the pulp is essential to achieve maximum enzyme efficiency. Addition of the enzyme after proper dilution at pulper near the beginning of the repulping process has been desirable. This allows the enzyme to react with waste paper furnish at higher consistency while undergoing optimum mechanical agitation.

## PULP CONSISTENCY

Results shown in the Figure 5 clearly indicate that pulping at medium consistency, 12% is advantageous for recovering toner through the combined effect of enzyme and mechanical action. However, the results shown in Table 1 clearly indicate that increasing the consistency to as high as 16% increases deinking efficiency. The results are consistent with the findings of these researchers (Jackson *et al.,* 1993; Borchardt, 1992; Park *et al.,* 1992; Sykes *et al.,* 1995) who showed that enzymatic deinking is most effective when used during high consistency fiberization. However, at higher deinking pulping consistencies and extended pulping time, the effect of mechanical action is predominant with only little additional benefit from the enzyme.

**Table 1: Effect of pulping consistency in ink removal[a]**

| *Particulars* | *Residual ink at different levels of consistencies (PPM)* | |
|---|---|---|
| | **12%** | **16%** |
| Control pulp | 681 | 343 |
| Enzyme treated pulp | 329 | 220 |

a: Residual ink on hand sheets counted in 10-2000 m range.

## SURFACTANTS

Addition of surfactants during enzymatic deinking process plays an important role in separating ink from pulp fibre during floatation. Surfactants make the cellulose more accessible to cellulase enzyme and facilitate enzyme dispersion, thus making the enzyme available to attach to cellulose sites. Under appropriate conditions, surfactants increase cellulase effectiveness. The addition of non-ionic surfactants to the pulp prior to addition of enzymes has been preferred to achieve better deinking efficiency as indicated from the results shown in Table 2.

**Table 2. Effect of adding surfactant via floatation cell or pulper.**

| *Specimen* | *Residual ink (PPM)* | |
|---|---|---|
| | *Floatation cell* | *Pulper* |
| Control | 326 | 231 |
| Enzyme treated | 278 | 168 |

Certain anionic surfactants containing sulphates or sulphonated functional groups can reduce the efficiency of the enzyme to hydrolyze cellulose and in turn reduce the deinking efficiency of the enzyme.

Moreover there should always be synergy of enzyme with a particular surfactant, which needs to be evaluated for individual enzyme preparation. Results of effectiveness of two of the identified enzymes and their synergy with the different surfactants is shown in Table 3.

**Table 3: Enzyme / surfactant synergy**

| *Variables* | *Residual ink (PPM)* |
|---|---|
| Control | 368 |
| Enzyme A, Surfactant A | 242 |
| Enzyme A, Surfactant B | 303 |
| Enzyme B Surfactant A | 278 |
| Enzyme B, Surfactant B | 214 |

## pH

The commercial cellulase enzyme preparations selected for deinking are most active, near to neutral pH range, around 6.5 to 7.5. Increasing the waste paper stream containing alkaline sized office waste paper place repulped waste at the upper end of pH range of approximately 8.5 - 8.9, however better deinking efficiencies have been obtained by adjusting the pH to 7.0. Results of the effect of pH on ink removal are shown in Table 4.

**Table 4: Effect of pH on ink removal**

| *Particulars* | *PH* | *Residual Ink (PPM)* |
|---|---|---|
| Control pulp | — | 139 |
| Enzyme treated pulp | 8.6 | 64 |
| Enzyme treated pulp | 7.0 | 47 |

Depending on performance of the mill, sulphuric or phosphoric acids can be used to maintaine proper pH.

## TEMPERATURE

The optimum temperature range for a conventional cellulase based enzyme has been 45-55°C. Having realized that the toner-based inks might begin to fuse and reattach at higher temperature, the temperature below 65°C is always preferable. The schematic flow sheet of the proposed process of the enzymatic deinking is shown in Figure 7.

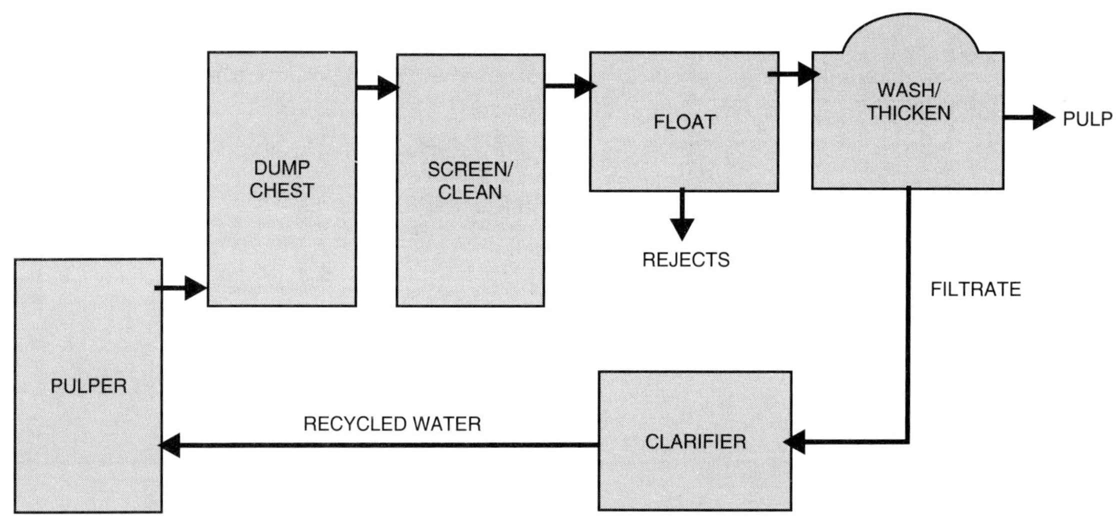


**Fig. 7.** Proposed process flow sheet for enzymatic deinking.

## CONCLUSION

1. Enzymatic deinking of laser and xerographic waste paper appears to be a promising technology for deinking of hard-to-remove toner inks.
2. Medium consistency pulping conditions of 11-13% appeared to be more effective than low consistency pulping conditions for toner removal.
3. Higher consistency and washing remove laser and xerographic toner from mixed office waste more effectively with a suitable blend of cellulases and xylanases enzymes.
4. Selection of an enzyme with optimum activity in the pH range of the pulped paper stock would minimize pH adjustment in the pulper and therefore simplify the process and lower the process cost.
5. Efforts are required to develop and evaluate enzyme preparations suitable for the kind of waste paper utilized by the Indian paper industry to make the technology techno-economically viable.

## REFERENCES

Borchardt, J.K. (1992). A Primer for Surfactants used in Deinking. *Progress in Paper Recycling,* (1), 55.

Freiermuth, B., Garrett, M. and Jokinen, O. (1994). The use of enzymes in the production of release papers. *Paper Technology,* 35(3), 21.

Jackson, L.S., Heitmann, J., and Joyce, T.W. (1993). *Enzymatic Modifications of Secondary Fiber. Tappi Journal,* 76(3), 147.

Jeffries, T., Klungness, J.H. Sykes, M., and Cropsey, K. (1993). TAPPI 1993 Recycling Symposium Notes, *"Preliminary Results of Enzyme Enhanced Deinking,"* TAPPI PRESS, Atlanta. Pp 183.

Park, J.W., Takahata, Y., Kajiuchi, T., and Akehata, T. (1992). Effect of Nonionic Surfactant on Enzymatic Hydrolysis of Used Newspaper. *Biotechnology and Bioengineering,* 39, 117.

Pommier, J.C., Fuentes, J.L., Goma, G. (1989). Using enzymes to improve the process and the product quality in the recycled paper industry, Part I: the basic laboratory work. *Tappi Journal,* 72(6), 187.

Pommier, J.C., Goma, G., Fuentes, J.L. and Rousset, C. (1990). Using enzymes to improve the process and the product quality in the recycled paper industry, Part II: Industrial applications. *Tappi Journal,* 73(12), 197.

Prasad, D.Y. (1993). Enzymatic deinking of laser and xerographic office wastes. *Appita,* 46(4), 289.

Sykes, M. Klungness, J., Jeffries, T., Cropsey, K., Abubakr, S. (1995). Tappi Recycling Symposium Proceedings, *"Enzymatic deinking of sorted mixed office waste Recommendations for scale-up,"* Tappi Press, Atlanta. Pp 61.

Yong, J. (1994). Enzone bleaching, Enzynk deinking advance to pilot plant trails. *Pulp and Paper,* 68(11), 81.

# 20

# Industrial Applications of Microbial Cellulases

**Ajay Singh[1], Ramesh C. Kuhad[2] and Owen P. Ward[1]**

[1]*Department of Biology, University of Waterloo, Waterloo, Ontario N2L 3G1, Canada*
[2]*Department of Microbiology, University of Delhi, South Campus, Benito Juarez Road, New Delhi–110 021, India*

Cellulases are hydrolytic enzymes, which have been commercially available for about the last three decades. Due to the complexity of these enzyme systems and the enormous potential for their application in a wide range of industries, cellulases have been a focus for research by both academic and industrial research groups. The potential applications of cellulases have been demonstrated in various industries including the food, animal feed, brewery and winery, agriculture, biomass refining, pulp and paper, and textile and laundry sectors. Customized cellulase preparations are currently being marketed by commercial producers catering to specific industry needs. The requirement for cellulases is growing rapidly and this demand is the driving force behind the cutting edge research, which is currently underway. In this chapter the scientific and technological developments and the future prospects for expanded use of cellulase biotechnology in different industries are discussed.

## INTRODUCTION

Microbial enzymes are commonly used in many industrial applications. The demand for highly active and stable enzymes is growing rapidly. For the year 2005, it is estimated that industrial enzyme products will represent a market of US$ 2.0 billion. Approximately 75% of the market for industrial enzymes is for hydrolases, especially polysaccharide-degrading enzymes and proteases. Cellulose is the most abundant polysaccharide on the planet and cellulases are the hydrolytic enzymes, which degrade these carbohydrates. Cellulases have been commercially available for about 30 years and these enzymes have represented a target for academic and industrial research due to both the complexity of enzyme system and to their potential applications in a wide range of industries (Singh, 1999; Vielle and Zeikus, 2001; Lynd *et al.,* 2002).

In what is perhaps an oversimplified description, the cellulase system in fungi has been considered to comprise three hydrolytic enzymes. The endo-(1,4)-β-D-glucanase (also known as endoglucanase, endocellulase, carboxymethyl cellulase [EC 3.2.1.4], cleaves β-linkages in cellulose in a random manner, commonly in the amorphous parts of cellulose molecule. The exo-(1,4)-β-D-glucanase (also known as cellobiohydrolase, exocellulase, microcrystalline cellulase, Avicelase [EC 3.2.1.91], releases cellobiose from either the nonreducing or the reducing end, generally from the crystalline parts of cellulose. The β-glucosidase (cellobiase [EC 3.2.1.21], releases glucose from cellobiose and short-chain cellooligosaccharides (Singh and Hayashi, 1995; Bhat, 2000; Perez *et al.,* 2002). A remarkable feature of the cellulase system is its synergism, in which the action of a mixture of two or more components of cellulase is generally found to be greater than the sum of the actions of each component.

Basic and applied research studies on cellulolytic enzymes have demonstrated the biotechnological potential of cellulases in various industries including food, animal feed, brewing and wine making, agriculture, biomass refining, pulp and paper, and textile and laundry (Table 1). As evidence of that interest and technological progress in the two and a half years up till October 2005 more than 100 US Patents were granted having claims related to cellulase production and/or applications. A selection from these patents is presented in Table 2. Recent developments and future prospects of cellulase technology are discussed here.

**Table 1. Applications of cellulolytic enzymes**

| *Industry* | *Applications* |
|---|---|
| Agriculture | Plant and fungal protoplasts; enhanced seed germination; enhanced plant growth and flowering; plant pathogen and disease control |
| Bioconversion | Conversion of cellulosic materials to ethanol, other solvents, organic acids and single cell protein, lipids |
| Detergents | Cellulase based detergents; superior cleaning action without damaging fibre |
| Environment | Wastewater treatment/purification, biosolids reduction |
| Fermentation | Macerating enzymes; improved pressing and color extraction of grapes; improved aroma of wines; improved filtration and stability; improved primary fermentation and quality of beer |
| Food | Improvement of yields in starch and protein extraction, improved maceration, pressing and color extraction of fruits and vegetables; clarification of fruit juices; improved texture and quality of bakery products |
| Pulp and Paper | Co-additive in pulp bleaching; bio-mechanical pulping; improved draining; enzymatic de-inking |
| Textile | Biostoning and biopolishing of jeans and other fibres; high quality fabrics |

## PULP AND PAPER INDUSTRY

Interest in the application of enzymes such as cellulases, xylanases and ligninases in the pulp and paper industry has increased considerably during the last decade (Mai *et al.,* 2004). Biomechanical pulping using white-rot fungi was originally applied as pretreatment of wood chips with fungal cultures to modify lignin and extractives (Akhtar *et al.,* 2004). Biopulping provides significant energy savings. Cellobiohydrolases have been applied either to chips or pulp reject fractions. Pilot and full-scale trials at the UPM-Kymmene Rauma mill in Finland (Pere *et al.,* 2002), indicated significant energy savings in the enzymatic process. Impregnation of wood chips with pectinase enzymes also requires lower energy input to reach the required freeness (water drainage from the fibers) and strength (Peng *et al.,* 2003), but the application is not yet commercialized. Cellulases and xylanases applications which improve the drainage from a pulp or paper machine have also been pursued by

**Table 2. Examples of Cellulases-Related US Patents Granted during the period April 2003-September 2005**

| *US Patent* | *Patent Title* |
|---|---|
| 6,943,144 | Concentrated stable, translucent or clear fabric softening compositions including chelants |
| 6,936,289 | Method of improving the properties of a flour dough, a flour dough improving composition and improved food products |
| 6,921,655 | Endoglucanases and cellulase preparations containing the same |
| 6,838,099 | Method for preparing soluble dietary fiber from corn hull |
| 6,835,707 | Laundry detergent compositions with a combination of cyclic amine based polymers and hydrophobically modified carboxy methyl cellulose |
| 6,835,558 | Beta-glucan compositions and process therefore |
| 6,830,655 | Method of modifying a xylan-containing carbohydrate substrate having hexenuronic acid groups attached to the xylan |
| 6,818,803 | Transgenic plants as an alternative source of lignocellulosic-degrading enzymes |
| 6,812,018 | Thermostable cellulase |
| 6,803,355 | Laundry detergent compositions with fabric enhancing component |
| 6,797,010 | Redeposition or backstain inhibition during stonewashing process |
| 6,790,822 | Detergent compositions having an anionically modified cellulose polymer |
| 6,777,383 | Solid detergents with active enzymes and bleach |
| 6,767,879 | Detergents comprising cellulases |
| 6,761,944 | Moulded articles of cellulose hydrate with enzymatically modified surface |
| 6,750,051 | Compositions and methods for enhancing fiber digestion |
| 6,740,508 | Fermentation-based products from corn and method |
| 6,730,653 | Method for manufacturing a molded detergent composition |
| 6,723,549 | Cellulases, the genes encoding them and uses thereof |
| 6,710,023 | Dishwashing detergent compositions containing organic polyamines |
| 6,685,748 | Enzymatic bleaching of natural non-cotton cellulosic fibers |
| 6,663,903 | Methods for using xyloglucan endotransglycosylase in baking |
| 6,635,465 | Mutant EGIII cellulase, DNA encoding such EGIII compositions and methods for obtaining same |
| 6,623,949 | Variant EGIII-like cellulase compositions |
| 6,623,750 | Enzyme composition adapted for application to ruminant feed to increase protein digestibility thereof and method of treating ruminant feeds relating thereto |
| 6,620,605 | Method and compositions for treating cellulose containing fabrics using truncated cellulase enzyme compositions |
| 6,617,268 | Method for protecting cotton from enzymatic attack by cellulase enzymes |
| 6,616,953 | Concentrated spent fermentation beer or *S. erythraea* activated by an enzyme mixture as a nutritional feed supplement |
| 6,602,842 | Cleaning compositions containing plant cell wall degrading enzymes and their use in cleaning methods |
| 6,596,678 | Laundry detergent compositions containing a polymer for fabric appearance improvement |
| 6,579,546 | Method of making microwavable yeast-leavended bakery product containing dough additive |
| 6,578,315 | Accelerated mulching paper degradation process and mulching paper for implementing the process |
| 6,569,646 | Process for the production of an enzyme preparation containing xylanase and carboxymethyl cellulase from *Termitomyces clypeatus* having accession no 11CB-411 |
| 6,566,112 | Cellulase producing actinomycetes, cellulase produced there from and method of producing same |
| 6,565,613 | Cellulase detergent matrix |
| 6,562,340 | Enzyme feed additive and animal feed including it |
| 6,558,666 | Chromium-containing carbohydrase enzyme food supplement composition |
| 6,555,147 | Methods for making popcorn kernels with improved flake production |
| 6,551,358 | Alkaline enzyme scouring of cotton textiles |
| 6,541,438 | Laundry detergent and/or fabric care compositions comprising a modified cellulase |

several mills with the objective to increase the production rate. It has been suggested that enzyme treatments removed some of the fines or peel off fibrils on the fibre surface and thus lower freeness.

Mixtures of cellulases and hemicellulases have been used for bio-modification of fibre properties with the aim of improving drainage and beatability in the paper mills before or after beating of pulp (Dienes *et al.,* 2004). Both endoglucanases I and II are useful in these fibre modification applications. Endoglucanase II can significantly reduce the pulp viscosity even at low concentrations without causing any structural damage to the fibres.

Cellulases alone, or used in combination with xylanases, are beneficial for deinking of different types of paper wastes. Conventional deinking methods use surfactants to float the toner followed by application of high temperatures to aggregate it, and then vigorous dispersion for size reduction. These processes are expensive and energy-consuming. Cellulases and xylanases release toner particles facilitating flotation and subsequent steps. Endoglucanases and endoxylanases have been shown to effectively deink selected old newsprint waste and to improve optical as well as strength properties of paper from enzymatically deinked pulp (Xia *et al.,* 1996). The enzymatic deinking process also improves freeness compared to chemically deinked paper. Cellulase, modified with the copolymer polyethylene oxide and maleic anhydride have been found to improve several physical properties of paper such as freeness, optical properties, and physical strength as compared to conventional treatment (Park and Park, 2001). Xylanases enhance the process of fibrillation and also facilitate enzymatic deinking of recycled fibres when used in combination with cellulases. Enzymatic deinking methods have been scaled up and represent one of the few promising biotechnological applications of cellulases in the pulp and paper industry (Heise *et al.,* 1996).

Xylanases can increase the viscosity of pulps by removing carbohydrates having a low degree of polymerization. Although the presence of cellulases generally causes a decrease in viscosity and impairs the pulp quality, endoglucanase I (EG I) from *T. reesei* can increase the bleachability of softwood kraft pulp and produces a final brightness score comparable to that of xylanase treatment (Suominen and Reinikainen, 1993).

## TEXTILE INDUSTRY

Traditional stone-washing of jeans involves amylase-mediated removal of starch coating (desizing) and treatment of jeans with pumice stone (1-2 kg/pair of jeans) in large washing machines (abrasion). A washing step is then performed to remove excess dye. However, the high loading of stones in washing machines not only decreases the capacity of jean loading, but also causes substantial mechanical wear to both the machine and the jeans. Cellulases have been successfully used for the biostoning of jeans and biopolishing of cotton and other cellulosic fabrics. Cellulases have been shown to effectively facilitate the abrasion of indigo dye from fibre surfaces. Major advantages in replacing stones with enzymes result from reduced mechanical wear and tear of washing machines, higher clothes loading capacities of the equipment, generally safer working conditions and improvements in process automation (Uhlig, 1998).

Enzymatic stone-washing or biostoning can produce up to a 50% higher jean load while causing the desired soft finish (Belghith *et al.,* 2001). Typically 100 kg of stones is replaced by 1 kg of enzyme. Two different types of cellulases, neutral and acidic, are generally used, but use of neutral cellulases is preferable because it results in the reduction in indigo back-staining and facilitates a better working pH profile. The conventional stoning process with cellulosic fibers causes problems with formation of fuzz (fibres protruding from the yarn) and pills (small fibre bundles adhering to the fabric) are often encountered.

Enzymatic biopolishing is a novel biological finishing process for textiles made from cellulosic fibres such as cotton, rayon, polynogic, ramie and tencel. This process is usually carried out during the wet processing stages which include desizing, scouring, bleaching, dyeing and finishing. The controlled partial hydrolysis treatment with acidic cellulases improve softness and water absorbance property of fibres, strongly reduces the tendency for pill formation and provides a clearer surface structure with less fuzz (Sreenath *et al.,* 1996). Cellulase preparations, rich in endoglucanases are best suited for biopolishing enhancing fabric look, feel, and colour without needing any chemical coating of fibres (Galante *et al.,* 1998). The action of cellulases removes short fibres, surface fuzziness, creates a smooth and glossy appearance, and improves colour brightness, hydrophilicity and moisture absorbance. Selection of suitable enzyme preparations is therefore a key in biopolishing.

Ultrasound energy can significantly intensify enzymatic activity on various types of cotton fabrics. A combined enzyme (cellulases and pectinases) and sonication treatment of cotton textiles offers significant advantages such as less consumption of enzymes, shorter processing time, less fibre damage and uniform treatment (Yachmenev *et al.,* 2002).

Iogen from Ottawa, Canada is one of the leading producers and suppliers of fibre-digesting enzymes used by the textile industry to improve the properties of its products (www.iogen.ca). Iogen produces two lines of enzymes for the textiles industry: the Denabraide® line (liquid and granulated products) is used in denim processing to substitute or enhance the performance of pumice stones which fade and soften fabric and the Fibrilase® line (liquid product) is used to soften fabrics and remove or prevent the formation of tiny balls or fuzz on garments.

## DETERGENT INDUSTRY

Enzymes such as proteases and lipases are now commonly used in the detergent industries, but the use of cellulases is a more recent innovation. Cellulase preparations capable of modifying cellulose fibrils can improve colour brightness, feel and dirt removal from the cotton blend garments. Removal of soil from the interfibre space in the fibre using common detergents is challenging. Cellulase supplementation can remarkably improve the cleaning power of the detergent (Horikoshi, 1999). The industrial application of alkaline cellulases as a potential detergent additive is being actively pursued with a view to selectively contacting the cellulose within the interior of fibers and removing soil in the interfibril spaces in the presence of the more conventional detergent ingredients.

## BIOETHANOL INDUSTRY

Enzymatic saccharification of lignocellulosic materials by cellulase produces monosaccharides, which can subsequently be fermented to a variety of products such as ethanol, other alcohols, organic acids, single cell protein, and lipids (Kuhad *et al.,* 1997a; Lynd *et al.,* 1999). Bioconversion of cellulose to glucose using cellulolytic enzymes, followed by production of bioethanol from glucose and other sugars produced from hydrolysis of agricultural residues, is evolving into a viable technology.

Bioconversion of lignocellulosic materials into useful, higher value products normally require multi-step processes (Ghosh and Singh, 1993; Sulbaran-de-Ferrer *et al.,* 2003; Wyman *et al.,* 2005). These processes include:

(i) pretreatment (mechanical, chemical or biological),
(ii) hydrolysis of the polymers to produce readily metabolizable molecules (e.g. hexose or pentose sugars),
(iii) bioconversion of these smaller molecules to support microbial growth or to produce chemical products, and
(iv) the separation and purification of the desired products.

Technologies are currently available for all steps in the bioconversion of lignocelluloses to ethanol and other chemical products (Kuhad *et al.,* 1993, 1997b; Beg *et al.,* 2001; Sun and Cheng, 2002; Mosier *et al.,* 2005). However, some of these technologies must be improved to produce renewable biofuel and other bioproducts at prices, which can compete with more conventional production systems.

Production of ethanol from renewable resources via fermentation represents an important process for production of alternative fuels. Ethanol is an advanced new transport fuel with a unique combination of attributes including low life-cycle greenhouse gas emissions, a high level of sustainability, and seamless integration into the existing transport system and has the potential to have a large-scale impact (Ward and Singh, 2002). Ethanol is a high fuel-blending component that is used widely by major oil companies and distributors. There is an established commercial market for ethanol; approximately 12% of all U.S. gasoline and 5% of all Canadian gasoline is currently blended with ethanol. Its manufacturing process does not consume fossil fuels, but rather uses plant byproducts to create the energy to run the process.

Industrial ethanol production in North America is currently based on corn starch that is first liquefied and saccharified. The oligosaccharide syrup is then used as a feedstock for ethanologenic yeast fermentation. The use of cellulases to increase the yields of starch liquefaction and saccharification has been discussed (Vielle and Zeikus, 2001). Since starch liquefaction is performed at high temperatures, using thermophilic endoglucanases during this step is an interesting option (Haki and Rakshit, 2003).

Considerable efforts are being directed towards development of an economically feasible ethanol production from cellulose. An efficient enzymatic saccharification step using a suitable mixture of cellulase enzymes is required to make cellulose and its degradation products suitable for ethanologenic yeast or bacterial fermentations. One of the main limitations to this three-step process of pretreatment, cellulose hydrolysis and sugar fermentation is the low activity (and high cost) of the cellulases used. Since alkaline pretreatment of cellulose is performed at high temperatures, hyperthermophilic cellulases may represent the best candidate catalysts for cellulose degradation (Maheswari *et al.,* 2000). However, the production of cellulases by hyperthermophilic microorganisms is rare. Thermostable endoglucanases and cellobiohydrolases were characterized in the *Thermotogales* (Haki and Rakshit, 2003). Among the enzymes characterized, pairings of endoglucanase and cellobiohydrolase, optimally active at around 100°C represent interesting enzyme combinations to be tested in cellulose processing.

It is expensive to produce cellulases, which accounts for as much as 60% of the total processing costs associated with enzymatic hydrolysis of cellulose to glucose (Nieves *et al.,* 1998). Thus 90 to 95% of the enzymes need to be recycled to obtain the enzymatic saccharification process economically competitive with acid hydrolysis. Development of a robust enzyme production system that can produce adequate amounts of highly efficient cellulase to make rapid and less expensive biomass conversion is the key challenge in lignocellulose biotechnology. Cellulase yield and productivity above 150 FPU per g of cellulose and 55 FPU per L per h, respectively, have been achieved using Solka Floc as a substrate (Himmel *et al.,* 1999). According to one estimate, bioethanol process would require about 11 million FPU (19 kg, 42 lbs) of cellulase to process 1 ton of biomass (1000 lbs cellulose) to 84 gallons of ethanol at an enzyme loading of 25 FPU/g cellulose (Himmel *et al.,* 1997). It costs US$ 3/gallon of ethanol ($0.8/L) for commercial enzyme formulations and $0.5/gallon ($0.13/L) for less processed formulations (Lynd *et al.,* 1999). To achieve total cost competitiveness, enzyme costs must be reduced to less than $0.07/gallon of ethanol, requiring a 10-fold increase in specific activity or production efficiency (Himmel *et al.,* 1999).

The U.S. Department of Energy and the National Renewable Energy Laboratory (NREL) are running major biomass to bioethanol programs using integrated approaches to achieve maximum process benefits at low costs in the following terms: consolidated bioprocessing < simultaneous saccharification and fermentation < separate hydrolysis and fermentation (Lynd *et al.,* 1999).

Iogen is the first commercial supplier of the cellulose-based biofuel (www.iogen.ca) with EcoEthanol™ as the trade-marked name of the cellulose ethanol process. The Canadian government has invested $21.1 million to help Iogen reach the commercialization stage. Petro-Canada and the Royal Dutch/Shell Group have invested about $24.7 million and $46 million, respectively, into the company. The process uses an enzyme hydrolysis step to convert the cellulose in agriculture residues into sugars. These sugars are fermented and distilled into ethanol fuel using conventional ethanol distillation technology. The yield of bioethanol is obtained in the range of 300 L per tonne of plant fibre. The technology for making ethanol from cellulose has been transferred from the laboratory and scaled-up to a commercial level. By 2008 it is predicted that a processing facility costing $300 million shall have an ethanol-producing capacity of 50 million gallons per year. Initially, production costs are expected to approximate to $1.30 a gallon, but should come down to under $1 a gallon with further advances.

## FERMENTATION INDUSTRY

Microbial glucanases and related polysaccharides play important roles in fermentation processes to produce alcoholic beverages including beers and wines (Galante *et al.,* 1998). These enzymes can improve both quality and yields of the fermented products (Nomachi *et al.,* 2002). Glucanases are added either during mashing or primary fermentation to hydrolyze glucan, reduce the viscosity of wort and improve the filterability. Endoglucanase II and cellobiohydrolase II enzymes of the

*Trichoderma* cellulase complex are responsible for the reduction in the degree of polymerization of the carbohydrates and wort viscosity (Canales *et al.,* 1988).

In wine production pectinases, glucanases and hemicellulases play an important role by improving colour extraction, skin maceration, must clarification, filtration and finally the wine quality and stability. β-Glucosidases can improve the aroma of wines by modifying glycosylated precursors. Macerating enzymes also improve pressability, settling and juice yields of grapes used for wine fermentation. A number of commercial enzyme preparations are now available to the wine industry.

Hemicellulases and engineered glycosynthases and β–glucosidase can be applied for the enzymatic synthesis of oligosaccharides (Palcic, 1999; Fairweather *et al.,* 2002). Specifically, high yields of different oligosaccharides can be obtained using engineered glycosynthases, in which the nucleophilic residue has been replaced. Currently a number of organic acids, amino acids, vitamins and a number of bacterial and fungal polysaccharides such as xanthan gums are produced by fermentation using glucose as the fermentation substrate mediated by the actions of cellulases and hemicellulases.

## FOOD INDUSTRY

Fungal cellulases have been shown to improve the yield of sugars from unmodified malts in the brewing process. *Trichoderma* cellulases improve ethanol yields from cassava. Fungal cellulases can improve total extractable solids from unmodified malts in the brewing process. The isolation of starch from tubers and cereals requires release of the non-degraded particles as cleanly as possible. Cereal starches are released to improve yield when amylase-free cellulases and xylanases are applied to the starch slurry prior to centrifugation and washing. Addition of extracellular microbial enzymes like amylases, proteases, xylanases, arabinases and esterases can improve the quality of dough, bread and other bakery products (Poutanen, 1997).

Cellulases and xylanases can improve and speed up colour extraction by pectinases from the skin of fruits. Cellulase has an important application as a part of an enzyme complex, macerating enzymes (cellulases, xylanases and pectinases) used for extraction and clarification of fruit and vegetable juices to increase the yield of juices (Minussi *et al.,* 2002). Occasionally α-amylases and amyloglucosidase are also added to the macerating enzymes to process starch containing fruits. The macerating enzymes are used to improve cloud stability and texture and decrease viscosity of the nectars and purees from tropical fruits such as mango, peach, papaya, plum, apricot and pear. Texture, flavour, aroma and volatile properties of fruits and vegetables can be improved and excessive bitterness of citrus fruits can be reduced by infusion of enzymes such as pectinases and β-glucosidases (Baker and Wicker, 1996). Enzyme mixtures containing pectinases, cellulases and hemicellulases are also used for improved extraction of olive oil.

White rot fungi are cultivated on lignin and cellulose-containing substrates such as cotton stalks and wheat or rice straw. As the white rot fungi can decompose lignocellulose efficiently without chemical or biological pretreatment, a large variety of lignocellulosic wastes can be recycled and utilized, for example, for the production of edible mushrooms. The approach provides well-proven and low-cost profitable technology for production of good quality human food (Cohen *et al.,* 2002). Currently, the most commercially viable process for bioconversion of agricultural lignocellulosic waste is composting to provide substrate for growth of edible fungal fruiting bodies.

Fungal fruiting bodies are rich in protein. *Agaricus, Lentinus, Volvariella, and Pleurotus* are the dominant taxa cultured for human consumption. In Europe, the major organism of interest is *Agaricus bisporus*. In the Far East, a wide variety of mushrooms, especially the oyster mushroom, *Pleurotus ostreatus*, and shitake are cultivated.

## ANIMAL FEED INDUSTRY

Applications of cellulases and hemicellulases in the feed industry have received considerable attention because of their potential to improve feed value and performance of animals (Dhiman *et al.,* 2002). Pretreatment of agricultural silage and grain feed by cellulases or xylanases can improve its nutritional value (Godfrey *et al.,* 1996; Beg *et al.,* 2001). The enzymes can also eliminate anti-nutritional factors present in the feed grains, degrade certain feed constituents to improve the nutritional value and provide supplementary digestive enzymes such as proteases, amylases and glucanases.

β-Glucanases and xylanases have been used in the feed of monogastric animals to hydrolyze non-starch polysaccharides such as β-glucans and arbinoxylans. Cellulases, used as feed additives alone or with proteases can significantly improve the quality of pork meat. Glucanases and xylanases reduce viscosity of high fibre rye- and barley-based feeds in poultry and pig. These enzymes can also cause weight gain in chickens and piglets by improving digestion and absorption of feed materials.

The complex ruminant diet contains mainly cellulose, hemicellulose, lignin and pectin. Most low-quality feedstuffs contain high concentrations of cellulose, small amounts of protein and fat and relatively high ash contents when compared with high-quality feedstuffs. Cellulases can be used to improve silage production for cattle feeding which involves enhancement of the digestibility of grasses, which contain large amounts of potentially total digestible nutrients and energy values together with only small amounts of water-soluble carbohydrates. The nutritional value of these low-quality feedstuffs can be increased significantly if they are effectively processed before feeding. It is possible to convert lignocellulosic substrate to more digestible and nutritious feedstuff for ruminants by upgrading animal feeds through solid-state fermentation processes. A large number of lignin-degrading white-rot fungi are known for solid-state bioconversion of lignocellulosic materials (Akin *et al.,* 1993). *P. chrysosporium, Cyathus* sp., *Coriolus hirsutus, Dichomitus squalens*, and *Pleurotus* spp. are the extensively studied organisms. Apart from improving digestibility and feed value of straw, the mycostraw by-product obtained after solid-state fermentation can potentially be used for biogas production. After microbial removal of lignin, these straw components are more accessible for anaerobic digestion.

Animal feedstock production processes include heat treatments that inactivate potential viral and microbial contaminants. Application of thermophilic cellulase in feedstock production has the potential to reduce pathogens as well as to enhance digestibility and nutrition of the feed, thereby facilitating a combination of heat treatment and feed transformation in a single step.

Lignocellulosic waste from the pulp and paper industries, as well as wastes from dairy and agricultural industries, provides supplies of substrate for use in the production of SCP for feed and food purposes (Kuhad *et al.,* 1997b). The potential to convert cellulose to SCP by growing the microbes directly on cellulosic substrate for microbial protein production and by facilitating production of extracellular cellulase enzyme that can be used for saccharification of cellulosic substrates represents an efficient waste recycling process. Spent sulfite liquor (available as a byproduct of paper pulping) is widely used as a carbohydrate source and contains about 3% fermentable sugars. The organisms used for SCP include bacteria *Cellulomonas* and *Lactobacillus*; the molds *Aspergillus, Chaetomium, Fusarium, Paecilomyces, Penicillium,* and *Polyporus*; the yeasts *Saccharomyces, Candida,* and *Rhodotorula*; and the algae *Spirulina* and *Chlorella.*

## AGRICULTURAL AND ENVIRONMENTAL INDUSTRIES

Various enzyme preparations consisting of different combinations of cellulases, hemicellulases and pectinases have potential applications in agriculture for enhancing growth of crops and controlling plant diseases (Chet *et al.,* 1998). Plant or fungal protoplasts produced using microbial hydrolases can be fused to produce hybrid strains with desirable properties. Fungal β-glucanases are capable of controlling diseases by degrading cell walls of plant pathogens. Cellulolytic fungal species of *Trichoderma*, *Geocladium*, *Chaetomium* and *Penicillium* are known to play a key role in seed germination, rapid plant growth, flowering and crop yields.

White rot fungi are gaining interest because of their capability to degrade a wide variety of natural and synthetic materials and environmentally persistent organopollutants (Pointing, 2001; Tango and Islam, 2002). Lignin and cellulose degrading cultures of *Phanerochaete chrysosporium* and *Trametes versicolor* are capable of degrading polychlorinated biphenyls (PCB), polycyclic aromatic hydrocarbons (PAHs) and dyes (Reddy, 1995). Fungi are capable of degrading a large spectrum of azo-, anthracene- and triphenylmethane-dyes (McMullan *et al.,* 2001; Kuhad *et al.,* 2004). The ability of these fungi to degrade such a range of organic compounds results from the relatively non-specific nature of their ligninolytic enzymes - lignin peroxidase, manganese peroxidase and laccase (Murugesan and Kalaichelvan, 2003; Ohmia *et al.,* 2003).

The possibility of converting spent casings, a co-product of the frankfurter/sausage processing industries, has been discussed by Sanders *et al.,* (2000). Greater than 90% degradation of spent casing to glucose has been achieved using a cellulase preparation. The traditional methods of disposal of spent casings are landfills and land application and their reuse as food for animals could provide an attractive environment-friendly disposal alternative.

Microbial enzymes can also reduce the disposable solid content of sewage sludge. A mixture of industrial cellulase, protease, and lipase, in equal proportion by weight can reduce total suspended solids (TSS) by 30-50% and improve settling of solids (Parmar *et al.,* 2001). Two-enzyme combinations of protease and cellulase resulted in better solid reduction than individual enzymes and that lipase further augmented this effect.

## CONCLUSIONS

Cellulases are being commercially produced by several companies worldwide (Table 3) and are widely used in food, animal feed, fermentation, agriculture, pulp and paper, and textile applications. One of the major potential applications of cellulases is for production of bioethanol, by conversion of plant fibres to glucose and its onward conversion by fermentation to ethanol. Although the economics of this process do not currently compete with fuel ethanol production from starch, the productivities are being improved through a significant decrease in cellulase manufacturing costs.

**Table 3. Some commercial producers of cellulases and related enzymes**

| | |
|---|---|
| Novozyme, Denmark, | – The largest producer of enzymes in the world; biggest market is detergent enzymes (35% of sales including proteases, cellulases and lipases). |
| Genencor, USA | – Another large producer of enzymes; research and production facilities in both Europe and North America |
| Iogen, Ottawa, Canada | – Primarily focused on fuel ethanol production from biomass, also produces cellulases and xylanases; developed a steam-explosion technology for pretreatment of cellulose prior to cellulase hydrolysis. |
| AB Enzymes, Darmstadt, Germany | – A range of enzymes for use in food, pulp and paper and textiles |
| Buckman Labs, Memphis, TN | – Cellulolytic enzyme producer; several other products under development to complement their existing chemical technologies |
| EDT, Norcross, Georgia | – Produces enzyme deinking and pitch control technologies |
| Tritex, Saint-Eustache, Canada | – Specializes in producing dyestuffs, pigment dispersions for the textile and paper industries; also sells enzyme deinking and pectinases products. |
| DSM, Netherlands | – Enzymes producers for other industries; produces enzymes for pulp and paper industry. |
| Diversa, USA | – Enzymes producers for other industries; produces enzymes for pulp and paper industry. |
| Kao (Japan). | – Enzymes producers for other industries; produces enzymes for pulp and paper industry. |
| Nalco, Kemira | – Primarily chemical suppliers with some hydrolytic enzyme product interest |
| Shanghai Simi | – A range of cellulases and amylases for food applications |

With modern biotechnology tools, especially in the area of microbial genetics, novel enzymes and new enzyme applications will become available for the pulp and paper industry (Wang *et al.*, 2005). Cellulose-binding domains of cellulases can be fused to heterologous proteins and used as an affinity tag purification or immobilization of fusion proteins (Pentilla, 1998). Discovery of extremophiles opens new opportunities for the development of enzymes having superior catalytic capabilities at high temperature and pH (Haki and Rakshit, 2003). So far most extremozymes in commercial use are altered little from their original state. In general, except for the thermostability, the molecular characteristics of cellulase components of thermophilic and mesophilic fungi are similar. However, for any new enzyme to gain commercial acceptance, its producers will have to reduce the costs of production.

Fungal cellobiohydrolase I promoter is particularly suitable for generating high yields of fusion proteins and antibodies (Ohmia *et al.*, 2003). Process solutions are being sought through aggressive R & D in this area. Potential enzyme applications under investigation include new enzymes for kraft pulp bleaching and new enzymatic approaches to reduce mechanical pulping energy. It is also anticipated that the newly developed genetic techniques will significantly reduce the cost for enzyme production and improve the characteristics of these biocatalysts.

## REFERENCES


Akhtar, M., Scott, G.M., Swaney, R.E., Shipley, D.F. (2000). Biomechanical pulping: a mill-scale evaluation. *Resource Conservation and Recycling* 28, 241-252.

Akin, D.E., Sethuraman, A., Morrison, W.H., Martin, S.A., Eriksson K.E. (1993). Microbial delignification with white rot fungi improves forage digestibility. *Applied Environmental and Microbiology,* 59, 4272-4282.

Beg, Q.K., Kapoor, M., Mahajan, L., Hoondal, G.S. (2001). Microbial xylanases and their industrial applications: a review. *Applied Microbiology and Biotechnology,* 56, 326-338.

Belghith, H., Ellouz-Chaabouni, S., Gargouri, A. (2001). Biostoning of denims by *Penicillium occitanis* (Pol6). cellulases. *Journal of Biotechnology,* 89, 257–262.

Baker, R.A., Wicker L. (1996). Current and potential applications of enzyme infusion in the food industry. *Trends Food Science Technology,* 7, 279-284.

Bhat, M.K. (2000). Cellulase and related enzymes in biotechnology. *Biotechnology Advances,* 18, 355-383.

Canales, A.M., Garza, R., Sierra, J.A., Arnold, R. (1988). The application of beta-glucanase with additional side activities in brewing. *MBAA Technology,* Q. 25, 27-31.

Chet, I., Benhamou, N. and Haran, S. (1998). Mycoparatism and lytic enzymes. In: *Trichoderma* and *Gliocladium – Enzymes, biological control and commercial applications,* Eds Harman, G. F. and Kubicek, C. P. Vol 2, Taylor & Francis, London, Pp 327-342.

Cohen, R., Persky, L., Hadar, Y. (2002). Biotechnological applications and potential of wood degrading mushrooms of the genus *Pleurotus. Applied Microbiology and Biotechnology,* 58, 582-594.

Dhiman, T.R., Zaman, M.S., Gimenez, R.R., Walters, J.L., Treacher, R. (2002). Performance of dairy cows fed forage treated with fibrolytic enzymes prior to feeding. *Animal Feed Science Technology,* 101, 115–125.

Dienes, D., Egyházi, A., Réczey, K. (2004). Treatment of recycled fiber with *Trichoderma* cellulases. *Industrial Crops Production,* 20, 11-21.

Fairweather, J.K., Faijes, M., Diguez, H., Planas, A. (2002). Specificity studies of *Bacillus* 1,3-1,4-β-glucanases and application to glycosynthase-catalyzed transglycosylation. *Chem Biochem,* 3, 866-873.

Galante, Y.M., DeConti, A., Monteverdi, R. (1998). Application of Trichoderma enzymes in food and feed industries. In: *Trichoderma* and *Gliocladium – Enzymes, biological control and commercial applications,* Eds Harman, G. F. and Kubicek, C. P. Vol. 2, Taylor & Francis, London, Pp 311-326.

Ghosh, P., Singh, A. (1993). Physicochemical and biological treatments for enzymatic/microbial conversion of lignocellulosic biomass. *Advanced and Applied Microbiology,* 39, 295-333.

Godfrey, T. and Reichelt, J. (1983). *Industrial Enzymology,* Nature Press, New York.

Haki, G.D. and Rakshit, S.K. (2003). Developments in industrially important thermostable enzymes: a review. *Bioresource Technology,* 89, 17–34.

Heise, O.U., Unwin, J.P., Klungness, J.H., Fineran, W.G., Sykes, M., Abubakr, S. (1996). Industrial scale up of enzyme-enhanced deinking of nonimpact printed toners. *Tappi Journal,* 79, 207–212.

Himmel, M.E., Adney, W.S., Baker, J.O., Elander, R., McMillan, J.D., Nieves, R.A., Sheehan, J.J., Thomas, S.R., Vinzant, T.B., Zhang, M. (1997). Advanced bioethanol production technologies: a perspective. *ACS Symposium Series,* 66, 2-45.

Himmel, M.E., Ruth, M.F. and Wyman, C.E. (1999). Cellulase for commodity products form cellulosic biomass. *Current Opinion in Biotechnology,* 10, 358-364.

Horikoshi, K. (1999). Alkaliphiles: some applications of their products for biotechnology. *Microbiol. Mol. Biol. Rev.,* 63, 735–750.

Kuhad, R.C. and Singh, A. (1993). Lignocellulose biotechnology: Current and future prospects. *Critical Review Biotechnology,* 13, 151-172.

Kuhad, R.C., Singh, A. and Eriksson, K-E. L. (1997a). Microorganisms and enzymes involved in the degradation of plant fibre cell walls, *Advanced Biochemical Engineering and Biotechnology,* 57, 45-125.

Kuhad, R.C., Singh, A., Tripathi, K.K., Saxena, R.K., Eriksson, K.-E.L. (1997b) Microorganisms as an alternative source of protein. *Nutrition Review,* 55, 65-75.

Kuhad, R.C., Sood, N., Tripathi, K.K., Singh, A. and Ward, O.P. (2004). Developments in microbial methods for the treatment of dye effluents. *Advanced Applied Microbiology,* 50, 185-213.

Lynd, L.R., Wyman, C.E., and Gerngross, T.U. (1999). Biocommodity engineering. *Biotechnol. Prog.,* 15, 777-793.

Lynd, L.R., Weimer, P.J., Van Zyl, W.H. and Pretorius, I.S. (2002). Microbial Cellulose Utilization: Fundamentals and Biotechnology. *Microbiol. Mol. Biol. Rev.,* 66, 506-577.

Maheswari, R., Bhardwaj, G., Bhat, M.K. (2000). Thermophilic fungi: their physiology and enzymes. *Microbiol Mol Biol Rev,* 64, 461–488.

Mai, C., Kües, U., Militz, H. (2004). Biotechnology in the wood industry. *Applied Microbiology and Biotechnology,* 634, 77–494.

McMullan, G., Meehan, C., Conneely, A., Kirby, N., Robinson, T., Nigam, P., Banat, I.M., Marchant, R., Smyth, W.F. (2001).. Microbial decolourisation and degradation of textile dyes. *Applied Microbiology and Biotechnology,* 56, 81-87.

Minussi, R.C., Pastore, G.M., and Duran, N. (2002).. Potential applications of laccase in the food industry. *Trends Food Science Technology,* 13, 205-216.

Mosier, N., Wyman, C.E., Dale, B., Elander, R., Lee, Y.Y., Holtzapple, M., Ladisch, M., (2005). Features of promising technologies for pretreatment of lignocellulosic biomass. *Bioresource Technology,* 96, 673–686.

Murugesan, K. and Kalaichelvan, P.T. (2003). Synthetic dye decolourization by white rot fungi. *Indian Journal of Experimental Biology,* 41,1076-1087.

Nieves, R.A., Ehrman, C.I., Adney, W.S., Elander, R.P. and Himmel, M.E. (1998). Survey and analysis of commercial cellulase preparations suitable for biomass conversion to ethanol. *World Journal of Microbiology Biotechnology,* 14, 301-304.

Nomachi, W., Urago, K.-I., Oka, T., Ekino, K., Matsuda, M., Goto, M., Furukawa, K. (2002). Molecular breeding of *Aspergillus kawachii* overproducing cellulase and its application to brewing barley Shochu. *Journal of Bioscience and Bioengineering,* 93, 382-387.

Ohmia, K., Sakka, K., Kimura, T., Morimoto, K. (2003). Application of microbial genes to recalcitrant biomass utilization and environmental conservation. *Journal of Bioscience and Bioengineering,* 95, 549-561.

Palcic, M.M. (1999). Biocatalytic synthesis of oligosaccharides. *Current Opinion in Biotechnology,* 10, 616-624.

Park, J. and Park, K. (2001). Improvement of the physical properties of reprocessed paper by using biological treatment with modified cellulase. *Bioresource Technology,* 79, 91-94.

Parmar, N., Singh, A. and Ward, O.P. (2001). Enzyme treatment to reduce solids and improve settling of sewage sludge. *Journal of Industrial Microbiology Biotechnology,* 26, 383-386.

Perez, J., Munoz-Dorado, J., de la Rubia, T., Martínez, J. 2002. Biodegradation and biological treatments of cellulose, hemicellulose and lignin: an overview. *International Microbiology.,* 5, 53–63.

Peng, F., Ferrtsius, R. and Angsas, U. (2003). Mechanical pulping with pectinase pretreatment of wood chips. *Proceedings International Mech. Pulp. Conf.,* pp. 335-340.

Pere, J., Ellmen, J., Honkasalo, J., Taipalus, P. and Tienvieri, T. (2002). Enhancement of TMP reject refining by enzymatic modification of pulp carbohydrates - A mill study. In: *Biotechnology in the Pulp and Paper Industry,* Eds Viikari, L. and Lantoo, R. Elsevier, Amsterdam, Pp 281-290.

Pentilla, M. (1998). Heterologous protein production in *Trichoderma.* In: *Trichoderma* and *Gliocladium – Enzymes, biological control and commercial applications,* Eds Harman, G.F. and Kubicek, C.P., Vol 2, Taylor & Francis, London. Pp 365-382.

Pointing, S.B. (2001). Feasibility of bioremediation by white rot fungi. *Applied Microbiology and Biotechnology,* 57, 20-33.

Poutanen, K. (1997). Enzymes: an important tool in the improvement of the quality of cereal foods. *Trends Food Science Technology,* 8, 300-306.

Reddy, C.A. (1995). The potential of white rot fungi in the treatment of pollutants. *Current Opinion in Biotechnology,* 6, 320-328.

Sanders, D.A., Belyeaa, R.L., Taylor, T.A. (2000). Degradation of spent casings with commercial cellulases. *Bioresource Technology,* 71, 125-131.

Singh, A. (1999). Engineering enzyme properties. *Indian Journal of Microbiology,* 39, 65-77.

Singh, A. and Hayashi, K. (1995). Microbial cellulases: protein architecture, molecular properties and biosynthesis. *Advanced Applied Microbiology,* 40, 1-44.

Sreenath, H.K., Shah, A.B., Yang, V.W., Gharia, M.M. and Jeffries, T.W. (1996). Enzymatic polishing of jute/cotton blended fabrics. *J. Ferment. Bioeng.,* 81, 18-20.

Sulbaran-de-Ferrer, B., Aristiguieta, M., Dale, B.E., Ferrer, A., Ojedade-Rodriguez, G., (2003). Enzymatic hydrolysis of ammonia-treated rice straw. *Applied Biochemistry and Biotechnology,* 105–108, 155–164.

Sun, Y., Cheng, J. (2002). Hydrolysis of lignocellulosic material from ethanol production: a review. *Bioresource Technology,* 83, 1-11.

Lange, N.K. (1993). Applications of cellulases in textile industry. In: *Trichoderma reesi cellulases and other hydrolases. Foundation for Biotechnical and Industrial Fermentation Research,* Eds Suominen, P. and Reinikainen T., Espoo, Finland. Pp 263-272.

Tango, M.S.A. and Islam, M.R. (2002). Potential of extremophiles for biotechnological and petroleum applications. *Energy Sources.* 24, 543-559.

Uhlig, H. (1998). Industrial enzymes and their applications. John Wiley & Sons, New York.

Vielle, C. and Zeikus, G.J. (2001). Hyperthermophilic enzymes: sources, uses, and molecular mechanisms for thermostability. *Microbiol. Mol. Biol. Rev.,* 65, 1-43.

Wang, T., Liu, X., Yu, Q., Zhang, X., Qu, Y., Gao, P., Wang, T. (2005). Directed evolution for engineering pH profile of endoglucanase III from *Trichoderma reesei. Biomol. Eng.,* 22, 89–94.

Ward, O.P., Singh, A. (2002). Bioethanol technology: developments and perspectives. *Advanced Applied Microbiology,* 51, 53-80.

Wyman, C.E., Dale, B.E., Elander, R.T., Holtzapple, M., Ladisch, M.R., Lee, Y.Y. (2005). Coordinated development of leading biomass pretreatment technologies. *Bioresource Technology,* 96, 1959-1966.

Xia, Z., Beaudry, A. and Bourbonnais, R. (1996). Effects of cellulases on the surfactant assisted acidic deinking of ONG and OMG. *Prog. Paper Recycl.,* 5, 46-58.

Yachmenev, V.G., Bertoniere, N.R., Blanchard, E.J. (2002). Intesification of the bio-processing of cotton textiles by combined enzyme/ultrasound treatment. *J. Chem. Technol. Biotechnol.* 77, 559-567.

# 21

# *Palo Podrido* Feed : A Pivotal Trigger for Biodegradation Research Round The Globe

**Raman Rao[1] and Ramesh C. Kuhad[2]**

[1]*Senior Research officer, Department of Animal Nutrition, College of Veterinary and Animal Sciences, G.B. Pant University of Agriculture and Technology, Pant Nagar-263 145; India.*
[2]*Professor, Department of Microbiology, University of Delhi, South Campus, Benito Juarez Road, New Delhi- 110 021; India.*

A perennial shortage of fodder and hay as well as ever increasing prices of conventionally used high quality feeds had culminated in a search for animal feeds of desired characteristics in the developing world. The lignocellulosics of agricultural crops offered high potentials in meeting increased demand of animal feed due to high content of cellulose and hemicellulose, constituting a staple-feed (basal) for the ruminant animals. In early nineties, the researchers reported the decomposition of plant residues to a varied extent and could not conclude on the available data due to the lack of simple and sensitive-assay techniques. Lately, white-rot fungi were observed to degrade lignin and significantly increased the crude-protein content as well as digestibility of these lignocellulosics manifold. In the past couple of decades, researchers all over optimized and standardized different cultural conditions of these organisms *in vitro*. This contributed the role of biotechnolgy in the different areas of pulp and paper manufacture—forestry, pulping, bleaching, waste-management etc. gaining importance world over. All these efforts led to the discovery of 'ligninase' enzyme in 1983, which became the pivotal point of biotechnolgical approach looking for the genes responsible for such solid-state fermentation process by way of immunoscreening, oligonucleotide probing and differential DNA-hybridization techniques. Cloning of the ligninase-gene to produce cellulose-less as well as hemicellulose-less mutants would pave-way for the new beginning and therefore, future breakthrough in the research.

## INTRODUCTION

The biodegradation process dates back to time immemorial when some wild ruminant animals were spotted feeding on the tree-woods in South Chile. The natives of South Chile called this wood as 'Palo-podrido (Huempe)' meaning–'animal feed' (Zadrazil *et al.,* 1982[c]). Later on, it was found that this wood was actually degraded by certain types of fungi under natural conditions and decomposed to Spent-wood, which turned out to be the animal feed. The scientists started working on these fungi as well as this type of process. Lately, the fungi were found to belong to higher-basidiomycetes class and called White-Rot Fungi. They mainly degraded lignin of the wood, however cellulose and hemicellulose to a limited extent only. But, surprisingly it was found that the digestibility and crude protein content increased to a considerable extent. Later on, the scientists started working on spent-straws too. The cereal straws (lignocellulosics) constituted a staple-feed (basal) in the developing countries for these ruminant animals. But, the presence of lignin (a biological plastic) limits the utilization of these cereal straws in the rumen eco-system. Thus, the concept of this type of Solid Sate Fermentation (SSF) came into this type of investigation. Some examples of such white-rot-fungi were – *Pleurotus ostreatus, Coprinus cineus, Sporotrichum thermophile, Phanerochaete chrysosporium, Polyporus berkeleyi, Trichoderma reesei, Neurospora crassa, Alternaria tenius* etc to name a few. Some examples of substrates upon which such types of fungi grew for Solid State Fermentation (SSF) were: wheat straw, paddy straw, paddy husks, ground nut shells, sugarcane bagasse, barley straw, sunflower stalk, aspen wood, birch wood, saw dust etc. The resurgence of interest in the use of straw from cereal and grain crops as animal feed had been witnessed in early nineties to maintain the performance of animals. Beckman (1923) used the specially designed extruders, pneumatic cyclones, and disk refiner for allowing intimate contact of straw particles with the chemicals. Such treatment decomposed the lignin moiety of the cohesive lignocellulosic complex. But, the treatment dictated certain limitations – the higher concentrations of chemicals were required under pressure for better efficiency, huge quantity of water was used for washing out the chemicals, loss of about 25% of the organics solubilized during hydrolysis, higher cost incurred during drying due to higher moisture content, generation of effluent and consequently, high expenses on waste-management (otherwise, environmental pollution). Waksman and Cordon (1939) worked on the decomposition of plant residues by pure cultures of thermophilic moulds and observed a lot of variation. Since, it was very difficult at that stage, to come to a conclusion about the degradability of lignin by these moulds based on the data, it had been strongly felt a need for the simple and sensitive assay-techniques for confirmation of the results. Bondy and Meyer (1948) in their estimations without the use of chemicals (e.g. sulphuric acid), found about 5% methoxyl in the lignin isolated from legumes and about 10% methoxyl in lignin isolated from various grasses and very little difference was found in the methoxyl content of the lignin isolated from the corresponding faces which indicated that there was very little breakdown of these linkages in the lignin molecule during passage through digestive tract. Schubert and Nord (1950) investigated the softwood lignin and lignification. They opined that alkali treatment did not decrease methoxyl content of lignin. High carbohydrate content accounted for the lower value of methoxyl groups in the lignin preparation. Investigations on lignin and lignification was further studied by different researchers – for the comparative evaluation of native lignins (Kutzin *et al.*,1951); for characterization of enzymatically liberated hard-wood lignins (Kutzin and Nord, 1951) as well as isolation and characterization of bagasse native lignin (de Stevens and Nord,1951). Waksman (1957) performed the growth-studies with *Streptomyces* cultures and tested the production of pigments in the Asparagine-glucose-agar and Tyrosine-glucose-agar media. After 10 d growth at 37°C, the plates were observed and score based on visual relative examination. Hungate (1957) studied microorganisms in the rumen of cattle, fed a constant ration and developed a culture-medium for carbon source for the inoculation of mixed anaerobes having a mixed gas atmosphere of $CO_2$ and $N_2$ (85:15). Farmer *et al.,* (1960) studied the aromatic alcohol oxidase activity in the growth medium of *Polysticus versicolor*, unravealing the direct involvement of enzymes. Ishikawa *et al.,* (1963) studied the enzymatic degradation of soft-wood lignin by white-rot fungi and advocated the possibility of direct role of extracellular and non-specific enzymes. Raghu and MacRae (1966 and 1967) studied the bio-degradation of isomers of benzene hexachloride (BHC) and observed faster breakdown under anaerobic conditions. They also reported that repeated addition of chlorinated Cyclo-Hexane (HCH) insecticide to flooded soil caused the enrichment of specific microorganisms resulted in the enhanced degradation of HCH. Other works indicated that white-rot fungi (Higher Basidiomycetes) could utilize lignin along with cellulose and other components of the substrate (Lawson and Still, 1957; Cochrane, 1958; Farmer, 1967). Reid (1979) showed studies on upgrading the straw quality as ruminant feed through biological treatment using *Phanerochaete chrysosporium* (a white-rot fungus). Increasing the lignolytic activity with a minimum loss of associated polysaccharides had been the prime objective of this study. Gradually, such type of interesting research area caught the attention of various other groups' world over. In the past couple of decades, this research-field gained tremendous momentum, so much so that, there was a lot of overlapping of work simultaneously and also, great communication among different researchers.

## SOLID STATE FERMENTATION (SSF) RESEARCH IN THE DEVELOPING WORLD

The research focussed mainly on *Pleurotus ostreatus* [commonly known as Oyster mushroom; the French version: champignon; (*Coprinus* sp tested for higher temperatures)] and prepared their 'spawn' on un-infected sorghum grains (alternatively, millet / wheat grains were also used by other workers). After thorough cleaning and boiling, these grains were filled in the 250 ml heat-resistant bottles with calcium oxide mixed (1% of grain quantity) and autoclaved at 15 lb pressure for 20 min. These bottles were then inoculated with the above fungi aseptically in the UV chamber separately and kept for incubation at 25°C for about a week. This spawn was used to treat the different straws in the poly-propylene bags to optimize the different cultural conditions – temperatures, moisture levels, incubation periods, light (in the BOD incubator) supplemental nutrients N and S (Raman and Naik, 1990, 1991, 1992, 1993) buffer concentrations (Raman, 1995[a]); heat-treatments (Raman, 1995 [b]); mineralization and composting as well as mineralization and moisturization (Raman, 1997[a,b,c]); anhydrous ammonia (Raman,1998); other substrates ( Raman, 1999[a]); mineralization alone (Raman and Chauhan, 1999 [b]), colonizing with different organic residues during initial phase (Raman and Hélène, 2000 ) as well as other work (Raman Rao, 2001; Raman Rao and Kuhad,2001; Raman Rao and Petit, 2002). The samples were analyzed for different nutritional parameters – crude protein (CP), neutral detergent fiber (NDF), acid detergent fiber (ADF), hemicellulose, permanganate-lignin, cellulose, pH, bio-mass, water solubles, *in vitro* dry matter digestibility (IVDMD), digestible dry matter recovery (DDMR) including nylon bag dry matter digestibility (NBDMD). During the SSF, in general, CP content increased ($P < 0.05$) with respect to their corresponding controls (ranging from 7.63–68.85% for the wheat straw and 7.36–14.30% for the paddy straw for different treatments), possibly due to proportionately lower N-loss compared to C-loss. The C-loss was found to be due to a loss of cellulose and/hemicellulose (Shivappa Shetty and Krishnamurty, 1981). The fungus in question, resulting into some DM-loss of about 8-10% might have utilized these carbohydrates.

These findings reflected that such fungi were known to lack the ability of fixing atmospheric nitrogen and therefore, tend to conserve available nitrogen for their growth. There was no effect of exposure to light in these studies too (Kamra and Zadrazil, 1986).

There were actually two phases during the colonization of such fungi. In the initial phase (of about a week / ten days), the fungus used soluble nutrients of the substrates for their own establishment and therefore, there was slow initial increase in digestibility (Raman and Hélène, 2000). And thereafter in the final phase, the fungus attacked the fibrous portion of substrate (digestibility ranging between 16.18 – 36.10 % for the wheat straw, while 9.5 – 40.13 % for the paddy straw for the different treatment groups). This was due to the weakening of ligno-carbohydrate complexes (the extent of weakening depended mainly on the substrate type and its preferential use by the fungus in question). The authors were more concerned for the fungal growth upto the mycelial-stage only and this took about three to four weeks' time in general and then the further growth was stopped. They did not want 'fruit-body formation' [i.e. mushroom formation (champignon)] as it saped more of the available nutrients. But, Zadrazil and his associates (1974, 1975, 1977, 1978, 1979, 1980[a,b], 1981, 1982[a,b,c], 1983, 1984, 1985, 1988; Kamra and Zadrazil, 1987) thought it a little commercially. These research groups too, worked extensively on optimizing the cultural conditions of different white-rot fungal species. But, they had profited by dual advantages – on the upper part of the substrates, they grew and harvested edible crop of mushrooms (in French = champignon) for commercial consumption of human beings and the lower spent-substrates were used as ruminant-feed. Kuhad with different co-workers (2004; 2005[a,b]) had attempted *in vitro* studies on Xylanase of *Ganoderma* sp. kk-02; Laccase from the birds' nest fungus *Cyathus bulleri* as well as soil microflora (which were responsible for the development of plants).

The ecological, physiological and biochemical aspects and economic importance of such thermophilic moulds had been reviewed by other researchers (Cooney and Emerson, 1964; Emerson, 1968; Sharma, 1989). Waksman and Cordon (1939) and Jain *et al.,* (1979) reported the decomposition of plant residues by pure cultures of *Aspergillus fumigatus, Chaetomium thermophile, Sporotrichum thermophile, Thermoascus aurantiacus, Humicola insolens* and *Torula thermophila* were observed to be very active decomposers of wheat and paddy straws. These organisms degraded cellulose (3.3 to 61.7 %) and lignin (1.0 to 37.3 %) to a varied extent. These wide differences were apparently due to different methods used for the analysis of the partially decomposed straws. Chahal *et al.,* (1979) reported that for the conversion of cellulose and hemicellulose into Single Cell Protein (SCP), *Chaetomium cellulolyticum* had been a good system. Lignocellulosic materials (with or without pre-treatments) were fermented with thermophilic fungus either by SSF or submerged fermentation. The above fungus was found to be fast growing on 0.09 $h^{-1}$; this rate increased to 0.3 $h^{-1}$ when 1% cellulose was added after the exhaustion of the first batch substrate. In the liquid medium, the growth rate was very rapid with a specific growth rate of about 5 h. This mould was observed to have 50-100 % faster growth rate than *Trichoderma viride*. The amino acid composition of *C. cellulolyticum* was better than that of *T. viride* and compared favourably with FAO reference protein and soyameal.

Singh and Singh (1989) aimed at isolation of lignin from wheat straw, its characterization and use in culture medium for degradation by an isolate of soil fungus, *Monilia* sp. Recovery of lignin from wheat straw was 8% on DM basis. It was a brown coloured powder, highly soluble in NaOH. Results of chemical analysis indicated negligible amounts of nitrogenous and inorganic impurities.

Carbon content of 60% lied in the range reported for lignin from various sources. Methoxyl group content of the lignin preparation of 8.5% was lower than that reported for soft wood lignin (14.4 to 16.8 %). Alkali treatment did not decrease methoxyl content of lignin (Schubert and Nord, 1950), therefore, either the WS lignin originally had lower content of methoxyl groups or the co-precipitating material devoid of this moiety had diluted its content in the preparation. High carbohydrate content of 47% accounted for the lower value of methoxyl groups in the lignin preparation. Chua and Wayman (1979) reported that absorption spectrum of lignin preparations in UV range (200-400 nm) along with the ionization spectrum, showed very strong and characteristic spectra of aromatic substituted structures. The wheat straw lignin showed peak absorption at 280 nm. However, the absorbance value of 110-190 was reported for various hard wood lignins, which was lower than the bagasse lignin, where the ionization spectrum showed peaks at 250-290 nm. These peaks of phenolic groups, suggested significant substitution of basic phenyl-propane structure in the lignin molecules by ionizable groups such as phenolics, carboxylic, hydroxyl and carbonyl moieties. Alkali soluble lignin from WS was used for studies on lignin degradation by *Monilia* sp. The fast growth was observed for first 5 d (2.2g DW/l) and slow growth continued till 10$^{th}$ day (3.3g DW/l). After an initial degradation of about 19% of lignin in first two days, slow degradation continued till 15$^{th}$ day (40%). This suggested non-growth associated degradation of lignin by the culture.

Yadav *et al.,* (1988) tested the performances of 12 white-rot,1 soft-rot and 6 other fungi for their bio-conversion abilities under SSF and evaluated for the DDMR. The biomass yield and CP content of the moulds grown on malt extract glucose broth was 22.6-215.4 mg/g glucose and 27.9-48.6%, respectively. All the fungi colonized WS very well during SSF resulting in DM losses of 13.0-28.84%. The DDMR of the fermented straw decreased (16.0 to 14.92%) with all fungi on 21-d fermentation at 30°C as compared to initial straw (41.6%). In the case of 3% urea treated WS - SSF, *Coprinus* sp. colonized very well, in both single and double stage. However, their performance with respect of DDMR showed a decline of 36.2-38.0% initial value in single stage and of 28.6-33.4% as compared to initial value of 43.7% after first phase of urea treatment in double stage during 10-d fermentation at 30°C. However, for the white-rot, soft-rot and other fungi was found to show moderate increase in DDMR in single stage SSF of urea treated straw and there was no significant increase in DDMR in double stage fermentation except in *Phanerochaete* fermented straw. They concluded that the fungi colonizing efficiently the urea treated straw decreased their DDMR due to OM losses. Hence efforts should be made to use mutants of fungi lacking cellulase or hemicellulase, so that part of the holocellulose may not be degraded, thus reducing OM losses.

Rai and Saxena (1988) opined that *Pleurotus* sp. colonized and degraded a variety of lignocellulosics to improve the feed value and also, biogas yield from 'mycostraw' found to be almost twice than that from untreated straw. Cultivation of oyster mushrooms was, therefore, one of the easiest biotechnologies for recycling agricultural wastes into feed and fuel. It was the biochemistry of lignin-degradation which remained the 'crux of the problem' and still in a flux. *Pleurotus* spp. were reported to produce extracellular degradative enzymes for degradation of cellulose, hemicellulose and lignin [Arora and Sandhu (1987); Bourbannais and Paice (1988); Hong *et al.,* (1985); Platt *et al.,* (1984); Sharma, (1987)]. It was established that they produce endoglucanase (EC 3.2.1.4) and β–glucosidase (EC 3.2.1.21). Lignin was reported to be degraded at a faster rate than cellulose during mycelial colonisation of lignocelluloses by *Pleurotus* but trend reversed during fruiting (Ginterova and Lazarova; 1987). There were some studies on lignin degrading enzymes, other than laccase and polyphenol oxidase in *Pleurotus* spp. Waldner *et al.,* (1988) compared lignolytic enzyme activity of six selected white-rot fungi including *P. ostreatus* and *P. chysosporium*. Peroxidase production was found to be a common feature of all the tested fungi. *P. chrysosprium* showed higher lignolytic activity than *P. ostreatus*. However, polyclonal antibodies raised against ligninase of *P. chrysosporium* reacted with extracellular peroxidases of *Coriolus versicolor*, but not with those of *P. ostreatus*. It had been suggested that oxidation route was, however, found to be different than that of lignin peroxidase of *P. chrysosporium* and its route in production of $H_2O_2$ during oxidation of lignin fragments. It had been indicated that initiation of mycelial growth was supported by uptake and utilization of soluble substances before the polysaccharides were degraded. Appearance of celluloses on 10$^{th}$ day could be due to depression caused by depletion of the soluble sugars during initial growth. In later stages, the content of free sugar did not rise significantly, indicating commensurate uptake to avoid catabolite repression of cellulases. Sugars were also reported to repress laccase production of *Pleurotus* spp. Gupta and Langar (1988) had also suggested that to reduce execessive substrate O. M. loss, an appropriate stage of specific fungus should be determined if spent-straw was intended to be used as feed. They found that spent-straw at primordium initiation stage (day 27) with minimum D.M. loss and maximum digestibility was best for animal feed. But, sacrificing the feed would be of great economic loss, as any substrate had the potential to produce mushrooms worth Canadian $ 20 per kg. It should be pragmatic and economically justifiable to fortify the spent-substrate to make-up for the loss of nutrient availability.

Gulati (1980), Gulati and Gaur (1988) studied the scale-up of biodegradation process related to growth and metabolism of *Trichoderma reesei* 2M 9414 and enzyme production with acid treated bagasse during different submerged and SSF conditions. The decrease in carbon content (56.2 – 25.6%) and increase in CP (5.5 – 26.3%), soluble protein (2.2 – 18.0) of biomass and enzyme activity (1.25 units / ml) were noticed with different environmental factors and methods. Highest CP (26.3 – 25.6%) and soluble protein (18.0 – 17.7%) were obtained in stirred fermentor with aeration and agitation in submerged method and the improvised fermentor with aeration in SSF method respectively, while the enzyme activity (1.25 units/ml) west par in shaker and improvised fermentor. The lowest yield of the CP (4.8 %), soluble protein (2.10 %) was observed in 500 ml volume of residual biomass or composting of SSF method. The results concluded that the rate of biodegradation of lignocellulosic residue and quality of microbial protein were markedly affected by environmental control and methods of production of biomass.

The SSF process for upgradation of the straw overcame all the limitations of submerged fermentation process. Moreover, it was also simple and waste-free method (Isrilides *et al.,*1978) with amenability and economics for operation at as low capacity as at farm scale (Gould *et al.,* 1989). In a typical process, the chopped straw was mixed with 3 parts of 0.5 N $H_2SO_4$, autoclaved at 121°C for 30 min., fermented with inoculum under standardized conditions, drying and then pelletization. The conditions opted for such processes were of achievable upgradation worked out by many investigators (Chahal *et al.,* 1983; Linden *et al.,* 1983; Zadrzil *et al.,* 1977, 1981, 1983). The technology was found to be feasible in the sense that assured future availability of straw in abundance and remote chances of its competition for direct human food. The combination of milder chemical pre-treatment and subsequent SSF process could also largely overcome the limitations of submerged fermentation process and lead to the product, which was adequately satisfactory in terms of digestibility and nutritive values. The design, cost estimation and feasibility considerations, as worked out for farm and village level plants to process 0.1, 1.2 and 10 ton straw/d, gave valuable data. The farm scale plant to process 0.1 ton straw/d, was not economically feasible. The process was, however, economical and the cost compared well with other animal feeds of similar quality when operated at 1.2 and 10 tons straw/d capacities. A co-operative village level venture was suggested for effective utilization of abundantly available straw and to overcome the perennial shortage of good quality animal feed in the developing countries.

Different co-workers (Langar *et al.,* 1980[a,b], 1986; Gupta *et al.,* 1986, 1988; Gupta and Langar 1983, 1984; Gupta *et al.,* 1984, 1986, 1987) tried modifications in the pre-compost mixture for *Agaricus* sp. and observed the monitoring of chemical and microbiological changes during 30 d indicated that urea could be used as a nitrogen source and 12 d composting with minimum OM and DM losses was sufficient to improve the nutritive value of straw as livestock feed. Subsequent standardization and evaluation showed that stacking urea-straw (3.5: 96.5) at 70% moisture without turning for 9 d provided a straw preparation capable to meet both the maintenance energy and protein requirements of an adult buffalo. Comparative nutritive evaluation of urea-straw stacked for 9 d with 40 to 70% moisture or for 28 d with 45% moisture further revealed that 9 d with 40% moisture provided the next best straw preparation after the one with 70% moisture, having better nutritive value than the one stacked for 28 d. These results indicated that the moisture level and stacking periods due to establishment of different microbial population and fermentation pattern influenced the substrate nutritive value in different ways. At 40 to 50% moisture level the mesophilic actinomycetes bacteria and fungi were present, where as at 70% moisture, their number was negligible after stacking for 9 d. Partial lignolysis with true protein enrichment within 9 d was seen only with 70% moisture, while only partial lignolysis was present after 28 d stacking at 45% moisture associated with high OM loss. The similar plate counts at 40 and 45% moisture level respectively after 9 and 28 d stacking also supported the view that 9 d stacking at 40% moisture level was sufficient for improving the nutritive value of urea-straw. There seemed to be no need to practice prolonged stacking at low moisture levels also. However, from the animal health point, the thermophilic-thermotolerant pattern attained here with only 70% moisture, appeared to be a desirable process than the mesophilic fermentation process established with low moisture (40-50%). The fixed treatment cost/kg available DM of a straw preparation after 9 d stacking with 70% moisture was less than a Canadian $ compared to untreated straw itself explains the importance of applying open stacking as an intermediate biotechnology process for the livestock development through better feeding of crop residue under the conditions of the developing world. Moreover, urea-straw with 70% moisture and 5%(fresh basis) fermented straw (as inoculum) after stacking for only 6 d provided a straw preparation of better nutrient availability with a minimum nutrient loss than 9 d stacking. Therefore, the open composting process could be further modified for improving the nutritive value of cereal crop residues as livestock feed in a shorter period.

Walli *et al.,* (1987[a,b], 1988) reported that WS treated with 3% urea, 1% single super phosphate, 0.1% CaO and 2% fungal culture of *Coprinus fimetarius*, with moisture level maintained at 65% was incubated in polythene bags with holes for 2 weeks till the appearance of white mycelial growth. For the urea treatment, all the ingredients except the fungal culture were used and the straw was placed under polythene cover for one month. The two types of straws, fungal treated (FTS) and urea treated (UTS) were fed to two groups of crossbred calves (4 in each group), supplementing the FTS and UTS groups with 400 and 585 g of Ground Nut Cake (GNC) respectively. The chemical composition revealed a slight increase in CP and hemicellulose content and a decrease in ADF content. The DM intake kg/100 kg body weight was significantly lower in FTS than in UTS. Except for ADF, which was significantly lower in FTS, the variation in the digestibility of other nutrients with respect to two types of straws was insignificant. Although, the N intake through cake was significantly lower on FTS, the N balance was similar on FTS and UTS diet. Fungal treated straw produced by 2 stage process (urea treatment followed by fungal treatment) using *Coprinus* showed a decrease in soluble N and $NH_3$-N and an increase in ADIN, insoluble N other than ADIN and fungal biomass as the fermentation proceeded from day 7. By the 5$^{th}$ d of fermentation, the DM loss was only 7.6 and the fungal biomass was 28.8%. Although, the potential degradability of the straw decreased with the increase in period of fermentation, the effective DM and OM degradability increased progressively.

Bhupinder *et al.,* (1988) studied chemical changes and DM digestion kinetic characterization during the course of fermentation of WS using *Pleurotus sajor caju.* The loss in OM at vegetative stage (VS) was 1.1% while it was 33.4% at harvesting stage (HS; just before first flush). Nitrogen content increased by 61.2% at VS and decreased by 14.8% at HS compared to that in control straw. The loss (on equal ash basis) in NDF, ADF, hemicellulose and cellulose at VS was 5.3, 6.3, 2.8 and 15.2% respectively, while at HS, these values were 35.6, 39.9, 25.2 and 42.9 respectively. The loss in lignin at HS was 18.8%. Digestion kinetic studies revealed a decreased rate constant (c) of DM digestion due to fungal growth being –0.0333 for VS and –0.0246 for the control. A lag (L) period of 4 hr was observed for control straw, whereas fungal treated straw at VS and HS did not show lag period. The rapidly soluble fraction (a) was absent in control straw, while it was 0.9% for VS and 3.6% for HS indicating solubilization of straw constituents due to the treatment. Degradable fraction (b) was 48.6, 42.3 and 37.4% and the effective degradability, p ($p = a+b.c/c+k$) was 31.8, 25.2 and 22.1% for control, VS and HS respectively. Further studies on digestion kinetics of fibre fraction after fungal growth on substrates (autoclaved and chemically sterilized) would provide a better insight into the degradation process.

Rai with other co-researchers (1984[a,b], 1986, 1987, 1989) used crop residues for SSF with various fungi (including *Coprinus* sp). The substrate converted into protein-enriched animal feed with some OM losses (about 15% in general) and other compositional changes *in vitro* or *in sacco* digestibility. Kishen *et al.,* (1989[a,b]) performed SSF of urea treated WS with *Coprinus fimetarius*, but under non-sterile conditions (at ambient temperature) and reported the significant increase in the chitin (8.8 to 24.9 mg/g) and biomass content (10.5 to 13.7 %) and the disappearance of lignin was 11.29%.

## BIOTECHNOLOGICAL PERSPECTIVE: DISCOVERY OF LIGNINASE ENZYME

In 1983, various laboratories announced the discovery of an extracellular $H_2O_2$ requiring enzyme in a white-rot fungus – *Phanerochaete chrysosporium* that catalyzes lignin degradation in the culture growth medium (Tien and Kirk, 1983; Glenn *et al.,* 1983; Higucchi *et al.,* 1983). These researchers had isolated and purified this enzyme, which was able to degrade lignin model compounds, in general and the c-c and ether bond cleavage reactions, in particular. This enzyme, however, was obligatorily depended on the presence of $H_2O_2$ for its activity *in vitro* (Gould *et al.,* 1989). Results from various laboratories had also shown that this enzyme was a peroxidase type metalloprotein. Further studies revealed that ligninase activity in *P. chrysosporium* might be associated with more than one enzyme and isozymes too (Walther *et al.,* 1988). The variation in the enzyme sizes may probably be due to post-translational modifications, strain differences, cultural conditions and age.

The ligninase enzymes had been further characterized and shown to have a molecular weight of 39-43 kDa showing similar catalytic properties. These enzymes, however, were glycoproteins. Immunological reactions and protease digestion patterns had suggested a considerable homology, though they differ in isoelectric points (Tien and Tu, 1987). Crude extractions of ligninase consist of a mixture of both active and inactive enzymes, which could not be separated. However, the enzyme was stable once isolated, though variation in specificity was observed, which probably might be due to cultural conditions and also possibly due to the enzyme mixture (Kirk, 1987).

Molecular Organisation of Ligninase gene(s): Various laboratories had isolated different cDNA and genomic sequences coding for ligninase. Based on these data, it was possible to generalize the molecular organisation of ligninase gene. The mature mRNA was approximately 1200 nucleotides in length having flanking regions of varying lengths on 5'and 3'ends (Walther *et al.,* 1988; Tien and Tu, 1987; Boer *et al.,* 1987). It had a poly (A) sequence at the 3' end similar to eukaryotic mRNAs. The G+C content of the gene was approximately 60% in the coding region(De Boer *et al.,* 1987). However, the noncoding 3' flanking region contained a lower G + C content (45%). Generally, it appeared that 5' untranslated region was shorter than the 3' untranslated region. The coding region started with the usual AUG codon.

The 3' end of the gene had a polyadenylation signal sequence helped in processing the transcript of adding poly(A) tail (De Boer *et al.,* 1987). Nucleotide sequence analysis of various cDNA sequences showed a heterogeniety (Walther *et al.,* 1988; Tien and Tu, 1987; De Boer *et al.,* 1987). This suggested the existence of more than one mRNA for ligninase. Furthermore, Southern blot analysis of digested genome DNA showed multiple bands when probed with cDNA probes. These results indicated the possibility of existence of a family of genes approximately ten that coded for ligninase. However, at present it was not possible to predict the exact number of genes in the ligninase gene family. Comparison between the genomic and cDNA sequences revealed the presence of intervening sequences of 50-60 nucleotides present in the coding region. The number of introns was about eight. The intron exon junctions of ligninase genes conformed to the consensus splice junctions found in other filamentous fungi too.

The coding region coded for ligninase protein of approximately 370 amino acids (Tien and Tu, 1987). However, the final size of ligninase was around 340 amino acids. It had been shown by various investigators that ligninase contained a leader sequence of about 25 amino acids at the amino terminal. Since ligninase was a secretory protein, it was natural to expect such a leader sequence, which played a role in transport across the membrane according to signal hypothesis. The leader sequence contained predominantly hydrophobic amino acids resembling eukaryotic and procaryotic signal peptides. The molecular weight was calculated to be about 37 kDa based on the amino acid composition. However, estimated molecular weight was more than the calculated one indicating post-translational modification, probably glycosylation.

Using $^{14}C$-labelled lignin, lignin model compounds or polymeric dyes (Poly R, Poly B), considerable progress had been made in understanding the microbial physiology associated with lignin degradation and in optimizing conditions *in vitro* under which lignolysis was expressed. Four key points emerged: First, although many organisms could degrade cellulose and hemicellulose, the ability to degrade lignin was very restricted. Secondly, low molecular weight aromatics could be degraded anaerobically, but the degradation of polymeric lignin required oxygen. This meant that approaches to introduce (genetically engineered) lignolytic organisms into rumen of herbivores to enhance straw digestibility, for example, were unlikely to succeed. Thirdly, lignolytic fungi required a second readily oxidisable substrate eg. cellulose or glucose in order to attack lignin. Fourthly, the extracellular fungal ligninase, important in biological lignin degradation was expressed widely amongst white-rot fungi inferring common mechanisms of lignin degradation (Eriksson *et al.,* 1980). Therefore, biotechnological approach to be developed to produce cellulose-less as well as hemicellulose-less mutants which might use only lignin for their growth without touching above carbohydrates (Eriksson *et al.,* 1980).

Searching for the Genes: Once it was evident that lignin degradation was catalyed by ligninase, many laboratories tried to isolate the ligninase genes. Rapid advancement in recombinant DNA technology was exploited not only to isolate the genes but also to over produce the ligninases for commercial use. The crux of the problem lied in the isolation of ligninase genes. By the diligent use of the molecular biological techniques, it might be possible to isolate any gene whose protein product was available. Following techniques were exploited to isolate such genes.

## IMMUNOSCREENING

This was tried as antigen-antibody reaction (Kirk *et al.,* 1978). Antibodies were raised against a protein whose gene was to be identified. Total mRNAs were isolated and full-length cDNAs were cloned in suitable expression vectors and transformed into suitable host such as *E. coli* or yeast. Cells harbouring recombinant DNA were replica plated on to nutrient agar. Following the growth, the cells within the colonies were lysed by exposure to chloroform vapour. The antibody was linked to a solid support such as DEM cellulose filter or PVC support and brought into gentle contact with the lysed cells to allow adsorption of antigen to antibody. This complex was detected by incubating the sheet with the second antibody protein (A) that was radiolabelled. Untreated antibody could be washed away and the position of the antigen-antibody complex could be determined by autoradiography. From the autoradiography, the position of the bacterial cells, which were synthesizing the antigen, on the master petriplate could be identified.

Oligonucleotide probing: Gene in question could also be identified by oligonucleotide sequence corresponding to the amino acid sequence of the protein in question was synthesized (Zhang *et al.,* 1986). The prerequisite was that one should know at least a part of the amino acid sequence of the protein, which was to be identified. The cDNA corresponding to the gene in question was also to be identified. The cDNA or the genomic library was screened using radiolabelled oligonucleotide probe if the mRNA of the gene in question was abundant, the isolation and identification of the clone harbouring cDNA corresponding to the gene in question would not pose any particular problem). From the signals on the autoradiograph, the colonies corresponding to them could be identified. These would harbour the gene that code for the protein in question.

## DIFFERENTIAL SCREENING

Chartrand *et al.,* (1986) and Trudel *et al.* (1986) had suggested this technique to isolate genes that were expressed at different stages of life cycle or environmental conditions such as developmentally regulated genes and heat shock genes. Similarly, lignin-modifying genes were expressed in stationary phase and not in exponentially growing cells. Differently expressed genes could be identified by DNA-hybridization techniques. mRNAs were extracted from exponentially growing and stationary phase cells ( lignin expressing and not expressing ). Both of these mRNAs were used to synthesize different radioactive cDNA-probes. Each probe was hybridized to identical blots from genomic library or cDNA constructed from ligninase expressing cultures. Since, ligninase mRNAs were present only in the induced cells, these genes would be visualised only with the induced probe (and not with the uninduced probe). Therefore, it would be possible to identify ligninase-coding genes using this technique and it was possible to isolate ligninase genes.

*Streptomyces* sp. could withstand a higher temperature, profusely sporulate and form secondary metabolites (Gottschalk, 1979). The pathway and mechanism of lignin by this organism was intensively studied in temperate isolates (Crawford *et al.,* 1981, 1982 and 1984). Kalaichelvan and Ramasamy (1988) studied this organism for its lignolytic ability and tested for its lignin degrading biocatalysts with eleven isolates. One of the isolates showed higher activity. Glucose was the best carbon source and urea as nitrogen source, inhibited the growth of this organism. The growth of this isolate was maximum at pH 6.0. It had secondary metabolic activities, like pigmentation and antibiotic activity. Lignolytic biocatalysts – laccase, ligninase, peroxidase and polyphenol oxidase were produced by this isolate. Its lignolytic character was confirmed by the production of $^{14}CO_2$ from 14C – DeHydrogenated Polymer (DHP) lignin. Its growth on insolubles indicated that its biocatalytic activity could be profitably used for the lignocellulose conversion.

Ramasamy with different co-researchers (1976,1985[a,b]) attempted biodegradation of coir-dust (a recalcitrant lignocellulosic waste) and observed that lignolysis started only after 8 d growth of *Pleurotus* sp. and laccase was growth associated. The evolution $^{14}CO_2$ from $^{14}C$ – lignin (DHP) was very less at initial stages, followed by an increase with increase in incubation period, whereas, as DHP supplemented with glucose, the release of $^{14}CO_2$ was very rapid. Qualitative separation for enzymatic proteins and thereafter, staining for laccase activity with diamino benzidine in the absence of $H_2O_2$, indicated that the activity was pronounced in the earlier colonization period and re-expressed to lesser degree in the final stage of incubation. Assay also indicated the presence of 3 peroxidase positive proteins in the crude cell-extract and their activity varied widely when electrophoretically tested for their mobility. Two proteins were positive for this enzyme as they showed high mobility.

Role of biotechnology in the different areas of pulp and paper manufacture, namely, forestry, pulping, bleaching, waste management etc. had gained importance world over. Constant efforts were needed by technologists and scientists to attain maximum efficiency through adopting the latest innovations and technologies. Eriksson(1985) suggested large silos as treatment chambers, where wood chips and fungal inoculum would be introduced at the top and the treated chips removed from the bottom after a suitable residence time. For hard wood chips, residence of one week or less should be adequate. Silos large enough to accommodate this residence time could be constructed to process 170,000 tons of pulp per year per silo, provided oxygen, moisture and nutrients were maintained.

## FUTURE PROJECTIONS

Lignocellulosics have the tremendous potential of recycling agricultural wastes into food, feed and fuel. Though, good beginning has been made in understanding the processes underlying the degradation of various crop-residues by white-rot fungi, but future-researches should address to the aspects of suitability of species for various substrates and in depth understanding of the enzymology of lignin degradation as well as mode of action. Ideal and optimum interaction between the substrate and white-rot fungi, the kinetic properties of the metabolic processes should also be worked out. This would lead to the production of active ligninase enzyme on commercial scale.

Molecular characterization of the cloned-DNA fragment in plasmid containing a new gene engineered for producing desired mutant is important. Therefore, efforts should be made to use mutants of white-rot fungi lacking in cellulase and /or hemicellulase activities during SSF of ligno-carbohydrate crop residues, so that holocellulose may not be degraded, thus reducing the extent of O.M. loss.

## HYPOTHESES

Penetration of fungal-hyphae may take place in these possible ways: (a) The fungus grows in the central cavity, the lumen of the cell and spreads in the neighbouring cells through pores/perforated zones, or (b) The enzymes of fungus facilitate penetration through the cell wall themselves.

By doing so, (i) either, lignin is removed, or (ii) modified, i.e. de-polymerized, but re-polymerized back!!!

Then, the desired mutants produced as a result of genetic-engineering, would not be stable for too long and might return back to original strain, where from it was originated.

(c) Fungus could also degrade modified-lignin, which leads to the concept of bio-bleaching (in that case, re-polymerization might not be possible).

Palmer *et al.,* (1987) opined that the discovery of ligninase in 1983 signified for many researchers, the end of a long search for the biological catalysts involed in lignin degradation and start of new biotechnological developments based on the enzyme. However, they had been unable to use the enzyme to degrade lignin with any degree of success. Under these circumstances, they might run a real risk of losing sight of the importance of ligninase to lignin breakdown and its potential for implementing either in its natural form, modified or mimicked. Accordingly, they suggested three main areas where, rationalization would contribute greatly to our understanding of how to use the enzyme *in vitro.*

First: How does ligninase attack lignin *in vivo* ? Is the attack made by a mediator, like veratryl alcohol or direct?

Second: Where is ligninase located *in vivo* ? Is it fixed in space, located on the fungal-hyphae or freely diffusing?

Third: How is further polymerization of phenolic compounds by ligninase with its attendant inhibitory effects on the enzyme controlled or avoided?

## ACKNOWLEDGEMENTS

The first author carried out Post Doctoral Research (June, 1[st] – Nov.30[th], 2000) at the Dairy and Swine Res. and Development Centre, Lennoxville(QC) – J1M 1Z3 (Canada) on a "Sir Stapledon's Common-wealth Research Fellowship of England".

## REFERENCES

Arora, D.S. and Sandhu, D.K. (1987). Decomposition of angiospermic wood, saw-dust and laccase production by two *Pleurotus* sp. *Journal of Basic Microbiology,* 27(4), 179-184.

Bakshi, M.P.S., Gupta, V.K. and Langar, P.N. (1985[a]): Acceptability and nutritive evaluation of *Pleurotus* harvested spent wheat straw in buffaloes. *Agricultural Wastes,* 13, 51-58.

Bakshi, M.P.S. and Langar, P.N. (1985[b]): Utilization of Agaricus biosporous harvested spent-straw in buffaloes. *Indian Journal of Animal Sciences,* 55, 1060-63.

Bakshi, M.P.S.; Gupta, V.K. and Langar, P.N. (1986). Fermented straw as complete basal ration for ruminants. *Agricultural Wastes,* 16, 37-46.

Bakshi, M.P.S., Gupta, V.K. and Langar, P.N.(1987): Effect of moisture level on the chemical composition and nutritive value of fermented straw. *Biological Wastes,* 21, 283-89.

Bakshi, M.P.S. and Langar, P.N. (1989). Effect of inoculum on the nutritive value of fermented straw. *Indian Journal of Animal Sciences,* 59, 846-50.

Beckmann, E. (1923). In: *Fodder from straw;* US Patent 1,442,321.

Bhupinder, S., Negi, S.S. and Makkar, H.P.S. (1988). Balance of various nurtrients and digestion kinetics of wheat straw at different stages of *Pleurotus* growth; (I): Rumen degradability of N in typical cultivated grasses and leguminous fodders. *Animal Feed Science and Technology,* 22, 79-89.

Bondy, A.H. and Meyer, J. (1948). Lignin in young plants. *Biochemical Journal,* 43, 248

Bourbannois, R. and Paice, M.G. (1988). Veratryl alcohol oxidase from the lignin degradation of basidiomycete *Pleurotus sajor caju. Biochemical Journal,* 255, 455-60.

Chahal, D.S., Moo-young, M. and Dhillon, G.S. (1979). Bioconversion of wheat straw and wheat straw components into single-cell protein. *Canadian Journal of Microbiology,* 25, 793-97.

Chahal, D.S., Valch, D. and Moo-Young, (1983). Upgrading the protein feed value of lignocellulosic materials using *Chaetomium cellulolyticum* in SSF. *Advances in Biotechnology,* 2, 327-32.

Chartland, P., Trudel, P., Courchesne, D. and Roy, C. (1986). In: *Search of lignin modifying genes.* Eds Moo-Young, Hasnain, S. and Lamptey, J. Elsevier Applied Science (Publishers), Amsterdam. Pp 112-117.

Chua, M.G.S. and Wayman, M. (1979). Characterization of autohydrolysis of aspen (*Pleurotus tremuloides*) lignins. *Canadian Journal of Chemistry,* 57, 2603-11.

Cochrane, V.W. (1958). In: *Physiology of fungi,* John Wiley and sons, Inc., New York. Pp 524.

Cooney, D.G. and Emerson, R. (1964). In: *Thermophilic Fungi,* Ed Freeman, W. H., Co., San Fransisco, Pp 188.

Crawford, R. (1981). *Lignin biodegradation and transformation.* John Wiley and Sons, Inc., New York. Pp 154.

Crawford, D.L., Barber, M.J., Pometto III, A.L. and Crawford, R.L. (1982). Chemistry of softwood lignin degradation by *Streptomyces virdosporous. Archives of Microbiology,* 131, 140-45.

Crawford, D.L. and Pometto III, A.L. (1984). Production of useful modified lignin polymers by bioconversion of lignocellulose with *Streptomyces. Biotechnology Advances,* 2, 217-32.

De Boer, A.A., Zhang, Y.Z., Collins, C. and Reddy, C. A. (1987). Analysis of nucleotide sequences of two ligninase cDNAs from a white-rot filamentous fungus *Phanerochaeto chrysosporium. Gene,* 60, 93-102.

De Stevens, G. and Nord, F.F. (1951). Investigation on lignin and lignification VIII. Isolation and characterization of bagasse native lignin. *Journal of American Chemistry Society,* 73, 4622-27.

Emerson, R. (1968). In: *The Fungi,* Eds Ainsworth, G. C. and Sussman, A. S. Academic Press, New-York, vol.3. Pp 106-28.

Eriksson, K.E. (1980). Production of $H_2O_2$ in *Phanerochaete chrysosporium* during lignin degradation. *Phil. Trans. R. Soc.,* London, A-321. Pp 455-50.

Eriksson, K.E. (1985). Swedish developments in biotechnology related to the pulp and paper industry; *TAPPI J.,* 68(7), Pp 46-55.

Farmer, R.H. (1967). In: *Chemistry in the utilization of wood,* Pergamon Press, New York, Pp 193.

Farmer, V.C., Henderson, M. E. K. and Rusell, J. D. (1960). Aromatic alcohol oxidase activity in the growth medium of *Polystictus versicolor. Biochemical Journal,* 74, 257-62.

Ginterova, A. and Lzarova, A. (1987). Degradation dynamics of lignocellulose materials by wood-rotting *Pleurotus* fungi. *Folia Microbiology,* 32, 434-37.

Glenn, J.K., Morgan, M.A., Mayfield, M.B., Kuwahara, M. and Gould, M.H. (1983). An extracellular $H_2O_2$ requiring enzyme preparation involved in lignin biodegradation by white-rot basidiomycetes. *B. B. Res. Comm.* 114(3), 1077-83.

Gottschal, K. (1979). Microbial metabolism and significance of glucose oxidase in lignin degradation. *Nature,* 268, 78-80.

Gould, J.M., Jasberg, B.K., Fahey, G.C. and Berger, L.L. (1989). Treatment of wheat straw with alkaline $H_2O_2$ in a modified extruder. *Biotech. Bioeng.,* 33, 233-236.

Gulati, S.L. (1980). Improvement in production of fungal protein by SSF. Zbl.Bkl.II; Ab. I, 135, 413-17.

Gulati, S.L and Gaur, A.C. (1988). Enzymatic hydrolysis of cellulose from agricultural and industrial wastes. *Biological Wastes,* 25, 309-13.

Gupta, B.N., Walli, T.K., Flagel, T.W. and Singh, K.(1986). Fungal treatment of wheat straw and its utilization as a sole feed for goats; In: *Second symposium on Animal Nutrition,* Animal Nutrition Society, Udaipur, India, Pp 43.

Gupta, B.N., Rai, S.N., Walli, T.K. and Singh, K. (1988). Evaluation of fungal treated rice-straw at different fermentation periods in relation to D. M. I. and nutritive utilization in goats. *Indian Journal of Animal Nutrition,* 5(4), 273-79.

Gupta, V.K. and Langar, P.N. (1983). Bioconversion of urea for enrichment of wheat straw. *Indian Journal of Animal Nutrition,* 53, 1324-27.

Gupta, V.K. and Langar, P.N. (1984). Fermentation for enrichment of wheat straw as livestock feed. *J. Res.,* P.A.U., India, 21, 141-42.

Gupta, V.K., Bakshi, M.P.S. and Langar, P.N. (1984). Use of natural inoculum for enrichment of wheat straw urea mixture by fermentation. *Indian Journal of Animal Nutrition,* 1, 41-44.

Gupta, V.K., Prasad, K.S., Bakshi, M.P.S. and Langar, P.N. (1986). Improving nutritive value of G.N. shells through fungal cultures. *Agril.Wastes,* 16, 161-169.

Gupta, V.K., Bakshi, M.P.S. and Langar, P.N. (1987). Microbiological changes during natural fermentation of urea-wheat straw. *Biological Wastes,* 21, 291-299.

Gupta, V.K. and Langar, P.N. (1988). *Pleurotus florida* for upgrading the nutritive value of wheat straw. *Biological Wastes,* 23, 57-64.

Higucchi, T., Cheng, H.M. and Kirk, J.K. (Eds.,1983); In: *Recent advances in lignin biodegradation research,* Tokyo University Pp 279.

Hong, J.S., Lee, J.B., Koh, M.S., Imm, J.S., Lee, K.R. and Kim, Mc. (1985). Studies on cellulolytic enzymes produced by *Pleurotus* sp. in synthetic medium: Effect of C and N sources. *Korean J. Mycol.,* 13, 213-220.

Hungate, R.E. (1957). Microorganisms in the cattle-rumen, fed a constant ration, *Canadian Journal of Microbiology,* 3, 389-91.

Ishikawa, H., Schubert, W.J. and Nord, F.F. (1963). Investigations on lignin and lignification XXVII. The enzymatic degradation of softwood lignin by white-rot fungi. *Archives of Biochemistry and Biophysics,* 100, 131-139.

Isrilides, C.J., Grant, G.A. and Khan, Y.W. (1978). Sugar level, fermentability and acceptability of straw treated with different acids. *Applied Environmental and Microbiology,* 36, 43-46.

Jain, M.K.; Kapoor, K.K.and Misra, M.M.(1979; Cellulase activity, degradation of cellulose and lignin, and humus formation by thermophilic fungi. *Transactions of the Brit. Mycol. Soc.,* 73, 85-89.

Kalaichevan, G. and Ramasamy, K. (1988). Lignolysis of pine needles for *Streptomyces* sp. In: *Assoc. Microbiol. India 28th Annual Conference,* (Golden-Jubilee year), Pune, India (Jan.16-18).

Kamra, D.N. and Zadrazil, F. (1986). Influence of gaseous phase, light and substrate pre-treatment on fruit body formation, lignin degradation and in vitro digestibility of wheat straw fermented with *Pleurotus* sp. *Agric. Wastes,* 18, 1-17.

Kamra, D.N. and Zadrazil, F. (1987). Factors affecting microbiological improvement of lignocellulosics for animal feeding; In: *Biological, Chemical and Physical treatment of fibrous crop residues for use as animal feed,* Eds Singh, K., Flagel T. W. and Schiere J. B. Indian Council of Agricultural Research (ICAR – publication), New Delhi.

Kirk, T.K. and More, W.E. (1972). Removing lignin from wood with white-rot fungi and digestibility of resulting wood. *Wood Fibre,* 4, 72-9.

Kirk, T.K., Connors, W.J., Bleam, R.D., Jackett, W.F. and Zeikus, J.G. (1975). Preparation and microbial decomposition of synthetic $^{14}C$-lignins. *Proceedings of National Academy of Sciences USA,* 72, 2515-2519.

Kirk, T.K., Schultz, F., Connors, W.J., Lorenz, L.F. and Zeikus, J.G. (1978). Influence of culture parameters on lignin metabolism by *Phanerochaete chrysosporium. Archives of Microbiology,* 117, 277-285.

Kirk, T.K. (1987). Enzymatic composition: The microbial degradation of lignin. *Annual Review of Microbiology,* 41, 465-505.

Kishen S., Neelkantan, S.N., Singh, G.P. and Gupta, B.N. (1989[a]). SSF of urea-ammo. treated wheat and paddy straws with *Coprinus fimetarius. Indian Journal of Microbiology,* 29, 371-376.

Kishen S., Neelkantan, S.N. and Han, Y.W. (1989[b]). Biochemical profiles of solid state fermented wheat straw with *Coprinus fimetarius.* Indian Journal of Animal Sciences (published), discussion cited In: *Proceedings of National Seminar on Biotechnology of lignin degradation,* IVRI, Izatnagar, India (20-21 Dec., 1989).

Kuhad, R.C., Kothamasi, D., Tripathi, K.K. and Singh, A. (2004). Diversity and functions of soil microflora in development of plants. In: *Plant surface microbiology,* Eds Verma, A., Abbott, I., Werner, D. and Hampp, R. Springer (Publication), Germany. Pp 71-98.

Kutzin, S.F., DeBraun, R.M. and Nord, F.F. (1951). Investigations on lignin and lignification VI. The comparative evaluation of native lignins. *Journal Am. Chem. Soc.* 73, 4615-4618.

Kutzin, S.F. and Nord, F.F. (1951). Investigations on lignin and lignification, VII. Characterization of enzymatically liberated hardwood lignins. *J. Am. Chem. Soc.,* 73, 4619-4621.

Langar, P.N., Sehgal, J.P., Rana, V.K. and Garcha, H.S. (1980[a]). Chemical changes in wheat and paddy straws after fungal cultivations. *Indian Journal of Animal Sciences,* 50, 942-946.

Langar, P.N., Sehgal, J.P., Rana, V.K. and Garcha, H.S. (1980[b]). Utilization of fungal treated spent-straw in ruminant diets. In: *Proceedings of Recycling Residues of Agro- industry symposium,* Ed Kalra M. S. Pp 37-49.

Langar, P.N. and Bakshi, M.P.S. (1986). Evaluation of *Volvariella* harvested spent-straw. *Agricultural Wastes,* 15, 309-311.

Lawson, L.R. Jr. and Still, C.N. (1957). The biological decomposition of lignin literature survey. *TAPPI J.,* 40(9), 56A – 58A.

Linden, J.C., Murphy, V.G. and Moreira, A.R. (1983). Wheat straw auto-hydrolysis. *Advanced Biotechnology,* 2, 41 – 46.

Palmer, J.M., Harvey, P.J. and Schoemaker, H.E. (1987). In: *The role of peroxidases radical cations and oxygen in the degradation of lignin. Phil. Trans. R. Soc.* London A-321, Pp 495-505.

Platt, M.W., Haldar, Y. and Chat, I.(1984). Fungal activities involved in lignocellulose degradation by *Pleurotus* sp. *Applied Microbiology and Biotechnology,* 20, 150-154.

Raghu, K. and Mac Rae, I.C. (1966). Biodegradation of isomer of benzene hexachloride (BHC) in submerged soils. *Science,* 154, 263-264.

Raghu, K. and MacRae, I.C. (1967). The effect of the isomer of benzene hexachloride (BHC) upon the microflora of submerged rice-soils, II ; Effect upon the nitrogen mineralization, fixation and selected bacteria. *Canadian Journal of Microbiology,* 13, 621-34.

Rai, R.D. and Saxena, S. (1988). How protein-rich are mushrooms. *Indian Journal of Horticulture,* 33(2), 2-3.

Rai, S.N. and Mudgal, V.D. (1984[a]). Utilization of poor quality roughages, 2: Enzymatic treatment of wheat straw. *Asian Journal of Dairy Resources,* 3, 193-200.

Rai, S.N., Gulati, S.L. and Mudgal, V.D. (1984[b]). Use of fungi for protein enriched bagasse as an animal feed. In: *National symposium on feeding system for maximising livestock production,* Hisar Agric. Univ., Hisar, India (Oct.29 – 31), Pp 37.

Rai, S.N., Gulati, S.L. and Mudgal, V.D. (1986). SSF of sugarcane bagasse with *Trichoderma reesei* Q.M. 9414 for animal feed. In: *Symposium of Animal Nutrition Society of India,* M. L. Sukhadia University, Udaipur, India (July 14 - 17).

Rai, S.N., Walli, T.K. and Gupta, B.N. (1987). Investigation on fungal treatment of rice-straw and its evaluation as sole feed for crossbread goats, In: *Biological, Chemical and physical treatment of fibrous crop residues for use as animal feed.,* I.C.A.R., New Delhi, India.

Rai, S.N., Walli, T.K. and Gupta, B.N. (1989). The chemical composition and nutritive value of rice straw after treatment with urea or *Coprinus fimetarious* in a SSF system; *Anim. Feed Sci. Technol.,* 26, 81-92.

Raman and Naik, D. (1990). Influence of two levels of N and S on the growth and lignolytic ability of *Pleurotus ostreatus* on wheat and paddy straws. *Indian Journal of Animal Nutrition,* 7(1), 71-74.

Raman and Naik, D. (1991). Influence of incubation period on the growth pattern of *Coprinus* sp. on the wheat and paddy straws at 35° C. *Indian Journal of Animal Sciences,* 61(5), 567-568.

Raman and Naik, D. (1992). Influence of moisture and incubation period on the quality of wheat and paddy straws colonized by *Pleurotus ostreatus. Indian Journal of Animal Nutrition,* 9(1), 21-25.

Raman and Naik, D. (1993). Evaluation of biological fermentation of straws. *Indian Journal of Animal Sciences,* 63(12), 1308-1309.

Raman (1995[a]). Compositional changes during the fermentation of buffered straws with *Pleurotus ostreatus* and *Sporotrichum thermophile. Indian Journal of Animal Sciences,* 65(6), 59-61.

Raman (1995[b]). Influence of heat and inoculation with *Pleurotus ostreatus* on degradation of wheat straw. *Indian Journal of Animal Sciences,* 12(1), 35-36.

Raman (1997[a]). Influence of wheat bran, limestone and inoculation with *Pleurotus ostreatus* on the degradation of wheat straw, In: *National symposium on feeding strategies for eco-friendly animal production in India,* Animal Nutrition Society, Ind. Vet. Res. Ind. (IVRI), Izatnagar, India (Feb.13-14).

Raman (1997[b]). Influence of wheat bran, calcium nitrate or mixer of limestone and ammonium nitrate with *Pleurotus ostreatus* on the degradation of wheat straw. In: *National symposium on microbial technologies for environmental management and resource recovery,* Deptt. Microbiol., University of Delhi South campus, New-Delhi (Oct.12-17).

Raman (1997[c]). Effect of rice bran, calcium nitrate and inoculation with Pleurotus ostreatus on the degradation of paddy straw, In: *38[th] Annual meeting of association of microbiologists of India and conf. on microbes in sustainable development,* Deptt. Microbiol., University of Delhi South Campus, New Delhi, India, (Nov.12-14).

Raman (1998). Influence of two levels of anhydrous ammonia with *Pleurotus ostreatus* on the degradation of wheat straw. In: *Golden jubilee national symposium on feeding strategies for ruminants,* IVRI – regional station, Palampur, India (June 19-20).

Raman (1999[b]). Influence mineralization with *Pleurotus ostreatus* on the degradation of wheat straw. In: International conference on sustainable animal production, *Animal Science,* Hisar Agric. Univ., Hisar, India (Nov.24-27).

Raman and Petit, H. V. (2000). Colonization of different organic residues by *Pleurotus ostreatus* in the early phase. In: *National conference on Biotechnology,* Deptt. Biotech., G. B. Pant Agric. Univ., Pant Nagar-263145, India (Apr.27 – 29).

Raman Rao (2001[a]); Solid-state fermentation of some organic residues by *P. ostreatus*; *Indian Journal of Animal Sciences* 71 (11), 1082.

Rao, R. and Kuhad, R.C. (2001[b]). SSF of composted wheat straw with *P. ostreatus* in early phase; In: *Oral presentation delivered in the technical session V, National Symposium on Lignocellulose and Biotech.,* Deptt. Microbiol., University of Delhi South Campus, New Delhi (Dec. 10-11).

Rao R. and Petit, H.V. (2002). Bio-degradation of cereal crop residues by white-rot fungi; In: *Oral presentation delivered in the technical session VIII, IV[th]* Biennial conference of Animal Nutrition, W. B. University of Animal and Fishery Sciences, Calcutta (Nov. 20-22).

Ramasamy, K. and Kandaswamy, T.K. (1976). Substrate colonization and degradation by Oyster mushroom, In: *Mushroom in India,* (Today and Tomorrow publication), New Delhi, Pp 528-536.

Ramasamy, K. and Reddy, C.A.(1985[a]); Glucose oxidase in lignin degradation, In: *85[th] American Society for Microbiology Meeting,* 87 K, pp.74.

Ramasamy, P.P., Sreeramulu, U. S. and Kothandaraman, G.V. (1985[b]). Efficient utilization of coir pith in farms. In: *Workshop on Coir Research,* T. N. Agric. Univ., Coimbatore, India, Pp 55-63.

Reid, I.D. (1979). The influence of nutrient balance on lignin degradation by the white-rot fungus – *Phanerochaete chrysosporium. Canadian Journal of Botany,* 57, 2050-58.

Schubert, W.J. and Nord, F.F. (1950). Investigation on lignin and lignification. (I) Studies on soft-wood lignification. *J. Am. Chem. Soc.* 72, 977-81.

Sharma, H.S.S. (1987). Comparative study of the degradation of flax by strains of *Pleurotus. Applied Microbiology and Biotechnology,* 25, 542-46.

Sharma, K.K.; Kapoor, M. and Kuhad, R.C. (2005). In vivo enzymatic digestion (IVED), in vitro Xylanase digestion (IVXD), metabolic analogues, surfactants and polyethylene glycol ameliorate laccase production from *Ganoderma* sp. kk- 02. *Letters in Applied Microbiology,* 56, 23-31.

Sharma, H.S.S. (1989). Economic importance of thermophilic fungi. *Applied Microbiology and Biotechnology,* 31, 1-10.

Shivappashetty, K. and Krishnamuthy, V. (1981). Possibility of protein enrichment of animal feed by *Pleurotus sajor caju.* In: *Recycling residues of agriculture and industry,* Eds Kalra, M. S., Bico printers, Ludhiana, India. Pp 363-367.

Singh, R. and Singh, D.P. (1989). Characterization of wheat straw lignin and its degradation in synthetic medium by *Monilia* sp. In: *Proceedings of National Seminar on Biotechnology of lignin degradation,* IVRI, Izatnagar, India. Pp 31-39.

Tien, M. and Kirk, T.K. (1983). Lignin degrading enzyme from hymenomycetic *Phanerochaete chrysosporium. Science,* 221, 661-663.

Tien, M. and Tu, D. (1987). Cloning and sequencing of a c-DNA for ligninase from *Phanerochaete chrysosporium. Nature,* 326, 522-523.

Trudel, P., Courcnesne, D., Roy, C. and Chartrand, P. (1986). Cloning of *Trametes versicolor* genes induced by N-starvation. *Applied Environmental and Microbiology,* 54, 1550-1556.

Vasdev, K.; Dhawan, S.; Kapoor, K.R. and Kuhad, R.C. (2005). Biochemical characterization and molecular evidence of a laccase from the birds' nest fungus *Cyathus bulleri. Fungal Genetics Biology,* 23, 20-30.

Waksman, S.A. and Cordon, T.C. (1939). *Soil Science,* 47, 83-113 (cited) In: Wakasman (1957).

Waksman, S.A. (1957). In: *Soil Microbiology,* John Wiley and sons Inc., New-York, Pp 356.

Waldner, R., Leisola, M.S.A. and Fiechter, A. (1988). Comparison of lignolytic activities of selected white-rot fungi. *Applied Microbiology and Biotechnology,* 29, 400-407.

Walli, T.K., Gupta, B.N. and Rai, S.N. (1987[a]). Effect of feeding fungal treated wheat straw to crossbread calves on DMI and nutrient utilization. In: *Indo-Dutch workshop on Bioconversion of crop residues,* New Delhi, India, Pp 28.

Walli, T.K., Gupta, B.N. and Rai, S.N. (1987[b]). Feeding trials with urea treated and fungal treated wheat straw. In: *Biological, chemical and physical treatment of fibrous crop residues for use as animal feed.* Eds Singh, K., T.W. Flagel and J.B. Shiere (ICAR-publication), India, Pp 103-111.

Walli, T.K., Orskov, E.R. and Bhargava, P.K. (1988). Rumen degradation of straw, III: Botanical fractions of two rice straw varieties and the effect of $NH_3$-treatment. *Animal Production,* 46(3), 346-352.

Walther, I., Kalin, M., Reiser, J., Suter, F., Fritsche, B., Salo heimo, M., Leisola, M., Teei, T., Knowles, J.K.C. and Feitchter, A. (1988). Molecular analysis of *Phanerochaete chrysosporium* lignin peroxidase gene. *Gene,* 61 (cited).

Yadav, J.S., Reddy, P.P. and Neelkantan, S. (1988). Lignolytic activity of fungi on wheat straw lignin. *Indian Journal of Microbiology,* 28, 133-35.

Zadrazil, F. (1974). The ecology and industrial production of *Pleurotus ostreatus, Pleurotus florida, Pleurotus cornupiae* and *Pleurotus enzngiv.* In: Mushroom sci. (part I), *Proceedings of 9[th] international sci. cong. on the cultivation of edible fungi,* (editor- Kisaku Mordi), Tokyo, Pp 621-652.

Zadrazil, F. (1975). Die Zersetzung des Stroh-Zellulose-Lignin-Komplexes mit Pleurotus florida und dessen Nutzung. *Z. Pflanzenernähr Bodenk,* 138, 263-278.

Zadrazil, F. (1977). The conversion of straw into feed by Basidiomycetes. *European Journal of Applied Microbiology and Biotechnology,* 4,273-281.

Zadrazil, F. (1978). Unwandlung von Pflanzenabfall in Tierfutter durch höhere Pilze. *Mushroom Science,* 10, 231-241.

Zadrazil, F. (1979). Screening of basidiomycetes for optimal utilization of straw, production of fruiting bodies and feed. In: *Straw decay and its effect on disposal and utilization,* Ed Grobard E., John Wiles, New York, Pp 139-146.

Zadrazil, F. and Brunnert, H. (1980[a]); The influence of ammonium nitrate supplementation on degradation and in vitro digestibility of straw colonized by higher fungi. *European Journal of Applied Microbiology and Biotechnology,* 9, 37-44.

Zadrazil, F. (1980[b]). Conversion of different plant wastes into feed by basidiomycetes. *European Journal of Applied Microbiology and Biotechnology,* 9, 243-48.

Zadrazil, F.and Brunnert, H.(1981). Investigation of physical parameters important for the SSF of straw. 11, 183.

Zadrazil, F. and Kurtzmann Jr., R.H. (1982[a]). The biology of *Pleurotus* cultivation in the tropics. In: *Tropical mushrooms – biological nature and cultivation methods,* Chinese Univ. press, Hong Kong, pp. 277-298.

Zadrazil, F. and Brunnert, H. (1982[b]). SSF of lignocellulose containing plant residue with *Sporotrichum pulveruentum* and *Dichomitus squalens. European Journal of Applied Microbiology and Biotechnology,* 16, 45-51.

Zadrazil, F., Grinbergs, J. and Gonzaler, A. (1982[c]). Palo-podrido decomposed wood, which was used as feed. *European Journal of Applied Microbiology and Biotechnology,* 15, 167-171.

Zadrazil, F. (1983). Factors determining lignin decomposition and in vitro digestibility of lignocellulose during SSF. *Advanced Biotechnology,* 2, 369-373.

Zadrazil, F. (1984). Microbial conversion of lignocellulose into feed. In: *Straw and other by-products as feed,* Eds Sundstol F. and Owen E. publishers- Elsevier publications, Amsterdam. Pp 276-291.

Zadrazil, F. (1985). In: *Angew. Botanik.* 59, 433.

Zadrazil, F. and Reiniger, P. (1988). In: *Treatment of lignocellulosics with white-rot fungi,* Elsevier Applied Science, London and New York, Pp 56.

Zhang, Y.Z., Zylstra, G.J., Olsen, R.H. and Reddy, C.A. (1986). Identification of c-DNA clones for ligninase from probes. *Biophy. Biochem. Res. Commun.,* 137, 649-56.

# 22

# Biotechnological Production of Xylitol from Hemicellulosic Materials

**R. Sreenivas Rao, Ch. Pavana Jyothi and L. Venkateswar Rao**

*Department of Microbiology, Osmania University, Hyderabad - 500 007, India*

Hemicellulose is the third most abundant polymer in nature. In some plants, it comprises upto 40% of the total dry material. The use of dilute acids to catalyze the hydrolysis of hemicellulose to its sugar constituents is well-known and effective. These sugars, of which D-xylose is the most abundant, can be converted to xylitol by yeasts. The conversion of hemicellulose for the production of xylitol is being studied intensively with a view to develop a technically and economically viable process. However, a major problem associated with acid pretreatment is the poor fermentability of the produced hydrolysate as a result of the presence of inhibitory compounds. In this paper, various substrates and various pretreatment options for xylitol production from hemicellulosic materials is reviewed. Biotechnological production of xylitol could be of economic interest and attractive, as crude hemicellulosic hydrolysates can be used as potential substrates, instead of pure xylose, to reduce the cost of production.

## INTRODUCTION

Xylitol is one of the most expensive polyol sweeteners and has specific health claims in the world market. It is suitable for diabetes, recommended for oral health and paranteral nutrition (Makinen, 2000; Sreenivas *et al.,* 2004). On an industrial scale, xylitol is currently produced through chemical reduction of xylose derived from birchwood chips and sugarcane baggase hemicellulose hydrolysate.

Xylitol price has decreased substantially over the last ten years (Table 1). This was due to of economies in the scale of production. The raw material for production is hemicellulose, which is abundantly available, with the main industrial application in food and pharmaceutical sector.

**Table 1. Price trends ($/kg) for xylitol, through to 2007.**

| *1995* | *2000* | *2003* | *2005* | *2007* |
|---|---|---|---|---|
| 7.2 | 4.5 | 3.8 | 3.5 | 3.3 |

Microbial production is lately becoming more attractive since the downstream processing is expected to be cheaper. Only certain groups of microbial species are known to metabolize the xylose as carbon source. Some yeast has ability to utilize and convert xylose to xylitol. In the last two decades a lot of effort had been put on investigating the fermentation of xylose by different microorganisms. A European project called XYLOPHONE has goals to the development of xylo-oligosaccharides and xylitol for use in pharmaceutical and food applications. USDA (United States Department of Agriculture) began to brighten prospects for making xylitol in the United States from corn nearly 10 years ago. The research is part of new uses, quality, and marketability of plant and animal products. USDA and ZuChem received $900000 fund from BRDC for commercialization of xylitol production from agricultural residues using biotechnological process. Recently funding agencies in India started encouraging research investigations on xylitol production process development using hemicellulosic material via biotechnological route. The main reason is that the xylose is one of the most abundant sugar monomers in lignocellulosic biomass. The possibility of large scale xylitol production from this sugar is therefore of great economic interest.

Several investigators have reported xylitol production using different yeast strains (Table 2). Some yeasts convert xylose to xylitol particularly species of *Pichia* and *Candida.* Naturally occurring strains have been reported to produce xylitol under aerobic and semi-anaerobic conditions upto 0.56-0.72g / g of xylose as a substrate.

**Table 2. Xylitol producing yeast strains**

| *Xylitol producing yeast sps.* | *References* |
|---|---|
| *Candida boidini* | Saha and Bothast, 1997 |
| *Candida peltata* | Saha and Bothast, 1999 |
| *Candida tropicalis* HXP 2 | Gong *et al.*, 1981 |
| *Debaromyces hansenii* CCMI 941 | Carvalheiro *et al.*, 2004 |
| *Candida sp.* 11-2 | Dominguez *et al.*, 1997 |
| *Candida mogii* ATCC 18364 | Mancilha and Karim, 2003 |
| *Candida parapsilosis* | Kim *et al.*, 1999 |
| *Debaromyces hansenii* | Cruz *et al.*, 2000 |
| *Debaromyces hansenii* UFV-170 | Sampaio et al 2005 |
| *Candida guilliermondii* FTI 20037 | Silva *et al.*, 1998 |
| *Pichia guilliermondii* NRRL Y-12723 | Leathers and Gupta 1997 |
| *Candida magnoliae* | Tada *et al.*, 2004 |
| *Candida sp.* (mutant) | Dominguez *et al.*, 2000 |
| *Candida tropicalis* | Sreenivas *et al.*, 2004 |
| *Candida tropicalis* OMV5 | Sreenivas *et al.*, 2006 |
| *Debaromyces hansenii* NRRL Y-7426 | Parajo *et al.*, 1996 |
| *Candida mogii ATCC* 18364 | Rizzi *et al.*, 1992 |
| *Pichia stipitis* NRRL Y-7124 | Delgenes *et al.*, 1990 |
| *Candida shehatae* NRRL Y- 12858 | Kuzmanova *et al.*, 1993 |
| *Pachysolen tannophilus* | Converti *et al.*, 1999. |

The key enzymes present in yeasts are D-xylose reductase which using either NADH or NADPH, reduce D-xylose to xylitol, and NAD-linked xylitol dehydrogenase which reoxidizes xylitol to D-xylulose. Xylitol accumulation in yeasts is sensitive to environmental conditions such as nutrition, temperature, pH, inoculum, substrate and aeration, the last two being critical for yeast growth and fermentation. Hemicellulosic hydrolysates derived from hardwood and particularly from agricultural residues, such as sugarcane bagasse, corn cobs, corn fiber, wheat and rice straw, are used as feedstock for xylitol production. Due to the presence of inhibitory components, some of the hydrolysates have to be treated prior to microbial utilization. The most investigated types of processes have been batch ones, although fed-batch and immobilized systems have been used. Apart from the naturally occurring yeasts, mutant strains, recombinant strains of *Saccharomyces cerevisiae* in free and immobilized form were also investigated for xylitol production.

Lignocellulosics are organic materials, which are abundant and renewable. Their major components, cellulose, hemicellulose and lignin, vary with plant species. The pentose fraction composed of D-xylose and L-arabinose is much higher in hardwoods than in softwoods. High amounts of pentosans up to 40% are also present in agricultural residues (Singh & Mishra, 1995). The fact that xylan is more easily hydrolyzed than cellulose offers the technical possibility of D-xylose and xylitol production (Jeffries, 1983).

Hemicellulose is a plant cell wall polysaccharide and the third most abundant polymer in nature. In some plants, it comprises upto 40% of the total dry material. Hemicelluloses are short branched chain heteropolysacharides of mixed hexosans and pentosans that are easily hydrolyzed. The most common form of hemicellulose is xylose polymer (xylan). As hemicellulose is abundant in nature and renewable, extensive research has been undertaken to convert hemicellulose-derived carbohydrates, particularly xylose, into useful products. Hydrolysis of hemicellulose yields glucose, D-xylose, L-arabinose and other minor sugars.

The chemical process is expensive because of the high temperature, pressure required for hydrogenation of xylose, and the xylose separation and purification before hydrogenation. A current industrial example is used by Roquette and Danisco companies are involved in the production of xylitol by chemical route. Availability and cost of production are the obstacles for the increased use of xylitol. Biotechnological production of xylitol could be of economic interest and attractive, as crude hemicellulosic hydrolysates can be used as potential substrates, instead of pure xylose, to reduce the cost of production (Sreenivas *et al.,* 2005).

During acid hydrolysis, xylose is degraded rapidly to furfural and other condensation byproducts. These degradation products are inhibitory to microorganisms. The inhibitory effect of different compounds like furfural, 5-hydroxymethyl furfural (HMF), acetate, hydroxybenzaldehyde (HBA), siringaldedyde (SGA) and vanillin on yeast growth is well documented. (Tsao *et al.,* 1999 ).

Various approaches are being considered to remove fermentation inhibitors or minimize their formation, CaO treatment (Silva *et al.,* 1998), use of activated charcoal (Pandey *et al.,* 2000), over liming (Dien *et al.,* 2000), ion exchange resins ( Mancilha and Karim, 2003; Sreenivas *et al.,* 2005), $Ca(OH)_2$ (Dien *et al.,* 2000), solvent extraction (Cruz *et al.,* 1999), intracellular acidification (Lohmeier-Vogel *et al.,* 1998), yeast strain variation ( Martin and Jönsson, 2003), laccase enzyme ( Cassland and Jönsson, 1999), recombinant strains (Dein *et al.,* 1997) and adaptation of the microbial strains (Rivas *et al.,* 2003; Sreenivas *et al.,* 2005) have been reported.

Many studies have been conducted utilizing the hemicellulose portion of agricultural residues like *Eucalyptus grandis* (Silva *et al.,* 1998), rice straw (Roberto *et al.,* 1994 ), aspen wood hemicellulosic hydrolysate (Preziosi-Belloy *et al.,* 2000), barley bran (Cruz *et al.,* 2000), hybrid polar wood chips (Dominguez *et al.,* 1997), corn cobs (Rivas *et al.,* 2003), corn stover (Mancilha and Karim, 2003) (brewery's spent grain hydrolyaste (Carvalheiro *et al.,*2004), jute stick and rice husk (Ahmed *et al.,* 2001), sorghum straw (Herrera *et al.,* 2003), wheat bran (Hollmann and Lindhauer 2005) for xylose and xylitol production.

Corn fibre represents a renewable resource that is available in significant quantities from the corn wet milling industries (Doner *et al.,* 2001). Corn fibre contains about 70% fermentable sugars (Dien *et al.,* 1999). Corn fibre is rich in pentose sugars (Saha, 2003) in the form of arabinoxylan (hemicellulose). Because of its relatively low commercial value and increasing supply, considerable interest exists in finding new, value-added uses for this material (Dien *et al.,* 1999). The low cost high carbohydrate content of corn fibre makes it an attractive feed stock for conversion to xylitol. About $3.4 \times 10^6$ dry tons of corn fiber are produced each year in the United States alone (Dien *et al.,*1999), and about 203.6 Tg of dry corn stover are globally available (Tsao *et al.,* 1999) for D-xylose and xylitol production. Nearly half of all fuel ethanol production in the USA employs a wet-milling process utilizing more than 300 million bushels of corn. Every bushel of maize processed for sweetener, oil or ethanol generate nearly 7 kg of protein and fibre rich residues.

Many research efforts were made to achieve the production of value added products from sugarcane bagasse (Pandey *et al.,* 2000). The annual global production of dry cut sugarcane is about 328 Tg and Asia (44%) is the primary production region (Kim and Dale, 2004). Each ton of milled sugarcane gives 180-280 kg of bagasse residues (Pessora *et al.,* 1997). The acid hydrolysate from bagasse contains xylose as the main component (Toit *et al.,* 1984). When sugarcane bagasse hemicellulose hydrolysate was evaluated for cultivation of 18 species of different yeasts, *Candida* sps. emerged as the best performer (Pessoa *et al.,* 1997).

## MICROBIAL PRODUCTION OF XYLITOL

Considerable efforts have been focused on the microbial production of xylitol from D-Xylose. The microbial production of xylitol has been always closely connected with that of ethanol and for years it was considered only a byproduct in ethanol fermentation processes from D-xylose (do Preez *et al.,* 1986; Slininger *et al.,* 1990; Vandeska *et al.,* 1993). Obviously this stems from the fact that ethanol production from D-xylose was the process that first attracted the attention of researchers. However, for a considerable period of time, views on the ability of yeasts to produce ethanol from aldopentoses were contradictory (Schneider, 1989). The matter was resolved in the beginning of the eighties when a number of different laboratories independently demonstrated the direct conversion of D-Xylose to ethanol (Gong *et al.,*(c) 1981; Slininger *et al.,* 1982; Rosenberg, 1980) as well as ethanol production from D-xylulose (Wang *et al.,* 1980; Gong(a) *et al.,* 1981). This discovery opened the possibility not only for ethanol but also for polyol production, primarily xylitol. The first significant papers on microbial production of xylitol refer to screening for suitable microorganisms (Gong[b] *et al.,* 1981; Barbosa *et al.,* 1988), but the real scientific interest began in the last few years, when xylitol, due to its unique properties as an alternative sweetener, began to be more frequently used (Mir, 1988; Isokangas *et al.,* 1988; Pepper & Olinger, 1988), and when awareness of protection of the environment grew.

The ultimate goal of all investigations regarding D-xylose fermentation is to acquire sufficient knowledge for establishing fermentation processes using the pentose fraction of the lignocellulosic hydrolysates.

Although the hydrolysis can be performed enzymatically most fermentation studies have focused on hydrolysates derived from acid hydrolysis. Due to its heterogeneous structure and relatively low degree of polymerization, hemicellulose is much easier to hydrolyze than the crystalline cellulosic components of biomass. In many cases even a simple steam treatment without the aid of acid catalysis has been found to be effective (Magee & Kosaric, 1985). However for the decomposition of hemicellulose, mild hydrolysis is the most suitable. Its advantages are that it prevents the formation of some degradation products, enhances the susceptibility of cellulose to subsequent enzymatic or acid hydrolysis, reduces the requirement for expensive corrosion-proof equipment and avoids the environmental problems incurred through the use of strong chemical treatments (Magee & Kosaric, 1985).

A critical feature of hydrolysates prepared using acid catalysis is the presence of inhibitors of microbial metabolism (Schneider, 1989) which makes them substantially more difficult for microbial utilization than the corresponding mixtures of pure sugars (Jeffries, 1983). During the acid hydrolysis of the lignocellulosics, different types of sugars (D-glucose, D-galactose, D-mannose, D-xylose, L-arabinose) and degradation products (furfural, 5-hydroxymethyl furfural, acetic acid, syringic acid, p-hydroxybenzoic acid, vanillin etc.) are formed. Knowledge regarding inhibitors and how to minimize their effects is of the utmost importance to achieve efficient fermentation processes (du Preez, 1994; Olsson & Hahn-Hagerdal 1996).

One of the most common inhibitors, particularly in ethanol fermentations, is acetic acid, the degree of toxicity of which is pH-dependent (Prior *et al.,* 1989). However, with regard to xylitol production in xylitol-producing process, no negative influence of acetic acid has been clearly identified. The concentration of acetic acid decreased continuously until its complete depletion when *Debaryomyces hansenii* NRRL Y-7426 (Parajo *et al.,* 1996) and *Candida guilliermondii FTI* 20037 (Roberto *et al.,* 1994) were grown on hemicellulosic hydrolysates. *Candida guilliermondii FTI* 20037 grown on rice straw hydrolysate exhibited the highest production rate of 0.56 g/l/h, whereas *Debaryomyces hansenii* NRRL Y-7426 grown on chips of *Eucalyptus globulus* had in fact the highest yield, 0.73 g/g.

The presence of inhibitory substances in hydrolysates very often imposes the necessity of purification of the hydrolysates prior to their utilization and/or adaptation of the microorganisms to the sugar which will be used. Thus in the process of xylitol production by *Candida guilliermondii* FTI 20037 from sugar cane bagasse hemicellulosic hydrolysate, the hydrolysate was treated in seven different ways (Roberto *et al.,* 1991). The best results were obtained following over-titration of the hydrolysate with $Ca(OH)_2$ and subsequent use of $H_2SO_4$.

A single hydrolysis stage in *Eucaliptus globulus* led to the formation of about 17 g/1 D-Xylose with a low concentration of inhibitors (Parajo *et al.,* 1996). However the relatively low substrate concentration limited both the productivity and yield of the subsequent fermentation step. Therefore the D-xylose content was raised by vacuum evaporation. In the comparatively high range of xylose concentrations studied (57-78 g/l) a single charcoal treatment was unable to reduce the amount of inhibitors to a satisfactory level and convert the hydrolysate to a suitable fermentation medium.

Because of this the combined strategy involving both charcoal adsorption of concentrated hydrolysates and a high initial cell concentration of up to 80 g/l was applied.

When the yeasts *Candida guilliermondii* FTI 20037 (Roberto *et al.,* 1991; Roberto *et al.,* 1994) and *Candida mogii* ATCC 18364 (Rizzi *et al.,* 1992) were grown on hemicellulose hydrolysates for xylitol production a sequential pattern of sugar consumption was observed. In addition, as a result of carbon source limitation, both yeasts started to consume the xylitol produced when D-xylose was almost exhausted.

There are other xylose-fermenting yeasts such as *Pichia* stipitis NRRL Y7124 (Delgenes *et al.,* 1990) and *Candida shehatae* NRRL Y-12858 ( Kuzmanova *et al.,* 1993) which produced xylitol when cultivated on wheat straw and corn cob hemicellulosic hydrolysate respectively. In these cases, however, xylitol was only a by-product of ethanol fermentation.

In our studies (Sreenivas *et al.,* 2005) corn fibre hydrolysate and sugarcane bagasse hydrolysate were neutralized with $Ca(OH)_2$ and treated with activated charcoal and ion exchange resins to reduce inhibitors and conductivity of the hydrolysate. Corn fibre has four-fold dissolved solids and less conductivity than sugarcane bagasse hydrolysate. In both cases conductivity was reduced up to 500 $\mu\Omega^{-1}.cm^{-1}$ by ion exchange resins. To improve *Candida tropicalis* growth and xylitol production from corn fibre and sugarcane baggase hydrolysates containing inhibitors, the cells were adapted in the hydrolysate medium by subculturing for 25 cycles . After every five cycles, the organism was tested for xylitol production. With corn fibre the xylitol production increased from initial 0.43 g/g to 0.46 g/g, 0.49 g/g, 0.52 g/g of xylose after 5,10,15$^{th}$ cycles respectively. After adaptation up to 20th cycle the xylitol production improved to 0.58g/g of xylose. With sugarcane bagasse, the xylitol production increased from initial 0.45 g/g to 0.48 g/g, 0.52 g/g, 0.57 g/g of xylose after 5,10,15$^{th}$ cycles respectively. After adaptation up to 20th cycle the xylitol production was improved to 0.65g/g of xylose. After 25$^{th}$ cycle, negligible increase in xylitol yield was observed. Xylitol accumulation was more in sugarcane bagasse hydrolysate than in corn fibre hydrolysate.

## CONCLUSION

In conclusion to produce xylitol by yeasts, in a cost-effective manner the fermentation of xylose from hemicellulosic material is important. The potential of corn fibre, corn cobs, wheat straw, wheat bran, rice straw and rice bran and sugar cane bagasse hemicellulosic hydrolyzates as substrates for xylitol production by natural, mutant, recombinant and adapted strains of yeasts are important.

## REFERENCES

Ahmed, Z., Banu, H., Akhter, F. and Izumori, K. (2001). A simple method for D-xylose Extraction from Jute Stick and Rice Husk. *Journal of Biological Sciences,* 1, 11, 1001-1004.

Barbosa. M.F.S., de Medeiros. M.B., de Mancilha. I.M., Schneider. H., and Lee. H. (1988). Screening of yeasts for production of xylitol from D-xylose and some factors which affect xylitol yield in *Candida guilliermondii. J. Ind. Microbiol,* 3, 241-251.

Carvalheiro, F., Duarte, L.C., Lopes, S.,Parajó, J.C., Pereira, H. and Girio. F.M. (2005). Evaluation of the detoxification of brewery's spent grain hydrolysate for xylitol production by *Debaryomyces hansenii* CCMI 941. *Process Biochemistry,* 40, 1215-1223.

Cassland, P. and Jösson, L.J. (1999). Characterization of a gene encoding *Trametes versicolor* laccase A and improved heterologous expression in *Saccharomyces cerevisiae* by decreased cultivation temperature. *Appl. Microbial Biotechnol,* 52, 393-400.

Converti, A., Perego, P. and Dominguez, J.M. (1999). Xylitol production from hardwood hemicellulose hydrolysates by *Pachysolen tannophilus, Debaryomyces hansenii,* and *Candida guilliermondii. Appl. Biochem. Biotechnol,* 82, 141-151.

Cruz, M.J., Domínguez, J.M., Domínguez, H. and Parajó, J.C. (1999). Solvent extraction of hemicellulosic wood hydrolysates: a procedure useful for obtaining both detoxified fermentation media and polyphenols with antioxidant activity. *Food Chem,* 67, 147-153.

Cruz, M.J., Domínguez, J.M., Domínguez, H. and Parajó, J.C. (2000). Xylitol production from barley bran hydrolysates by continuous fermentation with *Debaromyces hansenii. Biotechnol Letters,* 22, 1895-1898.

Delgenes. J.P., Moletta. R., and Navarro. J.M. (1990). Acid hydrolyzate of wheat straw and process consideration for ethanol fermentation by *Pichia stipitis* Y-7124. *Process Biochem, 25,* 132-135.

Dien, B.S., Hespell, R.B., Ingram, L.O. and Bothast, R.J. (1997). Conversion of corn milling fibrous co-products into ethanol by recombinant *Escherichia coli* strains K011 and SL40. *World J. Microbiol. Biotechnol,* 13, 619-625.

Dien, B.S., Iten, L.B., Bothast, R.J. (1999). Conversion of corn fiber to ethanol by Recombinant *E. coli* strain FBR3. *J. Industrial Microbiol, Biotechnol,* 22, 575-581.

Dien, B.S., Nichols, N.N., O' Bryan, P.J. and Bothast, R.J. (2000). Development of New Ethanologenic *Escherichia coli* strains for fermentation of lignocellulosic biomass, *Applied Biochem. Biotechnol,* 84, 181-196.

Domínguez, J.M., Cao, N.J., Krishnan, M.S., Gong, C.S. and Tsao, G.T. (1997). Xylitol Production from hybrid poplar wood chips pretreated by the ammonia steeping process. *Biotechnol. Techniques,* 11, 339-341.

Dominguez, J.M., Cao, N., C.S. Gong C.S. and Tsao., G.T. (1997). Dilute acid Hemicellulose hydrolysates from corn cobs for xylitol production by yeast. *Bioresource Technology,* 61, 85-90.

Doner, L.W., Johnston, D.B. and Singh, V. (2001). Analysis and properties of arabinoxylans from discrete corn wet-milling fiber fractions. *J Agr. Food Chem,* 49, 1266-1269.

do Preez. J.C., Bosch. M., and Prior. B.A. (1986). Xylose fermentation *by Candida shehatae* and *Pichia slipitis:* effects of pH, temperature and substrate concentration. *Enzyme Microb. Technol,* 8, 360-364.

do Preez. J.C. (1994). Process parameters and environmental factors affecting D-xylose fermentation by yeasts. *Enzyme Microb. Technol,* 16, 944-956.

Gong. C.S., Chen, L.F., Flickinger, C., Chiang, L.C., and Tsao, G.T. (1981). Conversion of hemicellulose carbohydrate. *Adv. Biochem. Eng. Biotechnol,* 20, 93-118.

Gong. C.S., Chen. L.F. and Tsao. G.T. (1981). Quantitative production of xylitol from D-xylose by a high-xylitol producing yeast mutant *Candida tropicalis* HXP2. *Biotechnol. Lett.* 3, 130-135.

Gong. C.S., Chen. L.F., Flickinger. C., Chiang. L.-C. and Tsao, G.T. (1981). Production of ethanol from D-xylose by using D-xylose isomerase and yeasts. *Appl. Environ. Microbiol,* 41, 430-436.

Herrera, A., S.J. TeÂllez-Luis, S.J., Ramõ Ârez, J.A. and VaÂzquez, M. (2003). Production of xylose from sorghum straw using hydrochloric acid. *Journal of Cereal Science,* 37, 267-274.

Hollmann, J., Lindhauer, M.G. (2005). Pilot-scale isolation of glucuronoarabinoxylans from wheat bran. *Carbohydrate Polymers,* 59, 225-230.

Isokangas. P., Alanen. P., Tiekso. J. and Makinen. K.K. (1988). Xylitol chewing gum in Caries prevention: a field study in children. *J. Am. Dent. Assoc,* 117, 315-320.

Jeffries. T.W. (1983). Utilization of Xylose by bacteria, yeasts and fungi. *Advances in Biochemical Engineering Biotechnology,* 27, 1-32.

Jeffries, T.W. and Jinn, Y.S. (2004). Metabolic engineering fro improved fermentation of Pentoses by yeasts. *Appl. Microbial. Biotechnol,* 63, 495-509.

Kim, S. and Dale, B.E. (2004) Global potential bio ethanol production from wasted crops and crop residues. *Biomass Bioenergy,* 26, 361-375.

Kuzmanova. S., Vandeska. E. and Georgievska. I. (1993). Selection of yeasts for fermentation of corn cob hemicellulosic hydrolyzates. *Microbiol,* 30, 19-30.

Leathers, T.D. (2003). Bioconversion of maize residues to value-added coproducts using yeast-like fungi. *FEMS Yeast Res,* 3, 133-140.

Lohmeier-vogel, E.M., Sopher, C.R. and Lee, H. Itracellular acidification as a mechanism for the inhibition by acid hydrolysis-derived inhibitors of xylose fermentation by yeasts. *J. Indus. Microbiol. Biotechnol.* 20, (1998), 75-81.

Magee. R. J., and Kosaric. N. Bioconversion of hemicelluloses. *Adv. Biochem. Eng. Biotechnol., 32,* (1985), 61-93.

Makinen, K.K. The rocky road of xylitol to its clinical application. *J. Den Res* 79, (2000), 1352-1355.

Mancilha, I.M., Karim, M.N. Evaluation of ion exchange resins for removal of inhibitory compounds from corn stover hydrolysate for xylitol fermentation. *Biotechnol. Prog.* 19, (2003), 1837-1841.

Martin, C., Jönsson, L.J. Comparision of the resistance of industrial and laboratory strains of *Saccharomyces* and *Zygosaccharomyces* to lignocellulosic-derived fermentation inhibitors. *Enzyme Microbial Technol.* 32, (2003), 386-395.

Mir. A. Caries prevention with xylitol. A review of the scientific evidence. *Wld. Rev. Ntr. Diet. 55*, (1988), 183-209.

Mussatto SI, Dragone G, Roberto IC. Kinetic behavior of *Candida guilliermondii* yeast during xylitol production from Brewer's spent grain hemicellulosic hydrolysate. *Biotechnology Prog.* 4, (2005), 1352-6.

Olsson. L., and Hahn-Hagerdal. B. Fermentation of lignocellulosic hydrolysates for ethanol production. *Enzyme Microb. Technol., 18*, (1996), 312-331.

Pandey A., Soccol, C.R., Nigam, P., Soccol, V.T. Biotechnological potential of agro-industrial residues. I; sugarcane bagasse, *Biores. Technol.* 74, (2000), 69-80.

Parajo. J.C., Dominguez. H., and Dominguez. J.M. Production of xylitol from concentrated wood hydrolysates by *Debaryomyces hansenii:* effect of the initial cell concentration. *Biotechnol. Lett., 18*, (1996), 593-598.

Pepper. T. and Olinger. P.M. Xylitol in sugar-free confections. *Food Technol., 10*, (1988), 98-106.

Pessoa J, A., de Manchilha, I.M., Sato, S. Evaluation of sugar cane hemicellulose hydrolyzate for cultivation of yeasts and filamentous fungi, *J. Ind. Microbiol. Biotechnol.* 18. (1997), 360-363.

Preziosi-Belloy, L., Nolleau, V., Navarro, J.M. Xylitol production from aspen wood hemicellulose hydrolysate by *Candida guilliermondii, Biotech. Letters* 22, (2000), 239-243.

Prior. B.A., Killian. S.G., and du Preez. J. C. Fermentation of D-xylose by the yeasts *Candida shehatae* and *Pichia stipitis. Proc. Biochem. 24*, (1989), 21-32.

Rivas, B., Torre, P., Domínguez, J.M., Perego, P., Converti, A., Parajó, J,C. Carbon material and bioenergestic balances of xylitol production from corncob by *Debaryomyces hansenii. Biotechnol. Prog.* 19, (2003), 706-713.

Rizzi. M., Sirisansaneeyakul. S., and Reuss. M. Microbial production of xylitol from wheat straw hydrolysates, p. 541544. In DECHEMA Biotechnology Conference 5. VCH Verlagsgemeinschaft, Weinheim (1992).

Roberto. I.C., Mancilha. 1.M., de Souza. C.A., Felipe. M.G.A., Sato. S., and de Castro. H.F. Evaluation of rice straw hemicellulose hydrolysate in the production of xylitol by *Candida guilliermondii. Biotechnol. Lett., 16*, (1994), 1211-1216.

Roberto. I.C., Felipe. M.G.A., Lacis. L.S., Silva. S.S., and de Mancilha. 1.M. Utilization of sugar cane bagasse hemicellulosic hydrolyzate by *Candida guilliermondii* for xylitol production. *Bioresource Technol., 36*, (1991), 271-275.

Rosenberg. S.L. Fermentation of pentose sugars to ethanol and other neutral products by microorganisms. *Enzyme Microbiol. Technol., 2*, (1980), 185-189.

Saha, B.C., Bothast, R.J. Microbial production of xylitol. In. Saha, B.C, Woodward, J. (eds) *Fuels and chemicals from biomass.* American Chemical Society, Washington, D.C. (1997). pp 307-319.

Saha, B.C., Bothast, R.J. Production of xylitol by *Candida peltata J. Ind Microbiol Biotechnol*, 22, (1999), 633-636.

Saha, B.C., Bothast, R.J. Pretreatment and enzymatic saccharification of corn fiber. *Applied Biochem. Biotechnol.* 76, (1999), 65-77.

Saha, B.C. Hemicellulose bioconversion. *J. Ind. Microbiol. Biotechnol.*30, (2003), 279-291.

Sampaio, F.C., Mantovani, H.C., Passos, F.J.V., Moraes, C.A., Converti, A., Passos F.M. Bioconversion of D-xylose to xylitol by *Debaryomyces hansenii* UFV-170: product formation versus growth. *Process Biochemistry.* 40, (2005), 3600-3606.

Schneider. H. Conversion of pentoses to ethanol by yeasts and fungi. CRC. *Crit. Rev. Biotechnol., 9*, (1989), 1-40.

Silva, S.S., Maria, G.A., Silva, B.A., Arnaldo, M.R. Acid hydrolysis of *Eucalyptus grandis* chips for microbial production of xylitol, *Process biochem.* 33, No.1, (1998), 63-67.

Singh. A., and Mishra. P. Microbial pentose utilization. *Adv. Appl. Microbiol. 39,* (1995), 91-152

Slininger. P.J., Bothast. R.J., van Cauwenberge. J.E., and Kurtzmann. C.P. Conversion of D-xylose to ethanol by the yeast *Pachysolen tannophilus. Biotechnol. Bioeng., 24*, (1982), 371-384.

Slininger. P.J., Bothast. R.J., Ladish. M.R., and Okos. M.R. Optimum pH and temperature conditions for xylose fermentation by *Pichia stipitis. Biotechnol. Bioeng., 35*, (1990), 727-731.

Sreenivas Rao, R., Prakasham, R.S., Krishna Prasad, K., Rajesham, S., Sharma, P.N., Venkateswar Rao, L. Xylitol production by *Candida* sp.: parameter optimization using Taguchi approach. *Process Biochem.* 39, (2004), 951-956.

Sreenivas Rao, R., Ch. Pavan Jyothi, R.S. Prakasham, P.N. Sarma, L. Vekateswar Rao. Xylitol production from corn fiber and sugar cane bagasse hydrolysates by *Candida tropicalis* – published online-*Bioresource Technology*, (2005), (In Press).

Sreenivas Rao, R., Ch. Pavan Jyothi, R.S. Prakasham, Ch. Subba Rao, P.N. Sarma, L. Vekateswar Rao. Xylitol Strain improvement of *Candida tropicalis* for the production of xylitol: Biochemical and physiological characterization of wild and mutant strain CT-OMV5. *The Journal of Microbiology*, 44, (2006), 113-120.

Tada. K., Horiuchi, J.I., Kanno, T., Kobayashi, M.Y. Microbial xylitol production from corn cobs using *Candida magnoliae. J of Bioscience and Bioeng.* 98, 2004, 228-230.

Toit, P.J., Olivier, S.P., Van Biljon, P.L. Sugarcane bagasse as a possible source of fermentable carbohydrates. I.Characterization of bagasse with regard to monosaccharide, hemicellulose and amino acid composition. *Biotech. and Bioeng.* 26, (1984), 1071-1078.

Tsao, G.T., Cao, N., Gong, C.S. Hemicellulose Conversion. Encyclopedia of Bioprocess Technology: Fermentation, Biocatalysis and Bioseperation. John Wiley & Sons.Inc., ISBN-0-471-13822-3, 3, (1999),1391-1400.

Vandeska. E., Georgievska. I., and Kuzmanova. S. Batch fermentation of xylose to ethanol by *Candida shehatae. J. Serb.Chem. Soc., 58*, (1993), 651-657.

Wang. V., Jonson. B.F., and Schneider. H. Fermentation of D-xylose by yeasts using isomerase in the medium to convert D-xylose to D-xylulose. *Biotechnol. Lett., 3*, (1980), 273-278.

# Index